# of South Dakota

THEODORE VAN BRUGGEN

Ames / 1976

Theodore **Van Bruggen** is Professor of Biology, College of Arts and Sciences, the University of South Dakota, Vermillion.

© 1976 The Iowa State University Press
Ames, Iowa 50010. All rights reserved

Printed by The Iowa State University Press

First edition, 1976

cmu-1073

Library of Congress Cataloging in Publication Data

Van Bruggen, Theodore, 1926-
  The vascular plants of South Dakota.

  Bibliography: p.
  Includes index.
  1. Botany—South Dakota. I. Title.
QK186.V36   581.9'783   76-3747

ISBN 0-8138-0650-X

CONTENTS

Basic Units of Measurement

10 millimeters (mm)

equals

1 centimeter (cm)

10 centimeters

equals

1 decimeter (dm)

METRIC 1| 2| 3| 4| 5| 6| 7| 8| 9| 1|0

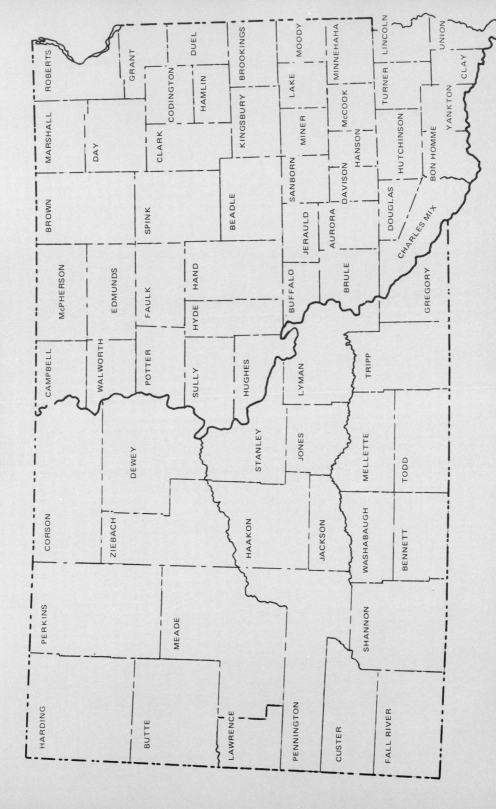

SOUTH DAKOTA COUNTIES

PREFACE

THIS EDITION is the result of approximately 17 years of study of the flora of this state. Previously published manuals used for the identification and distribution of plants in South Dakota inadequately covered this geographic area or are nomenclaturally outdated.

Many people have given much assistance in the preparation of this book. Special appreciation is due Dr. Ronald McGregor, Director of the State Biological Survey of Kansas and curator of the University of Kansas Herbarium. A foremost authority on the flora of the North and Central plains, he provided invaluable assistance on distribution of plant species, nomenclatural decisions, and sources of literature. Thanks are also due to former graduate and undergraduate students who participated by making extensive field collections in various parts of this state, and to The University of South Dakota, whose auspices made this work possible.

Acknowledgement is made to the South Dakota Geological Survey, Duncan McGregor, State Geologist, for use of the glacial, geologic, and physiographic maps.

I am grateful to Elise Parish who typed the manuscript and was helpful in many ways.

Finally, I owe a considerable debt for the patience and encouragement of my wife, Geraldine.

INTRODUCTION

THE EARLIEST known collections of plants from the area which is
now South Dakota were undoubtedly made by Lewis and Clark on
their famous expedition up the Missouri River in the summer of
1804. Following their first winter at Fort Mandan, 1804-1805,
they sent back 123 plant specimens of which some were no doubt
collected in South Dakota territory. Later expeditions by John
Bradbury and Thomas Nuttall in 1811 and 1812, by Thomas Nuttall,
1834-1836, F. V. Hayden, 1859, and the Custer Expedition, 1874,
provided several hundred collections from the state during the
19th century.

The discovery of gold in the Black Hills in 1875 stimulated
geological expeditions to the region and enhanced the collecting
of plants from both the Badlands and the Black Hills. Professor
W. P. Jenney, in the Newton-Jenney report of "The Geology and
Resources of the Black Hills" (1880) listed 166 species of plants
determined by Asa Gray of Harvard University.

In 1892 Per Axel Rydberg collected plants from the Black
Hills and in 1896 published "The Flora of the Black Hills of
South Dakota," listing some 700 species of plants. From 1908
through 1912, Stephen S. Visher made collections of South Dakota
plants while serving as a naturalist with the State Geological
Survey of South Dakota. On one of his trips to the northwestern
part of the state Visher met William H. Over, a homesteader in
Harding County who was also interested in natural history.
Visher encouraged Over's interest in natural history and later,
in 1913, after the Over family had moved to Vermillion, Over
joined the staff of the South Dakota Geological Survey as an
assistant curator of the museum. He was keenly interested in
the plant and animal life of South Dakota, and his hobby became
his lifetime occupation. From 1916 to 1924, Over made extensive
collecting trips throughout the state. During the years 1924
through 1930, Over collaborated with A. C. MacIntosh, professor
of biology at the School of Mines and Technology at Rapid City,

on a collection of plants from the Black Hills region. As a result of this field work, MacIntosh published "A Botanical Survey
of the Black Hills of South Dakota" as one of the issues of *The
Black Hills Engineer* (1931). It included over 1300 species of
plants. This publication was reprinted in 1949 by the South
Dakota School of Mines and Technology.

In 1932 William Over published *A Flora of South Dakota*, a
checklist of over 1500 species of vascular plants of the state.
It also included 75 illustrations of commonly encountered species. This work has been a standard reference for the flora of
the state since that time. Most of the binomials used by Over
were nomenclaturally similar to those used in Rydberg's *Flora
of the Rocky Mountains and Adjacent Plains* (1917).

Since the 1940's a number of botanists and their students,
as well as interested amateurs, have added to the collections
that are now incorporated in herbaria in South Dakota and elsewhere. Principal collections in South Dakota are located at
South Dakota State University, Brookings; University of South
Dakota, Vermillion; Black Hills State College, Spearfish; Northern State College, Aberdeen; and Augustana College, Sioux Falls.
Most of the Black Hills collections of A. C. MacIntosh, originally housed at the School of Mines and Technology, Rapid City,
were transferred to the University Herbarium in Vermillion.
Other known collections of South Dakota plants are located at
the University of Kansas, Lawrence; the Smithsonian Institution,
Washington, D.C.; the Missouri Botanical Garden Herbarium, St.
Louis; and North Dakota State University, Fargo. A number of
these collections were examined during the preparation of this
manual, or advice and suggestions were solicited from curators
of these collections pertaining to distributional data, annota-
tion of difficult taxa, and nomenclatural decision. Additional
help was obtained from monographers of certain groups.

GEOLOGY, PHYSIOGRAPHY AND CLIMATE

Geology

The state of South Dakota is essentially a rectangular area
which was determined in an arbitrary way without much regard to
natural boundaries. It is bounded on the north approximately on
the 46th parallel and on the south by the 43rd parallel, except
for the eastern part where the Missouri River forms the boundary.
On the west the 104th meridian forms the boundary. On the east
the state is bounded approximately halfway between the 96th and

SOUTH DAKOTA GEOLOGICAL SURVEY
Duncan J. McGregor, State Geologist

STATE OF SOUTH DAKOTA
Richard Kneip, Governor

EDUCATIONAL SERIES
Map Two

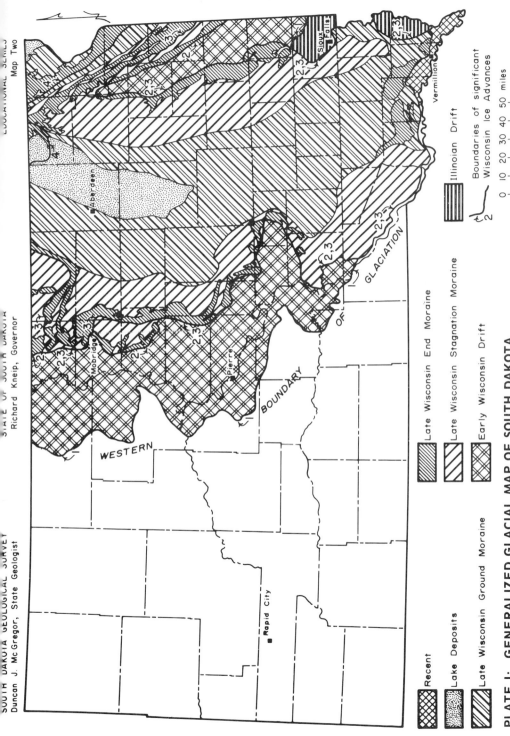

PLATE I: GENERALIZED GLACIAL MAP OF SOUTH DAKOTA

Recent

Lake Deposits

Late Wisconsin Ground Moraine

Late Wisconsin End Moraine

Late Wisconsin Stagnation Moraine

Early Wisconsin Drift

Illinoian Drift

Boundaries of significant
Wisconsin Ice Advances

0 10 20 30 40 50 miles

WESTERN   BOUNDARY   OF   GLACIATION

Aberdeen

Mobridge

Pierre

Rapid City

Sioux Falls

Vermillion

97th meridian. The northern part of the eastern boundary is
formed by Lake Traverse and Big Stone Lake. In the southern
part, just south of Sioux Falls, the Big Sioux River forms the
eastern boundary southward to where it flows into the Missouri
River at Sioux City, Iowa. Of the approximately 1440 miles of
total boundary, only 240 miles follow natural boundaries of
lakes or rivers. The dimensions of this approximate rectangle,
460 miles east and west by 260 north and south, include an area
of 77,047 square miles.

Most of the eastern part of the state was glaciated during
the Pleistocene. At least five major glaciations took place:
the Nebraskan, Kansan, Illinoian, Early and Late Wisconsin.
Deposits of the two earliest, Nebraskan and Kansan, were completely covered by the latter three. Outcrops from the first
two are known along Skunk Creek near Hartford and the Big Sioux
River south of Sioux Falls. If they advanced farther west of
these points, the later glaciations have destroyed the evidence.
The Illinoian glaciation only reached the southeastern corner of
South Dakota (Plate I). Only one glaciation, the Early Wisconsin, extended slightly beyond the Missouri River to the west.
The Early Wisconsin and the Late Wisconsin glaciations were re-
sponsible for forming the Missouri River sometime between 12,000
and 40,000 years ago.

The thickness of glacial drift in the eastern part of the
state averages between 50 and 100 feet. In places it is over
800 feet thick. The drift is underlain by flat-lying marine
Cretaceous strata, only locally capped by residual remnants of
Tertiary rocks. The position and direction of flow of the present major drainage systems of the eastern part were largely determined by the series of glacial sheets and their subsequent
melt water movement.

The country west of the Missouri River is one of canyons,
broad upland flats, buttes, and rolling plains. Much older geologic strata are exposed where extensive erosion has occurred.
The present Badlands is a classic case of irregular erosion of
the original Tertiary Tablelands. The alternating bands of
sand, ash, clay and gumbo, eroded by rapidly draining water
from torrential summer thunderstorms, combine to produce the
interesting earth sculpture. Perhaps the most interesting topographic feature of the state is the Black Hills, the *Paha Sapa*
(Hills of the Shadows) of the Sioux Indians, an elliptically
shaped dome that has been exposed to erosion for millions of
years. A series of concentric sedimentary formations is present. Although these sedimentary rocks persist around the rim
of the dome, the central region has had the sedimentary deposits
long removed, leaving a core of ancient Precambrian granites,
pegmatites, and quartz-mica schists. This core is much higher

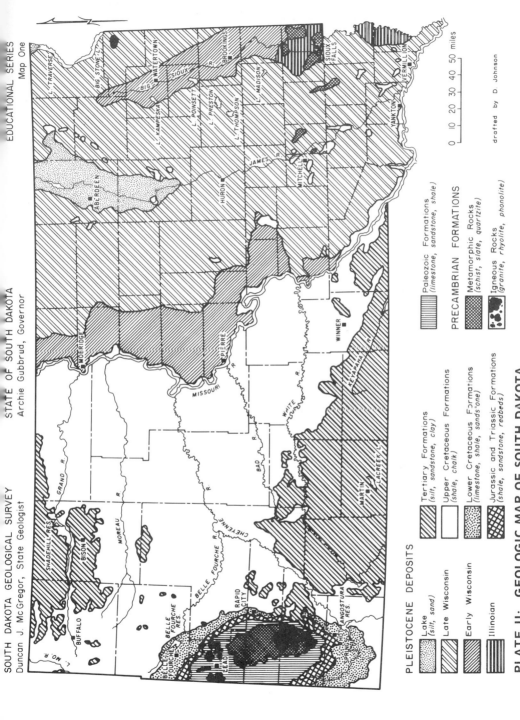

SOUTH DAKOTA GEOLOGICAL SURVEY
Duncan J. McGregor, State Geologist

STATE OF SOUTH DAKOTA
Archie Gubbrud, Governor

EDUCATIONAL SERIES
Map One

PLEISTOCENE DEPOSITS

Lake
(silt, sand)

Late Wisconsin

Early Wisconsin

Illinoian

Tertiary Formations
(silt, sandstone, clay)

Upper Cretaceous Formations
(shale, chalk)

Lower Cretaceous Formations
(limestone, shale, sandstone)

Jurassic and Triassic Formations
(shale, sandstone, redbeds)

Paleozoic Formations
(limestone, sandstone, shale)

PRECAMBRIAN FORMATIONS

Metamorphic Rocks
(schist, slate, quartzite)

Igneous Rocks
(granite, rhyolite, phonolite)

0 10 20 30 40 50 miles

drafted by D. Johnson

PLATE II: GEOLOGIC MAP OF SOUTH DAKOTA

in altitude than the surrounding, more recently deposited sedimentary rock. The limestone plateaus, ancient crystalline rocks,
rock fragments with mica flakes, and finer rock particles have
all been influential in determining the species of plants that
thrive there (Plate II).

Physiography

According to Fenneman (1930) and later Rothrock (1943), the
state of South Dakota lies in two physiographic provinces. The
western two thirds of the state is located in the Great Plains
province. The eastern one third, all of which was covered by
Pleistocene glaciation, is in the Central Lowlands. The line
dividing these two is approximately on the 100th meridian (Plate
III). Both of these physiographic units are comprised of several
identifiable divisions within the state. The Great Plains province, which extends east of the Missouri River, is substantially
higher in elevation than the Central Lowlands to the east. Rothrock's vivid description of the topography of the state follows:

To get a general picture of this topography let us consider
the surface of the state as a great table with about the pro-
portions of an ordinary office desk or kitchen table. The
northwest corner has been raised so that there is a general
slope from the northwest to the southeast. The highest part
of this slope, then, will be in Harding County on the divide
between the Little Missouri and the Grand Rivers where precise
level lines, run by the United States Coast and Geodetic Survey, give elevations of nearly 3,400 feet. The lowest point
on this general slope is in the southeast corner of the state
at the junction of the Big Sioux and Missouri Rivers with an
elevation of 1,100 feet.

This latter point is not the lowest point in the state,
however, for due to a glacial accident, a trench, now occupied
by Lakes Traverse and Big Stone, was cut in the Northeastern
corner, whose elevation (960 feet) is some 140 feet lower than
the elevation at the mouth of the Big Sioux River. This is
but a small chip off the corner of the state, however, since
less than twenty miles to the west of these lakes the surface
rises 1,000 feet in a great escarpment to an elevation of
2,000 feet above sea level.

The highest point in the state is not in the northwest corner of the state though this is the top of the general slope.
In the southwest corner is a tidy little mountain mass, the
Black Hills, whose highest peak, Harney Peak, reaches an elevation of 7,242 feet. (Rothrock, 1943)

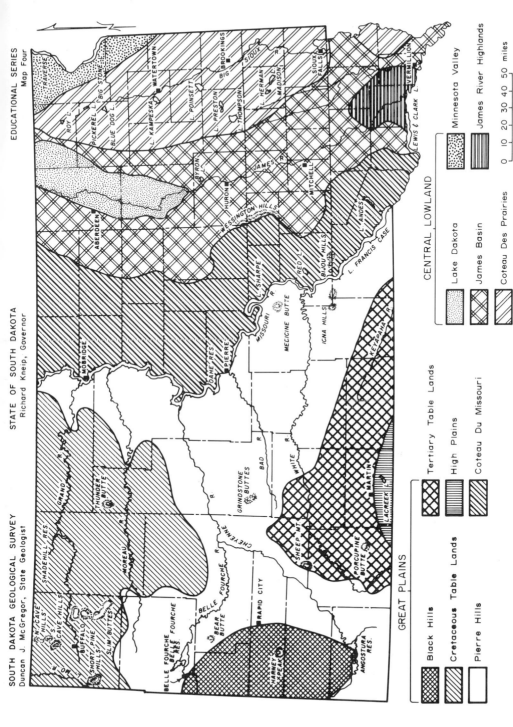

SOUTH DAKOTA GEOLOGICAL SURVEY
Duncan J. McGregor, State Geologist

STATE OF SOUTH DAKOTA
Richard Kneip, Governor

EDUCATIONAL SERIES
Map Four

0 10 20 30 40 50 miles

GREAT PLAINS

Black Hills

Cretaceous Table Lands

Pierre Hills

Tertiary Table Lands

High Plains

Coteau Du Missouri

CENTRAL LOWLAND

Lake Dakota

James Basin

Coteau Des Prairies

Minnesota Valley

James River Highlands

PLATE III: MAJOR PHYSIOGRAPHIC DIVISIONS OF SOUTH DAKOTA

The Great Plains province has four physiographic divisions.
At its eastern edge lie the *Coteau du Missouri*. They are partly
of glacial origin and partly a result of stream erosion by the
tributaries of the Missouri River. The surface is a 25 to 30
mile wide region of rolling glacial hills and rugged river
"brakes." At the southern edge of the state, in Shannon and
Bennett counties, lies a 12 to 15 mile wide northern extension
of the Sand Hills of Nebraska called the *High Plains* division.
They terminate in a pine covered, north-facing escarpment called
the Pine Ridge, which is 1,000 feet higher than the more northern plain in some locations. The *Missouri Plateau* is the largest division of the "west river" Great Plains province. It has
a butte and canyon topography with side stretches of hilly upland and a network of east moving main streams. In the southern
part of this division are the Tertiary Tablelands, whose western
edge along the White River has eroded to form the Big Badlands
of South Dakota. The fourth division of the Great Plains province represented in the state is the *Black Hills*. It is, without doubt, the most famous topographic region in the state.

The Central Lowlands province is divided in the *James Basin*,
a large, glacially eroded valley which is drained by the James
River. It runs the entire width of the state from north to
south. It is bordered on the east by a glacial highland of
prairie hills named by the early French traders, the *Coteau des
Prairie*. It reaches an elevation of over 2,000 feet above sea
level. Approximately as wide as the James Basin, it differs in
being a prairie pothole and lake-dotted highland. On its east
side the Coteau is marked by a 500 to 600 foot escarpment over-
looking the *Minnesota Valley*, a lowland that appears flat in
contrast to the prairie highland. In the Minnesota Valley are
the trench-like, elongate lakes of Big Stone and Traverse, which
form the eastern boundary of the state. An interesting feature
about these lakes is their position on the continental watershed.
Water from the southern part of the valley drains to Big Stone
Lake. Its outlet is the Minnesota River which flows ultimately
into the Gulf of Mexico. Water from the northern part of the
valley finds its way through Lake Traverse northward to the Red
River of the North and then into Hudson Bay.

Climate

South Dakota's location in the heart of the North American
continent, with its extremes of summer heat and winter cold, ex-
emplifies a characteristic continental climate. Rapid fluctua-
tions in temperature are common. Temperatures of 40° C. or
higher are experienced in some parts of the state each summer.

However, these high temperatures are usually accompanied by low
humidity, which reduces somewhat the oppressiveness of the heat.
Temperatures of -12° C. are frequent in the winter, and prolonged
cold spells where temperatures may not go above this mark for 6-8
days are not infrequent. The wind is from the north or northwest
during most of the winter, bringing masses of cold air from the
Arctic regions. Occasionally violent winds from the northwest,
accompanied by snow, will result in a raging blizzard which may
last for 48 to 72 hours. During the summer the wind prevails
from the south or southeast. The mean wind speed is slightly
over 11 miles per hour on a year-round basis, so the day is rare
when it is absolutely calm. During the nights of windy days,
however, the wind usually decreases, and it is not uncommon for
it to be calm at daybreak.

At Bison, in the northwest part of the state, the average
frost-free season extends from May 15 to September 25, a period
of 133 days. At Vermillion in the southeast the dates extend
from May 4 to October 7 for a frost-free period of 156 days.

The Black Hills have a climate all their own. During summer
the higher elevation results in cooler temperatures. Frost-free
periods in the summers are shortest high in the Black Hills, because brief freezing may occur at any time during the summer.
During the winter temperatures may be higher than on the surrounding plains because the dense cold Arctic air masses may not
extend to the higher elevations. Chinook winds and frequent
sunny skies also make the Black Hills the warmest part of the
state during the winter period.

Precipitation ranges from 25 inches per year in the southeast to less than 13 inches per year in the northwest. Usually
more than 50 percent of the yearly precipitation occurs from
April to August, which coincides with the crop season. Whenever
there is less than average rainfall during this period of time,
there is a threat of crop failure. Only 2 to 5 days of hot, dry
winds during one of these subnormal precipitation periods is suf-
ficient to seriously reduce yields of crops, especially corn, the
main crop in the eastern part.

Snowfall varies greatly from year to year. Less than one
third of the time is there excessive snowfall, and then it usually is accompanied by strong northwest winds resulting in high
drifts. The insulating effect of the snow, which is so apparent
in many temperate regions, is largely absent in South Dakota.
Frost usually penetrates the soil to depths of 4 feet or more
in most areas of the state.

FLORISTIC ELEMENTS AND VEGETATION

As the highway signs over the state indicate, South Dakota is the "Land of Infinite Variety." The flora of South Dakota, though not infinitely variable, exhibits considerable variation. Within the state there are areas representative of at least three of nine major floristic regions of North America. In addition, there are species that occur in isolated areas which have affinities to other floristic elements. At higher elevations of the Black Hills several species occur that are related to species commonly alpine, or at least circumboreal in their distribution. Representative species are *Adoxa moschatellina, Arnica rydbergii, Betula glandulosa, Calypso bulbosa, Carex aurea, Carex capillaris, Cornus canadensis, Cypripedium calceolus, Epilobium angustifolium, Linnaea borealis, Petasites sagittata, Potentilla fruticosa, Pyrola uniflora, Rubus pubescens, Salix phylicifolia,* and *Viburnum opulus.*

In the extreme southwestern part of the state, in Fall River and Shannon counties, certain species are found that are characteristic of the Great Basin to the southwest. Apparently similar ecological conditions exist here. Rocky soil with reduced rainfall and an alkaline soil has enhanced the migration of xerophytic species such as *Abronia fragrans, Artemisia cana, Artemisia tridentata, Asclepias arenaria, Cirsium ochrocentrum, Echinocereus viridiflorus, Ipomopsis longiflora, Redfieldia flexuosa, Sarcobatus vermiculatus,* and *Tidestromia lanuginosa.*

The three principal vegetation regions, along with their chief representatives, are outlined below.

The Eastern Deciduous Flora

This extensive, natural vegetation region reaches its most western limit in isolated areas along the eastern part of the state, especially in the southeast along the Big Sioux and Missouri Rivers and the wooded ravines and north-facing slopes in the northeast part. Several areas in these localities are threatened because of overgrazing; however, the woods of Newton Hills State Park and the adjoining game preserve in Lincoln County will hopefully be preserved. In the northeast the hard maple-basswood forest of Sieche Hollow State Park in Roberts County, the woods of Hartford Beach State Park along Big Stone Lake, and several deep wooded ravines west of Lake Traverse are representative of deciduous forested areas. The ranges of several species of dominant trees, as well as understory species, reach their western limit of distribution in these forests.

Typical trees and shrubs are:

Acer saccharum	Prunus americana
Amelanchier sanguinea	Quercus macrocarpa
Celtis occidentalis	Rhamnus lanceolata
Corylus americana	Ribes cynosbati
Euonymous atropurpurea	Tilia americana
Gymnocladus dioica	Ulmus thomasii
Juglans nigra	Viburnum lentago
Ostrya virginiana	Viburnum rafinesquianum

Herbaceous plants typical of these deciduous forests are:

Anemone quinquefolia	Hystrix patula
Arabis canadensis	Prenanthes alba
Asarum canadense	Solidago flexicaulis
Caulophyllum thalictroides	Trillium cernuum
Cystopteris bulbifera	Trillium nivale
Erythronium albidum	Uvularia grandiflora
Geranium maculatum	

Plains and Prairie Flora

The greatest area by far in South Dakota is that occupied by the Grassland Province of North America. The Great Plains grassland covers the western two thirds of the state east to about the 100th meridian, approximately at the physiographic juncture of the Central Lowlands and the Great Plains province. The eastern one third of the state is True Prairie, or Tall Grass Prairie. Although substantial amounts of grassland still occur west of the Missouri River, the True Prairie in the eastern part has largely been plowed and its potholes drained so that very few remnants remain. Those prairie remnants that remain are either heavily overgrazed or otherwise so disturbed that only indicator species can be found. Some of the most valuable sites of true prairie in the eastern part of the state are on railroad-roadside rights-of-way, state parks, and openings in wooded areas or heavily bouldered glacial till areas which were impossible to plow!

The transition from True Prairie in the eastern part of the state to Great Plains Grassland westward is largely a function of reduced effective precipitation. Many species are common to both of these vegetation regions. There is a general uniformity in the composition of the plant cover. A dominance of grasses, lack of trees, rolling topography, and a characteristic xerophytic flora are its main features. Along streams, lakes, and man-made bodies of water, however, trees are conspicuous. This apparently is in contrast to early historical accounts of the prairie which allude to an absence of trees except along permanent water courses or in ravines or below escarpments.

The perennial life habit is exhibited by all of the dominant prairie species and reproduction is largely by vegetative means. A few representative species of these two vegetational regions follow.

True Prairie

Grasses

Andropogon gerardi
Andropogon scoparius
Bouteloua curtipendula
Elymus canadensis
Koeleria pyramidata

Panicum virgatum
Poa pratensis
Sorghastrum avenaceum
Stipa spartea

Forbs

Amorpha canescens
Anemone canadensis
Anemone patens
Artemisia ludoviciana
Asclepias tuberosa
Aster ericoides
Aster sericeus
Astragalus crassicarpus
Calylophus serrulatus
Echinacea angustifolia
Erigeron strigosus

Kuhnia eupatorioides
Liatris aspera
Lithospermum canescens
Lygodesmia juncea
Onosmodium molle
Psoralea argophylla
Psoralea esculenta
Ratibida columnifera
Silphium laciniatum
Solidago missouriensis

Great Plains Grassland

Grasses

Agropyron smithii
Andropogon scoparius
Bouteloua curtipendula
Bouteloua gracilis
Buchloe dactyloides

Calamovilfa longifolia
Elymus canadensis
Koeleria pyramidata
Poa pratensis
Sporobolus cryptandrus

Forbs

Allium textile
Amorpha canescens
Artemisia frigida
Asclepias speciosa
Calylophus serrulatus
Carex eleocharis

Carex filifolia
Castilleja sessiliflora
Coryphantha vivipara
Erigeron strigosus
Eriogonum flavum
Grindelia squarrosa

Linum rigidum   Penstemon albidus
Lithospermum incisum   Penstemon grandiflorus
Lomatium orientale   Rosa arkansana
Opuntia compressa   Verbena stricta
Oxytropis lambertii   Viola nuttallii

Rocky Mountain Forest Flora

(Black Hills Montane Element)

Although geographically separated from the Rocky Mountains, the Black Hills have geologic affinities to this mountainous area which lies almost 300 miles to the west. During the glacial periods, when climates were more moist and cool, the alpine and montane vegetation zones were lower in altitude. This may have permitted migration of many Rocky Mountain species to the Black Hills area. This, however, is not the only source of the present Black Hills flora. Its composition is diverse and its origins are equally diverse. The accompanying table lists the percentage origin and composition of over 1,000 Black Hills species of plants based on Hayward (1924) and MacIntosh (1949).

Black Hills Flora

Percent Composition	Origin
8.0	North America and Eurasia
9.0	Eastern U.S. (Deciduous Forests)
17.0	Great Plains
30.0	Rocky Mountains and Plains
6.0	Northern Origin
8.0	Old World Weeds and Cultivars
4.5	South or Southwest
.5	Possible Endemics
1.0	Cosmopolitan Species
16.0	Wide Ranging Species (across North America)

It is evident that the Black Hills is not an area where endemics occur. If one assumes a less than conservative taxonomic interpretation of the species present, and their ranges, it is doubtful that any endemics are present in the state.

A list of principal western Rocky Mountain species represented in the Black Hills follows:

Trees and Shrubs

Berberis repens
Betula occidentalis
Betula papyrifera
Ceanothus velutinus
Cercocarpus montanus
Juniperus scopulorum
Picea glauca
Pinus contorta

Pinus flexilis
Pinus ponderosa
Populus acuminata
Populus balsamifera
Populus tremuloides
Ribes lacustre
Salix candida
Sorbus scopulina

Herbs

Aconitum columbianum
Adiantum capillus-veneris
Balsamorhiza sagittata
Calochortus nuttallii
Calypso bulbosa
Carex bella
Carex microptera
Carex obtusata
Carex praticola
Carex xerantica
Castilleja sulphurea
Chimaphila umbellata
Delphinium nuttallianum
Deschampsia caespitosa
Disporum trachycaulum

Gaillardia aristata
Goodyera repens
Hedysarum occidentale
Hesperochloa kingii
Lithophragma bulbifera
Perideridia gairdneri
Petrophytum caespitosum
Sagina saginoides
Saxifraga cernua
Swertia radiata
Telesonix jamesii
Triglochin concinuum
Vaccinium scoparium
Viola adunca

In addition to the preceding vegetation regions, there are several other plant communities of the state that are noteworthy.

Wooded Bottomlands

Several major drainage routes in the state, the most important of which are the Missouri, James, Big Sioux, Grand, Cheyenne, White, and the Moreau, have alluvial deposits that support a substantial number of trees and shrubs as dominant forms.

Acer negundo
Acer saccharinum
Celastrus scandens
Clematis virginiana
Cornus stolonifera
Elaeagnus angustifolia

Fraxinus pennsylvanica
Parthenocissus vitacea
Populus deltoides
Quercus macrocarpa
Salix amygdaloides
Salix exigua

Sambucus canadensis
Shepherdia argentea
Toxicodendron rydbergii

Ulmus americana
Vitis riparia

Lakes, Ponds, and Prairie Potholes

Although many of the natural ponds and potholes have been drained and are now in agricultural production, a significant number of them still remain, especially in the northeast Coteau region. Few are populated at their margins with trees and shrubs which are more typically in alluvial habitats. Most of them are open with herbaceous, marsh plants inhabiting their margins. Almost all of the state lies in the Central Flyway, a major migratory route for waterfowl. It is not unusual for seeds or other propagules of marsh and aquatic forms to be deposited in favorable habitats as the birds rest and feed at these aquatic or marshy places. Some of these marshes and swampy areas are quite extensive.

The large, impounded reservoirs along the Missouri River in South Dakota, Lake Oahe, Lake Sharpe, Lake Francis Case, and Lewis and Clark Lake, have widely fluctuating shoreline levels where typical marsh and aquatic forms are, for the most part, completely lacking. Here the upland clays and shales are constantly eroded by the changing waterline, both by the raising and lowering of the reservoir level and by wind and wave action.

A representative listing of the principal marsh and aquatic plants follows.

Alisma plantago-aquatica
Cicuta maculata
Cyperus ferruginescens
Eleocharis macrostachya
Equisetum hyemale
Juncus dudleyi
Lemna minor
Potamogeton nodosus

Potamogeton pectinatus
Sagittaria cuneata
Scirpus validus
Sparganium eurycarpum
Typha angustifolia
Typha latifolia
Utricularia vulgaris

REFERENCES, INTRODUCTION

Burrough, R. D. 1961. The Natural History of the Lewis and Clark Expedition. Michigan State Univ. Press, East Lansing.
Custer, G. A. 1874-1875. Expedition to the Black Hills. 43rd Cong., 2nd Sess. Ex. Doc. 32.

Fenneman, N. M.   1931.   Physiography of the Western United
       States.   McGraw-Hill Co., New York.
McIntosh, A. C.   1931.   A botanical survey of the Black Hills
       of South Dakota.   Black Hills Engineer 19 Reprint, 1949,
       28(4).   Rapid City, S.D.
McKelvey, S. D.   1955.   Botanical Exploration of the Trans-
       Mississippi West, 1790-1850.   Arnold Arboretum, Harvard
       Univ., Jamaica Plain, N.Y.
Newton, H., and W. P. Jenney.   1880.   Report of the geology and
       resources of the Black Hills of Dakota.   U.S. Geograph.
       Geol. Surv., Rocky Mountain Region.   Washington, D.C.
Over, W. H.   1932.   Flora of South Dakota.   Univ. of S. Dakota,
       Vermillion.
Rothrock, E. P.   1943.   A geology of South Dakota I The surface.
       S. Dakota State Geol. Surv., Bull. 13.
Rydberg, P. A.   1896.   Flora of the Black Hills of South Dakota.
       Contrib. U.S. Nat. Herb. 3:463-523.
_____.   1954.   Flora of the Rocky Mountains and Adjacent
       Plains, 2nd ed.   Reprint.   Stechert-Hafner Co., New York.
U.S. Dept. Commerce, ESSA.   1968.   Climatic Atlas of the United
       States.   U.S. Environmental Data Service.   U.S.G.P.O.

Systematic Treatment

THE PRIMARY purpose of the keys and descriptions in the next section is to provide the amateur, as well as the serious student, a means of identifying the vascular plants that grow in the state of South Dakota. Vascular plants are those plants which have an organized vessel system that extends from roots to stems to leaves. The three large taxa, or classification units, that make up the vascular plants are:

1. The Pteridophyta, including the club-mosses, horsetails and ferns. In other treatments these may be referred to as the ferns and fern-allies. The Pteridophyta reproduce by spores instead of seeds.
2. The Gymnospermeae, also commonly called the gymnosperms or conifers. These plants produce cones and reproduce by seeds which are borne on the surfaces of scales making up the cone.
3. The Angiospermeae, also commonly called the flowering plants. They produce flowers and reproduce by seeds. The seeds, however, are borne within ovaries, functional parts of flowers. The Angiospermeae are segregated into the Monocotyledoneae, having one seed-leaf in the developing embryo, and the Dicotyledoneae, having two seed-leaves in the developing embryo.

The keys to species have been kept as simple as possible using a strictly dichotomous, or paired, arrangement. The descriptions are brief but, it is hoped, adequate. Descriptions include diagnostic characteristics and a statement of the plant's habitat. The terms describing the plant's frequency are largely arbitrary, based on information presently available. The indicated location or distribution of where a species occurs is kept as general as possible except where it is known to be rare. The month or months listed at the end of each description indicate the time of flowering and fruiting for that species.

Insofar as possible the International Rules of Botanical Nomenclature are followed. Well-known synonyms are included in

3

brackets near the end of the description for that species. If
varieties or subspecies have been recognized by an authority for
a group, these are listed but not distinguished, that is, keyed
out. Common names, if known, are included at the lower part of
the description.

The sequence of families in the Angiospermeae is basically
according to Engler and Prantl. After each "Key to Genera" the
genera are listed alphabetically.

For the sake of brevity, distribution maps, illustrations,
and other detailed references to types, nomenclature, and synonyms so desirable in a definitive flora are omitted. Pertinent
bibliographic material, however, follows the systematic section.
A glossary of terms used to describe anatomic structures or conditions that may separate one group from another is found immediately preceding the index.

No distinction is made between plants that are known to be
native to South Dakota, those that have been introduced and now
are naturalized, or those that grow spontaneously. Those that
are known to have been planted and have escaped, or cultivated
forms that may be easily confused with closely related species
in the previously listed categories, are also included.

The following several suggestions may be helpful in using
this manual to identify unknown plants or to verify a determina-
tion.

1. If possible, work the plant through the key when the plant
    is living, or at least prior to drying and wrinkling. Be
    able to observe a number of plants of that same species to
    get a true picture of plant size, leaf structure and arrange-
    ment, flower color and texture, type of root system, and
    fruits in various stages of maturity. In summary, do not
    base your decision on a single observation!
2. Collect specimens of plants to be identified that are in
    flower *and* in fruit and show clearly a portion of the root
    system to determine whether the plant is annual or perennial.
    This is especially important if collections are pressed and
    dried with the intention of identifying them at a later date.
3. A record of the plant's location, habitat and date of collec-
    tion also is helpful in identifying or later verifying a de-
    termination.
4. Each contrasting pair of statements or couplets in the key is
    preceded by a number and its prime, indented at the same lo-
    cation on the page. Each couplet usually begins with the
    same word or phrase but is contrasting. Read both couplets
    before making a choice. The first couplet or phrase may
    seem acceptable but the second one may be even better.
    After making the choice that best fits the plant in ques-
    tion, go directly to the next couplet immediately below and

indented.   This procedure will ultimately lead to the description of the plant in question and its name.
5.   Finally, several basic instruments are almost indispensible
     for keying and accurately identifying most plants:
     a.   a 15 centimeter plastic ruler, divided into 1 millimeter
          units,
     b.   a hand lens that magnifies up to 10 diameters,
     c.   a pair of fine forceps,
     d.   several fine-pointed needles.

   Do not hesitate to ask assistance of the many biologists,
teachers, county extension agents, game, soil, or range conser-
vationists, or others who have an interest in and work in the
out-of-doors in South Dakota.   They usually will be eager to
stimulate your interest in an exciting and challenging experi-
ence.

Families of Vascular Plants
of South Dakota

52. Capparidaceae
53. Crassulaceae
54. Saxifragaceae
55. Rosaceae
56. Fabaceae
57. Geraniaceae
58. Oxalidaceae
59. Linaceae
60. Zygophyllaceae
61. Rutaceae
62. Polygalaceae
63. Euphorbiaceae
64. Callitrichaceae
65. Anacardiaceae
66. Celastraceae
67. Aceraceae
68. Balsaminaceae
69. Rhamnaceae
70. Vitaceae
71. Tiliaceae
72. Malvaceae
73. Hypericaceae
74. Elatinaceae
75. Tamaricaceae
76. Cistaceae
77. Violaceae
78. Loasaceae
79. Cactaceae
80. Elaeagnaceae
81. Lythraceae
82. Onagraceae
83. Haloragaceae

84. Araliaceae
85. Apiaceae
86. Cornaceae
87. Ericaceae
88. Primulaceae
89. Oleaceae
90. Gentianaceae
91. Apocynaceae
92. Asclepiadaceae
93. Convolvulaceae
94. Polemoniaceae
95. Hydrophyllaceae
96. Boraginaceae
97. Verbenaceae
98. Lamiaceae
99. Solanaceae
100. Scrophulariaceae
101. Orobanchaceae
102. Bignoniaceae
103. Lentibulariaceae
104. Martyniaceae
105. Phrymaceae
106. Plantaginaceae
107. Rubiaceae
108. Caprifoliaceae
109. Adoxaceae
110. Valerianaceae
111. Dipsacaceae
112. Cucurbitaceae
113. Campanulaceae
114. Lobeliaceae
115. Asteraceae

Statistical Summary

Principal Components of the Flora of South Dakota

	Species	Genera
Pteridophyta	38	21
Gymnospermeae	8	3
Monocotyledoneae	429	128
Dicotyledoneae	1110	427

Number of Families Represented: 115

The Ten Largest Families with Number of Species in Each:

Asteraceae	226	Rosaceae	55
Poaceae	187	Ranunculaceae	42
Cyperaceae	109	Scrophulariaceae	40
Fabaceae	92	Polygonaceae	37
Brassicaceae	65	Apiaceae	32

The Ten Largest Genera with Number of Species in Each:

Carex	74	Ranunculus	18
Astragalus	27	Potamogeton	17
Polygonum	20	Potentilla	16
Aster	19	Salix	16
Euphorbia	19	Chenopodium	15

Number of Vascular Plant Species in South Dakota: 1585

Descriptions of Groups

1   Plants lacking flowers.

    2   Reproduction by spores which are borne in sporangia on specialized sporophylls or on vegetative leaves.

    2' Reproduction by seeds which are borne on cones or modified cones, ours all evergreen trees or shrubs.

1' Plants with flowers produced (Angiosperms).

    3   Flower parts in 3's or multiples thereof. Embryo with a single seed leaf, the first leaves alternate. Leaves usually parallel veined.

    3' Flower parts in 4's or 5's or multiples thereof, rarely in 3's or 6's. Embryo with two seed leaves, opposite. Leaves usually net veined.

GROUP I. PTERIDOPHYTA

PTERIDOPHYTA--Plants without flowers or seeds, reproducing by spores and borne in sporangia (Ferns and Fern-allies).

1   Leaves whorled, fused by their lower margins to form a sheath around the stem at conspicuous nodes. Stems grooved and hollow, usually harsh-surfaced . . . . Equisetaceae   page 10

1' Leaves not whorled, stems usually solid and not conspicuously jointed. Sporangia borne on surfaces of normal or modified leaves or in their axils.

2   Leaves clover-like, with 4 leaflets, long-petioled. Spo-
    rangia borne in ovoid sporocarps at the base of the peti-
    oles. Plants perennial in marshy places . . . .
    . . . . Marsileaceae   page 20

2' Leaves not clover-like.

   3   Leaves narrow, sessile and simple, not fern-like.

      4   Stems elongated and leafy. Leaves not quill-like.
          Sporangia usually borne in terminal strobili.

         5   Strobili circular in cross-section. Leaves with-
             out a ligule. Spores of one kind (homosporus)
             . . . . Lycopodiaceae   page 13

         5' Strobili more or less 4-angled. Leaves with a
             short ligule. Spores of two kinds (heterosporus)
             . . . . Selaginellaceae   page 13

      4' Stems short, thick and corm-like. Leaves quill-like
          in a basal tuft. Sporangia borne in the spooned
          bases of fertile leaves. Plants of wet soil or
          marshy places . . . . . Isoetaceae   page 12

   3' Leaves usually broad and fern-like, petiolate and com-
       pound. If not compound at least petioled.

      6   Sporangia-bearing frond different in shape and color
          from sterile or vegetative frond.

         7   Sporangia large, sessile, borne in a stalked ter-
             minal panicle. Sporangial wall more than one cell
             thick . . . . Ophioglossaceae   page 13

         7' Sporangia small and naked or borne within rolled
             pinnae. Sporangial wall not more than one cell
             thick . . . . Polypodiaceae   page 14

      6' Sporangia-bearing frond not markedly different in
          shape and color from sterile or vegetative frond
          . . . . Polypodiaceae   page 14

Family Equisetaceae

EQUISETUM L.

1 Stem simple or with few branches not regularly arranged on the
   stem.

   2   Stems over 15 cm tall, with prominent central cavities.

3 Nodal sheaths longer than broad, widened upward. Tips
   of sheaths forming a dark band. Stems smooth to the
   touch, 2-4 dm tall, annual, from a creeping rhizome
   that is perennial. Strobili 1-2 cm long, on short
   peduncles. Common over the state in sterile sandy
   soil or in dry prairie or along railroad ballast.
   {*E. kansanum* J. H. Shaeffer}. July-Aug . . . .
   . . . . *Equisetum laevigatum* A. Br.

3' Nodal sheaths not much longer than broad, not prominently widened upward. Sheaths possessing two dark
   bands which extend around the stem, one at the tips
   of the sheath, the other at the base of the sheath.
   Stems rough to the touch, perennial and evergreen, a
   meter or more tall. Frequent to common on moist, sandy
   soils, flood plains and shores over the state. {*E.
   robustum* A. Br.}. June-Aug . . . . *Equisetum hyemale* L.

2' Stems low and diffuse, usually less than 15 cm tall and
   0.5-1 mm thick, lacking a central cavity. Stems all alike,
   6-grooved and with angles. Strobili small, 3-5 mm long, on
   short peduncles. Rare to infrequent in moist or marshy
   valleys at higher altitudes in the Black Hills and in the
   northeast. June-Aug . . . . . *Equisetum scirpoides* Michx.

1' Stem usually with branches in regular whorls.

4 Ridges of stem with a double row of recurved spinules.
   Teeth of nodal sheaths reddish brown, several teeth fused
   into a single lobe. Stems of two kinds, the sterile ones
   2-6 dm tall with whorled branches. Sheaths 4-10 mm long,
   scarious brown towards maturity. Both sterile and fertile
   stems pale green, annual. Strobili blunt, on short peduncles. Rare to infrequent in moist places in the Black
   Hills. June-Aug . . . . *Equisetum sylvaticum* L.

4' Ridges of stem with a single row of spinules or smooth.

5 First internode of the primary branches mostly shorter
   than the nodal stem sheath. Occasionally stems are
   found that are unbranched. Sterile and fertile stems
   alike and green, 2-10 dm tall, producing strobili in
   summer.

6 Rhizomes black. Stem sheaths 10-14 mm long, the
   teeth long, with scarious margins. The central
   cavity of the main stem small, less than one fourth
   of the total diameter. Stems 2-7 dm tall, 7-10
   ridged. The strobili long-peduncled, deciduous.
   Rare to infrequent in moist places in the north and
   eastern part. June-Aug . . . . *Equisetum palustre* L.

6' Rhizomes reddish. Stem sheaths 5-10 mm long, the
teeth narrow, brown and essentially without scarious
margins. The central cavity of the main stem large,
over three fourths of the total diameter. Stems 2-
10 dm tall, 9-15 ridged. The strobili 1-2 cm long,
on peduncles. Rare on sandy floodplains or in
meadows over the state. June-Aug . . . .
. . . . *Equisetum fluviatile* L.

5' First internode of the primary branches much longer than
the nodal stem sheath. Sterile and fertile stems unlike,
producing strobili in early spring.

7 Fertile stem unbranched, 5-20 cm tall, yellowish-
brown, appearing in April or May, soon withering.
Sterile stems with branches upcurved, bright green,
1-3 dm tall, diffuse. Sterile branches not as thick
as the main stem. Sheaths greenish, 5-7 mm long, the
teeth brown or black at their tips. Strobili 5-25 mm
long, with blunt tips. A commonly occurring horse-
tail in various soils over the state. Apr-May . . .
. . . . *Equisetum arvense* L.

7' Fertile stem becoming green and branched, persistent.
Branches not strongly upcurved, more or less at right
angles from the main stem. Stems annual, the sterile
ones 1-4 dm tall. Sheaths blackened with pale margins, mostly free, 2-6 mm long. Strobili 1-2 cm
long, on short peduncles, deciduous. Infrequent in
rich woods in the Black Hills. June-Aug . . . . . .
. . . . *Equisetum pratense* Ehrh.

Family Isoetaceae

ISOETES L.

Isoetes melanopoda Gay and Durieu. Perennial plants with deciduo-
ous leaves up to 5 dm tall. Stems short, mostly below soil level.
Leaves 15-30, erect, narrowly linear and pointed at the apex.
Sporangia oblong, about 6 mm long. Heterosporous, the megaspores
0.6 mm across. Rare in shallow water in the southern part of the
state in Mellette county, now probably extinct. June-Aug.
Quillwort.

Family Lycopodiaceae

LYCOPODIUM L.

Lycopodium obscurum L. Plants perennial and evergreen from
creeping rhizomes. Stems 1.5-2.5 dm tall, erect, only sparingly
branched. Leaves in 6-8 ranks, crowded, 3-5 mm long. Strobili
terminal, sessile, erect, 2-3.5 cm long. The sporophylls are
markedly different from the leaves. Sporangia 1-2 mm wide.
Rare in moist places in higher altitudes in the Black Hills.
June-July. Ground Pine.

Family Selaginellaceae

SELAGINELLA Beauv.

1 Plants densely branched, with a white hoary appearance due to
   dense bristle tips of individual leaves. Cilia along leaf
   margins less prominent than in the following species. Stems
   short, almost prostrate, forming mats up to 15 cm across.
   Sporophylls in 4 rows, usually less than 1 cm long. Frequent
   in exposed places in the western part of the state. June-July
   . . . . *Selaginella densa* Rydb.

1' Plants more openly branched, with a light green appearance.
   In comparison, cilia along leaf margins more prominent than
   the foregoing species. Awn tips of leaves also more prominently scabrous. Sporophylls in dense spikes 1-2.5 cm long.
   An eastern species that occurs in the extreme eastern part of
   the state on Sioux quartzite but ranges to the western part
   on rocky or exposed areas. June-July . . . .
   . . . . *Selaginella rupestris* (L.) Spring.

Family Ophioglossaceae

BOTRYCHIUM Sw.

1 Sterile leaf blades consisting of 1-4 pairs of pinnae, the
   entire blade not over 6 cm across. Pinnae rounded or irregularly lobed. The blade glabrous, with a petiole. Plants 3-
   10 cm tall, the fruiting spike 2-6 cm long, erect. Apparently
   rare in moist soil of valleys of higher altitudes of the Black
   Hills. June-Aug . . . . . . *Botrychium simplex* E. Hitchc.

1' Sterile leaf blades bipinnate to quadripinnatifid or more dissected, the entire blade often over 6 cm across.

2 Blade leathery or fleshy, evergreen, the ultimate segments
   rounded or obtuse. Petiole well developed, attached to the
   stipe axis at or below the middle. Plants 5-20 cm tall,
   stout, becoming glabrous toward maturity. Rare in rich
   woods and moist valleys at higher altitudes in the Black
   Hills. June-Aug. Leather Grape Fern . . . .
   . . . . *Botrychium multifidum* (Gmel.) Rupr.

2' Blade membranous, deciduous, much dissected, the ultimate
   segments toothed. Blade sessile, 10-30 cm across and almost as long, the surfaces slightly hairy. Plants 4 dm or
   more tall. The fertile spike bipinnate, 3-12 cm long, on
   a stalk 5-15 cm long. Infrequent in rich woods of the
   eastern part and in the Black Hills and Harding County
   in the northwest. June-Aug. Rattlesnake Fern . . . . . .
   . . . . *Botrychium virginianum* (L.) Sw.

Family Polypodiaceae

(Key to Genera)

1 Sori covered by the inrolled margin of the frond--a false
   indusium.

   2 Stipe and rachis of light color, without a polished appear-
      ance . . . . *PTERIDIUM*

   2' Stipe and rachis purple-brown to black and with a polished
      appearance.

      3 Ultimate segments with dense, brown hairs on the under-
         surface . . . . *CHEILANTHES*

      3' Ultimate segments glabrous or with few hairs on the
         undersurface.

         4 Margins of ultimate segments entire . . *PELLAEA*

         4' Margins of ultimate segments dissected . . . . . .
            . . . . *ADIANTUM*

1' Sori not covered by the inrolled margin of the frond.

   5 Sori without indusia. The frond pinnatifid . . . .
      . . . . *POLYPODIUM*

   5' Sori usually with indusia. The frond not pinnatifid.

      6 Frond dissected.

         7 Stipe and rachis of frond dark brown and polished
            in appearance . . . . *ASPLENIUM*

7' Stipe and rachis of frond green to brown, not polished in appearance.

  8 Sporangia bearing frond markedly different from the sterile frond.

    9 Fertile and sterile fronds broadest at their bases . . . . *ONOCLEA*

    9' Fertile and sterile fronds tapering to a narrow base . . . . *MATTEUCCIA*

  8' Sporangia bearing frond not markedly different from the sterile frond.

    10 Sori elongate and straight to crescent-shaped but never completely covering the lower surface . . . . *ATHYRIUM*

    10' Sori round to round reniform, or completely covering the lower surface at maturity.

      11 Indusia superior and rounded in shape. This is best determined on young or immature sori.

        12 Ultimate segments of pinnae entire or but slightly toothed. Sori medial, at maturity confluent. Indusia pale . . . . *THELYPTERIS*

        12' Ultimate segments of pinnae prominently toothed.

          13 Indusium like an umbrella with a central stalk, opening all around . . . . *POLYSTICHUM*

          13' Indusium horse-shoe shaped or reniform, opening along the sides and end . . . . . *DRYOPTERIS*

      11' Indusia superior and hood-like, inferior, or absent.

        14 Basal pinnae stalked, 5 mm or more long . . . . *GYMNOCARPIUM*

        14' Basal pinnae sessile or stalk much less than 5 mm long.

          15 Veins of ultimate segments reaching the margin of the leaflet. Indusium, when present, superior and hood-like . . *CYSTOPTERIS*

15' Veins of ultimate segments not
reaching the margin of the leaflet. Indusium, when present, inferior, with deeply lacerate segments . . . . . . *WOODSIA*

6' Frond simple, with an entire margin . . *ASPLENIUM*

ADIANTUM L.

1 Leaf blade longer than wide, the pinnae alternate on a continuous rachis. Pinnae with up to 6 pairs of pinnules or ultimate segments. Perennial from a short, creeping rhizome.
Fronds arched or lax-drooping, 1-4 dm tall, 1-3 times pinnate,
the pinnae alternate. Ultimate segments, pinnules, with
rounded lobes, the sori marginal, giving the lobes a crenate
appearance. Locally abundant on rocks along Cascade Creek in
Fall River county. July-Aug . . *Adiantum capillus-veneris* L.

1' Leaf blade as wide as long, the two primary divisions recurved-
spreading. Pinnae with up to 30 pairs of pinnules or ultimate
segments. Perennial from a short, creeping rhizome that is
stout. Leaves few, 1-3.5 dm long and almost or fully as wide,
the petiole purple-black. Main rachis divided sub-equally,
the divisions bearing progressively shorter pinnae. Sori
borne on the reflexed margins of the pinnules. Apparently
rare in north-facing woods at higher elevations in the northern Black Hills. July-Aug . . . . *Adiantum pedatum* L.

ASPLENIUM L.

1 Pinnae usually less than 4 in number on each frond, these narrowly linear with narrow, elongate sori. Fronds tufted, 4-15
cm long. Stipes tan or brownish on the lower part. Two or
three sori on each segment. A western species in rock crevices in the region of the Needles and Sylvan Lake area in the
Black Hills. June-Aug . . *Asplenium septentrionale* (L.) Hoffm.

1' Pinnae many, not narrowly linear or with narrow elongate sori.

2 Stipes of fronds brown to black only at the base, the upper
part of the rachis green like the pinnae. Fronds 5-15 cm
long, clustered from a perennial base, the old stipe bases
persisting for years. Fronds with 6-18 pairs of pinnae
each, with coarse, rounded lobes or teeth. Several sori
on each pinna, each one 1-2 mm long. Infrequent in crevices of limestone in Lawrence County in the Black Hills.
June-Aug. Green spleenwort . . . . *Asplenium viride* Huds.

2' Stipes and rachis brown to black their entire length.
Fronds spreading, up to 35 cm long, the pinnae glabrous

and evergreen. Stipe bases persisting on the perennial
base. Pinnae oval or oblong, the distal margins crenate.
Sori several per pinna, each 1-2.5 mm long. Indusium conspicuous. Infrequent in rock crevices at higher altitudes
in the Black Hills. June-Aug. Maidenhair Spleenwort . . .
. . . . *Asplenium trichomanes* L.

ATHYRIUM Roth.

Athyrium filix-femina (L.) Roth. Fronds 2-18 dm tall, several
from a short sub-erect rhizome, this covered by the bases of
previous year stipes. Stipes with chaffy scales that are tawny
brown. Blades twice pinnate, with 10-30 pairs of pinnae. Each
pinna with 12-30 pairs of offset pinnules. Sori rounded or elliptic, mostly 1 mm or so across. Infrequent in rich woods of
the Black Hills. June-Aug. Lady Fern.

CHEILANTHES Sw.

Cheilanthes feei Moore. Small ferns with fronds 5-13 cm long,
tufted, the fronds with a dense, brown, tomentose covering.
Blades tri-pinnatifid, rarely bi-pinnate. Stipes of fronds dark
brown or black. From 6-10 pairs of opposite pinnae, the ultimate
pinnules small, not much over 1 mm long. Sporangia covering the
entire lower surface of the pinnule. Frequent on limestone in
the western part. July-Aug. Lip Fern.

CYSTOPTERIS Bernh.

1 Basal pinnae slightly longer than the second pair of pinnae.
   Bulblets frequently borne on the undersurface of fertile
   fronds. Leaves in small clusters from a short rhizome, the
   blades 3-8 cm long, bipinnate, with 10-18 pairs of pinnae,
   the pinnules deeply pinnatifid. Rare in rich woods of
   Roberts and Marshall Counties in the northeast part. July.
   Bulblet Bladder Fern . . . . *Cystopteris bulbifera* (L.) Bernh.

1' Basal pinnae slightly shorter than the second pair of pinnae.
   Bulblets absent. Leaves from a creeping rhizome, the fronds
   scattered, 1-5 dm high. Blades 1-3 times pinnate, with 8-15
   pairs of pinnae. The pinnae deltoid-lanceolate or ovate, the
   ultimate segments toothed or cleft. Frequent in shady banks
   of upland or rich woods over the state. July. Common Bladder
   Fern . . . . *Cystopteris fragilis* (L.) Bernh.

DRYOPTERIS Adans.

Dryopteris filix-mas (L.) Schott. A large coarse fern with
leaves clustered from a short horizontal or ascending rhizome,

the stipes conspicuously scaly. Blades 2-10 dm long, evergreen,
bipinnatifid, with 15-25 pairs of pinnae. Pinnules finely serrate. Sori nearer the midvein than the margin, orbicular, in
two rows. Infrequent in moist rocky woods of the Black Hills.
June-Aug. Shield Fern.

GYMNOCARPIUM Newm.

Gymnocarpium dryopteris (L.) Newm. Small plants 1-4 dm tall, the
fronds delicate, from thin, scaly rhizomes. Blades 2-4 pinnate,
the basal pinna opposite, deltoid, with pinnatifid pinnules.
Sori scattered, rounded or broadly elliptical, indusium lacking.
Rare in rocky crevices of steep, north-facing gorge below Sylvan
Lake in Custer County and in deep canyons elsewhere in the Black
Hills. July-Aug. Oak Fern.

MATTEUCCIA Todaro

Matteuccia struthiopteris (L.) Todaro. Plants with coarse di-
morphous fronds, clustered from scaly, creeping rhizomes. Sterile fronds with numerous pinnatifid pinnae alternately arranged,
5-20 dm tall. Pinnae 5-15 cm long. Fertile fronds shorter, containing contracted revolute margins covering the sori. The sori
rounded, on slightly elevated structures. Infrequent in moist
wooded ravines and on stream banks in the Black Hills and the
extreme northeast part. June-Aug. Ostrich Fern.

ONOCLEA L.

Onoclea sensibilis L. Plants with dimorphic fronds, the sterile
ones 3-10 dm tall, the main rachis winged. Blade ovate, pinnatifid, the segments undulate or sinuate. Fertile fronds shorter,
bipinnate, the pinnules contracted into ball-like sporangia which
dehisce at maturity. Infrequent in damp springy soils and in
moist woods in the western part. June-Aug. Sensitive Fern.

PELLAEA Link.

1 Stipes and rachises glabrous. Fronds 5-15 cm long, pale blue-
   green. Leaves of one kind with the pinnules all quite narrow.
   Plants not more than 15 cm tall. Blades once-pinnate with 3-
   several pairs of almost opposite pinnae. The lowest ultimate
   segments 1-2 cm long. Sori continuous along the margin which
   is reflexed to form a false indusium. Frequent on calcareous
   rocks of the Black Hills. Ours var. *occidentalis* (E. Nels.)
   Butters. July-Aug. Cliff-brake . . . *Pellaea glabella* Mett.

1' Stipes and rachises with scurfy, jointed, appressed hairs.
   Fronds 10-50 cm long, grayish-green. Leaves of two kinds,

with some of the lower pinnules broader than those upward.
Plants with a short rhizome. Blades bi-pinnate below and
once pinnate above, the stalk and rachis over 15 cm long.
Sori as in the preceding species. Infrequent in rock crev-
ices of lower altitudes in the Black Hills. July-Aug . . . .
. . . . *Pellaea atropurpurea* (L.) Link.

POLYPODIUM L.

1 Paraphyses present among the sporangia. Rhizomes licorice-
flavored, the tissues very hard. Chaffy scales on the rhizomes uniformly brown. Sori medial on each side of the main
vein. Plants evergreen with simple, pinnatifid leathery
fronds 5-20 cm long. Lower pinnae usually shorter than the
middle ones. Frequent in shady rock crevices of sandstone
in the western part, including the Black Hills. {This species has been reported to be diploid whereas the following
species is tetraploid.} July-Aug. Western Polypody . . . . .
. . . . *Polypodium hesperium* Maxon

1' Paraphyses lacking among the sporangia. Rhizomes not sweet or
licorice-flavored, the tissues uniformly softer than in the
preceding species. Chaffy scales on the rhizomes dark on the
dorsal side. Sori located near the margins of the pinnae.
Plants evergreen, the pinnatifid fronds 4-25 cm long. Lower
pinnae usually as long as or longer than those at the middle.
Rare on Sioux quartzite in Minnehaha County in the extreme
eastern part. July-Aug. Common Polypody . . . .
Polypodium virginianum L.

POLYSTICHUM Roth.

Polystichum munitum (Kaulf.) Presl. Plants from stout rootstocks.
Fronds firm, evergreen, in clusters 2-12 dm long. Blades once-
pinnate, the pinnae offset 30-50 or more on each side of the
rachis. Rachis scaly. Upper side of the pinnae with a lobe or
tooth, the margins otherwise toothed. Sori on middle and upper
pinnae, in single rows on each side of the middle vein. The
indusium reniform, with threaded margins. Rare in rich woods
of the Black Hills. July-Aug. Christmas Fern.

PTERIDIUM Scop.

Pteridium aquilinum (L.) Kuhn. Coarse ferns of moist places, the
leaves 5-15 dm tall from extensive, creeping rhizomes. Blades
deltoid, decompound, the larger ones tri-pinnate below. Pinnules
entire, lobed or pinnatifid. Sori borne adjacent to the margin
connecting the veins, the margin of the pinnule curved back to
form a false indusium. A variable species found in rich, loamy

soils of the Black Hills. Ours var. *latiusculum* (Desv.) Underw. July–Aug. Bracken Fern.

THELYPTERIS Schmidel

Thelypteris palustris Schott. Plants from slender, creeping root stocks, the fronds 3–7 dm tall. Blades lanceolate, once-pinnate, the pinnae 3–6 cm long, broadly linear, pubescent beneath, pinnatifid. Sori median on the pinnae, 1–2 mm long, crowded. Rare but locally abundant along the Minnechadusa Creek in swampy soil in Bennett and Todd counties of the extreme south central part and in the Black Hills. June–Aug. Marsh Fern.

WOODSIA R. Br.

1 Lower frond surfaces glabrous (do not mistake segments of the laciniate indusium for hairs). Plants 4–25 cm tall, the blades irregularly lance-ovate, once pinnate. Pinnae oblong, triangular, the segments irregular. Teeth of the segments often curved back over the sori. Sori sub-marginal, the inferior indusium with narrow lobes. Infrequent in the western one half of the state. July. Oregon Woodsia . . . . *Woodsia oregana* D. C. Eaton

1' Lower frond surfaces bearing shiny, flexuous hairs. Plants 4–10 cm tall, the fronds lanceolate, pinnate. Pinnae oblong-ovate, deeply pinnatifid. Sori rounded, their inferior indusia with lacinate lobes. Frequent on rocks and rocky crevices at middle elevations of the Black Hills. July. Rocky Mountain Woodsia . . . . *Woodsia scopulina* D. C. Eaton

Family Marsileaceae

MARSILEA L.

Marsilea vestita Hook. & Grev. Marsh or aquatic ferns with long-petioled leaves that float. Rooted in shallow water at margins of ponds or lakes. Blades shaped like a four-leaved clover, the leaflets 4–15 mm across. Surfaces variously pubescent to glabrous. Spores borne in solitary sporocarps 4–6 mm long, from short axillary peduncles. Frequent in ponds and temporarily filled potholes of the eastern one-half. July–Aug. Pepperwort. {*M. mucronata* A. Br.}

GROUP II. GYMNOSPERMEAE

GYMNOSPERMS--Plants with seeds but not flowering, ours all evergreen trees or shrubs.

1 Leaves and cone scales opposite or whorled, the leaves small,
   scale-like or subulate . . . . Cupressaceae

1' Leaves and cone scales spirally arranged, fascicled or alternate, but never opposite or whorled . . . . . Pinaceae

Family Pinaceae

(Key to Genera)

1 Leaves quadrangular in cross-section, arranged singularly on
   the twig, not sheathed at the base. Cones without woody
   scales . . . . *PICEA*

1' Leaves hemispherical in cross-section, arranged in clusters
   of 2 to 3 or 5 on the twig, sheathed at the base. Cone
   scales woody . . . . *PINUS*

PICEA A. Dietr.

Picea glauca (Moench.) Voss. Erect tree up to 20 meters tall,
with a pyramidal crown. Leaves glaucous green, 4-sided, 12-20
mm long. Cones produced on one year-old twigs, the ovulate cones
maturing in one season, with thin, non-prickly scales, light
brown, 3-6 cm long. The Black Hills spruce of the horticulturist
has been referred to as *P. glauca* (Moench.) Voss var. *densata*
Bailey. Common in higher altitudes in the Black Hills. Apr-May.
White Spruce.

PINUS L.

1 Leaves in fascicles of 5, many times clustered at the ends of
   branches. Tips of cone scales not prickly. Trees 5-15 meters
   tall, not very symmetrical. Needles 4-6 cm long. Ovulate
   cones slender to slightly ovate, light brown, 10-13 cm long.
   Rare as an isolated stand of trees in the "Cathedral Spires"
   area south of Harney Peak on the Custer-Pennington County line
   in the Black Hills. May. Limber Pine . . *Pinus flexilis* James

1' Leaves in fascicles of 2 or 3. Tips of cone scales prickly.

   2 Leaves 8-20 cm long or longer, often in fascicles of 3 as
      well as 2. Cones over 7 cm long, falling from the tree
      after maturity. Trees up to 30 meters tall, the bark

becoming thick and scaly, reddish. Our common pine of the
state. From Tripp County westward to lower elevations of
the Black Hills. May. Ponderosa Pine . . . .
. . . . *Pinus ponderosa* Laws

2' Leaves less than 8 cm long, in fascicles of 2. Cones less
than 7 cm long, many staying on the tree for a year or more
after maturity. Trees 10-20 meters tall, with crowns
rounded to pyramidal. Bark not as thick, much furrowed,
gray-black. Infrequent in small patches throughout the
Black Hills. {*P. murrayana* Balf.}. May. Lodgepole Pine.
. . . . *Pinus contorta* Dougl.

Family Cupressaceae

JUNIPERUS L.

1 Leaves in whorls of 3, their bases joined, all subulate and of
one kind, without glands, 5-15 mm long, spreading. Plants
small arborescent trees, seldom over 2 meters tall, or spreading and shrubby. Foliage pungent, whitish on the upper surface, green on the lower. Plants dioecious, the ovulate cones
(berries) maturing the second season. Fruits blue-black, 5-8
mm long, 1-several seeded. Frequent on dry hillsides of the
western part. Apr-May. Common Juniper . . . .
. . . . *Juniperus communis* L.

1' Leaves mostly in 2's, opposite, of two kinds, most scale-like,
some subulate. Scale-like leaves with glands on the dorsal
surface.

2 Plants with stems spreading and prostrate, occasionally a
trailing shrub. Leaves not spreading, closely appressed,
blue-green to bright green. Foliage with a pungent odor.
Plants dioecious, the ovulate cones (berries) blue to
black, 6-8 mm long, maturing in one year, on recurved
peduncles. Frequent on rocky or sandy hillsides and banks
south of the White River and in the Black Hills. May-June.
Creeping Cedar . . . . *Juniperus horizontalis* Moench.

2' Plants upright, at least shrubby but more commonly arborescent. All leaves scale-like except for juvenile ones.

3 Leaves overlapping. Glands oval to elliptic, shorter
than the distance from the gland to the leaf tip. Tips
of leaves acute. Trees up to 10 meters, often with a
pyramidal or conic crown. Fruit maturing the first
season, 3-6 mm in diameter. Common on hillsides in the
southern part of the state and hybridizing with and

intergrading into the following species west of the central part of the state. May-June. Eastern Red Cedar . . . . *Juniperus virginiana* L.

3' Leaves not overlapping as much as the preceding species. Glands more elliptic than oval, longer than the distance from the gland to the leaf tip. Tips of the leaves obtusely rounded. Trees smaller, many times scraggly and with dome-like crowns. Fruit maturing the second season; therefore, female trees will usually possess two types of maturing cones (berries). Frequent on hillsides of the western one-half. Apr-May. Rocky Mountain Cedar. . . . . *Juniperus scopulorum* Sarg.

GROUP III. MONOCOTYLEDONEAE

MONOCOTYLEDONS--Flower parts in 3's or multiples thereof. Embryo with a single seed leaf, the first leaves alternate. Leaves usually parallel veined.

1 Plants grass-like with leaves narrow and parallel veined. Flowers not showy. Perianth, if present, green or brown, sometimes arranged in chaffy scales (grasses, sedges, and rushes) . . . . Section 1.   page 23

1' Plants not grass-like; however, the leaves may be narrow. {*Smilax* and *Trillium* in the Liliaceae have broad leaves and are treated in Section 3.}

   2 Plants growing on, in, or from out of the water, sometimes prostrate on the muddy or sandy margins of aquatic areas. . . . . Section 2.   page 24

   2' Plants growing on land, if at the margin of the water, their stems erect.

      3 Stems leafy, the leaves sometimes reduced to almost scale-like size . . . . . Section 3.   page 25

      3' Stems leafless or possessing a leaf or leaves immediately below the flower cluster only . . . . Section 4.   page 26

Section 1. Grasses, Sedges and Rushes

1 Perianth of 6 parts, green. Fruit a capsule with 3-many seeds . . . . Juncaceae   page 36

1' Perianth of bristles or lacking (leaf-like bracts or scales
   appearing as perianth parts in the grasses). Fruit a single-
   seeded caryopsis or achene.

   2 Stem cylindric. Nodes conspicuous. Leaf sheaths split.
      Leaves in two ranks on the stem. Fruit a caryopsis . . .
      . . . . Poaceae   page 70

   2' Stem triangular or cylindric. Nodes not conspicuous. Leaf
      sheaths closed. Leaves in 2 or 3 ranks on the stem. Fruit
      an achene . . . . Cyperaceae   page 39

                Section 2. Marsh or Aquatic Monocotyledons

1 Plants free-floating thalli less than 3 cm across, lacking
   normal stems and roots . . . . . Lemnaceae   page 122

1' Plants usually with leaves, stems, and roots.

   2 Plants aquatic with floating or wholly submersed leaves.
      When flowering, the flowers submersed, floating or raised
      above the water level.

      3 Leaves basal, long, tape-like, rooted on the bottom,
         Vallisneria, or leaves whorled, *Elodea* . . . .
         . . . . Hydrocharitaceae   page 28

      3' Leaves cauline or basal, alternate or opposite . . . . .
         . . . . Najadaceae   page 30

   2' Plants of shallow water or muddy margins. The leaves
      partly or all above water line.

      4 Perianth absent, or, if present, not having showy,
         petal-like inner segments.

         5 Flowers in an elongate spike.

            6 Spikes erect.

               7 Flowers perfect . . Juncaginaceae   page 29

               7' Flowers separated into a pistillate lower por-
                  tion and a staminate upper portion . . . . . .
                  . . . . Typhaceae   page 121

            6' Spikes lateral . . . Araceae   page 122

         5' Flowers in globose heads, unisexual, the pistillate
            heads below . . . . . Sparganiaceae   page 120

      4' Perianth more or less showy, the inner segments at least
         petal-like.

8 Many ovaries in each flower, leaves basal.

    9 Perianth of 3 green sepals and 3 white or almost white petals . . . . Alismataceae   page 26

    9' Perianth of 6 pink segments, not sharply differentiated . . . . . . Butomaceae   page 26

8' One ovary in each flower, leaves basal or on branched stems . . . . Pontederiaceae   page 125

Section 3. Terrestrial Leafy-stemmed Monocotyledons

1 Individual flowers borne on a fleshy spadix which are enclosed or arched over by a spathe. Leaves compound. *Arisaema* in . . . . Araceae   page 122

1' Individual flowers not borne on a fleshy spadix, the leaves simple.

  2 Ovary or ovaries superior.

    3 Sepals green, petals colored or white.

      4 Flowers irregular, leaves cauline. *Commelina* in . . . . Commelinaceae   page 124

      4' Flowers regular.

        5 Inflorescence a loose umbel. *Tradescantia* in . . . . Commelinaceae   page 124

        5' Inflorescence of solitary flowers or grouped other than in a loose umbel . . . . Liliaceae   page 125

    3' Sepals and petals not well-differentiated . . . . Liliaceae   page 125

  2' Ovary inferior, appearing below the perianth.

    6 Flowers regular, stamens 6 . . . . Amaryllidaceae   page 136

    6' Flowers irregular.

      7 Stamens 3, opposite the sepals, separate or fused above . . . . Iridaceae   page 136

      7' Stamens 1 or 2, fused to the style into a column . . . . Orchidaceae   page 137

Section 4. Terrestrial, Leafless Monocotyledons

1 Flowers densely crowded on a cylindric, lateral spike. *Acorus*
in . . . . Araceae   page 122

1' Flowers not as above.

  2 Carpels becoming separate at maturity . . . .
   . . . . Juncaginaceae   page 29

  2' Carpels fused into a compound ovary.

    3 Ovary superior . . . . . . Liliaceae   page 125

    3 Ovary inferior . . . . . . Orchidaceae   page 137

Family Butomaceae

BUTOMUS L.

Butomus umbellatus L. Perennial from a spreading rhizome system.
Leaves basal, narrow, erect, up to a meter or more tall. Flowering scapes equal or exceeding the leaves, up to 1.5 meter. Inflorescence umbellate, the flowers pink, 2 cm or more across.
Fruiting follicles in a whorl, the style terminal. Locally abundant at Lake Faulkton, Faulk County. June-July. Flowering Rush.

Family Alismataceae

(Key to Genera)

1 Carpels attached to the receptacle in a ring with the style on
the neutral margin. Flowers all perfect, stamens 6 . . . . .
. . . . *ALISMA*

1' Carpels in dense heads on convex receptacles.

  2 All flowers of the inflorescence perfect. Carpels plump.
   Head and achenes appearing bur-like when mature. Achenes
   ribbed or ridged. Plants annual . . . . . *ECHINODORUS*

  2' All of the upper flowers staminate, rarely pistillate,
   never perfect. Carpels flattened, achenes winged. Leaf
   blades sagittate, lanceolate, or linear. Plants perennial
   . . . . *SAGITTARIA*

ALISMA L.

1 Achenes with 2 grooves on the dorsal surface. Leaves tending
to be lance-shaped. Styles up to 5 mm long, curved. Flowering

scapes 1-3 dm tall, the inflorescence small. Flowers on short
pedicels on the main axis, the petals tending to be somewhat
pinkish. Infrequent in shallow water or muddy areas over the
state. {*A. geyeri* Torr.}. July-Aug. . *Alisma gramineum* Gmel.

1' Achenes with a single groove on the dorsal surface. Leaves
   ovate in shape. Styles up to 7 mm long, straight. Flowering
   scapes up to 7 dm or more, the inflorescence diffuse, large.
   Flowers in a compound panicle, the petals tending to be white.
   Common in moist areas over the state. Ours var. *americanum*
   Roem. & Schult. July-Sept. Water Plantain . . . .
   . . . . *Alisma plantago-aquatica* L.

ECHINODORUS Rich.

Echinodorus rostratus (Nutt.) Engelm. Annual or weak perennial,
2-5 dm tall. Plants without upright stems. Leaves from the soil
line, ascending, 1-4 dm tall, the blades ovate. Flowering scapes
taller than the leaves. Flowers in whorls on the scape, perfect.
Petals white, 7-9 mm long. Fruiting heads globose, 6-10 mm in
diameter. Infrequent in moist places in the southern part.
July-Sept. Burhead.

SAGITTARIA L.

1 Lower flowers with peduncles thickened and soft, often reflexed. The sepals appressed against the mature fruit.
   Plants with long petioled leaves, the blades hastate or
   sagittate, 10-50 cm long. Flowering scapes stout. Inflorescence in 3-8 whorls, the lower flowers pistillate. Achenes
   2-3 mm long, with narrow wings. Infrequent in marshes and on
   shores over the state. Ours ssp. *calycina* (Engelm.) Bogin.
   July-Sept . . . . *Sagittaria montevidensis* Cham. & Sch.

1' Lower flowers with peduncles not thickened or reflexed.
   Sepals spreading in fruit.

   2 Leaves with blades lanceolate, lacking basal lobes. Plants
      1-6 dm tall, often in water. Leaves flattened, the blades
      8-30 mm long, 3-5 ribbed. Inflorescence of 2-8 whorls, the
      upper staminate. Achenes 2-3 mm long, the beak short, less
      than 0.5 mm. Infrequent in ponds and on shores in the
      eastern part. Ours var. *graminea*. July-Aug . . . .
      . . . . *Sagittaria graminea* Michx.

   2' Leaves with sagittate basal lobes.

      3 Bracts of the inflorescence lanceolate with acute tips.
         Achenes with vertical or ascending style bases (beaks).

4 Achenes with erect beaks 0.2-4 mm long. Mature heads not depressed at the crown. The bracts at the base of the pistillate peduncle usually much shorter than the peduncle. Plants from a rhizome, the leaves long-petioled, sagittate or hastate. If present, floating or submersed leaves may be linear or ovate. Flowers on scapes up to 5 dm long. Achenes 2-3 mm long, with wings on the margins. Frequent to common over the state. July-Sept . . . . *Sagittaria cuneata* Sheld.

4' Achenes with ascending beaks 0.5-1.5 mm long. Mature heads depressed at the upper part. The bracts firm, often as long as the pistillate peduncles. Plants erect, the leaves petioled. Flowering scapes 3-7 dm tall, the flowers in 3-8 whorls. Sepals short, reflexed after flowering. Petals white, 8-10 mm long. Achenes 3-5 mm long, with marginal and facial wings. Rare to infrequent in the eastern part. Ours ssp. *brevirostra* (Mack. & Bush.) Bogin. July-Sept . . . . *Sagittaria engelmanniana* J. G. Smith

3' Bracts ovate to obtuse, thin, almost scarious. Achenes with a horizontal beak. Plants stout, 2-6 dm tall. Leaves with blades sagittate, 5-20 cm long. Flowering scapes with flowers in 3-12 whorls, the pedicels of staminate flowers longer than those of pistillate flowers. Mature achenes 2.5-4 mm long, the beak lateral. Our most common species in marshes and on shores over the state. July-Sept. Duck-potato . . . . *Sagittaria latifolia* Willd.

Family Hydrocharitaceae

(Key to Genera)

1 Stem elongate. Leaves whorled, less than 2 cm long. Plants rooted but easily broken and often found floating . . . . *ELODEA*

1' Stem short or lacking. Leaves basal, long and ribbon-like. Plants rooted in soil. Pistillate flowers very long stalked, the stalk becoming spiral in fruit. Staminate flowers released and floating free . . . . *VALLISNERIA*

ELODEA Michx.

1 Leaves elliptical or oblong, averaging 1.5-2.0 mm wide, often dense and overlapping on the stem, in whorls of 3. Stems

branched and forming large mats. Plants polygamo-dioecious.
Staminate flowers, when present, borne on a long capillary
peduncle that extends to the water surface. Petals of pistillate flowers 2.6 mm long, white. Fruits ovoid, maturing under
water. Infrequent in quiet water over the state. {*Anacharis
canadensis* (Michx.) Rich.}. July-Aug. Waterweed . . . . . .
. . . . *Elodea canadensis* Michx.

1' Leaves narrowly oblong to linear, averaging slightly less than
   1.5 mm wide, not dense or overlapping, in whorls of 3. Stems
   only sparingly branched. Plants polygamo-dioecious. Staminate flowers, when present, sessile on the stems and deciduous, floating to the water surface at anthesis. Pistillate
   flowers white, 2-4 mm across. Fruit narrowly ovoid, with few
   seeds, maturing under water. Infrequent in quiet water over
   the state. July-Aug . . . *Elodea nuttallii* (Planch.) St. John

VALLISNERIA L.

Vallisneria americana Michx. Submersed plants acaulescent,
rooted, with ribbon-like leaves 1-4 dm long and about 5 mm wide.
Pistillate flowers on long peduncles, reaching the surface of
the water. Staminate flowers crowded on a spadix, becoming detached at anthesis and rising to the surface before pollination.
Infrequent in the northeast and northern parts. July-Aug. Eel-
grass.

Family Juncaginaceae

TRIGLOCHIN L.

1 Fruits with 6 ovoid-shaped carpels, not tapering to the base.

   2 Ligules split nearly to their bases, appearing bi-lobed.
      Rhizomes slender. Plants 1-3 dm tall, their leaves rounded
      in cross-section. Flowering racemes 12-20 flowered, fewer
      than the following. Fruits 4-5 mm long. Rare to infre-
      quent in marshes in the extreme western part. Ours var.
      debile (M. E. Jones) J. T. Howell. June-July . . . .
      . . . . *Triglochin concinnum* Davy

   2' Ligules entire or only slightly toothed, not appearing bi-
      lobed. Rhizomes thick. Plants usually over 3 dm tall,
      their leaves somewhat compressed in cross-section. Racemes
      several to many flowered, their fruits about 5 mm long.
      Infrequent in saline or marshy soils over the state. June-
      July . . . . *Triglochin maritimum* L.

1' Fruits with 3 carpels, their bases slender and tapered.
   Plants slender stemmed, 2-5 dm tall, less than 2 mm in diameter. Ligules bi-lobed, parted to the base. Leaves sharp
   pointed. Fruits about 6 mm long. Rare in saline soils in
   the western part. June-July . . . . . *Triglochin palustre* L.

Family Najadaceae

(Key to Genera)

1 Leaves all or chiefly alternate, the uppermost sometimes sub-
  opposite.

   2 Leaves 1-3 cm long. Ovary 1, fruit solitary . . . .
     . . . . *NAJAS*

   2' Leaves 3-10 cm long. Ovaries usually 4. Fruits 2-4 . . .
      . . . . *ZANNICHELLIA*

1' Leaves all opposite. Flowers and fruits axillary.

   3 Plants of fresh or rarely brackish water. Flowers and
     fruit in an exsert spike or head . . . . . *POTAMOGETON*

   3' Plants of brackish water. Flowers 2, borne on a short
      spadix completely concealed at anthesis within the leaf
      sheath. Fruit becoming exserted due to the growth of the
      spadix base . . . . *RUPPIA*

NAJAS L.

1 Leaves opposite with inrolled margins and recurving, lanceolate tips, 1-2 cm long. Fruits lustrous, olive-brown at maturity, lance-shaped, 2-3.5 mm long and about one-third as
  thick. The surface of the fruit with 30-50 fine reticulate
  markings. Plants monoecious, submersed. Leaf bases triangularly enlarged in outline. Stems slender, forked, 1-5 dm
  long. Infrequent in shallow streams and at pond margins in
  the eastern part of the state. June-Aug. Shiny Najas . . .
  . . . . *Najas flexilis* (Willd.) Rostk. & Schmidt

1' Leaves opposite, flattened, the tip merely acute, 12-20 mm
   long. Fruits dull, 1.5-2.5 mm long, lance-shaped, with 15-20
   distinct reticulate markings on the surfaces. Plants monoecious, submersed. Leaves crowded, their bases much like the
   preceding. Stems branched, 2-6 dm long. Rare to infrequent
   in shallow water over the state. July-Sept . . . .
   . . . . *Najas guadalupensis* (Spreng.) Morong.

POTAMOGETON L.

1 All leaves narrowly linear or bristle-like, submersed.

  2 Lower or mature leaves thread-like, 1-nerved, tapering to
     points.

    3 Sheaths of leaves less than 2 cm long. Leaves not obvi-
       ously septate with cross partitions. Stems branched
       chiefly at the base. Style base flat, the achenes 2-3
       mm long, not beaked. Plants with stems mostly less than
       4 dm long. Flowering peduncles slender, 1 dm or more
       long. Frequent in shallow water of lakes and ponds over
       the state. July-Sept . . . *Potamogeton filiformis* Pers.

    3' Sheaths of leaves mostly more than 2 cm long. Leaves
       septate with cross partitions. Stems dichotomously
       branched from nodes. Achenes 3.5-4 mm long, with a
       short beak. Stems 2-4 dm long. Flowering spikes 1-3
       cm long, on peduncles 1 dm long. Quite common in shal-
       low water of ponds and lakes, especially alkaline ones.
       July-Sept. Sago Pondweed . . *Potamogeton pectinatus* L.

  2' Lower or mature leaves flattened, with 1-3 nerves, the tips
     acute or bluntly tapered.

    4 Stems flattened and winged, 2-3 mm wide. Leaves 2-5 mm
       wide, with tips abruptly pointed. Plants freely
       branched, to 5 dm long. Leaves linear, up to 15 cm or
       more long. Fruiting spikes 1-3 cm long on peduncles
       about 1 dm long. Infrequent in lakes of the eastern
       and northeastern part. July-Sept . . . .
       . . . . *Potamogeton zosteriformis* Fern.

    4' Stems rounded in cross section, not winged. Leaves less
       than 2 mm wide.

      5 Main stipules of lower leaves inflated and several
         times wider than the leaves, sheathing the stem and
         subtending 2 or more branches. Leaf tips rounded at
         the apex. Stems with 2 or more branches at a node,
         up to 5 m long. Flowering peduncles up to 8 cm long,
         with 4-8 whorls of flowers. Achenes obovate, about
         3 mm long, with an inconspicuous keel. Rare in deep
         water of Pennington County in the western part.
         July-Aug . . . . *Potamogeton vaginatus* Turcz.

      5' Main stipules of lower leaves not inflated or several
         times wider than the leaves. Leaf apices acute to
         bluntly tapered.

6 Achenes with a prominent, wavy, dorsal keel.
Leaves without basal glands. Stems branched, 3-8
dm long and 1-2 mm in diameter. Leaves 3-8 cm
long and 1-2 mm wide, their tips acute. Fruiting
peduncles 1-3 cm long, thick, the spikes short.
Achenes 2 mm long, the beak to the side. Locally
in ponds and lakes over the state. Ours var.
foliosus. July-Aug . . *Potamogeton foliosus* Raf.

6' Achenes without a prominent, wavy, dorsal keel.
Leaves usually with a pair of basal glands.

7 Basal glands 3-6 mm broad. Stipules strongly
fibrous, whitened. Achene with a dorsal keel
and obscure lateral keels. Stems much branched,
up to 8 dm long. Leaves linear, 5-9 cm long
and 2-3 mm wide, 3-7 nerved. Flowering spikes
terminal or in upper axils, the spike 1-2 cm
long. Achenes 1.5-2.5 mm long, the beak
slightly curved. Infrequent in lakes, more
so in the eastern part. July-Aug . . . . . .
. . . . *Potamogeton friesii* Rupr.

7' Basal glands very small, less than 1 mm broad.
Stipules not fibrous, greenish. Achene rounded
on the dorsal side. Stems slender, much
branched, up to 1 meter long, rounded in cross
section. Leaves linear, up to 2 mm wide and 6
cm long, mostly 3 nerved. Flowering spikes of
2 or 3 whorls of flowers, slender. Achenes
obovoid, 2 mm long. Frequent to common in
lakes and ponds over the state. July-Aug . .
. . . . *Potamogeton pusillus* L.

1' All or at least the upper or floating leaves ovate to elliptical.

8 Leaves sessile, with clasping, cordate bases. No specialized floating leaves present.

9 Tips of leaves cucullate (shaped like a boat). Stipules
conspicuous, rigid, usually persistent. Achenes 4-5 mm
long, with a prominent dorsal keel. Stems 1-3 meters
long, with very little branching, 2-4 mm in diameter.
Leaves 1-3 dm long, lanceolate to oblanceolate, with 3
or 5 prominent nerves. Flowering spikes 2-4 cm long,
on well-developed peduncles. Rare to infrequent in
lakes of the northeast part and in the Black Hills.
July-Aug . . . . *Potamogeton praelongus* Wulf.

9' Tips of leaves not cucullate. Stipules soon disinte-
grating into white fibers. Achenes 2-4 mm long, only
slightly keeled if at all. Stems only sparingly
branched, 1-2 mm in diameter and 2-6 dm long. Leaves
2 cm wide and 6-8 cm long, their margins characteristi-
cally undulate-wavy. Flowering spikes up to 4 cm long,
on stout peduncles. Frequent to common in lakes and
quiet running water over the state. July-Aug . . . . .
. . . . *Potamogeton richardsonii* (Benn.) Rydb.

8' Leaves with definite petioles, at least on the upper part
of the stem, their bases not auriculate or clasping.

10 All leaves of the same general shape. Submersed leaves
usually 4 mm wide or wider.

11 Stipules less than 3 cm long. Submersed leaves
with 3-9 distinguishable veins.

12 Leaves all submersed, 4-7 mm wide with a crisp,
wavy margin that is finely serrulate. Stems
branched, 3-8 dm long, forming mats underwater.
Stipules 4-9 mm long, deciduous early, their
margins free. Flowering and fruiting spike
about 1 cm long, dense, about as thick as the
stem. Achene body about 3 mm long, tapered
into a beak about as long as the body. Infrequent to rare in ponds and slow streams in the
southern part. July-Aug . . . .
. . . . *Potamogeton crispus* L.

12' Leaves floating in part, the blades oblanceolate, about 5 cm long and 1-2.5 cm wide. Submersed leaves lanceolate, 4-6 cm long and 1-3
cm wide, reddish tinged, with 3-9 distinguishable veins. Stipules 1-3 cm long, thin, broad
at their bases. Fruiting spike dense, 1-3 cm
long, with stout peduncles. Achenes 3-4 mm
long, the dorsal keel prominent. Infrequent
to rare in cold water of the Black Hills and
deep water in the northeast lake region. July-
Aug . . . . *Potamogeton alpinus* Balbis

11' Stipules 4-10 cm long. Submersed leaves with more
than 10 distinguishable veins.

13 Submersed leaves curved, with 25-50 veins.
Achenes 4-5 mm long. Stems up to 1 meter long.
Floating leaves with ovate to elliptic blades
and stout petioles, 5-10 cm long. Submersed

leaves narrower, on short petioles, the blades
lanceolate and up to 15 cm long. Spikes
densely flowered, up to 6 cm long, on stout,
well developed peduncles. Achenes with a
sharp, dorsal keel and 2 lateral keels. Infrequent in cold water lakes of the northeast
part. July-Sept . . . .
. . . . *Potamogeton amplifolius* Tuck.

13' Submersed leaves not curved, with fewer than 30
distinguishable veins. Achenes 3-4 mm long.

14 Underwater leaves sessile or with very
short petioles. Leaves 2-5 cm wide, their
tips pointed. Many times the floating
leaves not well developed. Stems mostly
simple, up to a meter or more long. Submersed leaves elliptic to oblanceolate,
10-20 cm long. Floating leaves, if present, shorter, 6-12 cm long and broadly elliptic. Flowering spikes 3-6 cm long,
densely flowered, on thick peduncles. Rare
in cool water of ponds from flowing streams
in the Black Hills. July-Aug . . . .
. . . . . *Potamogeton illinoiensis* Morong.

14' Underwater leaves with petioles up to 10 cm
long, the blades 4-20 mm wide, their tips
not sharply pointed. Stems simple or little branched, 1-2 meters long. Floating
leaves with ovate-elliptic blades 10-20 cm
long and long-petioled. Flowering spikes
2-5 cm long, on peduncles that are thicker
than the stem diameter. Achenes with a
sharp dorsal keel and a pair of lateral
keels not as sharp. Common in lakes,
ponds and river backwater areas over the
state. July-Aug . . . .
. . . . *Potamogeton nodosus* Poir.

10' All leaves not of the same general shape, the floating
ones with ovate shaped blades. The submersed leaves
linear, usually less than 4 mm wide.

15 Bases of submersed leaves free from their stipules.
Blades of floating leaves 3 cm long or more.

16 Floating leaves with petioles more than twice
the length of the blade. Juncture of the blade

and petiole forming an angle or flexibly attached. Stems up to 2 m long, only sparingly
branched. Floating leaf blades 5-10 cm long,
elliptic. Submersed leaves narrow, 1-2 mm
wide. Peduncles thick with a dense spike.
Achenes ovoid, 3-5 mm long, without a keel.
Infrequent in ponds and streams in the north
and eastern parts. July-Aug . . . .
. . . . *Potamogeton natans* L.

16' Floating leaves with petioles less than twice
the length of the blades. Blade-petiole junction not forming an angle. Submersed leaves
lacking bands on the midvein, or, if present,
very narrow. Stems 3-7 dm long, branched, only
slightly compressed. Floating leaves 2-5 cm
long, narrow to broadly elliptic. Peduncles
thicker than the stem, varying in length.
Achenes obovoid, 2-5 mm long. Infrequent
and wide-ranging in the state in cool water.
July-Aug . . . . . . *Potamogeton gramineus* L.

15' Bases of submersed leaves fused to the stipule for
at least one third of the length of the stipule.
Blades of floating leaves usually less than 3 cm
long, with blades obtuse to acute at their tips.
Achenes usually less than 1.5 mm in diameter, the
lateral keels prominent. Stems slender, branched,
up to 5 dm long. Submersed leaves linear, not exceeding 1.5 mm in width. Fruiting spikes numerous,
the submersed ones globose. Achenes circular,
flattened, the embryo coiled. Infrequent in shallow water in the western part. July-Aug . . . . .
. . . . *Potamogeton diversifolius* Raf.

RUPPIA L.

Ruppia maritima L. Submersed perennials from small rhizomes, the
stems branched, up to 6 dm long. Leaves flat, 6-11 cm long and
less than 1 mm wide. Flowers in axillary spikes with long peduncles reaching the surface. Petals and sepals lacking. Fruits
ovoid, in a cluster, 2-3 mm long. Infrequent in shallow water of
alkaline or brackish lakes and ponds of the eastern part. July-
Aug. Ditch Grass.

ZANNICHELLIA L.

Zannichellia palustris L. Submersed perennials from a creeping
rhizome. Stem slender, branching. Leaves needle-like, opposite,

3-6 cm long. Flowers monoecious, both kinds lacking a perianth,
the pistillate ones sessile in leaf axils and the staminate ones
on slender filaments. Achenes slightly curved, marginally winged
on the convex side, with a short beak. The fruit, including the
beak, 2-3 mm long. Frequent in fresh and brackish water over the
state. July-Aug. Horned Pondweed.

Family Juncaceae

(Key to Genera)

1 Fruiting capsule 3-seeded. Seeds attached basally in the
   capsule . . . . *LUZULA*

1' Fruiting capsule many-seeded. Seeds attached axially or
   parietally in the capsule . . . . *JUNCUS*

JUNCUS

1 Plants annual, low, usually 5-20 cm tall, branched from the
   base. Leaves linear, flat or inrolled, from mostly near the
   base. Inflorescence of single flowers at nodes in the upper
   one half of the plant. Perianth 4-6 mm long, subtended by a
   pair of bracteoles. Fruiting capsule 3-5 mm long, the seeds
   smooth or finely reticulate. Infrequent in moist places over
   the state. July-Sept. Toad Rush . . . . . *Juncus bufonius* L.

1' Plants perennial, usually exceeding 15 cm tall.

   2 Involucral bract erect, the inflorescence appearing lateral.
      Plants perennial from stout, spreading rhizomes, the culms
      solitary, 3-7 dm tall. Leaves reduced to bladeless sheaths.
      Inflorescence 3-5 cm long, greenish-brown, paniculate.
      Perianth 4-6 mm long, the segments lanceolate. Stamens 6.
      Fruiting capsule ovoid, slightly exceeding the perianth
      parts. Frequent to common in marshes and swales over the
      state. July-Aug. Bog Rush . . . . .*Juncus balticus* Willd.

   2' Involucral bract not a continuation of the stem, the in-
      florescence appearing terminal.

      3 Leaves laterally flattened so that one edge is toward
         the stem, the other edge away from the stem. Plants
         perennial, 3-6 dm tall, from creeping rhizomes. Leaves
         2-5 mm wide, the sheaths auricled and flattened. In-
         florescence of 2-10 heads, 7-10 mm in diameter, each
         many flowered. Perianth segments 2-3 mm long, brownish.
         Fruiting capsule about as long as the perianth. Infre-
         quent in wet meadows and valleys in the Black Hills.

{*J. saximontanus* A. Nels.}.   Ours var. *montanus*
(Engelm.) Hitchc.   July . . . . *Juncus ensifolius* Wikst.

3' Leaves terete or transversely flattened so that the flat
   edge is towards the stem.

   4 Leaf blades flat or inrolled but not hollow or sep-
      tate.

      5 Flowers borne in heads, lacking a pair of bracte-
         oles immediately below the perianth.

         6 Perianth and fruit less than 4 mm long. Cap-
            sule subglobose, not pointed at the summit.
            Groups of flowering glomerules numerous.
            Plants cespitose, the stems thickened at the
            base, 2-5 dm tall. Inflorescence of 5-20
            heads. Stamens 3, the perianth equaling the
            capsule. Rare in moist sandy places in the
            western part. July-Sept . . . .
            . . . . *Juncus marginatus* Rostk.

         6' Perianth and fruit exceeding 4 mm. Capsule
            oblong, with a pointed summit. Groups of
            flowering glomerules rarely more than 5.
            Plants from slender rhizomes, the stems aris-
            ing singly, 2-6 dm tall. Inflorescence of 2-5
            hemispherical heads. Stamens 6, the perianth
            5-6 mm long. Infrequent along streams and in
            moist places in the Black Hills and in the
            southern part. July-Aug . . . .
            . . . . *Juncus longistylis* Torr.

      5' Flowers borne singly on the inflorescence branches,
         each with a pair of bracteoles immediately below
         the perianth.

         7 Auricles of the leaf sheath scarious, thin,
            prolonged up to 3 mm above the base of the
            blade. Plants perennial, cespitose, 1-4 dm
            tall, rarely taller. Leaves mostly basal.
            Inflorescence commonly exceeded by the in-
            volucral leaf. Perianth often more than 4 mm
            long, the bracteoles obtuse to broadly acute.
            Fruiting capsule narrowly ovoid, shorter than
            the perianth. Infrequent in moist to dry
            soils over the state. June-Sept. Path Rush.
            . . . . *Juncus tenuis* Willd.

         7' Auricles of the leaf sheath hyaline to carti-
            laginous, not prolonged above the base of the
            blade.

8 Auricles of the sheath summit yellow and
cartilaginous. Bracteoles obtuse at the
apex. Plants perennial, the stems 2-8 dm
tall, clustered from a cespitose base.
Leaves basal, about one-half the length of
the stem. Inflorescence 3-5 cm long, the
panicle branches short and few-flowered.
Perianth 4-5 mm long, somewhat spreading.
Fruiting capsule obovoid, slightly shorter
than the perianth. Common in moist places
over the state. July-Aug . . . .
. . . . *Juncus dudleyi* Wieg.

8' Auricles of the sheath summit membranous
with hyaline margins. Bracteoles acuminate
at the apex. Plants perennial, 4-10 dm
tall, cespitose, the stems stout. Leaves
mostly basal, about one-third the length of
the stem. Inflorescence open, 3-8 cm long.
Perianth 3-4 mm long, the segments erect.
Fruiting capsules oblong, about the length
of the perianth. Frequent in dry to moist
sandy soils over the state. July-Aug . . .
. . . . *Juncus interior* Wieg.

4' Leaf blades hollow and septate.

9 Fruiting capsule subulate, about 5 times longer
than wide. Flowers in dense, globose heads on
the panicle branches.

10 Flowering heads 7-10 mm in diameter. Petals
and sepals about equal. Auricles of the leaf
sheaths yellow. Stems 2-5 dm tall, from
slender rhizomes. Leaves few, distributed
on the stem. Inflorescence somewhat congested, with 3-15 globose heads each with 10-
20 flowers. Fruiting capsule with a long
beak, 2.5-4 mm long. Frequent in marshy soil
or wet places over the state. Aug-Sept . . .
. . . . *Juncus nodosus* L.

10' Flowering heads 10-15 mm in diameter. Petals
much shorter than the sepals. Auricles of the
leaf sheaths white-scarious. Plants perennial, 4-9 dm tall, the stout stems from rootstocks with tubers. Leaves 2-4, elongate,
with prominent auricles. Inflorescence of 3-
10 heads, many flowered. Sepals 4.5-5.5 mm

long. Fruiting capsule slightly longer than
the perianth, tapered. Very common in moist
or marshy soils over the state. Aug-Sept . .
. . . . *Juncus torreyi* Coville

9' Fruiting capsules elliptic-oblong, not subulate,
about 3 times longer than wide. Flowers in 3-10
flowered glomerules on the panicle branches.

11 Branches of the inflorescence narrowly ascend-
    ing, over twice as high as wide. Plants per-
    ennial, the stems slender, 2-5 dm tall, from
    a slender rhizome. Leaves 1 or 2, elongate,
    almost to the inflorescence. Flowering heads
    4-10, forming a narrow inflorescence 5-10 cm
    long. Sepals obtuse with a mucronate tip, 2-
    3 mm long. Rare in wet, sandy soil in the
    southeast part and in the Black Hills. June-
    July . . . . *Juncus alpinus* Vill.

11' Branches of the inflorescence widely diver-
    gent, not more than twice as high as wide.
    Plants perennial, cespitose, the stems erect,
    2-6 dm tall. Leaves 2-4, the upper ones al-
    most reaching the inflorescence. Flowering
    heads 5-12, nearly hemispheric. Sepals
    lanceolate, 2-3 mm long. Fruiting capsule
    dark brown, longer than the perianth. Rare
    in marshy areas in the northeast part. July-
    Aug . . . . *Juncus articulatus* L.

LUZULA DC.

Luzula campestris (L.) DC. Plants perennial, tufted from a
cespitose base, the stems 2-5 dm tall. Leaves several, ascending, 2-4 mm wide, the blades pilose at the bases. Inflorescence
of small, congested clusters of 7-12 flowers on short peduncles
or sessile. Bracts of the inflorescence leaf-like. Perianth
segments 2-4 mm long, the capsule ovoid, with 3 seeds each about
1 mm long. Infrequent in open woods of the Black Hills. Ours
var. *multiflora* (Ehrh.) Celak. June-July. Wood Rush.

Family Cyperaceae

(Key to Genera)

1 Each flower subtended by a single scale. Flowers unisexual.
   Achene enclosed in a sac-like perignium, the perianth lacking
   . . . . *CAREX*   page 40

1' Each flower subtended by 1-several scales. Flowers perfect or
   unisexual. Achene not enclosed in a sac-like perigynium.
   Perianth present or lacking.

   2 Spikelets flattened with scales in 2 ranks . . . .
   . . . . *CYPERUS*   page 61

  2' Spikelets circular in cross-section. Scales spirally overlapped.

     3 Style base persistent as a tubercle on the summit of the
      achene. Perianth reduced to bristles.

       4 Plants with 1 terminal spikelet with many perfect
         flowers. Bristles 0-6 in each flower . . . .
         . . . . *ELEOCHARIS*   page 64

       4' Plants with several to many spikelets, each one 2-
         several flowered. Bristles 6 in each flower . . . .
         . . . . *RHYNCHOSPORA*   page 67

     3' Style deciduous. Achene pointed or blunt at apex but
      lacking a tubercle.

       5 Achene subtended by 0-6 bristles which do not exceed
         the scale length . . . . *SCIRPUS*   page 67

       5' Achene subtended by many elongate silky bristles that
         are white and cottony . . *ERIOPHORUM*   page 67

CAREX L.

(Keys to Species Groups)

1 Stigmas 2. Achenes lenticular or flattened.

  2 Spikes of two kinds on the same plant. Staminate spikes
      are usually terminal with the pistillate spikes below.
      (The dioecious species *Carex douglasii* will be found
      here.) . . . . *Group 1.*   page 41

  2' Spikes of one kind with staminate and pistillate flowers
      included.

     3 Staminate flowers at the apex of the spikes (Androgy-
         nous) . . . . *Group 2.*   page 43

     3' Staminate flowers at the middle or near the base of
         the spike (Gynecandrous) . . . . . . *Group 3.*   page 47

1' Stigmas 3. Achenes 3-sided or circular in cross section.

  4 Perigynia distinctly pubescent . . . . *Group 4.*   page 52

4' Perigynia glabrous (minutely scabrous or puberulent forms
   will be found here).

   5 Beak of perigynium sharply bidentate . . . .
      . . . . *Group 5.*   page 55
   5' Beak of perigynium entire or absent . . . .
      . . . . *Group 6.*   page 58

Group 1.

(Key to the Species)

1 Heads of one kind on a single plant, dioecious. Plants 1-3 dm
   tall, from slender rootstocks, the staminate and pistillate
   plants similar. Leaves 1-2.5 mm wide, mostly basal. Spikes
   several, sessile, 1-1.5 cm long, aggregated into a head about
   3.5 cm long. Perigynia pale brown, 3.5-4 mm long, plano-
   convex, with a well-developed bidentate beak, the latter 1-
   1.7 mm long. Infrequent in prairie swales over the state.
   May-July . . . . *Carex douglasii* Boott.

1' Heads of two kinds on a single plant, not dioecious.

   2 Perigynia with 4-6 prominent longitudinal ribs on each sur-
      face. Scales subtending the perigynium brownish-black ex-
      cept for the hyaline margins and lighter center. Plants
      stout, 2-9 dm tall, from spreading rhizomes. Pistillate
      spikes 2-5, the bract of the lowest spike sheathless.
      Staminate spikes 1-3, terminal. Periqynium flattened, 3-
      3.5 mm long, with a bidentate beak, the teeth 0.5 mm long.
      Frequent in moist places and in swales in the southwest
      part. May-July . . . . *Carex nebraskensis* Dew.

   2' Perigynia lacking longitudinal ribs. If these are inter-
      preted to be present, they will be many and due to shrink-
      age of the perigynium upon drying.

      3 Pistillate spikes of 15 or less perigynia, golden color
         at maturity. Plants with slender stems 5-20 cm tall,
         several in a clump from long rhizomes. Leaves narrow,
         flat, 1-4 mm wide, mostly basal and often surpassing
         the stems. Spikes 5-20 mm long, the staminate one
         terminal. Lower pistillate spikes usually without
         sheaths. Perigynium obovoid, beakless, 2-3 mm long,
         obscurely ribbed with 12-15 nerves, golden brown at
         maturity. Frequent in wet meadows of the northeast
         part and along open streams at higher altitudes in the
         Black Hills. May-July . . . . *Carex aurea* Nutt.

3' Pistillate spikes with many more than 15 perigynia, the
bracts and often the perigynia darkening at maturity.

   4 Pistillate scales near maturity obtuse to acute and
nearly equaling or slightly exceeding the perigynium
in length.

      5 Stems with basal leaves rudimentary or bladeless
(aphyllopodic), brown to brown-purple at their
bases. Perigynium elliptic, broadest at the middle. Bract of lowest pistillate spike not exceeding the entire inflorescence. Plants cespitose,
in tussocks, with short, scaly rhizomes, the culms
3-7 dm tall. Pistillate spikes erect, 2-4 cm
long, the lower ones with sheathless bracts.
Perigynium granular-roughened, broadest just at
or below the middle. Frequent in moist or marshy
places over the state. June-July . . . .
. . . . *Carex stricta* Lam.

      5' Stems with basal leaves fully developed (phyllo-
podic), reddish at their bases. Perigynium obovate, broadest above the middle. Bract of the
lowest pistillate spike elongate, often exceeding
the entire inflorescence. Plants cespitose, with
long scaly rhizomes, the culms 2-8 dm tall. Pistillate spikes appressed-erect, with 30-90 perigynia. Perigynium membranous, glandular-dotted,
2.5-3.0 mm long. Infrequent in swampy places in
the northern part. Ours var. *altoir* (Rydb.) Fern.
July-Aug . . . . *Carex aquatilis* Wahl.

   4' Pistillate scales near maturity acute to acuminate,
conspicuously exceeding the entire inflorescence.

      6 Perigynium 2 mm or longer, inflated, brownish at
maturity. Plants cespitose with short, ascending
stolons, not sod-forming, the culms 5-8 dm tall.
Pistillate spikes spreading, 1.5-3 cm long. Perigynium obovate, broadest above the middle, granular towards the summit, with a very short beak.
Infrequent in marshy places and swales in the
eastern part to the southwest. May-June . . . .
. . . . *Carex haydenii* Dewey

      6' Perigynium to 1.7 mm long, unequally bi-convex,
not inflated, greenish towards maturity. Plants
with horizontal stolons, forming sods, the stems
clustered, erect, 4-8 dm tall. Pistillate spikelets erect-ascending, 4-8 cm long. Perigynium

ovate, rounded at the base, broadest below the middle, granular-roughened at the summit, with a very short bidentate beak. Infrequent in swales and moist meadows in the east and southern part. June-July . . . . *Carex emoryii* Dewey

Group 2.

(Key to the Species)

1 Stems arising singly or few from creeping rhizomes.

  2 Perigynia 1-4 in each spike, the inflorescence slender and interrupted. Perigynium 2-3 mm long, the body abruptly contracted to a short, obliquely cleft beak. Plants 1-4 dm tall, the culms soft and weak, from slender rhizomes. Leaves 1-2 mm wide, borne at the base of the stem. Spikes 2-5, small and few-flowered, the entire inflorescence appearing depauperate. Infrequent in swamps and wet meadows in the Black Hills. July-Aug . . . . *Carex disperma* Dewey

  2' Perigynia 10 or more in each spike, the inflorescence not narrow and interrupted.

    3 Upper leaf sheaths green striate on the ventral surface. Leaves not clustered at the base of the culm. Perigynium thin-margined above, the beak 0.5 mm long, sharply bidentate. Plants with stout stems 3-7 dm tall, the culm bases at least 3 mm wide. Spikes numerous, about 1 cm long, sessile, the entire inflorescence 3-5 cm long. Perigynium 2.5-4.0 mm long, serrulate on the margin from the body to the beak, finely nerved on the dorsal side. Frequent in prairie swales in the northeast part and in the Black Hills. May-July . . . . *Carex sartwellii* Dewey

    3' Upper leaf sheaths hyaline on the ventral surface. Leaves clustered at the base of the culm. Perigynium sharp but not thin-margined from the body to the beak. Beak of the perigynium 1 mm long. Plants with stems 2-7 dm tall, the culm bases usually not more than 2 mm wide. Spikes 5-10, somewhat interrupted on the head which is up to 5 cm long. Perigynium 3-4 mm long, light to dark brown, with inconspicuous dorsal nerves. Common in moist prairie swales over the state. June-July . . . . *Carex praegracilis* W. Boott.

1' Stems from densely to loosely cespitose. Rootstocks sometimes prolonged but not long creeping.

4   Entire inflorescence not exceeding 1.5 cm in length.
     Plants usually not more than 3 dm tall, the culms ob-
     tusely angled. Slender rootstocks present, brownish,
     not more than 1 mm thick. Leaves narrow and involute,
     at least towards the ends. Spikes of 1-7 perigynia,
     each 2-3 mm long, gradually tapered to an obliquely
     cleft beak. Perigynium dark brown to blackened towards
     maturity. Common in dry prairies and exposed hillsides
     over the state. May-July . . . . . *Carex eleocharis* Bailey

4' Entire inflorescence commonly exceeding 1.5 cm. Plants
     exceeding 3 dm in height at flowering time.

   5   Perigynia 4-5 mm long, subulate, tapering to a long
        point. Stems below the inflorescence prominently
        winged on the angles, up to 3 mm wide. Plants cespi-
        tose, the stems 4-10 dm tall. Leaves flat, 4-8 mm
        wide, the sheaths transversely rugulose. Heads 3-9
        cm long with many androgynous spikes. Perigynia 4-8
        to a spike, obviously nerved on both faces, spongy
        below, the beak bidentate. Frequent in springy or
        marshy seepage slopes or swampy areas over the state.
        June-July . . . . *Carex stipata* Muhl.

   5' Perigynia ovate or ovate-lanceolate. Stems not promi-
        nently winged below the inflorescence.

     6   Spikes numerous, commonly exceeding 12, often decom-
          pound, the lower ones at times interrupted.

       7   Bracts subtending each spike elongate and needle-
            like. Ventral surface of leaf sheath cross-
            rugulose. Perigynia yellow-green. Plants 2-8
            dm tall, the leaves distributed on the culm. In-
            florescence 3-10 cm long, the spikes interrupted.
            Perigynia flattened, 2-3.5 mm long, with a gradu-
            ally tapered beak. Frequent in alluvial soils
            and on flood plains over the state. June-Aug . .
            . . . . *Carex vulpinoidea* Michx.

       7' Bracts subtending each spike scalelike, except for
            the lowest one which is short and setaceous. Ven-
            tral surface of the leaf sheath opaque, not cross-
            rugulose. Perigynia brownish-green. Plants 5-10
            dm tall, with 3-4 leaves on each culm. Inflores-
            cence 4-8 cm long, the spikes often decompound.
            Perigynia 5-10 on a spike, the edges sharply ser-
            rulate on the tapered beak. Infrequent in low
            meadows in the southeast part. {*C. prairea* Dewey}.
            June-July . . . . *Carex prarisa* Dewey

6' Spikes fewer than 12.

   8   Leaves 1.0-3.5 mm wide, the leaf sheaths tight around the culm.

      9   Perigynia wide-radiating on the few-flowered spikes, often interrupted by as much as 1 cm on the main axis.

         10   Principal leaves 1-2 mm wide. Stigmatic tips of the styles not twisted, slender. Stems 3-6 dm tall. Spikes 2-8, each spike 4-15 flowered. Perigynium ovoid to lanceolate, the beak stout, about one-fourth the length of the body, 2-4 mm long. Infrequent in rich woods in the eastern and western part. June-July . . . . *Carex rosea* Schk.

         10' Principal leaves 2-3 mm wide. Stigmatic tips of the styles stout and twisted. Stems slender, 3-5 dm tall. Spikes 4-7, each spike 4-15 flowered. Perigynia 2.5-4.5 mm long, green, flattened, abruptly contracted into the short beak. Frequent in rich woods of the eastern part, rare in rich woods of the Black Hills. May-July
Carex convoluta Mack.

   9' Perigynia ascending in the spike, not widely radiating. Spikes only scarcely interrupted on the main axis, if at all.

         11   Beak of perigynium obliquely cleft. Body of the perigynium essentially nerveless, abruptly tapering to the short beak. Principal leaves 1-2 mm wide. Plants cespitose with short, dark roots, the stems 2-5 dm tall. Inflorescence 1.5-3.0 cm long, the spikes few-flowered. Perigynia 3.5 mm long, becoming copper colored. Frequent on dry slopes of the Black Hills and in Harding County. June-July . . . .
Carex vallicola Dewey

         11' Beak of perigynium bidentate. The body of the perigynium with inconspicuous nerves on the dorsal surface and gradually tapering to the beak. Principal leaves 1.5-3.5 mm wide, borne on the lower part of the culms.

Culms 3-8 dm tall, densely clustered. Inflorescence 1-2 cm long, of 4-8 spikes,
compact. Perigynia 3.5-5.0 mm long, mostly
concealed by the brown, scarious scales.
Rare in meadows in the Black Hills. June-
July . . . . *Carex hoodii* Boott.

8' Leaves 3.5-8.0 mm wide, the leaf sheaths loose
around the culm.

12   Inflorescence short and compact, usually not
more than 3 cm long. Leaves 3-5 mm wide.

13   Perigynium body ovoid, tawny brown at maturity. The beak of the perigynium up to
one-third as long as the body. Plants
cespitose, the culms 3-5 dm tall. Spikes
6-10, compact and irregularly spreading,
some with elongate bracts. Perigynium 4
mm long, the beak serrulate and bidentate.
Common in alluvial woods or rich soil over
the state. Represented in South Dakota by
the two varieties *gravida* and *lunelliana*
(Mack.) Herm. June-July . . . .
. . . . *Carex gravida* Bailey

13' Perigynium body ellipsoid, deep green at
maturity. The beak of the perigynium one-
half the length of the body. Plants 4-6
dm tall, the culms slender. Leaves 3.5-6
mm wide. Spikes many, globose, densely
aggregated into a compact head. Perigynium 3 mm long, with ovate glumes having
hyaline margins. Apparently rare in woods
or thickets in the eastern part. May-July
. . . . *Carex aggregata* Mack.

12' Inflorescence elongate, interrupted, usually
3-5 cm long. Perigynia green at maturity.
Plants 3-7 dm tall. Leaves numerous, 5-8 mm
wide. Spikes 6-12, distinguishable, up to 50
flowered. Perigynia 3 mm long, nerveless,
with a bidentate beak. Rare in alluvial woods
in the eastern part. June-July . . . . . .
. . . . *Carex sparganioides* Muhl.

Group 3.

(Key to the Species)

1 Perigynia without winged margins, or if winged margins are present, then the perigynia at least 5-6 times longer than wide.

2 Inflorescence a dense head, the leaves overtopping the fertile culms. Perigynia 5-6 mm long and 1 mm wide, with a narrow wing. Stems 5-40 cm tall, the leaves 2-4 mm wide. Heads 1.5-3 cm long, composed of 6-12 compact spikes, sub-tended by long, leaf-like bracts. Perigynia numerous, slender, with a tapered serrulate beak. Frequent in prairie swales and sloughs in the eastern and northeast part. June-July . . . . *Carex sychnocephala* Carey

2' Inflorescence interrupted, the spikes few-flowered, as tall as or slightly overtopping the leaves. Perigynia without winged margins.

3 Perigynia less than 2 mm wide, with minute, whitish dots on the body surface. Plants cespitose, the stems 1-7 dm tall. Leaves clustered near the base, the blades 1-2.5 mm wide. Inflorescence narrow, the 4-9 spikes usually separated. Perigynia 5-10 in a spike, lightly nerved on the dorsal surface, 2.0-2.5 mm long, with a short beak. Rare in boggy places in the Black Hills. June-Aug . . . . *Carex brunnescens* (Pers.) Poir.

3' Perigynia exceeding 2 mm in width, not possessing minute, whitish dots on the body surface.

4 Perigynia 2-3 mm long, widely radiating in the spike. Leaves 1-3 mm wide. Stems 2-5 dm tall, stiff and wiry. Leaves mostly on the lower one third of the culm. Spikes 4-6 mm long, with 1-10 perigynia. Perigynium nerved on the dorsal surface, the beak serrulate and bidentate. Infrequent to rare in openings of woods and meadows in the Black Hills and in the eastern part. June-July . . . . *Carex interior* Bailey

4' Perigynia 4.5-5.5 mm long, appressed-ascending in the spike. Leaves 2-5 mm wide. Stems 2-10 dm tall, brownish at the base with old leaves. Spikes 3-4, in a head 2-5 cm long. Perigynia 3-14 in a spike, light green, only obscurely nerved on the dorsal surface. Body of perigynium appearing turgid with the achene. Infrequent in woods of the Black Hills and in the northeast. June-Aug . . . . . *Carex deweyana* Schw.

1' Perigynia usually with winged margins and not more than 3
   times longer than wide.

   5 Perigynium less than 2 mm wide, including the winged mar-
      gin, if present. {*C. microptera* and *C. festivella* rarely
      have perigynia exceeding 2 mm in width.}

      6 Sheaths of principal leaves green striate on the ventral
         surface. Spikes globose, the perigynia radiating in all
         directions. Plants cespitose, the stems 3-8 dm tall.
         Inflorescence 2-3 cm long, of 6-12 spikes, with one or
         more setaceous bracts on the lower spikes. Perigynia
         numerous, 3-4 mm long, with delicate nerves on both
         surfaces. Infrequent to rare in alluvial thickets or
         moist shady areas in the eastern part. July-Aug . . . .
         . . . . *Carex cristatella* Britt.

      6' Sheaths of principal leaves hyaline on the ventral sur-
         face.

         7 Spikes sessile, congested into an inflorescence usu-
            ally less than 3.5 cm long. Perigynia usually 4 mm
            long or less.

            8 Scales of the perigynia becoming yellow-brown with
               hyaline margins and a lighter 3-nerved midvein.
               Inflorescence narrow, 1 cm or less wide. Beak of
               the perigynium flattened and serrulate to the tip.

               9 Perigynia brown. Spikes rounded at their bases.
                  Plants 2-7 dm tall, the leaves 2-4.5 mm wide,
                  mostly on the lower one third of the culm.
                  Spikes 3-9, ovoid, each with many perigynia
                  closely aggregated. Perigynia 3.0-3.5 mm long,
                  obscurely nerved on the dorsal surface. Infre-
                  quent in swales and meadows over the state.
                  July-Aug . . . . *Carex bebbii* Olney

               9' Perigynia green, membranous. Spikes tapering
                  to their bases. Plants 3-12 dm tall, the
                  leaves 3-6 mm wide, on the lower one third of
                  the culm. Spikes 4-9, ovoid, the perigynia
                  appressed-ascending. Perigynia 3-4 mm long,
                  finely nerved on the dorsal side. Infrequent
                  in ditches and in open woods in the southeast
                  and in the Black Hills. June-July . . . . . .
                                         Carex normalis Mack.

            8' Scales of the perigynia chestnut brown with light,
               hyaline margins and midveins. Inflorescence
               bushy, 1-1.5 cm wide.

10   Spikes congested, not separated. The entire
     head ovoid to rounded. Perigynia spreading
     to ascending, the margins narrow, tapering
     gradually to the beak. Plants 3-8 dm tall,
     leafy. Spikes ovoid with 15-25 perigynia.
     Scales dull brown. Perigynia 3-5 mm long,
     dorsally nerved, lance-ovate. Infrequent in
     meadows and on open slopes at higher eleva-
     tions in the Black Hills. June-Aug . . . . .
     . . . . *Carex microptera* Mack.

10' Spikes separated and distinguishable. The
     entire head oblong to ovoid. Perigynia ap-
     pressed, the margins wide, their bodies ovate,
     abruptly tapering to the beak. Plants 3-8 dm
     tall, leafy, much like the preceding species.
     Spikes broadly oblong with 15-30 perigynia.
     Perigynia 3.5-5 mm long, membranous, lightly
     nerved on the dorsal surface. Frequent in
     meadows of the western part and in the Black
     Hills. June-July . . . *Carex festivella* Mack.

7' Spikes short pedicellate, the inflorescence 3-7 cm
    long. Perigynia 4 mm long or longer.

   11   Scales subtending the perigynia reddish-brown,
        longer than the perigynia. Perigynia lightly
        nerved on the dorsal surface. Plants cespitose,
        the stems 2-6 dm tall. Leaves mostly near the
        base, 1-3.5 mm wide. Inflorescence a flexuous
        head 2-5 cm long, of 3-7 spikes. Spikes of 7-18
        perigynia, the perigynia 5-6.5 mm long, mostly
        concealed by the silvery margined scales. Infre-
        quent in high meadows in the Black Hills. June-
        July . . . . *Carex praticola* Rydb.

   11' Scales light brown to green, shorter to almost as
        long as the perigynia.

      12   Perigynia lightly, if at all, nerved on the
           dorsal surface only, mostly 5-6 mm long and
           1.2 mm wide. Scales much shorter than the
           perigynia. Plants with slender, rough culms
           2-8 dm tall. Leaves 1-3 mm wide, on the
           lower part of the stem. Inflorescence 1-5
           cm long, the spikes crowded. Spikes 3-10,
           with many perigynia. Perigynia with a long,
           tapering serrulate beak. Frequent in moist
           or swampy places over the state. May-June .
           . . . . *Carex scoparia* Schk.

12' Perigynia strongly nerved on both surfaces, mostly 4 mm long and slightly less than 2 mm wide. Scales almost equaling the perigynia. Plants cespitose, the culms slender and flexuous, 4-7 dm tall. Leaves 1-2.5 mm wide, at the lower one third of the culm. Inflorescence 3-5 cm long, the main rachis flexuous, the spikes 4-7, not crowded. Spikes with 10-20 perigynia. Perigynia with gradually tapering beaks, shallowly bidentate. Infrequent in open woods of the northeast and in the Black Hills. June-July . . . . *Carex tenera* Dewey

5' Perigynium 2 mm wide or wider, including the winged margin, if present.

13 Body of the perigynium oval, gradually tapering to the beak. If the perigynium appears to taper abruptly, the widest part of the perigynium should exceed 2.5 mm.

14 Scales shorter than the perigynia as well as narrower above, making the perigynia conspicuous. Plants cespitose with slender rootstocks, the culms 3-8 dm tall. Leaves elongate, 2-3 mm wide, on the lower part of the stem. Inflorescence 2-3 cm long, of 4-7 spikes clustered but not crowded. Spikes globose, with 15-25 perigynia. Perigynium about 4.5 mm long, nerved dorsally, the beak contracted from the body but not abruptly. Infrequent to rare in thickets in the central and eastern part. May-July . . . . *Carex molesta* Macken.

14' Scales about the same length as the perigynia and about as wide at the top, the perigynia mostly concealed.

15 Perigynium 4-5 mm long, nerved on both surfaces. Scales with a green midrib having 3 nerves. Plants cespitose, the culms 4-10 dm tall, slender and weak. Leaves 4-6, elongate, 2-4 mm wide, on the lower one third of the culm. Inflorescence 4-7 cm long, the 4-8 spikes not crowded. Perigynia ascending-appressed in the spike. Rare on dry slopes in the Black Hills. June-July . . . . *Carex aenea* Fernald

15' Perigynia 5 mm long or longer, only obscurely nerved dorsally, the widest part of the perigynium at the top of or above the achene.

Plants with short, creeping rootstocks, the
culms 3-5 dm tall, narrow. Leaves mostly near
the base, 2-3 mm wide. Inflorescence 3-5 cm
long, of 3-5 spikes appressed and ascending.
Perigynia becoming reddish-brown at the beak,
mostly concealed by the scales. Frequent on
hillsides of the Black Hills. June-July . . .
. . . . *Carex xerantica* Bailey

13' Body of the perigynium orbicular, abruptly tapering to
the beak.

16 Scales as long as or longer than the perigynia.
Inflorescence moniliform and flexuous. Plants
cespitose, 4-10 dm tall. Leaves on the lower one
half of the culm, 2-4.5 mm wide. Spikes somewhat
separated with 5-20 perigynia. Perigynia 3-4.5 mm
long, about 2 mm wide, with a conspicuous wing,
both surfaces nerved. Infrequent in dry woods in
the Black Hills. {*C. siccata* Dewey}. June-Aug
. . . . *Carex foenea* Willd.

16' Scales not as long as the perigynia. Inflorescence
more compact, not moniliform.

17 Perigynium 3-5 mm long and usually less than 3
mm wide, the surface above coriaceous green
with prominent nerves. Green to white beneath,
the nerves inconspicuous. Plants cespitose,
3-9 dm tall, the leaves several, 2-4 mm wide,
on the lower one third of the culm. Inflorescence 2-5 cm long, with 3-6 spikes more or
less congested. Spikes with 8-20 perigynia,
ascending. Perigynium convex-curved, the beak
prominently bidentate. Very common in moist
meadows and thickets over the state. June-
July . . . . . . *Carex brevior* (Dew.) Macken.

17' Perigynium 5.5-6.5 mm long and more than 3 mm
wide, thin, with a broad, winged margin.
Strongly nerved on both surfaces. Plants
cespitose with short, black rootstocks, the
culms 4-11 dm tall and stout. Leaves elongate,
3-4 mm wide and borne on the lower one third of
the culm. Inflorescence 2-4 cm long, with 3-6
spikes, the lower ones with elongate bracts.
Spikes ovoid, the perigynia ascending. Perigynium flat, the beak with a prominent bidentate tip. Infrequent to rare in marshy areas

and prairie swales in the northeast part.
June-July . . . . *Carex bicknellii* Britt.

Group 4.

(Key to the Species)

1 Pistillate spikes with 20-50 perigynia, the spike usually exceeding 15 mm. Plants usually over 4 dm tall, from a stoloniferous base. Leaves 2-5 mm wide, well developed, distributed
on the lower one-half. Staminate spikes 2-4, the pistillate
ones 2 or 3, sessile or shortly pedunculate, bracted, up to 5
cm long. Perigynia 2.5-3.5 mm long, pubescent, the beak glabrous, bidentate. Common in moist meadows and marshy places
over the state. {incl. *C. lasiocarpa* of reports}. June-July
. . . . *Carex lanuginosa* Michx.

1' Pistillate spikes with 1-25 perigynia, the spike less than 15
mm long. Plants usually less than 4 dm tall.

   2 Perigynium 5-6 mm long, the pistillate spikes on slender
pedicels 2-6 cm long. Plants cespitose, the culms slender
and weak, 3-6 dm tall. Leaves thin and flat, 1-3 mm wide.
Inflorescence of usually a single terminal staminate spike
and 2 or 3 pistillate spikes below. Pistillate spikes up
to 3 cm long, few flowered. Perigynia hispid, erect, the
beak long tapered, the tip white, oblique. Infrequent in
rich upland woods in the east and northwest part. July-
Aug . . . . *Carex assiniboinensis* W. Boott.

   2' Perigynium 4.5 mm long or less.

      3 Leaves involute, stiff and wiry, less than 0.2 mm wide.
Spike solitary, androgynous. Plants densely cespitose
with many blackened sheaths and stem bases of previous
year's growth. Culms usually not more than 2 dm tall.
Inflorescence 1-3 cm long, the pistillate portion below
with 5-15 perigynia. Perigynium 3-4 mm long, membranous, elliptic-obovate with a short, cylindric beak.
Pistillate scales broadly obtuse, red-brown with a
hyaline margin. Common on dry prairie and plains over
the western two thirds of the state. May . . . .
. . . . *Carex filifolia* Nutt.

      3' Leaves over 1 mm wide. Staminate and pistillate spikes
distinguishable.

         4 Perigynia bidentately toothed, the cleft portion ex-
ceeding 0.5 mm.

            5 Spikes not widely separated, adjacent on well
developed culms, sessile.

6 Perigynia obtusely angled, 2-3 mm long and 1.5
mm in diameter. Plants stoloniferous, the
culms 1-4 dm tall, not wiry, the bases often
with old leaves. Leaves 1-3 mm wide, distributed on the fertile culms. Inflorescence of a
single staminate spike and several sessile,
few flowered pistillate spikes. Perigynia
yellow-green, pubescent, the beak bidentate.
Infrequent to rare in rich woods of the eastern
part. Apr-June . . . . *Carex pensylvanica* Lam.

6' Perigynia turgid to sub-orbicular, 3-5 mm long
and over 2 mm in diameter. Plants with long
stolons, the culms wiry, 1-3 dm tall. Leaves
1-2.5 mm wide, mostly basal on the fertile
culms. Inflorescence of a single staminate
spike and 1-3 few flowered pistillate spikes
below. Perigynium dull green, puberulent, the
beak deeply bidentate. Quite common in prairies and on open hillsides over the state.
June-July . . . . . *Carex heliophila* Macken.

5' Spikes widely separated, some of the pistillate
ones basal on short peduncles. Plants from stout
rootstocks, the stems wiry, 1-3 dm tall. Leaves
arising from near the culm bases, 1-2.5 mm wide.
Staminate spikes terminal but the pistillate ones
widely separated, some above and adjacent to the
staminate ones, others below. Perigynium 3-5 mm
long, the base spongy. Body of the perigynium
pubescent. The beak serrulate and bidentate at
the tip. Rare in dry soil of upland woods in the
Black Hills. May-June . . . . *Carex rossii* Boott.

4' Perigynia with oblique or erose beaks. If bidentate,
then the cleft portion less than 0.5 mm.

7 Beak of the perigynium toothed but the cleft portion very short. The body of perigynium obtusely
triangular, broadest above the middle and long
pubescent towards the beak. Plants stoloniferous,
the culms 2-5 dm tall. Leaves well developed
along the stems. Inflorescence of a terminal
staminate spike and 2-4 lateral pistillate ones
which are few flowered. Perigynia 3-4 mm long,
subtended by red-brown scales with white-hyaline
margins. Infrequent in open woods and prairie
openings in the eastern part and in the Black
Hills. May-June . . . . . *Carex peckii* E. O. Howe

7' Beak of the perigynium eroded or oblique but not
   toothed.

   8 Pistillate scale much shorter than the peri-
      gynium, obtuse, with a white-hyaline margin.
      Pistillate spikes sessile. Plants rhizomatous
      or cespitose, the culms 5-20 cm tall, wiry,
      curved. Leaves mostly basal, 2-2.5 mm wide.
      Inflorescence congested, short, the staminate
      spike terminal with 1-4 few flowered lateral
      pistillate spikes. Perigynia 2-ribbed, pubes-
      cent, with a spongy base. Beak almost obso-
      lete, with an oblique orifice. Rare in dry
      soils of the Black Hills. June-July . . . . .
      . . . . *Carex concinna* R. Brown

   8' Pistillate scale or awn of scale longer than
      the perigynium. Pistillate spikes peduncled.

      9 Perigynia 2-3 mm long, broadest near the
         middle. Pistillate scales obtuse, purplish
         with hyaline margins. Plants cespitose,
         with long rootstocks, the culms 2-4 dm tall.
         Leaves 2-3 mm wide, near the base of the
         culm. Terminal spike staminate, with usu-
         ally 2 lateral pistillate spikes that are
         short-peduncled. Perigynia 10-30 in each
         spike, obscurely triangular with a short
         pubescence above. Beak very short with an
         oblique opening. Infrequent in dry woods
         of the northeast and in the Black Hills.
         June . . . . . . *Carex richardsonii* R. Br.

      9' Perigynia 3.5-4.5 mm long, obovate, broadest
         near the top with a narrow, spongy base.
         Pistillate scales cuspidately awned. Plants
         with woody rootstocks, the culms 2-3 dm tall.
         Leaves mostly basal, 2-3 mm wide. Terminal
         spike staminate or androgynous. Lateral
         spikes 2-4, usually androgynous, on long
         peduncles. Perigynia few in each spike,
         triangular with a minute pubescence. Beak
         very short and bent with an entire opening.
         Rare in rich woods of the northeast part.
         May-June . . . . . . *Carex pedunculata* Muhl.

Group 5.

(Key to the Species)

1 Staminate spikes two or more (occasionally *C. retrorsa* and *C. sprengelii* may have but a single staminate spike).

  2 Teeth of perigynia 0.5 mm or less.

    3 Perigynia appressed-ascending in the spike, fusiform, 6-7 mm long, the beak short. Plants cespitose with scaly stolons, the culms 6-10 dm tall. Leaves 5-15 mm wide, elongate and well developed. Inflorescence of 2-3 terminal staminate spikes and as many pistillate spikes below. Pistillate spikes with 50-125 perigynia, each strongly nerved, with a short, bidentate beak. Rare in swampy places in the northeast part. July . . . . *Carex lacustris* Willd.

    3' Perigynia spreading or reflexed in the spike, ovoid and inflated, about 8 mm long, gradually tapering to a sharp, bidentate beak. Plants mostly cespitose, 2-8 dm tall, with several flat leaves 3-9 mm wide, elongate. Upper spikes staminate, the lower 3-5 pistillate, on slender peduncles. Perigynia many, broadly spreading, their bases inflated and crowded. Perigynium membranous with 8 or more nerves. Infrequent in swampy soils in the northeast part and in the Black Hills. July-Aug . . . . *Carex retrorsa* Schw.

  2' Teeth of perigynia exceeding 0.5 mm or occasionally the beak with an oblique orifice.

    4 Pistillate spikes on slender, spreading, elongate peduncles. Plants from dense, scaly and fibrous rootstocks, the culms 3-8 dm tall, slender and weak. Leaves 2-4 mm wide, distributed along the fertile stem. Inflorescence of 1-3 terminal staminate spikes and several pistillate spikes below. Perigynia 10-35 to a spike, each 5-6 mm long, widely spreading, the beak as long as or exceeding the body, narrow. Common in upland woods and thickets over the state. May-June . . . . *Carex sprengelii* Dewey

    4' Pistillate spikes sessile or on short, ascending peduncles.

      5 Bidentate beak of the perigynium with teeth 0.5-1.0 mm long. Style abruptly contorted above the achene.

        6 Perigynia in 8-12 rows on the pistillate spike, spreading towards maturity. Plants with heavy,

horizontal rhizomes, the stems 3-10 dm tall, with
obtusely triangular culms. Leaves many, elongate,
3-11 mm wide. Inflorescence of 2-3 staminate
spikes above and 2-5 pistillate spikes below on
very short peduncles or sessile. Perigynia many,
4-7 mm long, the body inflated, membranous. Beak
bidentate with stiff teeth. Frequent in moist or
marshy areas in the northeast part and in the
Black Hills. June-July . . . *Carex rostrata* Stokes

6' Perigynia in 6-8 rows on the pistillate spikes,
ascending-spreading towards maturity. Plants
with short rootstocks, lacking rhizomes, the
stems sharply triangular below the inflorescence.
Leaves 2-7 mm wide, evenly distributed on the
stem. Inflorescence of 2-3 upper staminate
spikes and 1-3 lower pistillate ones. Pistillate
spikes with many ascending perigynia. Perigynium
5-7 mm long, nerved, the body ovate with a narrow,
sharply bidentate beak. Rare in marshy places in
the lake region of the extreme northeast. Ours
var. *monile* (Tuck.) Fern. July-Aug . . . .
. . . . *Carex vesicaria* L.

5' Bidentate beak of the perigynium with teeth 1-2 mm
long. Style straight above the achene.

7 Leaf sheaths softly hairy, at least near the mouth
and on the dorsal side. Teeth of the beak exceeding 1 mm, spreading outward. Plants cespitose,
the culms spongy-enlarged, 4-12 dm tall. Leaves
well-developed, 4-10 mm wide. Inflorescence of
2-4 staminate spikes and 1-3 pistillate ones
below. Perigynia lanceolate, ribbed to the beak,
7-10 mm long, with prominent teeth. Frequent in
moist or marshy places over the state. May-June
. . . . *Carex atherodes* Spreng.

7' Leaf sheaths glabrous. Teeth of the perigynium
beak about 1 mm long, remotely curved. Plants
cespitose and with long stolons, the culms 4-10
dm tall. Leaves at the central and upper part,
2-7 mm wide. Inflorescence of 2-4 staminate
spikes and 1-3 lower pistillate spikes. Perigynia 25-40 in a spike, ascending-spreading.
Perigynium 6 mm long, strongly nerved, the beak
elongate and bidentate. Common in marshes and
bottom land in the eastern part. May-June . . .
. . . . *Carex laeviconica* Dewey

1' Staminate spikes one or lacking.

   8 Pistillate spikes up to 1.25 cm wide.

      9 Perigynia 5-7 mm long, with 15-20 nerves. Pistillate
         scales with a serrulate awn. Plants cespitose with
         narrow stolons, the culms 2-8 dm tall. Leaves few,
         3-8 mm wide, on the lower part of the fertile culm.
         Pistillate spikes 1-4, the perigynia many, spreading.
         Beak tapered to a bidentate tip with slightly spreading
         teeth 2 mm long. Frequent in marshes and wet soil over
         the state. June-July . . . . . . *Carex hystricina* Muhl.

      9' Perigynia 2-4 mm long. Pistillate scales not serrulate-
         awned.

         10 Spike solitary, androgynous. Plants from long,
             creeping rootstocks, the stems 6-20 cm tall.
             Leaves short, mostly basal, often from among pre-
             vious year's leaf bases. Spikes short, about 1 cm
             long or less, the upper part of staminate flowers,
             the lower part of 1-5 perigynia. Perigynium dark
             brown, ovate-lanceolate, somewhat angular, with a
             very short beak. Rare in sandy or rocky openings
             in woods in the Black Hills. May-June . . . .
             . . . . *Carex obtusata* Lilj.

         10' Spikes more than one, not androgynous.

             11 Terminal spike staminate, the lower ones pistil-
                 late, or the spikes only pistillate. Perigynia
                 2-3 mm long.

                 12 Plants cespitose. Scales much shorter than
                     the perigynia. Bracts subtending the lower
                     spikes long and leaf-like. Culms 1-4 dm
                     tall. Leaves short, near the base. Pistil-
                     late spikes 2-6, crowded, nearly sessile.
                     Perigynium obtusely triangular with a mi-
                     nutely bidentate beak. Rare in moist soil
                     of valleys in the Black Hills. July . . . .
                     . . . . *Carex viridula* Michx.

                 12' Plants strongly stoloniferous. Scales
                     longer than the perigynia. Bracts subtend-
                     ing the lower spikes short, only slightly
                     sheathing the culm. Culms 1-5 dm tall.
                     Leaves well-developed, borne near the base.
                     Pistillate spikes 1-4, erect, sessile.
                     Perigynium flattened, sharply nerved, the

beak very short, minutely bidentate. Infrequent in meadows in the eastern part and in
the Black Hills. June-July . . . .
. . . . *Carex hallii* Olney

11' Terminal spike gynecandrous, the lower spikes
usually pistillate. Perigynia 3-4 mm long.
Plants cespitose with fibrous rootstocks, the
culms 5-9 dm tall. Leaves elongate, on the
lower one half of the stem. Pistillate spikes
on flexuous and drooping peduncles. Pistillate
scales dark purple-brown with hyaline margins.
Perigynium ovate with a very short bidentate
beak. Apparently very rare in shaded ravines
in the Black Hills. July-Aug . . . .
. . . . *Carex bella* Bailey

8' Pistillate spikes exceeding 1.25 cm in width.

13 Perigynia 5-7 mm long, the beak with widely spreading
teeth. Perigynia many in the pistillate spike. Plants
5-12 dm tall, the stems stout and angled with wide,
well developed leaves. Terminal spike staminate with
2-5 lower pistillate spikes, the latter on slender peduncles. Perigynium many ribbed, the beak long-tapered.
Rare to infrequent in marshy soil of the northeast part.
July-Aug . . . . *Carex comosa* Boott.

13' Perigynia 10-20 mm long, the beak bidentate-spreading.
Perigynia 2-15 in the pistillate spike. Plants 2-10 dm
tall, the culms stout with many leaves. Pistillate
spikes 1-3, clustered on short peduncles or sessile.
Perigynium lance-ovoid, the beak tapered to a bidentate
beak, the teeth about 1 mm long, hispid on the inner
margin. Rare to infrequent in marshy soils in the
Black Hills. July-Aug . . . . *Carex intumescens* Rudge

Group 6.

(Key to the Species)

1 Leaves narrow or involute, less than 1 mm wide.

2 Spike solitary, androgynous. Perigynia 2.5-5 mm long,
strongly nerved, beakless, widest near the top. Plants
cespitose with slender rootstocks, the culms slender, 2-5
dm tall. Leaves mostly basal or near the base, linear.
Spike 4-15 mm long, the upper part of staminate flowers.
Perigynia loosely arranged, 2-10, ascending. Apparently
rare in rich woods and ravines in the Black Hills. June-
July . . . . *Carex leptalea* Wahl.

2' Spikes several, the terminal one staminate with 2-4 lateral pistillate spikes on slender peduncles. Perigynia about 2 mm long, ovoid, membranous, with a very short oblique beak. Plants with slender stolons, the stems tufted, only the fertile culms elongate to 3 dm tall. Leaves clustered at the base, involute and firm. Pistillate spikes with 2-6 perigynia, often overtopping the staminate spike. Infrequent in rich woods of the east part and in the Black Hills. May-June . . . . *Carex eburnea* Boott.

1' Leaves usually flat, narrow, exceeding 1 mm in width.

3 Tip of perigynium curved at the beak, sometimes only slightly so, with a circular, entire orifice.

4 Principal leaves less than 7 mm wide. Plants with deep, horizontal stolons.

5 Pistillate spikes 3-5 mm wide. Perigynia 1.5-2 mm wide, in 2-3 rows in the spike. Plants 2-6 dm tall, with leaves 2-4.5 mm wide. Pistillate spikes 1-3, linear, each 6-20 flowered, usually separated on the culm. Perigynia about 3 mm long, many-nerved, the beak minute, slightly curved. Infrequent in meadows and open woods in the eastern part. June . . . . *Carex tetanica* Schk.

5' Pistillate spikes 5-7 mm wide. Perigynia 2-2.5 mm wide, in 6 rows in the spike. Plants 2-5 dm tall from deep, horizontal stolons. Leaves 3-7 mm wide. Pistillate spikes 1-3, peduncled, 7-25 flowered. Perigynia 3-4 mm long, with many obscure nerves and two prominent ribs, tipped with a slightly bent beak. Infrequent in mesic prairies in the eastern one-half. May-June . . . . *Carex meadii* Dewey

4' Principal leaves wider than 7 mm, often to 15 mm. Plants cespitose, lacking horizontal stolons, the culms 1-5 dm tall. Leaves several to many, mostly basal. Terminal spike staminate or infrequently gynecandrous, the lateral spikes 2-4, pistillate. Perigynium ovoid-obtuse, widest above the middle, prominently nerved with a conspicuously bent beak. Pistillate scale often with a serrulate awn tip. Common in rich woods of the eastern part and in the Black Hills. June . . . . *Carex blanda* Dewey

3' Tip of perigynium not curved at the beak.

6 Mature perigynia with 7-many ribs or nerves.

7   Plants with deep, horizontal rhizomes. Lower pistil-
    late spikes on short, basal peduncles. Stems 1-3 dm
    tall, seldom taller. Leaves several, 1-3 mm wide,
    arising from near the basal part of the culm. Ter-
    minal spike staminate, with 2-4 lateral pistillate
    spikes, on short peduncles, the basal ones with
    longer peduncles. Perigynium ovoid to sub-orbicular,
    3-4 mm long, with a straight, hyaline beak. Frequent
    in mesic prairies in the northeast Coteau region.
    June-July . . . . *Carex crawei* Dewey

7' Plants cespitose with short roots. All pistillate
    spikes near the upper part of the fertile culms or
    at least above the middle.

    8   Leaves and leaf sheaths softly pubescent. Bases
        of stems purple-tinged. Pistillate spikes crowded,
        mostly sessile. Plants 2-4 dm tall, cespitose-
        tufted. Leaves flat, 2-3 mm wide, deep green.
        Terminal spike staminate, with 1-3 pistillate
        spikes immediately below, not interrupted. Peri-
        gynium sub-orbicular, inflated, with a very short
        beak. Infrequent in valleys and meadows in the
        Black Hills. July . . . . *Carex torreyi* Tuckerm.

    8' Leaves and leaf sheaths glabrous. Bases of stems
        brownish-tinged. Pistillate spikes separated, at
        least the lower ones peduncled.

        9   Leaves 1-2.5 mm wide. Pistillate spikes with
            capillary peduncles, drooping and elongate.
            Perigynia 2-nerved. Plants 0.5-6 dm tall,
            leafy at the base, very slender. Staminate
            spike terminal, up to 1 cm long. Pistillate
            spikes 2-4, with sheathing bracts. Pistillate
            spikelets 3-10 flowered. Perigynium 2-3 mm
            long, turgid, with a short, oblique beak.
            Rare in rocky woods in the Black Hills. Ours
            var. *elongata* (Olney) Fern. July-Aug . . . . .
            . . . . *Carex capillaris* L.

        9' Leaves 3-8 mm wide, glaucous. Pistillate spikes
            short-oblong, erect. Perigynia many-ribbed.
            Plants 2-8 dm tall, the stems clustered. Ter-
            minal spike staminate, the 2-5 pistillate
            spikes on interrupted peduncles. Perigynium
            2-4 mm long, ovoid, with a minute beak. Fre-
            quent in sloughs and swales of the northeast
            and infrequent in similar habitats in the

Black Hills. Ours var. *granularis*. June-July
. . . . *Carex granularis* Muhl.

6' Mature perigynia nerveless or, if nerves are present,
these very fine and numerous.

10 Spikes all androgynous, few flowered, the pistillate
portion subtended by leaf-like bracts. Plants with
stems 1-3 dm tall, the leaves elongate, overtopping
the fertile culms. Inflorescence of 1-3 spikes, the
lower pistillate portion with 2-5 perigynia. Perigynium 4 mm long, the base spongy. Beak 1 mm long,
with an oblique orifice. Infrequent in rich woods
of the Black Hills. June-July . . . .
. . . . *Carex saximontana* Macken.

10' Spikes separate, the terminal one staminate with 3-5
lateral pistillate spikes. Bracts subtending the
pistillate spikes sheathing them. Perigynia 7-12 in
each spike, loosely arranged. Pistillate scales of
the perigynium awn tipped. Perigynium 3.5-4.5 mm
long, with many fine, obscure nerves. Body of the
perigynium lance-ovoid, the beak obsolete with an
entire orifice. Infrequent in the east in low,
alluvial woods but locally abundant. Ours var.
turgida Fern. May-June . . . *Carex amphibola* Steud.

CYPERUS L.

1 Rachilla wings thick, curved and clasping the achene. Spike
lets breaking between the scales as 1-fruited joints. Plants
annual, the stems stout, 1-6 dm tall, reddish at the base.
Leaves 3-8 mm wide. Inflorescence of 3-6 principal rays that
are branched, with a congested appearance. Bracts elongate,
exceeding the rays. Spikelets 10-25 flowered, almost terete,
about 2 mm wide. Scales ovate with obtuse tips, 2-2.5 mm
long, the achenes ovoid, about 1 mm long. Common in wet or
sandy soils over the state. {*C. odoratus* of reports}. July-
Sept . . . . *Cyperus ferruginescens* Boeckl.

1' Rachilla wings, when present, very thin and transparent, not
clasping the achene. When mature, the entire spikelet falling
as a unit.

2 Styles 2-parted, the achenes lenticular.

3 Scales dull and membranous. Style conspicuously longer
than the scale, divided almost to the base. Plants annual, 5-20 cm tall, the leaves about as tall as the
culms. Inflorescence of 1-3 rays on peduncles up to

6 cm long. Scales ovate, 2-3 mm, red-purple at the
distal end. Achenes 1-1.3 mm long. Infrequent in
wet soil in the eastern part. Aug-Sept . . . .
. . . . *Cyperus diandrus* Torr.

3' Scales shiny and heavy textured. Style as long as or
shorter than the scales, divided to the lower one-third.
Plants annual, tufted, 5-20 cm tall. Leaves few, basal.
Bracts below the inflorescence leaf-like. Spikelets in
1-4 capitate clusters, the primary ray short, 3-12 mm
long. Scales ovate, 2-2.5 mm long, reddish at their
bases. Achenes 1-1.5 mm long. Frequent on sandy floodplains along the Missouri River and in the east and central part. Aug-Sept . . . . . *Cyperus rivularis* Kunth.

2' Styles 3-parted, the achenes triangular.

4   Scales 7-12 nerved, tipped with a noticeable recurved
awn. Plants annual, tufted, the culms slender, 5-15 cm
tall. Leaves few, arising from near the bases of the
stems. Involucral bracts leafy, surpassing the inflorescence. Spikelets in dense, capitate clusters, 4-10
mm long, the central cluster sessile. Scales 7-9
nerved, 1-1.5 mm long, recurved at the tip. Rachilla
with wings. The achenes triangular, less than 1 mm
long. Infrequent at muddy pond margins over the state.
Aug-Sept . . . . *Cyperus aristatus* Rottb.

4' Scales acute to obtuse, not awn tipped.

5   Spikelets one-half as wide as long, the scales wide-
spreading, their tips slightly recurved. Plants annual, tufted, with stems 5-35 cm tall. Leaves mostly
basal, 1-2 mm wide. Spikelets in dense, globose
heads on rays, the central one sessile. Involucral
bracts elongate, leaf-like. Spikelets 3-6 mm long,
the scales 1.5-2 mm long, 3-nerved. Achenes triangular, 0.5-0.9 mm long. Frequent in wet soils over the
state. Aug-Oct . . . *Cyperus acuminatus* Torr. & Hook.

5' Spikelets narrow, several times longer than wide.

6   Inflorescence arranged in a compound umbel, the
spikelets pinnately arranged on each rachis.

7   Plants annual. Scales 1-1.5 mm long. Achenes
less than 1 mm long. Stems tufted, 1-5 dm
tall. Leaves mostly basal, the blade 2-8 mm
wide, elongate. Spikes clustered cylindrically
on rays, the terminal one sessile, spikes up to
4 mm long. Spikelets 1-1.5 mm wide, compressed,

about 1 cm long. Scales keeled with apiculate
tips. Rachilla with a narrow, hyaline wing.
Achenes 0.7-1.0 mm long. Frequent in sandy
areas and on floodplains in the eastern one-
half. Aug-Oct . . . *Cyperus erythrorhizos* Muhl.

7' Plants perennial. Scales over 2 mm long.
Achenes exceeding 1 mm.

8 Scales 3.5 mm long or longer, the achenes
less than one-half the length of the scales.
Wings of the rachilla clasping the base of
the achene. Plants with culms few, sharply
angled, 1-5 dm tall. Leaves clustered from
near the base. Spikes cylindrical, 5-12
flowered, the spikelets many, 5-30 mm long.
Achenes with an obtuse, triangular shape.
Rare in moist meadows and along streams in
the eastern one-half. Aug-Sept . . . . . .
. . . . *Cyperus strigosus* L.

8' Scales about 2.5 mm long, the achenes 1.2-2
mm long. Wings of the rachilla narrow, not
clasping the achene. Stems triangular,
stout, 1-6 dm tall, the leaves arising from
near the base. Spikes cylindric, the terminal one sessile, the others with rays naked
near the base. Spikelets 8-25 flowered,
narrow, 1-3 cm long. Achenes obtusely triangular. Infrequent in moist soil in the
southeast part. Aug-Sept. Nut Sedge . . .
. . . . *Cyperus esculentus* L.

6' Inflorescence arranged in a sub-capitate cluster,
the spikelets unequally radiating or digitate, not
pinnately arranged on the rachis.

9 Plants annual. Scales 3-5 nerved, 1-1.5 mm
long. Stems slender, tufted, 1-3 dm tall.
Leaves narrow, 1-3 mm wide, on the lower part
of the stem. Involucral bracts leaf-like, much
exceeding the inflorescence. Spikes few, the
rays 1-3 cm long. Spikelets 3-5 mm long, reddish. Achenes about 1 mm long, almost equaling the scales. Rare in moist soils in the
southern part. July-Aug . . . *Cyperus fuscus* L.

9' Plants perennial, with hard, corm-like bases.
Scales 7-12 nerved, 2.0 mm or more long.

10 Spikelets crowded, causing them to radiate
in all directions, forming rounded heads.
Scales ovate, 2.5-3.5 mm long, ovate with
abruptly pointed tips. Plants rhizomatous,
the stems 1-6 dm tall. Leaves flat, 2-4 mm
wide. Spikelets 6-10 flowered, flattened,
the rachilla not winged. Achenes 1-2 mm
long. Rare to infrequent in dry to moist
sandy soils over the state. Aug-Sept . . .
. . . . *Cyperus filiculmis* Vahl.

10' Spikelets ascending in a 4-11 unequally
rayed umbel. Scales 2-4 mm long, the body
widely rounded.

11 Stems smooth. Achenes about 1.5 mm
long. Scales 2-2.5 mm long, many-
nerved. Plants 1-7 dm tall. Inflorescence of 1-2 sessile spikes in an irregular head, each with 2-5 rays up to
10 cm long. Spikelets 5-15, the scales
rotund and many-nerved. Achenes 1.5-2
mm long. Infrequent in sandy soil in
the eastern part and in the Black Hills.
Aug-Sept . . . *Cyperus houghtonii* Torr.

11' Stems roughened. Achenes 2-3 mm long.
Scales 3-4 mm long, 9-14 nerved. Plants
1-7 dm tall, the leaves crowded near the
base. Spikes on rays of unequal length,
the terminal ones central. Spikelets 5-
25 mm long, borne near the ends of the
rays. Rachilla very narrowly winged.
Common in sandy soils over the state.
Aug-Sept . . *Cyperus schweinitzii* Torr.

ELEOCHARIS R. Br.

1 Stigmas two. Achenes lenticular or bi-convex.

2 Plants annual. Tubercle flat, fitting closely to the summit of the achene. Achenes 0.5-1.5 mm long. Plants tufted,
cespitose, the stems 1-5 dm tall. Spikelets 5-12 mm long,
ovoid, commonly 35 flowered or more. Bristles 5-7, mostly
6, about as long as the achene. Achene light to dark brown.
Frequent in moist places in the eastern part. {incl. *E.*
engelmanii Steud., *E. ovata* (Roth) R. & S.} Ours var.
ovata (Roth.) Dr. & Mohl. July-Aug . . . .
. . . . *Eleocharis obtusa* (Willd.) Schult.

2' Plants perennial. Tubercle cone-like, constricted at the
base where it attaches to the achene.

   3 Scales at the base of the spikelet 2 or 3, each scale
not completely encircling the stem at the base of the
spikelet.

      4 Tubercle longer than wide. Scales firm, acute at the
apex. Stems stout, up to a meter tall and up to 5 mm
in diameter, often flattened. Spikelet linear to
ovoid, 5-20 mm long. Scales reddish-brown, about 2
mm long, with 4 bristles as long as or longer than
the body. Common in moist or marshy places over the
state. {*E. palustris* (L.) R. & S. in part.} July-
Aug . . . . *Eleocharis macrostachya* Britt.

      4' Tubercle as wide as long or wider. Scales firm,
acuminate at the apex. Stems wiry, terete, 2-8 dm
tall, about 2.5 mm wide. Spikelet slender to ovoid,
up to 2 cm long. Scales with spreading tips and 2
purple bands near the midrib. Achenes up to 1.2 mm
broad and 2 mm long. Bristles lacking or, if present, to longer than the achenes. Common in moist
places over the state. July-Aug . . . .
. . . . *Eleocharis smallii* Britt.

  3' Scale at the base of the spikelet solitary, completely
encircling the stem.

     5 Stem terete or nearly so, not rigid. Scales red to
purple. Culms 1-7 dm tall, slender. Spikelet lanceolate, 8-15 mm long, the scales obtuse and appressed.
Achenes narrowly obovoid, 1-1.4 mm long, the tubercle
cone-shaped. Bristles lacking or 4 bristles present,
as long as the achene. Common in wet places over the
state. {*E. calva* Torr.} June-Aug . . . .
. . . . *Eleocharis erythropoda* Steud.

     5' Stem compressed and often twisted, rigid, with a red-
dish sheath at the base. Scales light colored with
whitish margins, obtuse to rounded at the apex.
Culms 1-5 dm tall. Spikelets narrowly lanceolate,
1-2 cm long. Achenes obovoid, light chestnut in
color. Frequent in moist places in the western one-
half. June-Aug . . . .
. . . . . . *Eleocharis xyridiformis* Fern. & Brackett

1' Stigmas three. Achenes triangular to turgid. Plants perennial.

  6 Stems short, less than 1.5 dm tall, not flattened.

7 Achenes longitudinally ribbed and the tubercle distinguishable from the body. Stems slender, filiform, forming dense tufts of plants, 3-12 cm tall. Spikelets 2-7
   mm long, 3-10 flowered. Scales green with a hyaline
   margin. Achenes faintly yellow, about 1 mm long, with
   3 or 4 bristles. Frequent to common in moist places
   over the state. July-Sept . . . .
   . . . . *Eleocharis acicularis* (L.) R. & S.

7' Achenes smooth, the tubercle not forming a distinct cap
   at the apex. Stems slender, filiform, 2-5 cm tall,
   forming mats from slender rhizomes. Spikelets short,
   usually only 3 mm long, each one few to several flowered. Scales 1-5 mm long, yellowish. Achenes about 1
   mm long, the bristles, when present, as long as or
   longer than the body. Infrequent in slightly saline
   soils in the eastern part. Ours var. *anachaeta* (Torr.)
   Svens. July-Aug . . . *Eleocharis parvula* (R. & S.) Link

6' Stems exceeding 2 dm in height, more or less flattened.

8 Tubercle forming a distinct cap on the achene and distinguishable from it.

   9 Achenes with longitudinal ribs and with cross-
      striations. Bristles lacking. Spikelets 4-6 mm
      long, oblong to ovate. Plants 2-5 dm tall, from
      well-developed rhizomes, the culms flattened.
      Achenes obovoid, pale yellow to gray, with about
      9 prominent ribs. Tubercle depressed-conic, much
      smaller than the body of the achene. Not reported
      from South Dakota but should occur in the eastern
      part in marshes. June-Aug . . . .
      . . . . *Eleocharis wolfii* Gray

   9' Achenes granular-roughened, yellow-brown, but lack-
      ing ribs. Bristles 2-5, shorter than the achene.
      Spikelets 5-10 mm long, ovoid to oblong, the scales
      with pointed tips. Plants 2-6 dm tall, the culms
      compressed-flattened, up to 1.5 mm wide. Achenes
      obtusely triangular, the tubercle conic, minute.
      Frequent in moist places over the state. July-Aug
      . . . . *Eleocharis compressa* Sulliv.

8' Tubercle of the achene elongate-conic, not distinguishable from the body of the achene. Plants 4-10 dm tall,
   spreading with stolons or recurved culms that root.
   Spikelet 8-12 mm long, 10-20 flowered. Achenes obovoid,
   finely reticulate, with 4-8 bristles that exceed the
   achene. Rare in alkaline marshes in the southwest part.
   July-Aug . . . . *Eleocharis rostellata* Torr.

ERIOPHORUM L.

Eriophorum polystachion L. Perennial herbs with stems cespitose,
2-5 dm tall. Leaves 4-6 mm wide, not restricted to the lower
part of the stem. Inflorescence of 3-7 spikelets subtended by
leaf-like bracts 3-5 mm wide. Spikelets on short peduncles.
Scales thin, acutely pointed. Achenes brown, 2-3 mm long, the
bristles white or nearly so, 2-4 cm long. Rare in boggy places
in the northeast part and in the Black Hills. {*E. angustifolium*
Honckeny} Apr-June. Cottongrass.

RHYNCHOSPORA Vahl.

Rhynchospora capillacea Torr. Perennials with slender, triangular stems 1-3 dm tall, tufted from a cespitose base. Leaves involute, very narrow, mostly basal. Inflorescence of 2-10 spikelets, the lower most with a leaf-like bract. Spikelets few-
flowered, the achenes 1-2 mm long, obovate with a tubercle over
one-half as long. Bristles usually 6, slightly exceeding the
tubercle. Rare in marshy or boggy places in the southern sandhills area. July-Aug. Beakrush.

SCIRPUS L.

1 Inflorescence appearing to be lateral due to the erect position of the principal bract that seems to be a continuation
   of the culm.

   2 Spikelets 2-12, sessile or on short pedicels from one place
      on the culm.

      3 Plants annual. Stems terete to obtusely triangular,
         cespitose, 1-3 dm tall. Leaves small, reduced, the
         upper bract erect, appearing as a leaf-like extension
         of the culm. Inflorescence of 2-5 spikelets, sessile
         and peduncled, narrowly ovoid, 5-9 mm long. Achenes
         transversely rugose, 1-2 mm long. Bristles usually
         lacking. Infrequent on floodplains in the eastern and
         southern part. {*S. saximontana* Fern.} July-Sept . . .
         . . . . *Scirpus hallii* Gray

      3' Plants perennial. Stems strongly triangular, 1-8 dm
         tall, from creeping rootstocks. Leaves much reduced
         except for the upper ones. Inflorescence of 1-several
         spikelets, sessile, 6-10 mm long. Scales notched at
         the apex, scarious. Achenes smooth, 2-3 mm long.
         Bristles usually 4, slightly shorter than the achene.
         Common in wet places over the state. June-July . . . .
         . . . . *Scirpus americanus* Pers.

2' Spikelets many, in branched clusters or compound umbels.

    4 Stigmas 3, the achenes trigonous. Spikelets all on short
       peduncles. Culms terete, 1-2 meters tall. Spikelets in
       a loose or lax inflorescence, 6-11 mm long, few flowered.
       Scales reddish-brown, longer than the achene, with a
       mucronate tip. Achenes smooth, 2-3 mm long, with 2-4
       bristles almost as long as the achene. Infrequent to
       rare in shallow water of ponds in the eastern part.
       July-Aug . . . . *Scirpus heterochaetus* Chase

    4' Stigmas 2, the achenes lenticular.

       5 Achenes about 2.5 mm long, their scales 3-4 mm long,
          dull brown. Plants perennial, 1-3 meters tall, from
          stout rhizomes, the culms not as soft as in the fol-
          lowing species. Achenes with mostly 6 bristles,
          variable in length but about as long as the achene.
          Infrequent in marshy places in the northeastern part.
          July-Aug. Hardstem Bulrush. . . *Scirpus acutus* Muhl.

       5' Achenes 2 mm or less in length, their scales 2-3 mm
          long, reddish-brown. Plants perennial, 1-2 meters
          tall, from creeping rhizomes. Inflorescence of 20-
          40 spikelets arranged in an open or lax cluster, each
          one about 1 cm long. Achenes not entirely concealed
          by the scales. Bristles usually 6, about the length
          of the achene. Common in marshes and shallow water
          over the state. July-Aug. Giant Bulrush . . . . . .
          . . . . *Scirpus validus* Vahl.

1' Inflorescence appearing terminal, this due to the lateral
  spreading of the foliaceous principal bracts subtending the
  spikelets.

    6 Spikelets 1-3 cm long. Stems obtusely triangular.

       7 Stigmas 2, achenes lenticular. Spikelets on short pe-
          duncles. Leaf sheath straight or concave at the orifice,
          the nerves on the ventral surface gradually diverging to
          the sides. Plants rhizomatous and tuberous, 2-9 dm tall,
          the culms triangular. Spikelets 1-several, some sessile,
          others pedunculate. Scales scarious at the margins with
          a notched apex and a short, mucronate tip. Achenes 2.5-4
          mm long, the bristles 2-6, short or deciduous. Common at
          pond margins, in marshes, and on floodplains over the
          state. {incl. *S. paludosus* Nels.} Ours var. *paludosus*
          (Nels.) Kukenth. July-Aug. Prairie Bulrush . . . . . .
          . . . . *Scirpus maritimus* L.

7' Stigmas 3, achenes three-sided. Some of the spikelets
long-peduncled towards maturity in a loose, open inflorescence. Leaf sheath convex at the orifice, the nerves
on the ventral surface abruptly diverging to the sides.
Plants perennial from a thick rootstock, the culms 1-
1.5 meters tall, robust. Spikelets 10-40, 1-2.5 cm
long, some sessile. Scales brown, notched at the apex.
Achenes 3.5-5 mm long, the bristles equaling the achene.
Frequent to common along streams and ponds in the eastern
one-half. July-Aug. River Bulrush . . . .
. . . . *Scirpus fluviatilis* (Torr.) Gray

6' Spikelets 0.5-1 cm long. Stems terete to irregularly triangular.

8 Inflorescence of spikelets in loose clusters. Bristles
twice as long as the achenes or longer, conspicuously
exsert from between the scales.

9 Bristles smooth and flexuous, several times longer
than the achenes. Scales reddish-brown, acute to
broadly obtuse at their apices. Plants perennial,
cespitose, the culms 6-12 dm tall, leafy. Spikelets
3-7 mm long, many in a terminal, cymose cluster, some
short-peduncled. Bracts below the inflorescence
leafy. Achenes 1 mm long, obovate, sharply beaked.
Rare in moist places in the Black Hills. Aug-Sept.
Woolgrass . . . . . . *Scirpus cyperinus* (L.) Kunth.

9' Bristles very crooked, about twice the length of the
achene. Scales with a green midrib, the apices acuminate. Plants perennial, cespitose, up to 1.5
meters tall. Inflorescence of one terminal cluster
that is sessile and several other peduncled spikelets
that droop. Spikelets 6-10 mm long, the bristles
scarcely exceeding the scales. Achenes brown, 3-
angled, the beak short and sharply pointed. Rare
in swampy soil in the northeast part. July-Aug . . .
. . . . *Scirpus pendulus* Muhl.

8' Inflorescence of spikelets in dense glomerules. Bristles less than twice the length of the achenes, not conspicuously exsert from the scales.

10 Styles usually 3 or 3-cleft. Scales mucronate.
Leaf sheaths not tinged with red. Plants perennial,
cespitose, with slender stems 5-12 dm tall. Leaves
1-1.5 cm wide, on the lower part of the stem. Spikelets many, the principal rays branched. Spikelets
ovoid, 2-6 mm long. Achene about 1 mm long, the

bristles present, about as long as the achene, or lacking. Common in moist places over the state. {incl. *S. pallidus* (Britt.) Fern.} Represented in South Dakota by the two varieties *atrovirens* (rare), and *pallidus* Britt. (common). Aug-Sept . . . . *Scirpus atrovirens* Muhl.

    10' Styles usually 2 or 2-cleft. Scales obtuse to acute. Leaf sheaths tinged with red. Plants perennial from creeping rhizomes, the stems coarse, 5-12 dm tall. Leaves 8-13 mm wide, distributed on the stem. Spikelets numerous, sessile on branches of the main rays. Spikelets 4-8 mm long, dark green to gray-black. Achenes about 1 mm long, with 4-6 bristles slightly longer than the achene. Rare in moist places along streams in the Black Hills. Ours var. *rubrotinctus* (Fern.) Jones. July-Aug . . . . *Scirpus microcarpus* Presl.

Family Poaceae

(Key to Tribes)

1 Spikelets 2 to many flowered.*

    2 Inflorescence sessile in a usually continuous rachis to form a spike.

        3 Spikelets solitary or in groups of 2 to 6, these arranged alternately on opposite sides of the spike. Spike terminal and solitary . . . . *Tribe 6. Hordeae*   page 106

        3' Spikelets on one side only. Spikes usually more than one, forming a digitate or racemose head . . . . *Tribe 4. Chlorideae*   page 88

    2' Inflorescence pedicellate in an open or contracted panicle, sometimes contracted enough to appear spike-like.

        4 Glumes shorter than the lowest fertile lemma . . . . *Tribe 5. Festuceae*   page 91

        4' Glumes as long as or longer than the lowest fertile lemma, excluding the awn.

*Members of the genus *Hordeum* in the Tribe Hordeae and *Beckmannia, Bouteloua, Chloris, Schedonnardus,* and *Spartina* in the Tribe Chlorideae are exceptions, having one functionally perfect flower. These are properly treated in the second section of this key.

Tribe 1. Agrostideae

(Key to Genera)

1 Lemma of more delicate texture than the glumes.

  2 Inflorescence an open or spike-like panicle.

    3 Lemma and palea not easily separated from the grain.

      4 Stamen 1. Palea 1-nerved and 1-keeled. Tall perennials with nodding heads . . . . . . *CINNA*

      4' Stamens 3. Palea 2-nerved and 2-keeled.

        5 Lemma surrounded with a tuft of hairs up to 5 mm long, these arising from the callus.

          6 Lemma and palea membranous. Rachilla bristle-like, prolonged beyond the palea . . . . *CALAMAGROSTIS*

          6' Lemma and palea papery. Rachilla not as above. . . . . *CALAMOVILFA*

        5' Lemma glabrous or with hairs; if with hairs, these less than 2 mm long.

          7 Lemma with a terminal awn or with an abrupt point . . . . *MUHLENBERGIA*

          7' Lemma awnless or awn dorsal . . . . *AGROSTIS*

    3' Lemma and palea easily separated from the grain . . . . *SPOROBOLUS*

  2' Inflorescence very dense and spike-like. Glumes keeled.

    8 Spikelets articulating below the glumes, the entire spikelet falling or breaking off.

      9 Glumes with awns 3-7 mm long . . . . *POLYPOGON*

      9' Glumes without awns . . . . *ALOPECURUS*

    8' Spikelets articulating above the glumes . . . . *PHLEUM*

1' Lemma more hardened than the glumes when mature, closely enveloping the grain.

  10 Lemmas without awns . . . . *MILIUM*

  10' Lemmas with awns.

    11 Lemmas with 3 awns . . . . *ARISTIDA*

11' Lemmas with 1 awn.

    12 Awn of lemma twisted or bent . . . *STIPA*

    12' Awn of lemma not twisted but sometimes bent.

        13 Lemma broad, the awn deciduous . . . .
. . . . *ORYZOPSIS*

        13' Lemma narrow with an abrupt tip . . . .
. . . . *MUHLENBERGIA*

AGROSTIS L.

1 Palea evident, about one-half as long as the lemma. Perennial
with rhizomes, the stems ascending to erect, 5-12 dm tall.
Leaves flat to somewhat folded, 2-8 mm wide. Inflorescence
a pyramidal panicle, 5-20 cm long, often purple or reddish.
Panicle branches verticillate with numerous spikelets. Spikelets 2-4 mm long, the lemmas two-thirds the length of the
glumes, awnless. Common in yards, on roadsides and in pastures over the state. {*A. alba* L. misapplied.} June-July.
Redtop . . . . *Agrostis stolonifera* L.

1' Palea minute or lacking.

    2 Panicle narrow and contracted, with some of the lower
branches bearing spikelets near the base. Plants perennial, usually without rhizomes, the culms 2-10 dm tall.
Leaves flat, scabrous, with a ligule about 6 mm long.
Inflorescences narrow, interrupted to dense, 5-25 cm long.
Glumes about 3 mm long, the lemmas usually awnless. Infrequent in valleys and along streams in the Black Hills.
Ours ssp. *minor* (Hook.) C. L. Hitchc. July-Aug. Spike-
bent . . . . *Agrostis exarata* Trin.

    2' Panicle open, sometimes diffuse. Branches slender, smooth
or scabrous, the lower ones not bearing spikelets near
their bases.

        3 Inflorescence very diffuse, 15-60 cm long and purplish
towards maturity. Branches scabrous, forked near the
ends. Plants perennial, tufted, usually without rhizomes, the culms 2-5 dm tall. Leaves flat, 2-4 mm
wide, scabrous. Flowering panicle ovoid to pyramidal.
Spikelets 2-3 mm long, the glumes scabrous on the keel.
Frequent in light or sandy soil over the state. Represented in South Dakota by the two varieties *hyemalis*
and *tenuis* (Tuckerm.) Gl. May-June. Ticklegrass . . .
. . . . *Agrostis hyemalis* (Walt.) B.S.P.

3' Inflorescence open but not diffuse, 10-20 cm long and greenish towards maturity. Branches smooth, forked at or below the middle. Plants perennial, cespitose, the culms erect to decumbent, 3-8 dm tall. Leaves flat, 3-6 mm wide. Spikelets 2-2.7 mm long, the glumes about equal. Infrequent to rare on dry or sandy soil in the central and eastern part. Aug-Oct. Autumn Bent . . . . *Agrostis perennans* (Walt.) Tuckerm.

ALOPECURUS L.

1 Awns exsert from the glumes less than 1.5 mm, arising from near the middle of the lemma. Plants perennial, the culms slender, in tufts of 1-several, 2-4 dm tall. Leaves 2-4 mm wide, the sheaths as long as the blade. Panicle compact, spike-like, 2-7 cm long and about 4 mm wide. Frequent in low or wet places over the state. May-July. Short Awn Foxtail . . . . *Alopecurus aequalis* Sobol.

1' Awns exsert from the glumes 2.0 mm or more.

2 Plants annual. Anthers 0.5 mm long. Culms ascending to erect, 1-5 dm tall. Leaves narrow, pointed, 3-6 cm long. Panicle slender, 2-6 cm long, the spikelets 2-2.5 mm long. Awn exserted 2-3 mm beyond the glumes. Rare to infrequent in marshy soils over the state. June-July . . . . *Alopecurus carolinianus* Walt.

2' Plants perennial. Anthers exceeding 1.0 mm in length. Culms decumbent to ascending, rooting at the nodes, 3-5 dm tall. Leaves 1-4 mm wide, flat. Panicle spike-like, bristly due to the exsert awns. Spikelets about 2.5 mm long, purple-brown. Rare in moist or marshy places and appearing sporadically over the state without any specific distribution pattern. July-Aug. Water Foxtail . . . . *Alopecurus geniculatus* L.

ARISTIDA L.

1 Plants annual.

2 Central awn of the lemma spirally coiled at the base, its distal end projecting laterally about 1 cm. Lateral awns erect, less than 4 mm long. Plants cespitose, 2-5 cm tall. Leaves narrow, involute, about 1 mm wide, erect. Panicle terminal, mostly unbranched, 3-7 cm long. Infrequent on dry, sandy soil in the southeast part. Ours var. *curtissi* Gray. Sept-Oct . . . . *Aristida dichotoma* Michx.

2' Central awn of the lemma not spiralled at the base.

3 Lateral awns of the lemma 4-7 cm long, divergent, about
   equal to the central one. Plants branched from the base,
   the culms 3-5 dm tall. Leaves narrow, mostly involute,
   about 1 mm wide. Panicle loose, branched, 10-18 cm long,
   the spikelets on short pedicels. Glumes 2-4 cm long,
   tapering to a short awn. Infrequent on dry soil in the
   eastern part. July-Sept. Prairie Threeawn . . . . . .
   . . . . *Aristida oligantha* Michx.

3' Lateral awns of the lemma 1-2 cm long, unequal to equal,
   irregularly divergent. Plants annual, branched below,
   the culms 2-5 dm tall. Leaves flat or involute, less
   than 2 mm wide. Panicles narrow, elongate, up to 20 cm
   long. Lemmas 4-6 mm long. Glumes about equal, 6-8 mm
   long, ending in a short awn. Infrequent on sandy, sterile soil along the Missouri River in the southeast part.
   {*A. intermedia* Scribn. & Ball.} . . . .
   . . . . *Aristida longespica* Poir.

1' Plants perennial, 2-3 dm tall, much branched from the base.
Flowering culms spreading overtopping the many narrow, inrolled leaves. Panicle loosely branched, 4-6 cm long. Awns
divergent, about equal in length, varying from 4-8 cm long.
Second glume usually over 18 mm long. Frequent to common in
prairies and dry soil over the state. Aug-Sept. Red Threeawn . . . . *Aristida longiseta* Steud.

CALAMAGROSTIS Adans.

1 Awn of the lemma longer than the glumes, twisted. The glumes
about 7 mm long. Inflorescence pink to lavender, dense, contracted, the panicle 5-10 cm long. Plants usually cespitose,
the culms several, 4-7 dm tall. Leaves thick, 2-4 mm wide,
tending to roll inward. Spikelets 6-9 mm long, the callus
, hairs one-third to one-half the length of the lemma. Rare
to infrequent in rocky woods of the Black Hills. July-Aug.
Purple Reedgrass . . . . . . *Calamagrostis purpurascens* R. Br.

1' Awn of the lemma shorter than the glumes, mostly included.
The glumes 7 mm or shorter.

2 Plants usually less than 4 dm tall. Awn of the lemma
   twisted. Stems stiff, erect, from slender rhizomes.
   Leaves more or less involute, 2 mm wide, erect. Inflorescence dense, 5-8 cm long, with spikelets about 5 mm
   long. Lemmas nearly as long as the glumes, the awn of
   the lemma inserted near the base. Rare in dry soils of
   plains and prairies. July-Aug. Plains Reedgrass . . . . .
   . . . . *Calamagrostis montanensis* Scribn.

2' Plants usually exceeding 4 dm tall. Awn of the lemma
straight or only slightly twisted.

    3 Principal leaf blades 4-8 mm wide, flattened. Inflores-
cence lax and open, 10-20 cm long. Callus hairs as long
as the lemma. Plants perennial from creeping rhizomes,
the culms 6-12 dm tall. Spikelets 3-6 mm long, the
lemma with a delicate straight awn inserted near the
middle. Frequent to common in wet meadows and along
streams and ponds over the state. June-July. Blue-
joint . . . . . *Calamagrostis canadensis* (Michx.) Beauv.

    3' Principal leaf blades 1-4 mm wide, becoming involute on
drying. Inflorescence contracted and spike-like, 5-14
cm long, with some side branching. Callus hairs one-
half to three-fourths as long as the lemma.

        4 Ligules 1-3 mm long. Leaf blades soft, smooth on the
undersurfaces. Plants rhizomatous, the culms clustered, 3-8 dm tall. Panicle narrow and dense, often
with few erect branches. Spikelets 2.5-3.5 mm long,
the awn almost reaching the tip of the lemma. Rare
to infrequent in moist places in the north and eastern part. {*C. neglecta* (Ehrh.) Gaertn.} July-Aug . .
. . . . *Calamagrostis stricta* (Timm) Koeler

        4' Ligules 4-6 mm long. Leaf blades 2-4 mm wide, firm,
scabrous on the undersurfaces. Plants from slender
rhizomes, the culms clustered, 4-10 dm tall. Panicle
narrow and dense, often with several erect and appressed branches. Spikelets 3-5 mm long, the awn
about as long as the lemma. Infrequent in marshes
and wet meadows in the east and northern part and in
the Black Hills. June-Aug. Northern Reedgrass . . .
. . . . *Calamagrostis inexpansa* Gray

CALAMOVILFA Hack.

Calamovilfa longifolia (Hook.) Scribn. Tall perennial from stout,
scaly rhizomes, the culms usually solitary or few, 5-15 dm tall.
Leaves flat, rigid, 4-8 mm wide, tapering to fine, flexuous
points. Panicle erect, narrow and contracted to open, 1-3 dm
long, with ascending branches. Spikelets 6-7 mm long, the callus
hairs one-half the length of the lemma. Common in sandy prairie
over the state. Aug-Sept. Sandreed.

CINNA L.

1 Spikelets 4-6 mm long. Panicle contracted, ascending. Plants
perennial, the culms up to 1 meter tall. Leaves flat, less

than 1 cm wide with glabrous sheaths. Inflorescence gray-
green, nodding, 10-25 cm long, the panicle dense and many
flowered. Rare in alluvial woods in the eastern part. July-
Aug. Woodreed . . . . *Cinna arundinacea* L.

1' Spikelets 3-4 mm long. Panicle open, drooping. Plants perennial, the culms up to 1 meter tall. Leaves shorter than the
   preceding, the blades wider, to 1.5 cm. Inflorescence green,
   the panicle loosely flowered with the branches drooping. Infrequent in moist or alluvial woods of the Black Hills. July-
   Aug. Drooping Woodreed . . . *Cinna latifolia* (Trevir.) Griseb.

MILIUM L.

Milium effusum L. Erect perennial, the culms glabrous, 6-10 dm
tall. Leaves 10-16 mm wide, the blades lax and flattened. Inflorescence an open panicle 10-18 cm long, the branches spreading, flowered towards their distal ends. Spikelets one-flowered,
3-4 mm long. Lemmas white and smooth. Rare in rich loamy woods
in Marshall and Roberts counties in the extreme northeast part.
May-July.

MUHLENBERGIA Schreb.

1 Inflorescence diffuse and open, 3 cm or more wide. Spikelets
   on thin pedicels.

   2 Spikelets less than 2 mm long, without awns.

      3 Plants annual. Culms spreading to erect, 1-3 dm tall.
         Leaves flat, about 1 mm wide and up to 8 cm long. In-
         florescence one-half as long as the plant is tall, with
         many flowers. Spikelets 1-1.8 mm long, the glumes
         sparsely pubescent, one-half to two-thirds as long as
         the lemma. Rare in sandy soil of the southwest part.
         Aug-Sept . . . *Muhlenbergia minutissima* (Steud.) Swallen

      3' Plants perennial, from slender rhizomes, the culms
         branched from the base, 1-5 dm tall. Leaves flat, 1-2
         mm wide and up to 5 cm long. Inflorescence diffuse, up
         to 15 cm long, the pedicels slender. Spikelets 1.5-2 mm
         long, the glumes nearly as long as the lemmas, scabrous
         on the dorsal keel. Infrequent in moist, sandy soil and
         alkali areas over the state. July-Aug. Scratchgrass
         . . . . *Muhlenbergia asperifolia* (Nees and Mey.) Parodi

   2' Spikelets 4 mm or more long, the lemma with a short awn.
      Leaf blades involute, basal, with sharp pointed tips.
      Plants perennial with stout, creeping rhizomes, the culms
      decumbent to erect, tufted, 2-4 dm tall. Inflorescence

erect, the branches ascending to erect, with long spreading
capillary branches. Spikelets brown-purple, 4-5 mm long.
Infrequent in sandy soils of the southern part. Aug-Sept.
Sand Muhly . . . . *Muhlenbergia pungens* Thurb.

1' Inflorescence dense and contracted, less than 3 cm wide.
Spikelets on short pedicels, appearing sessile.

  4 Lemmas glabrous to minutely pubescent near their bases.

    5 Ligule not more than 0.5 mm long. Glumes more than one-
    half as long as the lemmas. Plants perennial with scaly
    bases, the culms in dense clusters 2-4 dm tall. Leaves
    short, wiry, the blades narrow. Inflorescence narrow,
    spikelike, 4-9 cm long. Spikelets 3-5 mm long. Fre-
    quent to common in dry or sandy soil over the state.
    July-Aug. Plains Muhly . . . .
    . . . . *Muhlenbergia cuspidata* (Torr.) Rydb.

    5' Ligule 1.5-3.0 mm long. Glumes less than one-half as
    long as the lemmas.

      6 Plants perennial, forming mats from robust rhizomes.
      Leaf sheaths and culms immediately below the nodes
      roughened with small, warty areas. Leaves narrow,
      inrolled and curved, 1-5 cm long. Flowering culms
      3-6 dm tall, the inflorescence 2-10 cm long, spike-
      like. Spikelets 2-3 mm long, the lemmas minutely
      awn-tipped. Infrequent in sandy or gravelly soils
      in the south and western part. July-Aug. Mat Muhly.
      . . . . *Muhlenbergia richardsonis* (Trin.) Rydb.

      6' Plants annual, but forming mats from above-ground
      runners that may persist for more than one year.
      Leaf sheaths and culms not roughened. Leaves flat,
      1-5 cm long. Flowering culms spreading-ascending,
      3-10 cm long. Inflorescence narrow and interrupted,
      1-4 cm long. Spikelets up to 2.5 mm long, with an
      awned tip. Rare in moist places of middle altitudes
      in the Black Hills. July-Aug . . . .
      . . . . *Muhlenbergia filiformis* (Thurb.) Rydb.

  4' Lemmas pilose near their bases.

    7 Glumes with stiff awn tips, longer than the awnless
    lemma. Flowering panicle compact but interrupted, 7-10
    mm wide. Plants perennial from creeping rhizomes, the
    culms 1-several, 3-8 dm tall. Leaves flat, 2-5 mm wide,
    the ligule about 1 mm long. Inflorescence 3-9 cm long,
    the branches erect and appressed to the main axis.
    Lemmas 3 mm long. Frequent to common in moist meadows

and alluvial areas over the state. Aug-Sept . . . . . .
. . . . Muhlenbergia racemosa (Michx.) BSP.

7' Glumes without stiff awn tips, shorter than the lemma
(excluding awn). Flowering panicle less than 7 mm wide.

8 Flowering culms glabrous just below the nodes.
Plants usually sprawling and top heavy due to many
upper branches. Plants perennial, from scaly rhizomes, the stems 4-9 dm tall. Leaves flat, 3-5 mm
wide, 6-12 cm long. Inflorescences numerous, many
exsert from upper sheaths, up to 10 cm long. Panicle
branches slightly spreading but erect, densely flowered. Lemmas 2-3 mm long. Infrequent in alluvial
thickets in the eastern part. Aug-Sept. Wirestem
Muhly . . . . . Muhlenbergia frondosa (Poir.) Fern.

8' Flowering culms pubescent or puberulent below the
nodes. Plants bushy branching from the lower part.

9 Inflorescences compactly flowered. Lemmas awnless
or with awns less than 5 mm. Plants from scaly
rhizomes, the stems 4-10 dm tall. Leaves 2-4 mm
wide and flat with glabrous sheaths. Panicles
narrow, 10-15 cm long, with branches mostly appressed. Spikelets 2-3 mm long. Infrequent in
low, alluvial woods and thickets of the eastern
part. Aug-Sept. Common Muhly . . . .
. . . . Muhlenbergia mexicana (L.) Trin.

9' Inflorescences not compactly flowered, the panicle
branches slender, lax. Lemmas with awns to 10 mm
long. Plants from creeping, scaly rhizomes, the
stems 4-10 dm tall. Leaves flat, the blades 3-6
mm wide, the sheaths glabrous. Inflorescence nodding or erect on slender branches. Rare to infrequent in alluvial woods of the southeast part.
Aug-Sept . . . Muhlenbergia sylvatica (Torr.) Rydb.

ORYZOPSIS Michx.

1 Lemmas smooth. Leaves involute, less than 2 mm wide. Plants
perennial and tufted, the culms 3-7 dm tall. Panicle open,
10-15 cm long, the branches spreading, irregular, 2-6 dm long.
Spikelets arranged at the ends of the branches. Lemma rarely
puberulent, with an awn 6-8 mm long. Infrequent in thickets
and ravines in the western one half of the state. June-July.
Littleseed Ricegrass . . . Oryzopsis micrantha (T. & R.) Thurb.

1' Lemmas variously pubescent. Leaves involute to over 1 cm wide.

  2 Hairs on lemma dense and silky, up to 3 mm long. Plants
     tufted, the culms 3-6 dm tall. Leaves slender, involute,
     1 mm wide, almost as long as the stems. Panicle 8-15 cm
     long, the branches loose, divaricate and dichotomously
     branched. Pedicels filiform, ending in solitary spikelets.
     Lemmas 3 mm long, thick, blackened towards maturity. Fre-
     quent to common on plains and ridges in the western part.
     June-July. Indian Ricegrass . . . .
     . . . . *Oryzopsis hymenoides* (R. & S.) Ricker

2' Hairs on lemma short and appressed.

    3 Spikelets, excluding the awn, 7-9 mm long, leaves flat.

       4 Lower leaf blades elongate, the upper blades of culms
         short and stiff, not more than 1 cm long. Plants
         densely tufted, the flowering culms decumbent, 2-7 dm
         long, usually with short leaves. Principal leaves 3-
         8 mm wide, flat. Inflorescence narrow, 5-8 cm long,
         slightly pubescent, the awn 6-9 mm long. Frequent
         on rich, wooded slopes in the northeast part and in
         the Black Hills. May-June . . . .
         . . . . *Oryzopsis asperifolia* Michx.

       4' Lower leaf blades short, the upper ones elongate.
         Plants perennial from an irregular rhizome, the culms
         3-10 dm tall. Principal leaves flat, 5-15 mm wide.
         Inflorescence 10-20 cm long, irregularly branched,
         the spikelets appressed at the ends of the branches.
         Lemma pubescent, shiny black towards maturity, 5-7
         mm long, with an awn 1-3 cm long. Frequent in rich
         upland woods of the eastern part. June-July . . . .
         . . . . *Oryzopsis racemosa* (J. E. Smith) Ricker

    3' Spikelets, excluding the awn, 5 mm long or less. Leaves
      narrow or involute. Awn of lemma 1-2 mm long, straight.
      Plants tufted, the stems slender, erect, 2-5 dm tall.
      Leaves flat or inrolled, up to 2 mm wide. Inflorescence
      3-6 cm long, the branches few, spreading. Spikelets
      solitary at the ends of filiform pedicels. Lemmas 3-4
      mm long, densely pubescent. Rare in dry, rocky places
      of the Black Hills. June-July. Mountain Ricegrass . .
      . . . . *Oryzopsis pungens* (Torr.) Hitchc.

PHLEUM L.

1 Inflorescence cylindric, 5 or more times longer than wide.
  Leaf sheaths not inflated. Plants perennial, tufted, 3-8 dm

tall, the culms tending to be bulbous at the base. Leaves
flat, with scabrous margins, 4-8 mm wide. Spikelets crowded
in the spike-like panicle, the glumes 3-4 mm long, pectinate-
ciliate on their keels. Common in fields and on roadsides
over the state. June-July. Timothy . . . *Phleum pratense* L.

1' Inflorescence ovoid, not more than 4 times longer than wide.
   Leaf sheaths somewhat inflated. Plants perennial, tufted, 1-4
   dm tall, the culms not bulbous at the base. Leaves flat, 4-7
   mm wide, the blades usually less than 10 cm long. Panicle
   spike-like, bristly, the spikelets to 5 mm long. Glumes
   hispid on the keels. Awns of the glumes stout, to 2 mm long.
   Rare in meadows at higher altitudes in the Black Hills. June-
   July. Alpine Timothy . . . . *Phleum alpinum* L.

POLYPOGON Desf.

Polypogon monspeliensis (L.) Desf. A tufted annual with culms
3-7 dm tall, the stems often rooting at the nodes. Flowering
panicle dense and bristly, 1-2 cm wide and 3-10 cm long. Spikelets yellowish towards maturity. Glumes with awns 6-9 mm long,
arising from between the two terminal lobes. Infrequent in moist
places above and below the main stem dams of the Missouri River
and at Angustora Dam in Fall River County. June-July. Rabbit-
foot Grass.

SPOROBOLUS R. Br.

1 Plants annual. Inflorescence almost always contracted and
   enclosed in the upper sheath. Plants branching from the base
   with ascending to erect culms 2-5 dm high. Leaves slender,
   flat to inrolled. Inflorescence 1-3 cm long, terminal or in
   axils of upper leaves. Lemmas 2.5-4.5 mm long. Ours represented by the following two varieties:
   1 Lemmas glabrous, 2-3 mm long. Frequent on dry sandy soil
      over the state . . . . . . var. *neglectus* (Nash) Scribn.
   2 Lemmas pubescent, 3-5 mm long. Rare in dry soils of the
      eastern part . . . . var. *vaginiflorus*
   Aug-Sept. Annual Dropseed . . . .
   . . . . *Sporobolus vaginiflorus* (Torr.) Wood

1' Plants perennial. Inflorescence variously contracted to open.

   2 Spikelets 1-2.5 mm long.

      3 Inflorescence partly enclosed in the upper sheath.
         Leaves with a conspicuous tuft of hairs at the blade-
         sheath juncture. Stems solitary or few, erect and
         stout, 3-8 dm tall. Leaves 2-5 mm wide, tapered to a
         long point. Panicle 1-2 dm long, the branches appressed,

bearing spikelets near their bases. Lemma 2-2.5 mm
long, glabrous. Common in sandy soil over the state.
Aug-Sept. Sand Dropseed . . . .
. . . . *Sporobolus cryptandrus* (Torr.) A. Gray

3' Inflorescence loose and open. Leaf sheaths naked at
the summit. Plants stout, tufted from the cespitose
base, culms 3-8 dm tall. Leaves 2-4 mm wide, involute.
Flowering panicle 10-30 cm long, with slender branches.
Spikelets 1.5-2.5 mm long, glabrous. Infrequent in
sandy soil and on floodplains in the western part.
July-Aug. Alkali Sacaton . . . .
. . . . *Sporobolus airoides* (Torr.) Torr.

2' Spikelets 3-7 mm long.

4 Second glume shorter than the lemma. Inflorescence contracted, more or less included in the upper sheath.
Culms cespitose, erect, 6-9 dm tall. Leaves 1-4 mm
wide, tapered to a long point, becoming involute at
maturity. Panicle 5-14 cm long, the spikelets 4-6 mm
long, glabrous. Frequent on sandy prairie over the
state. Ours var. *asper.* July-Sept . . . .
. . . . *Sporobolus asper* (Michx.) Kunth.

4' Second glume as long as the lemma. Inflorescence open,
gray or lead-colored, 5-15 cm long. Plants densely
tufted, the stems erect, 3-7 dm tall. Leaves narrow
and involute, 1-2 mm wide. Panicle branches ascending,
3-5 cm long. Spikelets 3-6 mm long. The caryopsis
globose, nut-like, exposed at maturity. Infrequent in
sandy prairie over the state. Aug-Sept. Prairie Dropseed . . . . *Sporobolus heterolepis* A. Gray

STIPA L.

1 Awns usually more than 6 cm long. Lemma 11-25 mm long, excluding the awn.

2 Lemmas pale, becoming light brown or tan at maturity, 10-
12 mm long, pubescent over the entire surface. Plants
cespitose, erect, the culms 3-7 dm tall. Leaves flat to
slightly inrolled, 1-2 mm wide, the ligule 3-4 mm long.
Inflorescence 10-20 cm long, the bases included in the
uppermost sheath. Spikelets crowded, the awn of the lemma
flexuous and twisted. Common in prairies over the state.
May-July. Needle and Thread . . *Stipa comata* Trin. & Rupr.

2' Lemmas dark brown at maturity, 14-25 mm long, pubescent
below and above only at the margin. Plants tufted, erect,

the culms 6-9 dm tall. Leaves 3-5 mm wide, the ligules 4-6 mm long. Inflorescence narrow, paniculate, the branches ascending, 10-20 cm long. Spikelets 1 or 2 on each branch, the awns twisted below. Common in prairies over the state. May-June. Porcupine Grass . . . . . . *Stipa spartea* Trin.

1' Awns usually less than 6 cm long. Lemmas less than 10 mm long.

    3 Inflorescence open and diffuse, the branches spreading. Awns about 2 cm long. Plants tufted, perennial, the culms 5-10 dm tall. Leaves narrow, filiform, mostly basal. Panicle 10-20 cm long, the spikelets near the tips of the branches. Lemmas brownish, with appressed pubescence, the body about 5 mm long. Rare in open woods of the Black Hills. July-Aug. Richardson's Needlegrass. . . . . *Stipa richardsonii* Link.

    3' Inflorescence narrow, contracted, the branches appressed-ascending.

        4 Leaf sheaths villous at the throat. The lower nodes of the panicle also villous. Plants perennial, tufted, the culms 5-10 dm tall. Leaves 2-5 mm wide, flat, the ligule about 1 mm long. Inflorescence 10-20 cm long, narrow, the ascending branches each bearing 2-4 spikelets. Lemma pale brown, the body 4-6 mm long, with an appressed pubescence. Awns 2-3 cm long. Very common on prairies over the state. July-Aug. Green Needlegrass. . . . . *Stipa viridula* Trin.

        4' Leaf sheaths glabrous at the throat. The lower nodes of the panicle also glabrous. Plants 3-6 dm tall. Leaves 1-3 mm wide, the ligules 1-2 mm long. Inflorescence 7-15 cm long, narrow and dense, often purplish. Lemmas 6-7 mm long, pubescent throughout, their awns 2-3 cm long. Rare in open woods in the Black Hills. {*S. columbiana* Macoun.} Ours var. *minor* (Vasey) Hitchc. July-Aug. Columbia Needlegrass . . . . *Stipa occidentalis* (Thurb.) ex. Wats.

Tribe 2. *Andropogoneae*

(Key to Genera)

1 Spikelets paired in slender racemes, the racemes digitate or racemose on the stem . . . . *ANDROPOGON*

1' Spikelets in open or contracted panicles, the racemes reduced to a single pair of spikelets.

2 Pedicellate spikelet male or neutral. The panicle open.
   Awns deciduous . . . . *SORGHUM*

2' Pedicellate spikelet reduced to a hairy projection. The
   panicle narrow. Awns persistent . . . . . . *SORGHASTRUM*

ANDROPOGON L.

1 Racemes arranged racemosely on the culm, not clustered at the
   same level. Plants densely tufted, the stems 5-15 dm tall.
   Leaves 3-7 mm wide, flat. Inflorescence branched, the
   branches consisting of several solitary racemes each 3-6 cm
   long. The sessile spikelet 6-8 mm long, with an awn 8-14 mm
   long. The pedicellate spikelet reduced and short-awned, sterile. Common on dry soils and prairies over the state.
   {*Schizachryium scoparium* (Michx.) Nash} Aug-Sept. Little
   Bluestem . . . . *Andropogon scoparius* Michx.

1' Racemes arranged digitately on the culm, 2-6 parted, and at
   about the same level.

   2 Rhizomes long creeping, yellowish. Hairs of rachis joints
      more than 3 mm long, gray to yellowish. Plants 1-2 meters
      tall. Leaf blades 4-9 mm wide, the ligules 3-4 mm long.
      Fertile, sessile spikelets 8-9 mm long, with awns 2-4 mm
      long, giving the racemes a villous appearance. Infrequent
      to rare on sandy prairie in the sandhills of the southern
      part and western one-half. July-Aug. Sand Bluestem . . .
      . . . . *Andropogon hallii* Hack.

   2' Rhizomes short or wanting. Hairs of rachis joints less
      than 3 mm long, purple. Plants glaucous, 1-2 meters tall,
      the stems mostly unbranched. Leaf blades flat, 5-10 mm
      wide, their sheaths often villous. Branches of the inflo-
      rescence 2-5, each raceme 3-9 cm long. Fertile, sessile
      spikelet 8-10 mm long, its awn twisted, 9-15 mm long.
      Common in dry soil and moist to dry prairies over the
      state. July-Aug. Big Bluestem . . *Andropogon gerardi* Vit.

SORGHUM Moench.

1 Plants annual, 2-3 meters tall, cespitose. Leaves up to 1 cm
   or more wide. Panicle elongate and loose, 2-3 dm long, the
   branches of 2-5 nodes with spikelets. A variable cultivated
   form which occasionally escapes but does not persist. The
   common sorghum races of milo, kafir and broom corn are included here. {*Sorghum vulgare* Pers.} Aug-Sept. Sorghum.
   . . . . *Sorghum bicolor* (L.) Moench.

1' Plants perennial, with extensive creeping rhizomes that are
   scaly. Stems 5-15 dm tall, not as robust as the preceding.
   Leaf blades 1-2 cm wide. Panicle up to 3 dm long, branched,
   open. The fertile (sessile) spikelet 4-6 mm long, silky, the
   awn 1-2 cm long. Occasionally cultivated for forage and escaping in the eastern part. July-Aug. Johnson Grass . . . .
   . . . . *Sorghum halepense* (L.) Pers.

SORGHASTRUM Nash.

Sorghastrum avenaceum (Michx.) Nash. Tall perennials with erect
stems, from short, scaly rhizomes. Leaves flat, the blades 5-10
mm wide. Inflorescence terminal, narrow, the panicle dense with
ascending branches. Spikelets 6-8 mm long, hirsute, becoming a
rich tan or tawny brown at maturity. Awns 1-1.5 cm long, twisted,
geniculate. Common in prairies over the state. {*S. nutans* (L.)
Nash} Aug-Sept. Indian Grass.

Tribe 3. Aveneae

(Key to Genera)

1 Spikelets more than 6 mm long.

    2 Lemmas awned from the back, or awns absent.

        3 Plants annual. Spikelets more than 1 cm long . . . . .
          . . . . *AVENA*

        3' Plants perennial.

            4 Spikelets less than 1 cm long . . . . *ARRHENATHERUM*

            4' Spikelets more than 1 cm long . . . . *HELICTOTRICHON*

    2' Lemmas awned from between a bifid tip . . . *DANTHONIA*

,1' Spikelets less than 6 mm long.

    5 Articulation below the glumes. The second glume much wider
      than the first one . . . . *SPHENOPHOLIS*

    5' Articulation above the glumes.

        6 Inflorescence contracted or spike-like. Glumes unequal.

            7 Lemmas with a twisted or bent awn 5-8 mm long . . . .
              . . . . *TRISETUM*

            7' Lemmas awnless or slightly awn-tipped . . . .
               . . . . *KOELERIA*

        6' Inflorescence open, panicled. Plants of moist or marshy
           places . . . . *DESCHAMPSIA*

ARRHENATHERUM Beauv.

Arrhenatherum elatius (L.) Presl. Tall perennial grass from
cespitose bases, the culms often exceeding 1 meter. Leaves flat,
5-10 mm wide, scabrous and pilose. Panicle narrow, 1-2 dm long,
with short branches. Spikelets 7-8 mm long, the first lemma with
an awn 1-2 cm long. Rare as an escape in the eastern part. June-
July. Tall Oatgrass.

AVENA L.

1 Lemma with a stiff, brown awn that becomes twisted towards
   maturity. Spikelets usually 3-flowered. Plants annual, the
   culms 3-7 dm tall. Leaves flat, 4-8 mm wide. Panicle loose
   and open, the spikelets with glumes 2.5 cm long. Lemma 2 cm
   long, with tan hairs, the awn stout, 3-4 cm long. Occasional
   as a weed of fields in the eastern part. June-July. Wild
   Oat . . . . *Avena fatua* L.

1' Lemma with a short awn or awn lacking. Spikelet 2-flowered.
   Plants annual, glabrous, the culms to 8 dm tall. Leaves flat,
   up to 8 mm wide. Panicle open, lax, 6-10 cm long. Spikelets
   with glumes 2.5 cm long, the florets not readily separating.
   Lemmas glabrous, the awn straight or lacking. Cultivated and
   escaping but not persisting. June-July. Oat . . . .
   . . . . *Avena sativa* L.

DANTHONIA Lam.

1 Lemma 7-8 mm long, glabrous on the back but pilose on the
   margins. Plants perennial, the culms 1-5 dm tall. Leaves
   variously flattened or rolled. Panicle narrow with few
   flowers, the spikelets solitary on the branch, purplish.
   Glumes 13-15 mm long, the lemmas with an awn 5-7 mm long.
   Infrequent to rare in meadows in the Black Hills. July-Aug.
   Timber Oatgrass . . . . *Danthonia intermedia* Vasey

1' Lemma 3-5 mm long, pilose on the back. Plants from a cespitose base, perennial, the culms 2-5 dm tall. Leaves mostly
   basal, curled, narrow or involute. Panicle 2-5 cm long,
   narrow, each branch with 1 or 2 spikelets. Glumes up to 1
   cm long. Awn of the lemma about 5 mm long. Infrequent in
   dry woods in the western part. June-July. Poverty Oatgrass
   . . . . *Danthonia spicata* (L.) Beauv.

DESCHAMPSIA Beauv.

Deschampsia caespitosa (L.) Beauv. Densely tufted perennial,
cespitose, the culms 2-9 dm tall. Leaves mostly basal, stiff,
narrow or involute, 1-3 mm wide. Panicle open, 8-20 cm long,

the branches spreading. Spikelets shiny, purplish or tan, borne at the ends of the branches. Lemmas smooth, 2-4 mm long, with short callus hairs and awns from their bases. Rare in moist meadows at higher altitudes in the Black Hills. July-Aug. Hairgrass.

HELICTOTRICHON Besser

Helictotrichon hookeri (Scribn.) Henr. Plants perennial, the stems erect, tufted, 3-4 dm tall. Leaves 1-3 mm wide, flattened or folded. Panicle erect, 6-8 cm long, the branches ascending. Spikelets 4-6 flowered, 1-2 cm long, the lowest lemma about 1 cm long. The awn twisted and geniculate, up to 1.5 cm long. Infrequent to rare in dry places of the northern part. June-July. Spike Oat.

KOELERIA Pers.

Koeleria pyramidata (Lam.) Beauv. Perennial cespitose plants with culms 3-6 dm tall, tufted. Leaves mostly near the base, narrow, 1-3 mm wide. Panicle dense, spike-like, 4-12 cm long, the lower branches sometimes slightly open. Spikelets 4-5 mm long, the glumes 3-4 mm long. Lemmas scabrous, awn-tipped or awnless. Common in prairies over the state. {*K. cristata* (L.) Pers.} June-July. Junegrass.

SPHENOPHOLIS Scribn.

Sphenopholis obtusata (Michx.) Scribn. Plants cespitose, perennial, the culms 2-8 cm tall. Leaves flat, scabrous, 3-5 mm wide. Panicle 5-12 cm long, the branches short, erect, with few flowers. Spikelets usually with 2 flowers, the second glume nearly as wide as long. Lemma 2.5 mm long, usually awnless. Common in low, open woods and low places over the state. {incl. *S. intermedia* (Rydb.) Rydb.} Represented in South Dakota by the two varieties *obtusata* and *major* (Torr.) Erd. May-June. Prairie Wedgegrass.

TRISETUM Pers.

Trisetum spicatum (L.) Richt. Plants perennial, the stems tufted, erect, 1-5 dm tall. Leaves puberulent to pilose, the blades 2-4 mm wide. Panicle contracted, spikelike, often interrupted, 2-12 cm long. Spikelets mostly 2 flowered, 4-6 mm long. Lemma 4-5 mm long, the awn divaricate and geniculate, 5-6 mm long, attached below the bifid lemma. Rare in rocky soil in the Black Hills. July-Aug. Spike Trisetum.

Tribe 4. *Chlorideae*

(Key to Genera)

1 Plants dioecious and stoloniferous. Male spikelets one-sided,
   the female ones bur-like . . . . *BUCHLOE*

1' Plants bisexual, not stoloniferous.

   2 Spikelets with more than one perfect flower. Plants annual.

      3 Inflorescence a few-flowered head or capitate panicle
         hidden among the sharp pointed leaves . . *MUNROA*

      3' Inflorescence exsert from the leaves.

         4 Spikes numerous, slender and racemose on an elongate
            axis . . . . *LEPTOCHLOA*

         4' Spikes few and digitate . . . . *ELEUSINE*

   2' Spikelets with only one perfect flower.

      5 Imperfect florets above the perfect one.

         6 Spikes digitate. Plants annual . . . *CHLORIS*

         6' Spikes racemose. Plants perennial . . *BOUTELOUA*

      5' Imperfect florets lacking above the perfect one.

         7 Rachilla articulated below the glumes, the spikelet
            falling entire.

            8 Spikelets globose. Glumes about equal . . . . . .
               . . . . *BECKMANNIA*

            8' Spikelets narrow. Glumes unequal in length . . .
               . . . . *SPARTINA*

         7' Rachilla articulated above the glumes, the latter
            remaining . . . . *SCHEDONNARDUS*

BECKMANNIA Host.

Beckmannia syzigachne (Steud.) Fern. Cespitose annual plants
with coarse, hollow culms 3-10 dm tall. Leaves flat, 5-10 mm
wide. Inflorescence narrow, 1-3 dm long, contracted, the panicle branches erect, full-flowered. Spikelets 1-flowered, laterally flattened, appearing circular, in two rows on one side of
the rachis. Glumes and lemma about 3 mm long, with apiculate
tips. Frequent in muddy or swampy areas over the state. July-
Aug. American Sloughgrass.

BOUTELOUA Lag.

1 Spikes 10-40, arranged racemosely on a common axis. Spikelets
   not pectinate on the rachis. Plants perennial, with rhizomes,
   the culms tufted, 5-8 dm tall. Leaves flat, 3-4 mm wide. Inflorescence of short-pedicellate spikes 10-15 mm long, each
   5-8 flowered. Spikelets with fertile lemmas up to 5 mm long,
   awn-tipped. Common in prairies and plains over the state.
   July-Aug. Side-oats Grama . . . .
   . . . . *Bouteloua curtipendula* (Michx.) Torr.

1' Spikes 1-4, with spikelets arranged pectinately on the axis.

   2 Rachis of spike prolonged as a stiff awn beyond the upper-
      most spikelet. Spikes usually less than 2 cm long. Peren-
      nial with densely tufted stems, the culms 2-6 dm tall.
      Leaves flat to involute, flexuous, mostly basal. Inflores-
      cence of 1-4 spikes with 30-40 spikelets, each 3-5 mm long.
      Glumes with spreading hairs giving the sides of the spike a
      fuzzy appearance. Infrequent in the southern part. July-
      Aug. Hairy Grama . . . . *Bouteloua hirsuta* Lag.

   2' Rachis of spike not exceeding the uppermost spikelet.
      Spikes often exceeding 2 cm. Plants densely tufted, peren-
      nial, the culms 2-5 dm tall. Leaves 1-2 mm wide, mostly
      basal, flat or involute. Inflorescence of 1-2 spikes,
      each 2-5 cm long, often curled backward towards maturity.
      Spikelets numerous, each up to 5 mm long. Fertile lemma
      hairy, three-cleft and awn tipped. Very common on prairies
      and plains over the state. July-Aug. Blue Grama . . . . .
      . . . . *Bouteloua gracilis* (H.B.K.) Griffiths.

BUCHLOE Engelm.

Buchloe dactyloides (Nutt.) Engelm. Plants perennial, dioecious,
forming stoloniferous mats. Flowering culms of staminate plants
5-15 cm tall. Pistillate plants prostrate, the inflorescence at
the ends of very short culms and in the axils of upper leaves.
Staminate spikelets 2-flowered, in 2 rows on a slender rachis.
Pistillate spikelets 3-5 in a head-like spike, 3-4 mm wide, 1-
flowered. Frequent in prairie in the western two thirds of the
state. June-July. Buffalo Grass.

CHLORIS Swartz.

Chloris verticillata Nutt. Plants perennial with tufted stems
1-4 dm tall. Leaves mostly basal, the blades 2-4 mm wide. Inflorescence of 1-3 whorls of spreading spikes, 7-12 cm long.
Spikelets about 3 mm long, the fertile lemma 2-3 mm long, with

an erect awn 5-8 mm long. Infrequent weed of lawns and waste places of the southeastern part but spreading westward. June-July. Windmill Grass.

ELEUSINE Gaertn.

Eleusine indica (L.) Gaertn. Plants annual, basally branched with spreading culms 3-6 dm long. Leaves flat, occasionally folded, 3-7 mm wide. Inflorescence of 2-4 spikes digitately arranged with 1 or 2 additional spikes below. Spikes 4-12 cm long. Spikelets mostly sessile in two rows on one side of the rachis, the lemmas awnless. Infrequent weed of waste places in the southeastern part. July-Aug. Goosegrass.

LEPTOCHLOA Beauv.

1 Spikelets 2 mm long or less. Leaf sheaths pilose-hairy. Plants annual, the culms purple-pigmented, 3-6 dm tall. Leaves flat, thin, up to 1 cm wide. Inflorescence of many racemes 5-12 cm long, the central axis 1-4 dm long. Spikelets 1-2 mm long, 3-4 flowered. Lemmas awnless, pubescent. Rare as a weed in gardens, apparently not persisting. Aug-Sept. Red Sprangletop . . *Leptochloa filiformis* (Lam.) Beauv.

1' Spikelets 6-10 mm long. Leaf sheaths glabrous. Plants annual, the culms spreading, 3-9 dm tall. Leaves flat to involute, 1-3 mm wide. Inflorescence of panicled racemes 10-20 cm long, the lower ones having their bases included in the upper leaf sheaths. Racemes 4-8 cm long, with overlapping spikelets. Spikelets 7-11 mm long, their lemmas 4-5 mm long, with awns about 5 mm long. Infrequent on flood plains and alkali swamps over the state. July-Sept . . . . *Leptochloa fascicularis* (Lam.) Gray.

MUNROA Torr.

Munroa squarrosa (Nutt.) Torr. Annual with matted stems, mostly prostrate or weakly ascending, 3-8 cm long. Leaves fascicled in clusters at the nodes, 3 cm long, stiff and pointed. Inflorescence of clustered spikelets at the nodes in the leaf axils, each spikelet 6-8 mm long. Lemmas with pilose hairs on the margins. Infrequent on dry plains and sandy soil in the western part. July-Aug. False Buffalo Grass.

SCHEDONNARDUS Steud.

Schedonnardus paniculatus (Nutt.) Trel. Plants perennial, tufted, with culms 1-3 dm tall. Leaves mostly basal, flexuous with whitened margins, the blades 1-2 mm wide, flat or involute. Inflorescence of remotely and irregularly arranged 1-sided spikes,

each spike 3-8 cm long. Spikelets sessile, remote, with pointed,
awnless lemmas. Infrequent in waste places and overgrazed areas
over the state. June-Aug. Tumble Grass.

SPARTINA Schreb.

1 Principal leaves involute, less than 5 mm wide. Glumes awnless. Plants perennial, the culms 3-5 dm tall. Inflorescence
   of 4-8 spikes closely overlapping and appressed, each 2-4 cm
   long. Spikelets borne on one side of the spike, each 1-
   flowered, flattened, 6-8 mm long. Glumes ciliate on their
   keels but awnless. Frequent on saline soils of the western
   part. July-Aug. Alkali Cordgrass . . *Spartina gracilis* Trin.

1' Principal leaves flat, 5-12 mm wide. Glumes awned. Plants
   perennial from extensive rhizomes, the culms 1-2 meters tall.
   Inflorescence of 10-15 spikes, each 4-8 cm long, appressed
   and ascending. Spikelets crowded on the rachis. The glumes
   scabrous on the keels. Awns of the glumes 2-4 mm long. Common along streams and ponds and in sloughs over the state.
   July-Aug. Prairie Cordgrass . . . . *Spartina pectinata* Link.

Tribe 5. Festuceae

(Key to Genera)

1 Rachilla hairy. Tall, stout perennials usually 2-4 meters
   tall. Lemmas glabrous . . . . *PHRAGMITES*

1' Rachilla not hairy. Plants usually less than 2 meters tall.

   2 Plants dioecious, perennial.

      3 Stems densely tufted, erect from short rhizomes. Plants
         of dry slopes . . . . *HESPEROCHLOA*

      3' Stems not densely tufted, from extensively creeping
         rhizomes. Plants of alkaline soil . . . *DISTICHLIS*

   2' Plants monoecious or with perfect florets.

      4 Lemmas 3-nerved, these prominent.

         5 Callus or nerves of lemma hairy.

            6 Nerves of lemma glabrous. Panicle diffuse, 18-25
               cm long . . . . *REDFIELDIA*

            6' Nerves of lemma silky hairy. Panicle short, 3-5
               cm long . . . . *TRIPLASIS*

         5' Callus and nerves of lemma not hairy.

7 Lemmas papery. Grain large, beaked, at maturity
   forcing the lemma and palea open . . . .
   . . . . *DIARRHENA*

7' Lemmas membranous. Grain neither large nor beaked.

   8 Spikelets 2-flowered, the lemmas broad at their
      tips . . . . *CATABROSA*

   8' Spikelets 3-many flowered, the lemmas acutely
      pointed . . . . *ERAGROSTIS*

4' Lemmas 5-many nerved, many times the intermediate or
   lateral nerves obscure.

   9 Lemmas awned or awn tipped.

      10 Apex of lemma with paired teeth, the awn just
          below the apex or behind the teeth.

          11 Callus bearded . . . . *SCHIZACHNE*

          11' Callus not bearded . . . . . . *BROMUS*

      10' Apex of lemma entire with an awn tip.

          12 Spikelets strongly flattened, crowded into
              1-sided clusters at the ends of branches.
              Keels of glumes hispid . . . . *DACTYLIS*

          12' Spikelets not strongly flattened or crowded
              into 1-sided clusters. Keels of glumes
              smooth . . . . *FESTUCA*

   9' Lemmas awnless.

      13 Callus bearded or cobwebby.

          14 Lemma irregularly notched or eroded at the
              summit. Beard of the callus short and
              lateral . . . . *SCOLOCHLOA*

          14' Lemma not eroded. Callus with cobwebby
              hairs . . . . *POA*

      13' Callus not bearded or cobwebby.

          15 Sterile lemmas present above the fertile
              florets . . . . *MELICA*

          15' Sterile lemmas lacking above the fertile
              florets.

              16 Spikelets strongly flattened, crowded
                  into 1-sided clusters at the ends of
                  long branches. Keels of the glumes
                  hispid . . . . *DACTYLIS*

16' Spikelets not strongly flattened or crowded. Glumes not hispid.

    17 Lemmas with parallel nerves.

        18 Nerves prominent. Plants tall, of woods or marshes . . . . GLYCERIA

        18' Nerves faint. Plants of alkaline soils . . . . . . PUCCINELLIA

    17' Lemmas with nerves converging toward the apex.

        19 Spikelets 14 mm long or more, many-flowered. Lemmas 8-11 mm long . . . . BROMUS

        19' Spikelets shorter than 14 mm, few flowered. Lemmas less than 8 mm long.

            20 Lemmas keeled on the back, the apex obtuse or acute . . . . POA

            20' Lemmas rounded on the back, acute or awn tipped . . . . FESTUCA

BROMUS L.

1 Spikelets strongly flattened, the lemmas keeled. Plants annual to weakly perennial, the culms 5-10 dm tall. Leaves scabrous to pubescent, 3-10 mm wide, the sheaths also pilose-pubescent. Inflorescence 15-30 cm long, the branches erect or spreading. Spikelets compressed, 5-9 flowered. Lemmas sparsely to densely pubescent, 15-20 mm long, the awns 5-12 mm long. Infrequent in meadows and on open hillsides in the Black Hills. {B. marginatus Nees} June-July. California Brome . . . . Bromus carinatus Hook. & Arn.

1' Spikelets not strongly flattened nor lemmas strongly keeled.

    2 Plants perennial.

        3 Creeping rhizomes present. Spikelets awnless or awn less than 4 mm long. Lemmas glabrous. Plants 5-10 dm tall, the culms erect. Leaves 5-9 mm wide, flat, the ligules 1-2 mm long. Panicle narrow to open, the branches ascending, 10-20 cm long. Spikelets 2-3 cm

long, 6-8 flowered. Lemmas 9-12 mm long. Very common
throughout the state as cultivated plant and naturalized. Represented in South Dakota by the two subspecies
inermis and *pumpellianus* (Scribn.) Wagnon. May-June.
Smooth Brome · · · · · · · · · · *Bromus inermis* Leyss.

3' Creeping rhizomes lacking. Spikelets with obvious awns.

   4 Lemmas pubescent on the margins and on the back near
the base. Plants tufted, the culms 7-11 dm tall.
Leaves soft, lax, up to 1 cm wide, their sheaths
glabrous to pilose. Inflorescence open, 10-20 cm
long, the panicle branches drooping, flowered towards
the ends. Spikelets 4-10 flowered, 1-2 cm long.
Lemmas 10-11 mm long, the awn 3-5 mm long. Infrequent on hillside thickets and woods over the state.
July-Aug. Fringed Brome · · · · · *Bromus ciliatus* L.

   4' Lemmas evenly pubescent on the back.

     5 Panicle 10-25 cm long. The first glume 1-nerved,
the second glume 3-nerved.

      6 Nodes of mature culms numbering 4-6. Upper
leaf sheaths shorter than the internodes. Base
of leaf at sheath juncture without an auricle.
Plants 7-11 dm tall, the culms solitary or few.
Leaves flat, 5-13 mm wide. Inflorescence with
open and spreading branches. Spikelets 2-3 cm
long, the lemmas 9-12 cm long. Awn of lemma
2-8 mm long. Infrequent in rich woods of the
eastern part and in the Black Hills. {B.
purgans L.} July-Aug. Canada Brome · · · · ·
· · · · · · · · · · · *Bromus pubescens* Willd.

      6' Nodes of mature culms numbering 10-12. Upper
leaf sheaths longer than the internodes and
overlapping. Base of leaf at sheath juncture
auricled, often times pubescent. Plants 8-11
dm tall, the culms few. Inflorescence open,
branched. Spikelets 2-3 cm long, the lemmas
about 1 cm long. Awn of lemma 3-9 mm long.
Infrequent in alluvial areas in the eastern
part. July-Aug · · · · · · · · · · · · · · · ·
· · · · · · *Bromus latiglumis* (Shear) Hitchc.

    5' Panicle 7-10 cm long. The first glume 3-nerved,
the second glume 5-7 nerved.

      7 Leaf sheaths and blades sparsely pilose-hairy.
Leaves 2-4 mm wide, scabrous. Plants 3-6 dm

tall, the culms slender. Leaves mostly basal.
Inflorescence about 10 cm long, the panicle
branches drooping. Spikelets 8-10 flowered,
the lemmas rounded on the back, 10-12 mm long.
Awn of the lemma 2-3 mm long. Infrequent on
hillsides in the Black Hills. {*B. anomalus*
of reports} July-Sept. Nodding Brome . . . .
. . . . *Bromus porteri* (Coult.) Nash

7' Leaf sheaths and blades conspicuously pubescent. Leaves 5-10 mm wide, flat. Plants 5-10
dm tall, the culms slender and usually solitary.
Inflorescence 5-10 cm long, the branches elongate and nodding. Spikelets 6-10 flowered, 15-
20 mm long. Lemmas 8-10 mm long, the awn 2-3
mm long. Rare to infrequent in open woods of
the Black Hills. June-July . . . .
. . . . *Bromus kalmii* Gray

2' Plants annual or biennial.

8 Spikelets about 1 cm wide, usually awnless. Lemmas
broadly ovate with scarious margins. Plants 2-6 dm
tall, the culms solitary or few. Leaves 2-5 mm wide,
the blades and sheaths pilose pubescent. Inflorescence
5-15 cm long, with lax, open panicle branches. Spikelets usually solitary on the branch, 1.5-3 cm long,
compressed. Lemmas usually glabrous, essentially awnless. Rare in waste places in the western part. June-
July. Rattlesnake Chess . . . .
. . . . *Bromus briziformis* Fisch. & Mey.

8' Spikelets less than 1 cm wide, with conspicuous awns.
Lemmas not broadly ovate.

9 Lemma sharply acuminate, its teeth 2 mm long. The
first glume 3 nerved. Plants with stems spreading,
3-6 dm tall. Leaves 2-4 mm wide, the blade and
sheath pubescent. Inflorescence 5-15 cm long,
densely branched, the spikelets nodding. Spikelets
3-6 flowered, slender, glabrous to villous-hairy.
Very common weed of waste places. May-June. Cheat
Grass . . . . *Bromus tectorum* L.

9' Lemma not sharply acuminate, the teeth shallow, less
than 2 mm long. The first glume 3 nerved, the second
one 5-7 nerved.

10 Panicle contracted, rather dense. Spikelets
turgid, the awns erect. Plants 3-7 dm tall, the

stems softly pubescent. Leaves flat, 2-4 mm
wide, the sheaths pubescent. Inflorescence
erect, 3-8 cm long. Spikelets 1-2 cm long, 5-7
flowered. Lemmas 6-8 mm long, pubescent to
pilose, the awn 6-10 mm long. Rare as a weed
in the western part. May-July. Soft Chess . . .
. . . . *Bromus mollis* L.

10' Panicle open, the branches spreading.

11 Leaf sheaths glabrous. Awns shorter than
the lemmas, undulate-divergent, 3-5 mm long.
Plants 3-6 cm tall, the panicle branches
drooping in fruit. Spikelets 10-18 mm long,
5-9 flowered. Lemma 6-8 mm long, obtuse,
the margins incurved in fruit. Infrequent
to rare as a weed in the southern part.
June-July . . . . *Bromus secalinus* L.

11' Leaf sheaths pilose. Awns 8-12 mm, as long
as or longer than the lemmas.

12 Leaves 1-3 mm wide, their sheaths not as
retrorsely pilose as the following species. Plants 2-5 dm tall, the inflorescence 7-14 cm long. Panicle branches
spreading and drooping, the pedicels
longer than the spikelets. Lemmas 8-10
mm long, their awns twisted and flexuous
at maturity. A common weed of waste
places over the state. June-Aug.
Japanese Chess . . *Bromus japonicus* Thunb.

12' Leaves 3-5 mm wide, their sheaths retrorsely pilose. Plants with tufted
stems 2-4 dm tall. Panicle branches
shorter than the preceding species, the
pedicels short and usually bearing single
spikelets. Lemmas tending to be obovate
with twisted, divergent awns that are
scabrous. Infrequent as a weed along
roadsides and in dry places in the western one-half. July-Aug . . . .
. . . . *Bromus squarrosus* L.

CATABROSA Beauv.

Catabrosa aquatica (L.) Beauv. Plants perennial from a creeping
base, often rooting at the lower nodes, the stems weakly ascending, 1-4 dm long. Leaves flat, 2-8 mm wide, distributed along

the culm. Inflorescence an oblong panicle, 10-20 cm long, the
branches spreading and evenly flowered. Spikelets 2 flowered,
about 3 mm long. Lemmas 2-3 mm long, the broad apex scarious,
awnless. Infrequent in slow moving streams and ponds in the
western part. June-Aug. Brookgrass.

DACTYLIS L.

Dactylis glomerata L. Plants perennial, from strongly cespitose
bases, the culms 6-9 dm tall. Leaves elongate, flattened, 2-8
mm wide, mostly glabrous. Inflorescence 5-15 dm long, the panicle branches stiff, open, flowered at their ends. Spikelets
compressed, few flowered, in 1-sided clusters. Lemmas 7 mm
long, hispid on the keel, the awn tip short. Commonly cultivated over the state and persisting. May-June. Orchard Grass.

DIARRHENA Beauv.

Diarrhena americana Beauv. Plants perennial from slender rhizomes, the culms soft, decumbent to erect, 6-8 dm tall. Leaves
flat, 1-2 cm wide, pubescent on their undersurfaces. Inflorescence of few panicle branches that are elongate and arching, few
flowered. Spikelets 10-16 mm long, 3-5 flowered. Lemmas 6-9 mm
long, pointed, the beaked caryopsis turgid, yellowish. Rare in
alluvial woods in the extreme eastern part. Ours var. *obovata*
Gleason. July-Aug.

DISTICHLIS Raf.

Distichlis spicata (L.) Greene. Plants perennial, dioecious.
Male and female plants similar in appearance, the culms 1-3 dm
tall, from rhizomes. Leaves distributed along the stem, pointed,
flat to folded. Inflorescence a contracted and crowded panicle,
the branches ascending. Spikelets of staminate plants yellowish,
8-12 flowered, the pistillate spikelets dull green, 7-9 flowered.
Frequent in alkali soils over the state. {incl. *D. stricta*
(Torr.) Rydb.} Ours var. *stricta* (Torr.) Beetle. July-Aug.
Saltgrass.

ERAGROSTIS Beauv.

1 Principal stems creeping, rooted at the nodes. Plants forming
   mats of various sizes.

   2 Lemmas and stems glabrous. Lemmas less than 2 mm long.
      Plants annual, the stems with few leaves at the nodes.
      Leaves short, scabrous, 1-3 mm wide. Panicle open, few
      flowered, the spikelets 5-10 mm long. Florets perfect,
      the lemmas 1-2 mm long, awnless. Frequent in moist

alluvial soil and on sandbars of the Missouri River in the
eastern part. July-Aug . . *Eragrostis hypnoides* (Lam.) BSP.

2' Lemmas and stems pubescent. Lemmas about 3 mm long. Plants
annual, the stems with tufts of short leaves at the nodes.
Leaves 1-3 cm long, pointed, pubescent. Panicles short,
ovoid in shape, of many clustered branches, 1-2 cm long.
Plants dioecious, growing in close proximity. The staminate spikelets 10-20 flowered, in a more open panicle than
the pistillate spikelets. Infrequent in moist alluvial
soils in the eastern part. July-Aug . . . .
. . . . *Eragrostis reptans* (Michx.) Nees.

1' Principal stems erect or ascending, not matted.

3 Glumes and lemmas conspicuously glandular on their keels.
Plants annual, the culms several from the base, 1-5 dm
tall. Vegetation glandular with an ill-smelling odor.
Leaves flat, 2-5 mm wide, the sheaths pilose. Panicle
erect and much branched, 5-15 cm long. Spikelets dull
green, 5-12 mm long, with 12-35 flowers. A common weed
of waste and rich or moist areas over the state. {*E.
megastachya* Link.} July-Aug. Stinkgrass . . . .
. . . . *Eragrostis cilianensis* (All.) E. Mosher

3' Glumes and lemmas sparsely ciliate or glabrous, but not
glandular.

4 Plants perennial.

5 Branches of the panicle scabrous, stiff and erect.
Spikelets usually 4-6 flowered. Plants 3-6 dm tall,
in dense clusters from a short rhizome. Leaves flat,
3-8 mm wide, harsh and ascending. Inflorescence
purple-colored, varying much in length. Spikelets
long-pediceled, 4-8 mm long. Lemmas 1-2 mm long.
Rare to infrequent on sandy soil in the southern
part. Aug-Sept. Purple Lovegrass . . . .
. . . . *Eragrostis spectabilis* (Pursh) Steud.

5' Branches of the panicle glabrous and flexuous. Spikelets usually 6-15 flowered. Plants 5-12 dm tall, the
stems tufted. Leaves elongate, 2-5 mm wide, pointed.
Inflorescence large and diffuse, up to one-half the
height of the plant. Branches capillary, the spikelets long pediceled. Spikelets 4-7 mm long, the
lemmas 2-3 mm long. Rare in sandy places and blow-
outs in the southern part. July-Sept. Sand Lovegrass . . . . . . *Eragrostis trichodes* (Nutt.) Wood

4' Plants annual.

6 Spikelets 1 mm wide, slender, the grains 0.5-0.7 mm
   long. Plants slender, erect from a spreading base,
   1-5 dm tall. Leaves flat, 1-3 mm wide. Inflorescence open, diffuse, 5-15 cm long, the panicle
   branches flexuous. Spikelets 3-5 mm long, dark
   gray, 3-9 flowered. Lemmas 1-1.5 mm long. Occasional as an escape in the southern and western
   part. June-July. India Lovegrass . . . .
   . . . . Eragrostis pilosa (L.) Beauv.

6' Spikelets 1.5 mm wide or wider, linear to ovate, the
   grains about 1 mm long. Plants densely cespitose,
   the stems 1-5 dm tall, branched. Leaves 1-3 mm wide.
   Inflorescence much branched, up to one-half the
   height of the plant. Panicle branches erect, the
   spikelets not terminal, appressed along the branches.
   Spikelets 5-11 flowered, 5-8 mm long. A common weed
   of waste places over the state. July-Sept . . . . .
   . . . . Eragrostis pectinacea (Michx.) Nees.

FESTUCA L.

1 Leaf blades flattened and more than 2.5 mm wide.

  2 Spikelets loosely scattered in an open panicle with
    branches 2-4 cm long, usually not more than 5 flowered.

    3 Lemmas 3-4 mm long, essentially awnless. Plants peren-
      nial, the culms 5-8 dm tall, solitary or few. Leaves
      flat, 4-7 mm wide. Inflorescence 12-20 cm long, open,
      with drooping branches, the spikelets borne towards
      their ends. Lemmas rather turgid, with acute to obtuse
      apices, awnless. Infrequent in upland woods in the
      eastern part. May-July. Nodding Fescue . . . .
      . . . . Festuca obtusa Biehler

    3' Lemmas about 7 mm long, awned from the tip, the awn from
       as long as the lemma to up to 17 mm. Plants perennial,
       cespitose, 7-9 dm tall, the stems leafy. Leaves flat,
       4-10 mm wide, thin and lax. Inflorescence lax, 15-35
       cm long, the branches in 2's or 3's, drooping. Lemmas
       tapered to an acuminate tip, the awn prominent. Rare
       in woods of the Black Hills. May-June. Bearded Fescue.
       . . . . Festuca subulata Trin.

  2' Spikelets aggregated in a contracted, almost spike-like
     panicle, 8-10 flowered and more than 10 mm long. Plants
     perennial, the culms 5-8 dm tall. Leaves flat, dark green,
     4-8 mm wide, glabrous. Inflorescence 10-20 cm long, the
     branches erect. Spikelets 6-9 flowered, 7-11 mm long.

Lemmas 5-7 mm long, rarely awn-tipped. Frequent in open
woods and meadows over the state. Introduced with seed
and persisting. {*F. elatior* L.} June-July. Meadow
Fescue . . . . *Festuca pratensis* Huds.

1' Leaf blades involute or less than 2.5 mm wide.

    4 Plants annual. Culms 1-3 dm tall, slender and erect, not
densely tufted. Leaves narrow, involute, tapering to fine
points. Inflorescence narrow, 3-10 cm long, the spikelets
6-8 mm long, turning brown towards maturity. Lemmas glabrous, 4-5 mm long, the awn 3-7 mm long. Frequent in dry,
sandy soils over the state. May-June. Six-weeks Fescue.
. . . . *Festuca octoflora* Walt

    4' Plants perennial, densely tufted and cespitose.

        5 Culms 2-4 dm tall, the leaves usually less than 10 cm
long, narrow and involute. Inflorescence 4-9 cm long,
narrow, the branches few and short, erect. Spikelets
4-6 flowered, the lemmas about 4 mm long. Awnless or
awned to 2 mm long. Frequent on dry hillsides over the
state. Ours var. *rydbergii* St.-Yoes. May-June. Sheep
Fescue . . . . *Festuca ovina* L.

        5' Culms 4-10 dm tall, some of the leaves usually more than
10 cm long, mostly basal, narrow. Panicle narrow, 10-20
cm long, the branches erect. Spikelets 5-8 flowered,
the lemmas about 7 mm long. Awns 2-5 mm long, stout.
Rare in sandy or rocky woods in the Black Hills. June-
July. Blue Bunchgrass . . . . *Festuca idahoensis* Elmer.

GLYCERIA R. Br.

1 Spikelets 10 mm or more long.

    2 Lemmas 4 mm long or less, glabrous between the nerves.
Culms 6-10 dm tall, slender. Leaves 2-4 mm wide, flat or
folded. Inflorescence 2-4 dm long, narrow, the branches
elongate, erect. Spikelets linear, appressed to the panicle branch, 10-15 mm long, 6-10 flowered. Lemmas 3-4 mm
long, scarious at the apex. Infrequent to rare in moist
places in the Black Hills. June-Aug. Northern Mannagrass.
. . . . *Glyceria borealis* (Nash) Batch.

    2' Lemmas usually exceeding 5 mm, slightly scabrous between
the nerves. Plants up to 1 meter tall, the culms decumbent
at the lower part. Leaves 6-10 mm wide, the blades flat,
with smooth sheaths. Inflorescence 2-4 dm long, narrow to
open, the panicle branches ascending to open after anthesis. Spikelets 8-14 flowered, 1.5-3 mm long. Lemmas 5.5-

6.5 mm long, white scarious at the apex. Rare in moist
places in the Black Hills. July-Aug. Mannagrass . . . . .
. . . . *Glyceria fluitans* (L.) R. Br.

1' Spikelets not more than 5 mm long.

   3 First glume 1 mm long or less. Plants in large clumps, the
     stems erect, slender, 3-9 dm tall. Leaves flat, 2-6 mm
     wide, lax and thin. Inflorescence 1-3 dm long, the panicle
     branches numerous, spreading, elongate. Spikelets 4-6 mm
     long, 5-7 flowered. Lemmas prominently nerved, narrowly
     scarious at the apex, 2 mm long. Frequent to common in
     wet places over the state. June-July. Fowl Mannagrass . .
     . . . . *Glyceria striata* (Lam.) Hitchc.

   3' First glume 1.5 mm long. Plants tufted from a rhizomatous
      base, the culms stout, erect, 8-14 dm tall. Leaves 6-13 mm
      wide, flat. Inflorescence 2-4 dm long, much branched, the
      panicle loose. Spikelets 5-8 flowered, 5-6 mm long. Lem-
      mas about 2.5 mm long, lavender-pigmented with a very nar-
      row scarious margin. Frequent in wet places over the state.
      June-July. American Mannagrass . . . .
      . . . . *Glyceria grandis* S. Wats.

HESPEROCHLOA (Piper) Rydb.

Hesperochloa kingii (S. Wats.) Rydb. Plants perennial, dioecious,
the stems 5-7 dm tall, in dense clusters from the base. Leaves
mostly flat, 3-6 mm wide, often glaucous. Inflorescences of male
and female plants similar, 7-15 cm long. The staminate spikelets
slightly larger, 7-11 mm long. Lemmas 5-7 mm long. Rare in canyons and on hillsides of the Black Hills. July-Sept.

MELICA L.

Melica smithii (Porter) Vasey. Plants perennial, the culms
slender and erect, 6-11 dm tall. Leaves flat, 5-10 mm wide,
scabrous. Inflorescence 10-20 cm long, with few remote panicle
branches. Spikelets 17-19 mm long, borne towards the ends of the
branches, each 3-6 flowered. Lemmas purplish, 8-10 mm long, with
a short dorsal awn 3-5 mm. Apparently rare in moist shady places
in the Black Hills. May-July. Melicgrass.

PHRAGMITES Trin.

Phragmites australis (Cav.) Trin. Coarse perennial grasses, the
culms reed-like, 2-4 dm tall, from stout, horizontal rhizomes.
Leaves flat, 1-5 cm wide, glabrous. Inflorescence plume-like,
the panicle 10-30 cm long, densely flowered. Spikelets 3-6 flowered, 12-14 mm long. Lemma 9-11 mm long, exceeded by long

rachilla hairs. Common in sloughs and marshy places over the
state. {*P. communis* Trin.} Aug-Sept. Plume Reed.

POA L.

1 Spikelets not compressed. Keels, if present, obscure. Panicle usually narrow, the branches erect and appressed.

2 Lemmas puberulent to pubescent on the lower one-half.

3 Flowering culms usually over 5 dm tall and having leaves
1-3 dm long. Panicles about 10 cm long or longer, narrow and dense, the branches erect. Spikelets 5 mm long,
3-5 flowered, the lemmas pubescent over the lower one-
half. Infrequent in dry soil in the western part.
June-July. Canby Bluegrass . . . .
. . . . *Poa canbyi* (Scribn.) Piper

3' Flowering culms usually less than 4 dm tall and having
short, involute basal leaves. Panicles usually less
than 10 cm long, loosely flowered, the branches erect.
Lemmas scabrous-pubescent below, scarious at the apex.
Plants often with a reddish pigment (anthocyanic). Infrequent on dry plains or woodland openings in the
western part. May-June. Sandberg Bluegrass . . . . . .
. . . . *Poa sandbergii* Vasey

2' Lemmas glabrous to short scabrous on the lower one-half.
Plants perennial, tufted, 4-10 dm tall. Leaves involute
to flat and up to 3 mm wide. Panicle narrow, the branches
erect. Spikelets 3-7 flowered. Lemmas 4-6 mm long. Not
reported from South Dakota but present in surrounding
states. Should be looked for on dry plains or alkaline
places in grassland in the western part. May-June . . . .
. . . . *Poa juncifolia* Scribn.

1' Spikelets compressed, the glumes and lemmas keeled. Panicle
erect to variously spreading.

4 Plants annual, tufted, usually less than 10 cm tall. Culms
glabrous, spreading, sometimes forming mats. Leaves soft,
2-3 mm wide. Inflorescence open, 2-6 cm long, with few
crowded branches. Spikelets 3-5 flowered, only 2-5 mm
long. Lemmas naked at the base but pubescent on the
nerves and keel. Infrequent in waste places and edges
of gardens over the state. April-July. Annual Bluegrass.
. . . . *Poa annua* L.

4' Plants perennial.

5 Creeping rhizomes present. These not to be confused
   with offset stems which are *not* rhizomes.

   6 Culms strongly flattened and 2-edged immediately
      below the inflorescence. Lemmas only sparsely
      webbed at their bases. Plants perennial, from
      prominent rhizomes, the culms usually solitary,
      2-4 dm tall. Panicle narrow, 3-8 cm long, with
      irregular branches. Spikelets 3-5 flowered, the
      lemmas 2-3 mm long, the keel and margins slightly
      pubescent toward their bases. Common in a variety
      of habitats over the state. June-July. Canada
      Bluegrass . . . . *Poa compressa* L.

   6' Culms not strongly flattened below the inflorescence,
      mostly terete.

      7 Inflorescence strongly contracted, the panicle
         branches erect and stiff. Plants tufted with
         short rhizomes 2-5 dm tall. Leaves chiefly basal,
         1-3 mm wide, flat or often folded. Panicle 2-8 cm
         long, the spikelets crowded on the erect branches.
         Lemmas 3-4 mm long, villous. The keel and margi-
         nal nerves also villous. Lemmas not cobwebby at
         their bases. Frequent in prairie and swales over
         the state. May-June. Plains Bluegrass . . . . .
         . . . . *Poa arida* Vasey

      7' Inflorescence open, the panicle branches spread-
         ing.

         8 Lemmas webbed at the base, pubescent on the
            nerves. Plants green, strongly rhizomatous,
            culms tufted, erect, 3-9 dm tall. Leaves soft,
            flattened, 2-4 mm wide. Panicle pyramidal, 3-9
            cm long, the branches spreading. Spikelets 3-5
            flowered, 3-5 mm long. A very common grass of
            almost every habitat in the state. June-July.
            Kentucky Bluegrass . . . . . . *Poa pratensis* L.

         8' Lemmas not webbed at the base but pubescent on
            the nerves below. Plants glaucous, tufted from
            short rhizomes, the culms 3-8 dm tall. Leaves
            flat, 2-4 mm wide. Panicle open, elliptic, 6-
            14 cm long, the branches in distant whorls of
            3's. Spikelets 3-4 flowered, 6-8 mm long.
            Lemmas 3-4 mm long. Infrequent in meadows in
            the western one-half. May-June . . . .
            . . . . *Poa glaucifolia* Scribn. & Will.

5' Creeping rhizomes lacking; however, offset stems may be
   present.

   9 Florets swollen into bulblets with dark purple bases.
   Plants with tufted stems, their bases also swollen.
   Culms 3-6 dm tall. The leaves inrolled, 1-2 mm wide.
   Inflorescence narrow, 5-8 cm long, the branches as-
   cending. Spikelets 3-5 flowered, up to 1.5 cm long.
   Lemmas 2-3 mm long, webbed at their bases. Infre-
   quent in meadows in the Black Hills. May-June.
   Bulbous Bluegrass . . . . *Poa bulbosa* L.

  9' Florets and culm bases not swollen.

    10 Bases of lemmas webbed with a tuft of hairs.

      11 Lower panicle branches reflexed and flexuous,
      3 or more in a whorl. Plants cespitose, the
      culms slender, 4-7 dm tall. Leaves soft,
      lax, 2-6 mm wide. Inflorescence elongate,
      open, 10-20 cm long. Spikelets 2-4 flowered,
      3-4 mm long. Lemmas 2-3 mm long, pubescent
      on the nerves as well as webbed. Rare in
      rich woods in the eastern part. May-July.
      Woodland Bluegrass . . . *Poa sylvestris* Gray

      11' Lower panicle branches ascending, 4 or more
      in a whorl.

        12 Inflorescence usually exceeding 10 cm
         in length, large and open at maturity.
         Ligule of principal leaves 2-7 mm long.
         Plants of wet places.

          13 Intermediate nerves of the lemma not
            prominent. The marginal nerves of
            the lemma hairy. Plants tufted, 4-
            10 dm tall. Leaves 1-3 mm wide,
            elongate, the ligules 3-5 mm long.
            Panicle 10-30 cm long, large and
            open at maturity. The spikelets 4
            mm long, 2-4 flowered. Lemmas 2-3
            mm long, the lower parts of the
            nerves pubescent as well as webbed.
            Frequent to common in meadows and
            moist thickets over the state. July-
            Aug. Fowl Bluegrass . . . .
            . . . . *Poa palustris* L.

          13' Intermediate nerves of the lemma
            prominent. The marginal nerves of

the lemma glabrous. Plants 4-10
dm tall, tending to be decumbent.
Leaves 2-4 mm wide, the ligules
3-7 mm long. Panicle 8-14 cm long,
loosely flowered. Spikelets 2-3
flowered, the lemmas about 3 mm
long. Rare in moist woods of the
Black Hills. May-July. Rough Bluegrass . . . . *Poa trivialis* L.

12' Inflorescence usually less than 10 cm
long, somewhat contracted. Ligule of
principal leaves less than 2 mm long.
Plants erect, tufted, the culms stiff,
2-5 dm tall. Leaves narrow, 1-2 mm wide.
Panicle with ascending branches, the
spikelets 2-4 flowered, about 4 mm long,
borne at the ends of the branches. Lemmas villous on the nerves as well as
webbed at the base. Frequent in open
meadows and swales in the western part.
June-July. Inland Bluegrass . . . . . .
. . . . *Poa interior* Rydb.

10' Bases of lemmas not webbed with a tuft of hairs.
Plants cespitose, the culms erect, 2-5 dm tall.
Leaves narrow, involute, 1-3 mm wide. Inflorescence contracted and dense, 2-7 cm long, the
spikelets congested. Spikelets 6-8 mm long,
4-5 flowered. Lemmas 4 mm long, villous on the
3 principal nerves. Infrequent in woods in the
western part. May-June. Muttongrass . . . . . .
. . . . *Poa fendleriana* (Steud.) Vasey

PUCCINELLIA Parl.

Puccinellia nuttalliana (Schult.) Hitchc. Plants perennial with
cespitose bases, the culms pale green, erect, 3-6 dm tall.
Leaves flat, 1-3 mm wide. Inflorescence open, the panicle 10-20
cm long, flowered at the ends of the branches. Spikelets 5-6 mm
long, 4-6 flowered. Lemmas 2-3 mm long, erose at the apex. Frequent in alkaline swales over the state. {*P. airoides* (Nutt.)
Wats. and Coult.} June-July. Alkaligrass.

REDFIELDIA Vasey

Redfieldia flexuosa (Thurb.) Vasey. Perennial grass from creeping rhizomes, the culms coarse and wiry, 6-10 dm tall. Leaves
elongate, involute, tapering to a long point. Inflorescence

diffuse, pyramidal, 20-40 cm long. Spikelets compressed laterally, 3-4 flowered, 5-7 mm long. Lemmas 4-5 mm long, densely
hairy on the nerves. Infrequent on sandy soil in the southwestern part. Aug-Sept. Blowout Grass.

SCHIZACHNE Hack.

Schizachne purpurascens (Torr.) Swallen. Perennial grass from a
cespitose base, the stems clustered, 5-9 dm tall. Leaves flat,
2-3 mm wide. Inflorescence of irregularly paired panicle
branches, about 10 cm long, few flowered. Spikelets purplish
at their bases, 2-2.5 cm long, 2-4 flowered. Lemmas strongly
nerved with two prominent apical teeth, with an awn 1 cm long
from the dorsal side. Callus with short hairs. Infrequent in
open, rocky woods in the Black Hills. June-July. False Melic.

SCOLOCHLOA Link.

Scolochloa festucacea (Willd.) Link. Tall, perennial grass from
thick, creeping rhizomes, the culms stout, up to 1 meter tall.
Leaf blades flat, elongate, 5-8 mm wide, pointed. Panicle loose
and open, 15-25 cm long, the branches ascending. Spikelets 7-9
mm long, the lemmas about 6 mm long, villous hairy at the base
and purple nerved and erose at the distal part. Frequent in
sloughs and swales in the eastern one-half. July-Aug. Sprangletop.

TRIPLASIS Beauv.

Triplasis purpurea (Walt.) Chapm. Plants annual with tufted
stems 2-6 dm tall, purplish, the nodes with axillary spikelets
in the lower sheaths. Leaves short, flat or involute, 1-3 mm
wide. Inflorescence a panicle 3-5 cm long, the branches few-
flowered. Spikelets 6-8 mm long, 2-4 flowered. Lemmas densely
villous on the nerves, the palea conspicuously so. Rare but
locally abundant on sandy soil of the southern part. Aug-Sept.
Sand Grass.

Tribe 6. Hordeae

(Key to Genera)

1 Spikelets solitary at each node of the rachis.

  2 Spikelets placed edgewise (one rank of florets adjacent)
     to the rachis. First glume of lateral spikelets lacking
     . . . . *LOLIUM*

  2' Spikelets placed flatwise to the rachis. Glumes 2.

    3 Plants annual.

        4 Spikelets cylindrical, their bases embedded in the rachis. Glumes awned . . . . *AEGILOPS*

        4' Spikelets flattened.

           5 Glumes 1-nerved, linear-subulate. Spikelets with 2 perfect flowers . . . . *SECALE*

           5' Glumes 3-several nerved, broad. Spikelets 2-5 flowered . . . . *TRITICUM*

    3' Plants perennial . . . . *AGROPYRON*

1' Spikelets 2-6 at each node of the rachis.

    6 Spikelets 3 at a joint, 1 flowered, the lateral pair aborted and reduced to awns . . . . *HORDEUM*

    6' Spikelets alike, 2-6 flowered.

        7 Glumes lacking or reduced to 2 short bristles. Spikes loosely flowered, the spikelets widely spreading . . . . *HYSTRIX*

        7' Glumes almost the size of the florets or longer. Spikelets ascending.

           8 Rachis continuous. Glumes broad or narrow, entire . . . . *ELYMUS*

           8' Rachis disarticulating at maturity. Glumes extending into long awns, making a bristly spike . . . . *SITANION*

AEGILOPS L.

Aegilops cylindrica Host. Annual, weedy plants with erect stems 4-6 dm tall. Leaves flat, 2-3 mm wide, with long hairs. Inflorescence a narrow, cylindrical spike 4-8 cm long. Spikelets 2-5 flowered, 8-10 mm long, embedded in the curvature of the rachis. Glumes with awns 1-3 cm long. Lemmas awned or lacking awns. Rare along railroads and roadsides in the eastern part but spreading westward. June-July. Jointed Goatgrass.

AGROPYRON Gaertn.

1 Plants with creeping rhizomes.

    2 Principal leaves flat, 5-10 mm wide. Lemmas glabrous on the back. Stems clustered from a yellowish rhizome system, the culms 5-10 dm tall. Spike 5-10 cm long, the spikelets 4-6 flowered, 1-2 cm long. Lemmas awnless to awns almost

as long as the lemma. Frequent in disturbed soil or on
roadsides and along fencerows over the state. June-July.
Quackgrass . . . . *Agropyron repens* (L.) Beauv.

2' Principal leaves involute or if flat, less than 5 mm wide.
Lemmas glabrous to copiously pubescent on the back.

　　3 Glumes obtusely blunt. Lemmas not acutely pointed.
　　　Leaf sheaths ciliate-margined. Plants 5-7 dm tall,
　　　with stiff, involute leaves. Spikes slender and stiff,
　　　10-20 cm long. Glumes 7-9 mm long, 5-7 nerved. Lemmas
　　　usually awnless but may be awn-tipped. Introduced in
　　　the western part and persisting. June-Aug. Intermedi-
　　　ate Wheatgrass . . . *Agropyron intermedium* (Host) Beauv.

　　3' Glumes attenuate to awn-tipped. Lemmas also acutely
　　　pointed. Leaf sheaths not ciliate-margined.

　　　　4 Glumes 3-5 nerved, lanceolate, gradually tapered from
　　　　　the base to an attenuate awn tip. Lemmas glabrous to
　　　　　pubescent but not densely so. Plants rhizomatous,
　　　　　the culms usually glaucous, 4-8 dm tall. Spike
　　　　　erect, 6-12 cm long, the spikelets 4-10 flowered.
　　　　　Lemmas awn tipped or with an awn to 5 mm long. Com-
　　　　　mon in prairie soils over the state. June-July.
　　　　　Western Wheatgrass . . . . . *Agropyron smithii* Rydb.

　　　　4' Glumes 5-7 nerved, not rigid, abruptly tapering from
　　　　　the middle to the awn tip. Lemmas usually hairy over
　　　　　the entire back. Plants often glaucous, 4-8 dm tall.
　　　　　Spike 6-14 cm long, the spikelets overlapping, 1-1.5
　　　　　cm long, 4-8 flowered. Lemmas 1 cm long, usually
　　　　　pubescent. Infrequent in dry, sandy soils in the
　　　　　western part. {*A. albicans* Scribn. and Smith}
　　　　　July-Aug. Thickspike Wheatgrass . . . .
　　　　　. . *Agropyron dasystachyum* (Hook.) Scribn. and Smith

1' Plants usually without creeping rhizomes.

　　5 Spikes pectinate, the spikelets crowded and divergent from
　　　the rachis. Lemmas and glumes short-awned.

　　　　6 Spikes densely hairy, especially on the flattened sides.
　　　　　Plants tufted, perennial, 5-10 dm tall. Leaves 2-5 mm
　　　　　wide, the blades tending to be pilose on the ventral
　　　　　surface. Spikelets spreading, the glumes gradually
　　　　　tapering into awns. Awns curved, 2-5 mm long. Widely
　　　　　planted and occasionally escaping over the state.
　　　　　{incl. *A. desertorum* (Fisch.) Schult.} June-July.
　　　　　Crested Wheatgrass . . *Agropyron cristatum* (L.) Gaertn.

6' Spike nearly glabrous but sometimes glabrate. Plants
   tufted, perennial, 4-9 dm tall. Leaves scabrous on the
   ventral surface. Spikelets appressed and ascending, the
   glumes abruptly awned. Awns straight to curved, 2-6 mm
   long. Frequently escaping from cultivation and persisting. This species becomes established more often than
   the preceding one. June-July . . . .
   . . . . *Agropyron pectiniforme* Roem. & Schult.

5' Spikes not distinctly pectinate; narrow and ascending.
   Lemmas and glumes awnless to strongly awned.

   7 Lower internodes of the flowering spike 1.5-3.0 cm long,
     becoming shorter at the upper portion of the spike.
     Plants 4-10 dm tall, cespitose, the culms stiff and
     erect. Leaves 3-5 mm wide, mostly on the lower one
     half of the culms. Spikes 18-30 cm long, the spikelets
     widely spaced below. Glumes and lemmas mostly awnless.
     Introduced and escaping in alkaline areas in the central
     and western part. July-Aug. Tall Wheatgrass . . . . .
     . . . . *Agropyron elongatum* (Host) Beauv.

   7' Lower internodes of the flowering spike not longer than
      the upper ones.

      8 Awns of the lemma strongly divergent when mature, 1-2
        cm long, bent at right angles. Spikes open, most of
        the spikelets shorter than the internodes. Plants
        4-8 dm tall, leaves many, glabrous to puberulent.
        Flowering spikes 8-15 cm long, the spikelets 6-8
        flowered, rarely more than one per node. Lemmas
        8-11 mm long, usually strongly awned. Occasionally
        rhizomatous forms are found. Infrequent in the
        western part. {*A. inerme* (Scribn. & Smith) Rydb.}
        June-July. Bluebunch Wheatgrass . . . .
        . . . . . *Agropyron spicatum* (Pursh) Scribn. & Smith.

      8' Awns of the lemma not strongly divergent, or the
         lemmas awnless. Spikes not loose and open, the
         spikelets usually overlapping. Plants strongly
         perennial and cespitose, 5-9 dm tall, leafy. Flow-
         ering spikes 4-15 cm long, mostly compact, the spike-
         lets 4-5 flowered. Lemmas awnless, or the awn up to
         2.5 cm long, straight to curved. A wide-ranging and
         common species of varied habitats that produces vari-
         ation in its physical appearance. {incl. *A. subse-
         cundum* (Link.) Hitchc.; *A. trachycaulum* of authors}
         July-Aug. Slender Wheatgrass . . . .
                              Agropyron caninum (L.) Beauv.

ELYMUS L.

1 Lemmas awnless or with awns 1-5 mm long.

2 Body of the lemma glabrous to sparsely pubescent, awnless.

3 Plants with short rhizomes, the stems clumped, 1-1.5
meters tall. Leaf blades 10-15 mm wide. Inflorescence
12-18 cm long, the spikelets dense, with 3 or more at a
node. Spikelets 4-6 flowered, the lemmas 10-15 mm long.
Infrequent on open wooded slopes and in ravines over the
state. June-July. Giant Wildrye . . . .
. . . . Elymus cinereus Scribn. & Merr.

3' Plants without rhizomes, the stems 5-7 dm tall, blue-
green. Leaf blades 4-8 mm wide. Inflorescence 10-15
cm long, the spikelets usually 2 at a node, 3-5 flowered. Lemmas awnless, 8-12 mm long. Introduced in
drier portions of the western part and persisting in
certain places. June-July. Russian Wildrye . . . . . .
. . . . Elymus junceus Fisch.

2' Body of the lemma densely villous. Awns 1-4 mm long.
Spikelets usually not more than 2 at a node. Leaf blades
5-8 mm wide. Plants with rhizomes, forming clumps, the
stems 4-8 dm tall. Inflorescence densely villous, 5-11
cm long, the spikelets 3-5 flowered. Lemmas 7-9 mm long,
awn tipped or awn to 4 mm. Infrequent in rocky woods at
higher elevations in the Black Hills. June-July . . . . .
. . . . Elymus innovatus Beal

1' Lemmas awned at least 5 mm or more. Plants not noticeably
rhizomatous.

4 Main rachis of the spike disarticulating towards maturity.
Spike narrow, less than 7 mm wide. A commonly occurring
hybrid between, most likely, Agropyron caninum and Hordeum
jubatum, which most conveniently keys out in the genus
Elymus. The plants are 5-8 dm tall, with involute leaves
2-5 mm wide. The inflorescence is 4-11 cm long, with
appressed spikelets about 1 cm long. The glumes are narrow and awned as well as the lemmas. {Elymus macounii
Vasey} . . . . . . X Agrohordeum macounii (Vasey) LePage

4' Main rachis of the spike not disarticulating towards maturity. Spikes usually broader than 7 mm.

5 Glumes subulate-tapering, not broadened above the base.

6 Awns of the lemmas straight. Plants with tufted
stems 6-9 dm tall. Leaves flat, pubescent on the
upper surfaces. Inflorescence ascending to drooping,

5-10 cm long, the spikelets somewhat divergent.
Lemmas scabrous to hirsute, 8 mm long, with the awn
1-3 cm long. Frequent in woods and thickets, mostly
in the eastern part. June-Aug . . . .
. . . . *Elymus villosus* Muhl.

6' Awns of the lemmas flexuous and divergent. Plants
with erect culms 7-13 dm tall. Leaves flat and scabrous, 5-10 mm wide. Spikes 8-18 cm long, nodding
at maturity. Lemmas glabrous to hirsute, 1 cm long,
with a flexuous or divergent awn up to 3 cm long.
Rare in rich, moist soil in the western part. July-
Aug . . . . *Elymus interruptus* Buckl.

5' Glumes lanceolate, broadened above the base, 3-several
nerved.

7 Bases of glumes not hardened-indurate, relatively
thin and flattened, several nerved. Plants cespitose, 6-10 dm tall. Leaves flat, 5-10 mm wide.
Inflorescence stiff and erect, 5-13 cm long. Spikelets 2 at a node, each 3-5 flowered. Lemmas 10-12
mm long, the awn erect, 1-2 cm long, or occasionally
only awn-tipped. Rare in open and dry woods in the
Black Hills and in the northeast part. July-Aug.
Blue Wild Rye . . . . *Elymus glaucus* Buckl.

7' Bases of glumes hardened-indurate, stiff and hardened
on the back.

0 Awns straight towards maturity of the spike, the
lemmas glabrous or nearly so. Plants 6-10 dm
tall, erect. Leaves flat, scabrous, 5-12 mm wide.
Inflorescence erect, sometimes exsert from the
uppermost sheath, 5-12 cm long, closely flowered.
Glumes about 13 mm long, curved, awn tipped.
Lemmas with a straight awn 1-2 cm long. Frequent
in thickets and along alluvial streams over the
state. July-Aug. Virginia Wild Rye . . . .
. . . . *Elymus virginicus* L.

8' Awns curved out towards maturity of the spike, the
lemmas usually pubescent. Plants 8-15 dm tall,
forming large clumps. Leaves flat, coarse, 7-13
mm wide. Inflorescence flexuous, drooping after
anthesis, 10-20 cm long, densely flowered. Glumes
scarcely bowed or curved, with an awn as long as
the body. Lemmas 10-15 mm long, tapered to a
slender curved awn 2.5-4 cm long. Common in

prairies and on roadsides over the state. July-
Aug. Canada Wild Rye . . . . *Elymus canadensis* L.

HORDEUM L.

1 Plants annual. Awns of the lemmas less than 2 cm long.

  2 Leaf blades without prominent auricles at the sheath juncture. Plants 1-4 dm tall, mostly erect. Leaves flat, 2-4
mm wide. Inflorescence 3-6 cm long, dense, the spikelets
erect-ascending. Glumes awnlike, the lemmas with an awn
10-15 mm long. Frequent on dry, sterile or eroded soils
over the state. May-June. Little Barley . . . .
. . . . *Hordeum pusillum* Nutt.

  2' Leaf blades with prominent auricles at the sheath juncture.
Plants 6-10 dm tall, with smooth, erect culms. Leaves 10-
14 mm wide, flat. Inflorescence 2-11 cm long, stout, the
spikelets divergent with erect awns. Lemmas usually awned;
however, certain varieties are short-bearded or lacking an
awn. Cultivated but not persisting. June-July. Barley.
. . . . *Hordeum vulgare* L.

1' Plants perennial. Awns of the lemmas 2-5 cm long, slender,
flexuous. Plants tufted, the culms 2-4 dm tall. Leaves 2-5
mm wide, scabrous. Inflorescence erect to nodding, 4-8 cm
long. The lateral spikelets reduced to awns. Perfect spikelet with awned glumes and lemmas 2-4 cm long. Common as a
weed in waste places over the state. June-July. Foxtail
Barley . . . . *Hordeum jubatum* L.

HYSTRIX Moench.

Hystrix patula Moench. Perennial cespitose plants with stems
solitary or few, 5-8 dm tall. Leaves flat, 7-12 mm wide. Inflorescence long-exsert, erect and later nodding, 8-12 cm long,
the spikelets horizontal from the vertical rachis. Spikelets in
pairs, 1-2 cm long. Lemmas glabrous to hirsute, with awns 1-4 cm
long. Rare in rich upland woods of the eastern part. June-July.
Bottlebrush Grass.

LOLIUM L.

1 Glumes longer than the entire spikelet. Plants annual, erect,
4-8 dm tall. Leaves flat, 3-10 mm wide. Inflorescence of
alternately arranged spikelets, 10-20 cm long. Spikelets 5-7
flowered, the glumes 15-20 mm long, exceeding the upper floret.
Lemmas 6-8 mm long, awnless or with a short awn. Infrequent
as a weed of lawns and gardens in the eastern part. Ours var.

leptochaeton A. Br.   June–July.   Darnel . . . .
. . . . *Lolium temulentum* L.

1' Glumes shorter than the rest of the spikelet. Plants perennial, 4-10 dm tall. Leaves 4-7 mm wide, often folded. Inflorescence 10-20 cm long, the spikelets 6-10 flowered, usually awnless. Lemmas 5-7 mm long. Introduced and occasionally appears as an escape from lawns or grass plantings where
   it is used in seed mixtures. Ours var. *perenne*. June–July.
   . . . . *Lolium perenne* L.

SECALE L.

Secale cereale L. Annual with stems 6-12 dm tall, glabrous.
Leaves 4-10 mm wide, flat. Inflorescence 8-14 cm long, the
spikelets erect, 2-flowered. Glumes slender, rigid, with subulate tips. Lemmas curved, long awned. Cultivated but not persisting. May–June. Rye.

SITANION Raf.

Sitanion hystrix (Nutt.) J. G. Smith. Plants perennial, tufted,
1-5 dm tall. Leaves flat or folded to involute, 1-3 mm wide.
Inflorescence exsert from the upper sheath, 3-12 cm long, disarticulating when ripe. Glumes rigid, subulate. The lower
floret of both spikelets of each node reduced to a structure
resembling a glume. Lemmas with awns as long as the glumes.
Frequent on dry hillsides and plains in the western one-half.
Represented in South Dakota by the two varieties *hystrix* and
brevifolium (Sm.) Hitchc. June–July. Squirreltail.

TRITICUM L.

Triticum aestivum L. Plants annual, or winter annuals, the stems
5-15 mm wide. Inflorescence 5-11 cm long, the spikelets plump.
Spikelets 2-5 flowered, the glumes firm, subulate. Lemmas awnless or with awns. Cultivated but not persisting. May–June.
Wheat.

Tribe 7. Oryzeae

LEERSIA Soland.

1 Spikelets more than 4 mm long. Panicle branches of the lower
  part of the rachis clustered. Plants spreading and decumbent,
  the culms several from a rhizomatous base, 7-12 dm long.
  Leaves flat, 6-9 mm wide, the surfaces sharply retrorse-
  scabrous. Inflorescence 10-15 cm long, with open, flexuous
  branches. Spikelets 5 mm long, with bristly lemmas, also

about 5 mm long. Frequent in marshy soil or in thickets at
the borders of ponds and lakes in the eastern part. Aug-Sept.
Rice Cutgrass . . . . *Leersia oryzoides* (L.) Sw.

1' Spikelets less than 4 mm long. Panicle branches of the lower
part of the rachis solitary. Plants with slender weak stems
5-8 dm long. Leaf blades short and flat, 6-10 mm wide. Inflorescence an open panicle, 10-20 cm long, with irregular,
capillary branches. Spikelets appressed to the branches,
each about 3 mm long. Lemmas ovate on the margins, keeled
on the back. Infrequent to rare in alluvial woods in the
eastern part. July-Aug. Whitegrass . . . .
. . . . *Leersia virginica* Willd.

Tribe 8. Paniceae

(Key to Genera)

1 Spikelets subtended by an involucre of bristles or spines.

  2 Involucre of bristles, these persisting when the spikelets
fall. Inflorescence a dense, spike-like panicle . . . . .
. . . . *SETARIA*

  2' Involucre of spines, these attached to the spikelet and
separating as spiny burs . . . . *CENCHRUS*

1' Spikelets not subtended by bristles or spines.

  3 Glumes or sterile lemmas awned or awn pointed . . . .
. . . . *ECHINOCHLOA*

  3' Glumes not awned.

    4 Spikelets in slender, one-sided racemes digitately ar-
ranged at the summit of the culm . . . . *DIGITARIA*

    4' Spikelets in panicles . . . . *PANICUM*

CENCHRUS L.

Cenchrus longispinus (Hack.) Fern. Annual matted plant with
short, prostrate or ascending stems to 6 dm long. Leaves flat,
2-6 mm wide. Inflorescence a short, crowded raceme of 5-9 burs
within which are the spikelets. Spikelets 4-6 mm wide, usually
2 enclosed by the involucre with sharp, radiating spines. Very
common weed of cultivated and sandy or thin soils over the state.
{*C. pauciflorus* of reports} July-Sept. Field Sandbur.

DIGITARIA Heis.

1 Leaf sheaths glabrous. Fertile lemmas brown, the sterile ones
   with glandular hairs. Plants annual, tufted from the base,
   the stems prostrate-ascending, rooting at the lower nodes.
   Racemes 2-7, digitate, 4-9 cm long. Rachis one-sided, the
   spikelets 2 mm long. Common in waste places and lawns over
   the state. Aug-Sept. Smooth Crabgrass . . . .
   . . . . *Digitaria ischaemum* (Schreb.) Muhl.

1' Leaf sheaths pilose. Fertile lemmas pale green, the sterile
   ones without glandular hairs. Plants annual, tufted and
   spreading, the stems prostrate or decumbent, 2-8 dm long.
   Leaves flat, 5-10 mm wide, pubescent. Inflorescence of
   digitate racemes 2-5, up to 10 cm long. Spikelets about 3
   mm long. Very common in waste places and lawns over the
   state. July-Sept. Crabgrass . . . .
   . . . . *Digitaria sanguinalis* (L.) Scop.

ECHINOCHLOA Beauv.

1' Lemma of fertile floret broadly acute to obtuse with a sharply
   differentiated, withering tip. Dorsal surface of the lemma
   with a line of minute hairs. Some hairs on the panicle
   branches as long as or longer than the spikelets. Annual
   plants with coarse, pithy stems, erect to spreading, the
   culms 4-10 dm tall. Leaves soft and flat, 4-12 mm wide.
   Spikelets 3-4 mm long, the sterile lemma awned, the fertile
   ones without awns. Common in alluvial soil, waste areas and
   otherwise fertile soil over the state. July-Sept. Barnyard
   Grass . . . . *Echinochloa crusgalli* (L.) Beauv.

1' Lemma of fertile floret narrowly acute with a persistent,
   stiff, mucronate tip. Dorsal surface of the lemma lacking
   lines of hairs. Hairs on the panicle branches absent or
   shorter than the spikelets. Plants coarse annuals with culms
   8-10 dm tall, from spreading bases. Leaves broad, with scabrous margins. Spikelets 2.5-3.0 mm long. Lemma of lower
   floret awnless or with an awn up to 1 cm long. Frequent to
   common in waste soil over the state. Ours var. *microstachya*
   Wiegand. July-Sept . . . *Echinochloa muricata* (Beauv.) Fern.

PANICUM L.

1 Plants annual.

   2 Spikelets more than 4 mm long, the panicle nodding. Lower
      glume one-half as long as the entire spikelet. Plants
      stout, the culms erect, 2-9 dm tall. Leaves 2 cm wide,
      glabrate to pilose. Inflorescence 10-30 cm long, with the

basal part of the panicle included, compact. Spikelets
ovate, many-nerved, 3-4 mm long. Occasional as an escape
in waste soil. July-Aug. Broomcorn Millet . . . .
. . . . *Panicum miliaceum* L.

2' Spikelets less than 4 mm long, the panicle erect.

   3 Leaf sheaths glabrous. The lower glume much less than
one-half the length of the entire spikelet. Plants irregularly spreading from the base, the culms 4-8 dm
tall, thick. Leaves up to 2 cm wide. Inflorescence
10-40 cm long, mostly terminal but smallar panicles may
be axillary. Spikelets 2-3 mm long, on diffuse panicle
branches. Infrequent in waste soil, on roadsides and
in farmyards in the eastern part. Aug-Oct. Fall Panicum . . . . *Panicum dichotomiflorum* Michx.

    3' Leaf sheaths pilose-pubescent. The lower glume almost
one-half the length of the entire spikelet. Plants
tufted, the stems coarse and fleshy, 2-8 dm tall.
Leaves 1-2.5 cm wide, hairy on both surfaces. Inflorescence diffuse, usually exsert, 10-30 cm long, the
panicle branches densely flowered. Spikelets 2-2.5 mm
long. Common in fields and on roadsides over the state.
July-Sept. Witchgrass . . . . . . *Panicum capillare* L.

1' Plants perennial.

  4 Spikelets 3 mm long or more.

    5 Plants with culms up to 1 meter or more, stout, from
scaly rhizomes. Panicle 1.5-4 dm long, open and
branched. Leaves 1-1.5 cm wide, mostly glabrous.
Spikelets 3.5-6 mm long, ovoid, the first glume over
one-half the length of the spikelet. Lemmas nerved,
acutely pointed. Common in low to open prairie and
on roadsides over the state. July-Sept. Switchgrass.
. . . . *Panicum virgatum* L.

    5' Plants much less than 1 meter tall, lacking stout,
scaly rhizomes.

      6 Leaf blades glabrous, especially on the upper sur-
face. Spikelets glabrous to short pubescent.

        7 Panicle spreading, 3-7 cm long, almost as wide as
long. Spikelets obovoid, 3 mm or more long, glabrous to short pubescent. Plants 2-5 dm tall.
Leaves flat, 6-10 mm wide, smooth above, sparingly pilose beneath. Panicle 4-7 cm long, the
branches ascending or later spreading. Common

in prairies and other grassy places over the
state. Ours var. *scribnerianum* (Nash) Fern.
May–July. Small Panic Grass . . . .
. . . . *Panicum oligosanthes* Schultes

7' Panicle erect, 5–12 cm long, much narrower than
long. Spikelets elliptical, up to 4 mm long,
usually short pubescent. Plants 2–6 dm tall.
Leaves 8–15 mm wide, erect, the blades glabrous
on both surfaces except pilose at their bases.
Leaf sheaths loose, overlapping each other near
the bases of the stems. Rare to infrequent in
sandy soil in the Black Hills. June–July . . . .
. . . . *Panicum xanthophysum* Gray

6' Leaf blades papillose-hairy on both surfaces, sometimes sparse on the upper surface. Spikelets pilose,
the hairs to 1 mm long. Plants 2–7 dm tall, the
stems branched from the base. Leaves flat, 7–13 mm
wide. Panicle 8–13 cm long, diffuse, the spikelets
3–4 mm long. Infrequent in open thickets and prairie
openings over the state. June–July . . . .
. . . . *Panicum leibergii* (Vasey) Scribn.

4' Spikelets less than 3 mm long.

8 Spikelets 1.0–2.0 mm long.

9 First glume broadly angular to rounded at the apex,
about one-third the length of the spikelet. Leaf
sheaths glabrous to softly pubescent. Hairs on the
stems not copiously villous, less than 4 mm long.
Plants in clumps, the culms 4–7 dm tall. Panicle
6–12 cm long, mostly surpassed by the leaves. Spikelets 1.5–1.9 mm long, short pubescent. Frequent in
moist to dry woods and openings in the eastern and
far western part. Ours var. *fasciculatum* (Torr.)
Fern. {incl. *P. huachucae* Ashe and *P. tennesseense*
Ashe} June–July . . . . . *Panicum lanuginosum* Ell.

9' First glume triangular-ovate with an acute apex, over
one-third the length of the spikelet. Leaf sheaths
and culms with long, pilose hairs 4–5 mm long,
spreading. Plants 2–5 dm tall, in clumps but later
spreading. Leaf blades erect to strongly ascending,
with papillose-pilose hairs on both surfaces. Panicle 4–6 cm long, ovoid. Spikelets 1.8 mm long, with
long, pilose hairs. Rare in dry soil of the eastern
part. June–July . . . .
. . . . *Panicum praecocius* Hitch. & Chase

8' Spikelets 2.5-3 mm long.

   10 Leaves densely pilose on both surfaces. Panicles at
      maturity up to 4 cm long. Plants cespitose, erect,
      the stems 1-3 dm tall, densely pilose. Leaves 3-6
      mm wide, stiff, ascending. Panicle with flexuous
      branches, not densely flowered. Spikelets papillose-
      pubescent. Infrequent in prairie and open areas
      over the state. June-July . . . .
      . . . . *Panicum wilcoxianum* Vasey

   10' Leaves scabrous above, pilose beneath. Panicles at
      maturity 4-8 cm long. Plants cespitose, the culms
      stiff, slender, 1-4 dm tall. Leaves tending to be
      basal due to the short internodes of the culms at
      the lower part. Leaves 2-5 mm wide. Panicle sur-
      passing the leaves, 3-8 mm long. Spikelets minutely
      pubescent. Infrequent in prairies and on hillsides
      over the state. May-June . . . .
      . . . . *Panicum perlongum* Nash.

SETARIA Beauv.

1 Clustered bristles below each spikelet retrorsely scabrous.
  Panicle branches of the inflorescence verticillate. Plants
  annual, the stems branched from the base, 3-8 dm tall. Leaves
  4-9 mm wide, scabrous or sparsely pilose. Inflorescence a
  dense panicle 5-10 cm long. Spikelets with 1 or 2 bristles
  4-6 mm long. Common as a weed in waste places or in rich soil
  over the state. July-Sept. Bristly Foxtail . . . .
  . . . . *Setaria verticillata* (Lam.) Beauv.

1' Clustered bristles below each spikelet antrorsely scabrous.
   Panicle branches not verticillate.

   2 Leaves loosely pubescent on the upper surface. Panicle
     nodding, 2-3 cm thick. Plants erect, 2-5 dm tall. Leaves
     flat, 8-15 mm wide, elongate. Inflorescence dense, 8-13 cm
     long, nodding. Bristles 1-3 below each spikelet, about 3
     times the length of the spikelets which are 2-2.5 mm long.
     A rare weed of cultivated soil in the eastern part but
     spreading westward. July-Sept. Chinese Foxtail . . . . .
     . . . . *Setaria faberii* Herrm.

   2' Leaves glabrous on both surfaces.

      3 Bristles 5-15 at the base of each spikelet. Spikelets
        3 mm long. Plants annual, branched from the base, the
        stems ascending, 5-9 dm tall. Leaves about 10 mm wide,
        elongate. Inflorescence dense, usually erect, 5-11 cm

long, the main rachis pubescent. Spikelets yellowish,
the enclosed grain transversely ridged at maturity.
Very common in waste places over the state. {*S.
lutescens* (Weig.) Hubb.} July-Sept. Yellow Foxtail . .
. . . . *Setaria glauca* (L.) Beauv.

3' Bristles 1-3 at the base of each spikelet. Spikelets
2-3 mm long.

4 Spikelet maturing and falling entire with the grain
enclosed in the glumes and the sterile lemma. Culms
4-6 dm tall, the panicle cylindric and green. Plants
annual, with spreading stems. Leaves soft, flat, 10-
15 mm wide. Panicle dense, erect to nodding, 8-10 cm
long. Spikelets with enclosed grain finely rugose.
A common weed of waste places and cultivated soil
over the state. June-Aug. Green Foxtail . . . . . .
. . . . *Setaria viridis* (L.) Beauv.

4' Spikelet maturing with grain falling free from the
glumes and sterile lemma. Culms up to 1 meter,
robust, the panicle lobed or interrupted. Plants
annual, the culms thick. Leaves very broad in some
races, up to 4 cm wide. Panicle up to 25 cm long,
purple to yellow. Spikelets 2-3 cm long, the grain
separating from it at maturity. An occasional weed
in the eastern part. July-Sept. Foxtail Millet . .
. . . . *Setaria italica* (L.) Beauv.

Tribe 9. *Phalarideae*

(Key to Genera)

1 Spikelets bronze or brown-colored, in an open panicle. Lower
florets staminate. The first lemma ovate . . . *HIEROCHLOE*

1' Spikelets green or yellow, in a contracted, spike-like panicle.
Lower florets reduced to bristle-like lemmas . . *PHALARIS*

HIEROCHLOE R. Br.

Hierochloe odorata (L.) Beauv. Plants perennial from rhizomes,
the culms 3-5 dm tall, with a fragrant odor. Leaves mostly basal,
pointed, 3-5 mm wide. Panicles bronze-colored, open, 5-8 cm long.
Spikelets 5-7 mm long, with 1 perfect floret. Fertile lemma awnless but pubescent towards the summit. Frequent in meadows and
prairie swales in the northeast and the western part. May-June.
Sweetgrass.

PHALARIS L.

1 Plants annual. Panicles ovoid. Glumes with dorsally winged
   keels. Stems 3-6 dm tall, erect from a cespitose base.
   Leaves flat, 4-9 mm wide. Inflorescence short, 2-5 cm long.
   Spikelets 6-8 mm long, green and white striped. Fertile lemma
   5-6 mm long, pubescent. Rarely occurring but not persisting.
   Seed of canary feed and other feed for pets. July. Canary
   Grass . . . . *Phalaris canariensis* L.

1' Plants perennial. Panicle contracted, cylindric. Glumes only
   scarcely winged on the keel. Plants from creeping rhizomes,
   the stems stout and erect, 6-13 dm tall. Leaves flat, 7-15 mm
   wide. Inflorescence dense, green, becoming yellow-white towards maturity, 8-13 cm long. Spikelets glabrous, the fertile
   lemma 4 mm long, with hairs towards the apex. Sterile lemmas
   reduced to 2 opposite tufts of hairs 1-1.5 mm long. Common in
   marshy places and ditches over the state. May-July. Reed
   Canary Grass . . . . *Phalaris arundinacea* L.

Tribe 10. Zizanieae

ZIZANIA L.

Zizania aquatica L. Plants annual, monoecious, the stems 1-3
meters tall, stout and slender. Leaves long, 1-4 cm wide, softly
scabrous. Panicle separated into a staminate upper portion which
is narrower and shorter than the lower pistillate portion. The
entire structure 30-50 cm long. Pistillate spikelets 2 cm long,
the lemma awn tipped, pubescent. Rare in shallow water of the
northeast part. July-Sept. Wild Rice.

Family Sparganiaceae

SPARGANIUM L.

1 Pistils with two stigmas. Mature achenes obpyramidal, truncate at the summit. Plants without floating leaves, 5-9 dm
   tall, erect. Leaves flat, 3-6 dm long and 9-12 mm wide. Inflorescence branched, with 4-10 staminate heads and 1 or 2
   pistillate heads. Fruits 2-seeded, up to 1 cm long. Common
   at borders of ponds and lakes and in marshy areas over the
   state. June-July . . . . . . *Sparganium eurycarpum* Engelm.

1' Pistils with one stigma. Mature achenes fusiform or ellipsoid, tapering to the summit. Plants less than 6 dm tall,
   usually with floating leaves.

2 Fruiting heads 2 cm or more across towards maturity. Principal leaves strongly triangular-keeled. Plants erect, up to 6 dm tall, the leaves also erect, not floating, 3-7 mm wide. Pistillate heads 1-4, crowded, sessile or on short peduncles. Achenes about 4 mm long, fusiform. Infrequent to rare in swampy soil or muddy shores in the eastern part. June-July . . . . *Sparganium chlorocarpum* Rydb.

2' Fruiting heads less than 2 cm across towards maturity. Principal leaves rounded on the back, or flat, often floating.

3 Leaves rounded on the back, 2-5 mm wide, the upper leaves dilated at the base. Stems 3-6 dm tall. Inflorescence usually with 2-4 pistillate heads and 2-5 staminate ones. Fruits with an abruptly pointed tip and stalked, 5-7 mm long. Infrequent in marshy soils and ponds in the northeast part. June-July . . . . *Sparganium angustifolium* Mx.

3' Leaves flat, 5-10 mm wide, not dilated at the base. A slender marsh plant with leaves 2-5 dm long. Inflorescence simple, or only a little branched. Staminate heads 3-5, clustered, the pistillate ones 2-5. Fruits pointed, on stalks at least 2 mm long. Infrequent in shallow ponds locally in the Black Hills. June-July. . . . . . . *Sparganium multipedunculatum* (Morong) Rydb.

Family Typhaceae

TYPHA L.

1 Pollen grains single. Pistillate and staminate portion of the spike usually separated. Leaves 3-8 mm wide. Plants up to 2.5 m tall. Leaves convex on the dorsal surface. Stigmas in pistillate portion of inflorescence with subtending bracts. Pistillate spike becoming 0.5-1.5 cm in diameter. Locally abundant in alkaline water over the state. July. Narrow-leaved Cattail . . . . *Typha angustifolia* L.

1' Pollen grains in tetrads. Pistillate and staminate portions of the spike usually confluent, pistillate flowers without subtending bracts. Leaves flat, 0.5-2.5 cm broad. Stems often over 1.5 m tall. Common in marshy places over the state. June-July. Broad-leaved Cattail . . . . *Typha latifolia* L.

Family Araceae

(Key to Genera)

1 Leaves simple. The spathe an extension of the scape. Flowers covering the spadix . . . . *ACORUS*

1' Leaves compound with 3 leaflets. The spathe conspicuous, enveloping the fleshy spadix. Flowers on the basal part of the spadix . . . . *ARISAEMA*

ACORUS L.

Acorus calamus L. Perennial aromatic herbs from stout rhizomes which form large colonies. Upright stems or scapes 1-2 meters tall, triangular. Leaves mostly basal, erect and elongate, 1-2 cm wide. Flowers yellow-brown, crowded on a spadix 3-8 cm long, the perianth reduced to 6 short projections. Ovary maturing to a hardened, obpyramidal fruit with 1-3 seeds. Rare in marshy areas below springs in the eastern part. July-Aug. Sweet Flag.

ARISAEMA Mart.

Arisaema triphyllum (L.) Schott. Perennial herbs with fleshy stems 2-6 dm tall, from a deep-seated corm. Plants functionally dioecious. Leaves mostly 2, the petiole 2-4 dm long. Leaflets 3, the central one larger and rhomboidal. Flowers reduced, crowded towards the base of the spadix, mostly dioecious. Fruiting berries on pistillate plants becoming red, globose, up to 8 mm across. Frequent in rich, loamy woods in the eastern part. Apr-May. Jack in the Pulpit.

Family Lemnaceae

(Key to Genera)

1 Thallus exceeding 1 mm in length, nerved, with rootlets from the ventral surface.

  2 Each thallus with 2-5 rootlets from the ventral surface. Ventral surface of the thallus red to purple . . . . *SPIRODELA*

  2' Each thallus with 1 rootlet from the ventral surface. Ventral surface not red pigmented . . . . . *LEMNA*

1' Thallus less than 1 mm long, without rootlets. Nerves lacking . . . . *WOLFFIA*

LEMNA L.

1 Thallus tapering to the base, the stipe 6-10 mm long, often
   remaining attached together to form matted, T-shaped submersed
   colonies. Thallus often rootless. Body of the thallus 5-10
   mm long, with faint nerves. Flowers, when produced, in a
   pouch or cleft of the thallus, with only anthers and ovary
   formed. Flowering specimens are rare and should be looked
   for in August. Frequent to common in quiet water in the
   eastern one-half. Aug . . . . *Lemna trisulca* L.

1' Thallus not tapering to a base, asymmetrically oval to falcate, less than 6 mm long, solitary or in groups.

   2 Individual frond 2-3 mm long, often reddish underneath,
      with an oblique or falcate shape. Thallus with one root-
      let, floating on the surface, often in colonies of 1-6.
      When flowers develop, these are reduced to 2 anthers and
      1 ovary in a small pouch. Only rarely do they reproduce
      in this way in our area. Reproduction usually by vegeta-
      tive buds. Infrequent in quiet water in the eastern part.
      Aug . . . . *Lemna perpusilla* Torr.

   2' Individual frond 3-4 mm long, not reddish underneath, with
      an oval to rounded shape, not tapering to the base. Entire
      thallus usually solitary, with 1 rootlet, floating on the
      surface. Reproduction mostly by budding. When flowers
      develop, they are reduced to 2 anthers and 1 ovary in a
      pouch on the upper surface. Common in quiet water over
      the state. Aug. Duckweed . . . . *Lemna minor* L.

SPIRODELA Schleiden

Spirodela polyrhiza (L.) Schleid. Small floating colonies attached in groups of 2-8, each segment ovate, 4-8 mm long, with
several rootlets. Under surface of the thallus red or purple
pigmented, the upper surface nerved. Reproduction mostly by
budding, only rarely flowering. When present, flowers reduced
to 2 anthers and 1 ovary in a cleft or depression at one side
of the thallus. Rare to infrequent in quiet water in the eastern part. July-Aug. Big Duckweed.

WOLFFIA Horkel.

Wolffia columbiana Karst. Thalli globose to ellipsoid, or
slightly flattened on the upper side with 3 papillae, 0.2-1 mm
long, usually floating just under the surface of the water at
the air-water interface. Flowers rarely produced, but if they
are, in a small pouch on the upper surface. Usually reproducing
by vegetative budding. Occasional in the eastern part in quiet
water. Aug.

Family Commelinaceae

(Key to Genera)

1 Petals unequal in length. Fertile stamens 3 . . . .
. . . . COMMELINA

1' Petals equal in length. Fertile stamens 6 . . . .
. . . . TRADESCANTIA

COMMELINA L.

1 Stems erect, 2-6 dm tall, from a perennial base of fibrous
roots. Leaves linear-lanceolate, 4-11 cm long, their bases
sheathing the stem. Flowering spathes arising near the terminal part of the plant on short peduncles. The margins of the
spathe fused at the base. Upper petals 1-2 cm long, blue, the
lower one short and white. Infrequent in sandy soils in the
southwestern part. Ours var. *angustifolia* (Michx.) Fern.
July-Aug . . . . *Commelina erecta* L.

1' Stems decumbent, spreading and rooting from the nodes, annual.
Leaves ovate to broadly lanceolate, 4-8 cm long. Flowering
spathes broad, the margins open to the base. Upper petals
blue, 8-12 mm long, the lower one white. Infrequent in moist
or shaded soil and in yards and thickets in the eastern part.
July-Sept. Dayflower . . . . *Commelina communis* L.

TRADESCANTIA L.

1 Sepals less than 10 mm long, sparsely pubescent with glandular
hairs. Vegetation usually glaucous. Plants perennial with
thickened, succulent roots. Stems often branched, 2-5 dm
tall. Leaves narrowly linear, folded, usually glabrous. Inflorescence a solitary flower or a few-flowered cyme, the
bract leaf-like. Petals pink or light blue, 1-1.5 cm long.
Infrequent in dry prairie or sandy roadside areas in the western two-thirds. Rare in the eastern part. May-June . . . . .
. . . . *Tradescantia occidentalis* (Britt.) Smyth.

1' Sepals exceeding 10 mm, pubescent with glandular and non-
glandular hairs. Vegetation glabrous or sparsely pubescent
but not glaucous. Plants 2-4 dm tall, perennial, the stems
stout and usually not branched. Leaves 8-13 mm wide, not
strongly folded. Flowers in a solitary terminal cyme, the
petals very light blue to deep lavender, 1-1.5 cm long. Common in sandy prairie and especially on railroad embankments
over the state. May-June. Common Spiderwort . . . .
. . . . *Tradescantia bracteata* Small.

Family Pontederiaceae

HETERANTHERA R. & P.

1   Leaves ovate. Flowers white or blue. Inflorescence 1-several
flowered. Partially submersed aquatic plants rooted in the
mud, the stems up to 2 dm tall, branched. Leaves 2-5 cm long,
the petioles up to 15 cm long. Leaves floating or submersed.
Inflorescence sheathed by a spathe 2-4 cm long, the flowers
solitary. Flowers with tubes 2-4 cm long, with 6 narrow perianth lobes. Stamens 3, unequal in length. Fruit a 3-valved
capsule, dehiscent, with many seeds. Infrequent in ponds and
ditches over the southern part of the state. July-Sept. Mud
Plantain . . . . *Heteranthera limosa* (Sw.) Willd.

1' Leaves narrowly linear. Flowers yellow, solitary. Partially
submersed aquatic plants with slender stems usually rooted in
the mud. Leaves grass-like, 2-5 mm wide and up to 10 cm long.
Flowers enclosed in a leaf-like spathe 2-4 cm long. Petals
forming a tube 2-4 cm long, the free portion spreading. Fruit
about 1 cm long, indehiscent, rarely set in our region. Infrequent in quiet water in the northeast lake region. July-
Aug. Water Stargrass . . . . *Heteranthera dubia* (Jacq.) MacM.

Family Liliaceae

(Key to Genera)

1   Stems leafy above the base of the plant at flowering time.

2   Petals exceeding 3 cm long. Flowers large and showy.

3   Perianth segments unlike, the 3 outer narrow and sepaloid. Petals pale yellow . . . . *CALOCHORTUS*

3' Perianth segments all alike. Petals orange or orange-
red . . . . *LILIUM*

2' Petals less than 3 cm long. Flowers smaller and not as
showy.

4   Flowers or flower branches borne laterally on the stem
or its branches.

5   Leaves scale-like, branchlets compounded into filiform segments . . . . *ASPARAGUS*

5' Leaves broad and flat.

6   Flowers in umbels. Leaves distinctly petioled
. . . . *SMILAX*

6' Flowers 1-3 on peduncles. Leaves sessile on the stem.

    7 Perianth segments separate to the base . . . . *STREPTOPUS*

    7' Perianth segments fused to form a tube . . . . *POLYGONATUM*

4' Flowers or flower branches borne terminally.

    8 Plants with 3 leaves arranged in a single whorl . . . . *TRILLIUM*

  8' Plants with leaves arranged alternately.

    9 Leaves linear, more than 10 times longer than broad.

       10 Flowers brown to purple, over 1 cm long. Plants 1 to several flowered . . . . *FRITILLARIA*

       10' Flowers greenish-white to yellow, about 1 cm long or less. Inflorescence a panicle or raceme, several to many flowered . . . . *ZIGADENUS*

    9' Leaves broader, less than 10 times longer than broad.

      11 Principal stem branched.

        12 Anthers longer than their filaments. Fruit a capsule . . . . . . *UVULARIA*

        12' Anthers shorter than their filaments. Fruit a berry . . . . *DISPORUM*

      11' Principal stem not branched.

        13 Leaves 2 or 3, cordate at their bases. Perianth segments 4 . . . . *MAIANTHEMUM*

        13' Leaves several to 10 or more, tapering to their bases. Perianth segments 6 . . . . *SMILACINA*

1' Stems with leaves at the base of the plant or lacking leaves at flowering time.

  14 Flowers bright orange, the perianth segments over 6 cm long. Plants at flowering time almost a meter high . . . . *HEMEROCALLIS*

14' Flowers not orange-colored, the perianth segments shorter than 6 cm.

    15 Inflorescence a solitary flower which is nodding . . . . *ERYTHRONIUM*

    15' Inflorescence several to many flowered.

        16 Plants with a large woody base. Leaves sword-like, harsh, with a spiny tip . . . *YUCCA*

        16' Plants herbaceous with woody bases. Leaves not spine-tipped.

            17 Perianth segments with a broad green stripe on the dorsal side. Flowers in an open umbel-like corymb . . . . *ORNITHOGALLUM*

            17' Perianth segments without a broad, green stripe on the dorsal side.

                18 Inflorescence 1-3 flowered, borne on short scapes less than 7 cm from ground level . . . . *LEUCOCRINUM*

                18' Inflorescence many-flowered or with bulb-lets in a dense umbel. Flowering scapes usually exceeding 7 cm in height . . . . *ALLIUM*

ALLIUM L.

1 Leaves absent at flowering time. Leaves formed early in spring usually as two flat lance-ovate blades 2-3 dm long and 2-4 cm wide, from a deep seated bulb. Flowering occurring 2 months later, the scape naked, 2-5 dm tall, the inflorescence immediately subtended by 1 or 2 foliaceous bracts. Flowers in tight umbels, the petals white, 4-7 mm long. Infrequent in rich, loamy woods in the eastern part. June-July. Wild Leek . . . . *Allium tricoccum* Ait.

1' Leaves present at flowering time, not broad and flat.

    2 Scape recurved immediately below the umbel at flowering time. Stamens slightly exceeding the perianth. Plants from elongate bulbs, the scapes 3-5 dm tall. Outer bulb layers with longitudinal parallel nerves. Leaves 2-4, narrow, at the soil level. Flowers white to pink, the perianth segments 4-5 mm long. Frequent in prairie over the state. July-Aug . . . . *Allium cernuum* Roth.

2' Scape and umbel erect at flowering time. Stamens usually
   included in the perianth.

   3 Outer bulb coats persisting as reticulated fibers form-
      ing diamond shaped areas.

      4 Ovary and fruit with 3 or more crests at the summit.
         Flowers rarely replaced by bulblets.

         5 Usually 2 leaves on each scape, about as long as
            the scape. Flowers white, the tips of the inner
            perianth whorl spreading. Plants 1-3 dm tall,
            seldom taller, from 1 or 2 reticulate-coated
            bulbs. Flowers in open umbels, 15-25, the pedi-
            cels unequal. Perianth segments about 6 mm long,
            usually white. Common in dry to moist prairie
            over the state. May-June . . . .
            . . . . *Allium textile* Nels. & Mac Br.

         5' Usually 3 leaves on each scape, not reaching the
            inflorescence. Flowers pink-rose, seldom white,
            the tips of the inner perianth whorl erect.
            Plants 2-5 dm tall, from reticulate-coated bulbs
            that are often paired. Umbels 10-25 flowered,
            sometimes with bulblets. Perianth segments 4-7
            mm long, usually pink, the stamens included.
            Rare to infrequent in low meadows and valleys
            in the western part. May-June . . . .
            . . . . *Allium geyeri* S. Wats.

      4' Ovary and fruit without crests at the summit.

         6 Plants 3-7 dm tall. Flowers often replaced by
            bulblets. Bracts of the spathe subtending the
            umbel 3-7 nerved. Basal bulb with coarse, well-
            developed reticulations. Leaves several on the
            lower one half of the scape, flattened, 2-4 mm
            wide. Flowers, when present, white to pink, in
            open umbels, the pedicels 1-3 cm long. Frequent
            to common in moist prairies in the eastern part
            and in the Black Hills. June-July. Wild Onion.
            . . . . *Allium canadense* L.

         6' Plants 1-3 dm tall. Flowers rarely replaced by
            bulblets, the umbels erect.

            7 Bracts of the spathe subtending the umbel 1-
               nerved. Perianth segments spreading, becoming
               papery and rigid. Reticulations of the outer-
               bulb coats very fine and mesh-like. Plants 1-3
               dm tall, the scapes usually solitary. Leaves 3

or more from each bulb, 2-3 mm wide. Umbel 5-
20 flowered, the pedicels about 2 cm long.
Flowers white to rose, the perianth parts 5-7
mm long. Rare in dry prairie in the southern
part. {*A. nuttallii* Wats.} June . . . .
. . . . *Allium drummondi* Regel.

7' Bracts of the spathe subtending the umbel 3-7
nerved. Perianth urn-shaped, permanently surrounding the fruiting capsule. Reticulations
of the outer bulb coats coarse meshed and
fibrous. Plants 1-2 dm tall, the scapes terete, with 3 or more leaves. Umbel 5-25 flowered, erect, the pedicels equaling the perianth. Flowers deep rose to fading purple, the
perianth segments 7-10 mm long. Infrequent in
prairies in the central and eastern part. June-
July . . . . . . *Allium perdulce* S. V. Fraser

3' Outer bulb coats membranous with longitudinal, parallel
nerves. Plants with scapes 3-6 dm tall, from a deep-
seated bulb. Leaves several, folded, arising at about
the soil surface, 1-3 mm wide. Flowers in open umbels,
the pedicels 1-2 cm long. Perianth segments usually
pink or rose-colored, 4-6 mm long. Stamens about as
long as the perianth. Frequent in prairies and open
places in the eastern part. July-Aug . . . .
. . . . *Allium stellatum* Ker.

ASPARAGUS L.

Asparagus officinalis L. Perennial herbs from deep spreading
rhizomes, the branched stems 5-15 dm tall. Leaves reduced to
scales, the branch endings filiform. Flowers solitary at nodes
and at paniculate branch endings, the perianth green or white.
Perianth segments about 4 mm long. Fruiting berries less than
1 cm across, green, maturing red. Frequently escaping and persisting in waste places over the state. May-July. Asparagus.

CALOCHORTUS Pursh.

1 Anthers acutely pointed. Gland on the petal face half-moon
shaped, wider than long, with a lavender band above it and a
purple spot at the lower part of the petal. Plants perennial,
from deep seated bulbs, stems 2-5 dm tall. Leaves erect, involute, glabrous. Flowers solitary or few at the stem terminus. Petals cream-colored or white, broadly obovate, about
3 cm long. Fruiting capsule 2-4 cm long, narrowly lanceolate.

Frequent in meadows and on hillsides in the western part.
July-Aug. Sego Lily . . . . . *Calochortus gunnisonii* S. Wats.

1' Anthers obtusely pointed. Gland on the petal face rounded,
not broader than long, surrounded by a hairy fringed area.
Plants perennial, 2-5 dm tall, from bulbs. Leaves 2-4,
erect, inrolled and grass-like. Flowers showy, white to
cream-colored, the petals tending to be narrower and more
acutely tipped than the preceding. Fruiting capsule lanceolate, tapered at both ends, about 3 cm long. Frequent to
common in prairies and hillsides in the western one-half.
June-July. Mariposa Lily . . . . *Calochortus nuttallii* T. & G.

DISPORUM Salisb.

Disporum trachycarpum (Wats.) B. & H. Plants perennial, herbaceous, the stems branched and flexuous, 2-5 dm tall, from a rhizomatous base. Leaves sessile and clasping, broadly ovate, 4-8
cm long, with acute tips. Vegetation finely pubescent. Flowers
white or greenish white, solitary or more usually paired in upper
axils, the perianth segments narrow and separate, 1.0-1.5 cm long.
Fruiting berry maturing red or red-orange, lobed, with few seeds.
Infrequent in rich woods in the Black Hills and Harding County.
May-June. Fairybells.

ERYTHRONIUM L.

Erythronium albidum Nutt. Perennial herbs with deep seated bulbs,
growing in colonies, the stems 1-3 dm tall. Leaves glaucous, mottled with lavender, lance-shaped, 5-12 cm long, arising near soil
level. Flowers usually solitary, the scape reflexed immediately
below the flower. Perianth parts not well differentiated, white
to lavender, the segments separated, about 3 cm long. Rare to
infrequent in rich woods of the southeastern part. Apr-May.
Dogtooth Violet.

FRITILLARIA L.

Fritillaria atropurpurea Nutt. Perennial herbs from scaly bulbs,
often with bulblets, the stems 2-4 dm tall. Leaves several, linear and ascending, 3-6 mm wide. Flowers usually solitary, the
scape and flower usually recurved and nodding. Perianth purple
or brown, spotted white or yellow, the segments distinct, 1.5-2.0
cm long. Fruiting capsule erect, angled, 1-1.5 cm long, with
many seeds. Infrequent on grassy slopes and prairies in the
western one-half. May-June. Leopard Lily.

HEMEROCALLIS L.

Hemerocallis fulva L.   Perennial herbs from fibrous, spreading
rhizomes, the leaves folded and arched, 3-6 dm long.   Scapes
leafless, 4-10 dm tall.   Flowers several at the ends of the
scapes, bracted.   Perianth segments fused below, the free portion recurved, yellow-orange, the entire flower up to 10 cm
across.   Commonly cultivated and occasionally persisting along
roadsides and at abandoned homesteads in the eastern part.   June-
July.   Day Lily.

LEUCOCRINUM Nutt.

Leucocrinum montanum Nutt.   Plants tufted from a perennial,
fleshy rooted base, cespitose.   Stems short, hardly above ground.
Leaves several to many, mostly at soil level, linear and arching,
up to 15 cm long.   Flowers several, clustered irregularly at the
apex of the stem, the perianth segments white, about 2 cm long.
Perianth not differentiated, with a fragrant aroma.   Fruiting
capsule 6 mm long.   Frequent on hillsides and plains in the
western part.   May.   Mountain Lily.

LILIUM L.

1 Flowers erect at anthesis.   Perianth distinctly clawed, the
   parts not recurved.   Plants 3-8 dm tall, from thick, scaly
   bulbs, the stems erect.   Leaves alternate except for an upper
   whorl or two.   Leaf blades linear to narrowly lanceolate, 5-
   10 cm long.   Flowers usually solitary, erect, the perianth
   segments separate, 5-7 cm long, orange to orange-red with
   purple spots.   Frequent to common in open woods and meadows
   in the eastern and far western part.   {L. *umbellatum* Pursh}
   Ours var. *andinum* (Nutt.) Ker.   June-July.   Wood Lily . . . .
   . . . . *Lilium philadelphicum* L.

1' Flowers nodding at anthesis.   Perianth segments prominently
   recurved.   Plants 6-12 dm tall, stout and erect, from a scaly
   bulb.   Leaves whorled at the middle of the stem, becoming opposite upward.   Leaf blades lanceolate, 6-12 cm long.   Flowers
   usually solitary or few on recurved pedicels.   Perianth segments fused below, strongly recurved above, lanceolate, 6-10
   cm long.   Flowers a strikingly beautiful orange-red with purple spots.   Infrequent to rare in moist meadows and openings
   of woods in the eastern part.   Ours ssp. *michiganense* (Farw.)
   Boivin & Cody.   June-July.   Turk's Cap Lily . . . .
   . . . . *Lilium canadense* L.

MAIANTHEMUM Weber

Maianthemum canadense Desf. Perennial herbs from creeping rhizomes, the plants low, not over 2 dm tall. Leaves 3-6 cm long,
usually two, the blades ovate to cordate with prominent nerves.
Inflorescence a short raceme of 5-30 flowers. Perianth segments
4, alike and spreading, white or nearly so, about 5 mm across.
Frequent in rich woods in the northeast and in the Black Hills.
Ours var. *interis* Fern. June-July. Wild Lily of the Valley.

ORNITHOGALUM L.

Ornithogalum umbellatum L. Perennial herb from a scaly bulb, the
scape erect, 2-4 dm tall. Leaves mostly basal, erect, 2-5 mm
wide, dark green with a lighter midrib. Flowers in terminal
corymbs, erect, the perianth parts scarcely differentiated.
Petals white or greenish white, 1-2 cm long. Sepals greenish
with white margins. Rare on roadsides and persisting at abandoned dwellings in the eastern part. May-June. Star of Beth-
lehem.

POLYGONATUM Mill.

Polygonatum biflorum (Walt.) Ell. Perennial herbs from creeping,
horizontal rhizomes, the stems simple, arching, 5-9 dm tall.
Leaves alternate, almost sessile, the blades ovate to oblong,
6-12 cm long. Flowers in drooping axillary clusters, the peri-
anth up to 2 cm long, the segments forming a greenish-white tube.
Fruiting berry blue-black, 6-8 mm across, with several seeds.
Frequent in rich woods over the state. June-July. Solomon's
Seal.

SMILACINA Desf.

1 Inflorescence a many flowered panicle. Stems arched ascending, 6-8 dm tall, from horizontal rhizomes, mostly unbranched.
   Leaves elliptic, sessile, 7-13 cm long, finely pubescent on
   the under surface. Flowers numerous, on short pedicels, the
   panicle 4-12 cm long. Perianth parts 1-2 mm long, inconspicuous, the stamens more prominent. Fruiting berries maturing
   reddish, 6-7 mm across, with several seeds. Infrequent in
   rich, loamy woods in the eastern part and in the Black Hills.
   May-June. False Solomon's Seal . . . .
   . . . . *Smilacina racemosa* (L.) Desf.

1' Inflorescence a few-flowered raceme. Stems mostly zig-zag,
   erect, 2-6 dm tall, from horizontal rhizomes. Leaves sessile,
   somewhat folded, broadly lanceolate, 4-8 cm long. Flowers
   several in a terminal raceme, the perianth white, spreading.

Perianth segments 4-6 mm long, starlike. Fruiting berries
yellow-green with black stripes, 8-9 mm across, with several
seeds. Common in woods and thickets over the state. May-
June. Spikenard . . . . *Smilacina stellata* (L.) Desf.

SMILAX L.

1 Stems herbaceous, without thorns. Flowers ill-scented.

 2 Flowering peduncles from the axils of foliage leaves. Tendrils usually present. Stems usually climbing, up to 3
  meters tall. Inflorescence often with more than 25 flowers. Leaves alternate, 4-8 cm long, petioled, the blades
  ovate with entire margins. Perianth poorly differentiated,
  green. Fruiting berries blue or black, 6-9 mm in diameter,
  with 2-5 seeds. Frequent to common in woods over the state.
  Ours var. *lasioneuron* (Small) Rydb. June-July. Carrion
  Flower . . . . *Smilax herbacea* L.

 2' Flowering peduncles in the axils of bracts borne below the
  foliage leaves, in some instances a few in the upper leaf
  axils. Tendrils usually lacking. Stems erect, not climbing, up to 6 dm tall. Inflorescence usually with less than
  25 flowers. Leaves broadly elliptic, glaucous-pubescent on
  the ventral surface. Flowers green, polygamo-dioecious.
  Fruiting berries globose, 9-11 mm across, blue-black. Rare
  in rich woods in the southeast part. May-July . . . . . .
  . . . . *Smilax ecirrhata* (Engelm.) Wats.

1' Stems woody, at least at the base. The older wood green-
 barked and thorny. Plants twining or at least climbing, often
 3-4 meters high. Leaves thin, ovate, the blades 4-8 cm long,
 dark green. Inflorescence of 10-25 flowers in short, umbellate clusters at axillary positions. Perianth weakly differentiated, the segments greenish-white, 4-6 mm long. Plants
 dioecious. Fruiting berries dark blue or black, about 6 mm
 across. Infrequent in thickets and rich woods in the southeastern part. May-June. Greenbrier . . . .
 . . . . *Smilax hispida* Muhl.

STREPTOPUS Michx.

Streptopus amplexifolius (L.) DC. Perennial herbs from creeping
rhizomes, the stems branched and appearing twisted, 5-10 dm tall.
Leaves alternate, ovate with acuminate tips, 4-10 cm long, the
bases sessile and clasping the stem. Flowers mostly solitary in
the axils of leaves, the peduncles 1-2 cm long. Perianth greenish white, poorly differentiated, the segments nearly separate,
1-1.5 cm long. Fruiting berries red or yellow, many-seeded,

1-1.5 cm across. Frequent in rich, shady woods in the Black
Hills. June-July. Twisted-stalk.

TRILLIUM L.

1 Stem 2-4 dm tall at anthesis. Leaves broadly rhombic, 8-15 cm
   long. Ovary 6-angled.

   2 Staminal filaments more than one-half as long as the
      anthers. Flowering peduncle 1-4 cm long, sharply reflexed
      and recurved below the leaves. Leaves broadly rhombic, 6-
      10 cm long. Flower solitary, the petals white, 1.5-2.5 cm
      long. Sepals equaling the petals. Ovary white or pink,
      sharply 6-angled. Rare in rich woods in the northeast
      part. May-June. Nodding Trillium . . . .
      . . . . *Trillium cernuum* L.

   2' Staminal filaments less than one-half as long as the
      anthers. Flowering peduncle 4-12 cm long, horizontal to
      somewhat recurved. Leaves broadly rhombic, 8-15 cm long.
      Flower solitary, the petals white with obtuse tips, spread-
      ing. Sepals lanceolate, equaling the petals. Ovary white
      to pink, sharply 6-angled. Rare in rich woods of the north-
      east part. {*T. flexipes* Raf.} May-June . . . .
      . . . . *Trillium gleasoni* Fern.

1' Stem less than 1 dm tall at anthesis. Leaves ovate, 3-5 cm
   long. Ovary scarcely 3 lobed. Plants perennial, from fleshy
   rhizomes, the stems hardly reaching above soil level. Leaves
   3, in a single whorl, dark green. Flower usually solitary, on
   an erect or arching peduncle 1-3 cm long. Petals oblong with
   obtuse tips, white, 2-4 cm long. Rare in rich, loamy woods in
   the eastern part. Apr-May. Dwarf White Trillium . . . . . .
   . . . . *Trillium nivale* Riddell

UVULARIA L.

1 Leaves distinctly perfoliate, broadly ovate to oblong, up to
   12 cm long. Perianth segments 3-5 cm long. Plants perennial
   from creeping rhizomes, the stems leafy, branched above, 2-5
   dm tall. Leaves about 9 cm long, scarcely puberulent beneath.
   Flowers 1-3, the perianth parts yellow, nodding and twisted.
   Fruiting capsules obtusely 3-angled. Infrequent in rich woods
   in the eastern part. May-June. Bellwort . . . .
   . . . . *Uvularia grandiflora* Sm.

1' Leaves sessile but not perfoliate, elliptic-oblong, about 8 cm
   long, glabrous. Perianth segments 1.5-2.5 cm long. Plants
   perennial from rhizomes, the stems usually forked once above,
   1-3 dm tall. Leaves glaucous or nearly so, with acute tips.

Flowers 1 or 2 on terminal branches but appearing axillary, nodding, the perianth parts yellow. Apparently rare in rich woods in the eastern part and now may be extinct. May. Small Bellwort . . . . *Uvularia sessilifolia* L.

YUCCA L.

Yucca glauca Nutt. Perennial herbs with deep woody taproots, the stems short. Leaves stiff, pointed, arranged in a dense rosette and ascending-spreading, 2-6 dm long. Inflorescence a stout, elongate, paniculate raceme, the flowering stem often 1-1.5 meters tall. Flowers greenish-white, not well differentiated, mostly separate, 3-5 cm long. Fruiting capsules woody, 3-angled, 4-6 cm long. Frequent to common on dry prairie knolls and exposed ridges over the state except not in the northeast. June-July. Soapweed.

ZIGADENUS Michx.

1   Perianth parts 8-11 mm long, not obviously clawed. The gland on the inside of the petal bilobed or obovate. Stamens perigynous with the ovary half inferior. Plants 2-7 dm tall, from a deep seated bulb. Leaves mostly basal, grass-like. Inflorescence racemose, from few to many flowered. Flowers green to white, the segments ovate and distinct, equal in length. Frequent in moist meadows of prairie in the eastern part and to the southwestern part. June-July. White Camas . . . . . .
. . . . *Zigadenus elegans* Pursh

1' Perianth parts 4.5-8 mm long. Glands on the petals not well-defined. Stamens hypogynous, the ovary superior.

2   Inner whorl of perianth parts clawed, the gland not well-defined. Staminal filaments fused to the perianth near their bases. Plants perennial, the scapes 2-5 dm tall. Leaves narrow, elongate. Inflorescence racemose to paniculate, the perianth parts yellow-green, 3-5 mm long. Infrequent to locally frequent on dry slopes of the western part. A potentially poisonous plant for range animals. Ours var. *gramineus* (Rydb.) Walsh. May-June. Death Camas.
. . . . *Zigadenus venenosus* Wats.

2' Inner whorl of perianth parts not clawed, the gland tending to be obovate. Staminal filaments free from the perianth near their bases. Plants perennial, 3-7 dm tall. Leaves 1-5 dm long, arching. Inflorescence racemose to paniculate, the perianth yellow-white, 6-8 mm long. Rare to infrequent in dry prairies in the extreme southern part. May-June . .
. . . . *Zigadenus nuttallii* S. Wats.

Family Amaryllidaceae

HYPOXIS L.

Hypoxis hirsuta (L.) Cov. Perennial herbs from corms, the flowering scapes 1-5 dm tall at anthesis. Leaves mostly basal, grass-
like, overtopping the flowers. Inflorescence an irregular umbel
of 2-5 flowers, the pedicels elongate and unequal. Perianth
poorly differentiated, yellow, the segments spreading, 1.5-2.5
cm across. Fruiting capsule 3-6 mm long, with black, roughened
seeds. Rare in low meadows in the eastern part and in the southern sand hills area. Apr-May. Star Grass.

Family Iridaceae

(Key to Genera)

1 Leaves more than 1 cm wide. Flowers large, more than 2 cm
   wide or long.

   2 Flowers light blue to purple. Style branches broad, concealing the stamens . . . . *IRIS*

   2' Flowers orange with mottled purple. Style branches thread-
   like, not concealing the stamens . . . . . *BELAMCANDA*

1' Leaves less than 1 cm wide. Flowers less than 2 cm wide or
   long . . . . *SISYRINCHIUM*

BELAMCANDA Adans.

Belamcanda chinensis (L.) DC. Perennial herbs from wide-spreading
horizontal rhizomes, the stems 3-5 dm tall. Leaves erect to arching, sword-shaped, 2-4 cm wide. Inflorescence paniculate to cymose, much branched, the flowers at the ends of branches. Perianth 3-5 cm wide, bright orange with purple mottling. Fruiting
capsule up to 3 cm long, dehiscing to expose black fleshy seeds
attached to a column resembling a large blackberry. Infrequent
as an escape in the southeast part. July. Blackberry Lily.

IRIS L.

Iris missouriensis Nutt. Perennial herbs from thick, horizontal
rhizomes, the flowering scapes 2-6 dm tall. Leaves mostly basal,
erect and fleshy, 1-3 cm wide. Inflorescence 1-4 flowered, the
flowers in a 1 or 2 valved spathe. Outer 3 sepals spreading,
lavender to blue, the inner 3 petals shorter, erect or nearly so,
usually of a lighter color. Each of the perianth segments clawed.

Frequent in open meadows and prairie in the Black Hills. June-
July. Blue Flag.

SISYRINCHIUM L.

1 Outer bract of the inflorescence with margins free to the base.
   Scape narrowly winged, 1-2 mm wide. Perianth white to pale
   blue. Plants pale green, 1-4 dm tall, tufted from fibrous
   roots. Leaves grassy, basal, 1-3 mm wide. Flowers 1-3 in an
   umbel-like cluster from a 2-valved spathe, the perianth radially symmetrical, the segments 8-10 mm long. Fruiting capsule
   orbicular, 2-3 mm in diameter. Frequent in prairies in the
   eastern part. Represented in South Dakota by the two varieties *campestre* and *kansanum* Bickn. May-June. White-eyed
   Grass . . . . *Sisyrinchium campestre* Bickn.

1' Outer bract of the inflorescence with margins united 2-5 mm
   above the base. Scape broadly winged, 3-4 mm wide. Perianth
   pale to deep purple.

   2 Plants bright green, drying dark. Fruit dark. Scapes 1-5
      dm tall, from fibrous roots. Leaves several, 2-4 mm wide.
      Inflorescence an umbellate cluster of 1-4 flowers, appear-
      ing terminal from the 2-valved spathe. Flowers radially
      symmetrical, the perianth segments 8-11 mm long. Fruits
      globose to obovoid, 4-6 mm long, dark. Frequent in prai-
      ries and openings of woods over the state. May-June. Blue-
      eyed Grass . . . . . . *Sisyrinchium angustifolium* Miller

   2' Plants light green, tending to be glaucescent, drying to a
      pale color. Fruit pale green to straw-colored. Scapes 1-5
      dm tall, clustered from fibrous roots. Leaves several, 2-4
      mm wide. Inflorescence an umbel of bright, violet flowers,
      the spathe light or tinged with purple. Fruits 4-6 mm long,
      globose to obovoid, pale green. Frequent in meadows and
      openings of woods over the state. Certainly not easily
      distinguished from the preceding species. May-June . . . .
      . . . . *Sisyrinchium montanum* Greene

Family Orchidaceae

(Key to Genera)

1 Plants lacking chlorophyll. Stem arising from a coral-like
   cluster of rhizomes. Leaves reduced to scales . . . .
   . . . . *CORALLORHIZA*

1' Plants with green color in at least some of their parts.

2   Lower lip of corolla a large inflated sac at least 1.5 cm
    long.

    3   Stem from a corm.   Leaf 1 . . . . *CALYPSO*

    3' Stem from coarse fibrous roots.   Leaves 2 . . . .
       . . . . *CYPRIPEDIUM*

2' Lower lip, if sac-like, small, not over 1 cm long.

    4   Spur conspicuous at the base of the lip.   Leaves cauline,
        3 or more   . . . . *HABENARIA*

    4' Spur lacking at the base of the lip.

        5   Foliage leaves basal or with a single pair of oppo-
            site leaves.

            6   Leaves basal, spreading in a rosette, with whitish
                veins.   Cauline leaves reduced to scales . . . .
                . . . . *GOODYERA*

            6' Leaves two, opposite or nearly so, at the middle
                part of the stem . . . . *LISTERA*

        5' Foliage leaves basal and cauline, linear to ovate,
            the cauline leaves not reduced to scales.

            7   Flowers 1.5-3 cm long, green-purple.   Inflores-
                cence few flowered, not congested . . . .
                . . . . *EPIPACTIS*

            7' Flowers less than 1.5 cm long, white or nearly so.
                Inflorescence spike-like or twisted, the flowers
                congested . . . . *SPIRANTHES*

CALYPSO Salisb.

Calypso bulbosa (L.) Oakes.   Low growing perennial from a fleshy
corm, the scape 1-2 dm tall.   Leaf 1, basal, 2-4 cm long, ovate
to cordate, petioled.   Flower usually solitary and subtended by
a single linear bract.   Sepals and petals similar and of approximate equal length, 1-2 cm long, purple or lavender.   Sac about
2 cm long, yellow-brown, spotted with purple, and with several
rows of hairs.   Rare in rich woods of the northern Black Hills.
June-July.   Venus' Slipper.

CORALLORHIZA Chat.

1   Sepals and petals 1 cm long or longer, with 3-5 prominent
    purple stripes.   Stems often lavender colored, 2-5 dm tall.
    Flowers 8-15, in racemes.   Petals and sepals arching, yellow-
    white, the lip not spurred or lobed.   Fruiting capsule

ellipsoid, 1-2 cm long. Infrequent under pines in the northern Black Hills. June-July. Striped Coral-root . . . .
. . . . *Corallorhiza striata* Lindl.

1' Sepals and petals less than 1 cm long, without prominent
   stripes.

   2 The enlarged lower petal, the lip, laterally lobed or
      toothed.

      3 Flowers yellow green or yellowish. The lip only toothed
         near the base. Sepals 5 mm long. Plants 1-2.5 dm tall,
         the stems slender, from a creeping rhizome. Inflores-
         cence a 6-15 flowered raceme, the flowers rarely purple-
         tinged. Petals and sepals 1-nerved, the lip white or
         nearly so. Infrequent to rare in rich woods of the
         Black Hills. June-July. Pale Coral-root . . . .
         . . . . *Corallorhiza trifida* Chat.

      3' Flowers white or pink with purple spots. The lip with a
         pair of prominent lateral lobes. Sepals more than 6 mm
         long, usually 3-nerved. Plants 2-5 dm tall, often
         tinged with purple. Inflorescence a 10-25 flowered
         raceme about 10 cm long. Flowers 8-12 mm long, the
         lateral sepals diverging and 3-nerved. Fruiting cap-
         sule 1.5-2 cm long, tapered. Frequent in woods in the
         Black Hills. June-July. Spotted Coral-root . . . . . .
         . . . . *Corallorhiza maculata* Raf.

   2' The enlarged lower petal, the lip, entire or wavy but not
      lobed or toothed.

      4 Lip 3-5 mm long, broadly ovate with an erose margin,
         often with two purple spots. Lateral petals 3-5 mm
         long. Plants flowering in late summer. Stems 1-2 dm
         tall, purple or brown. Inflorescence 3-5 cm long, of
         4-12 flowers. Fruiting capsules 6-7 mm long. Rare to
         infrequent in rich woods of the Black Hills. Aug-Sept.
         Late Coral-root . . . .
         . . . . *Corallorhiza odontorhiza* (Willd.) Nutt.

      4' Lip 5-7 mm long, irregularly notched on the upturned
         margin with an indentation at the apex, white, or with
         purple spots. Lateral petals 7-8 mm long. Plants
         flowering in spring. Stems 2-4 dm tall, pink to lav-
         ender. Inflorescence a few-flowered raceme 2-6 cm
         long. Lateral sepals and petals slightly flared or
         spreading, 2-3 mm wide. Fruiting capsules 9-11 mm
         long. Infrequent in rich woods of the Black Hills.
         Apr-May. Wister's Coral-root . . . .
         . . . . *Corallorhiza wisteriana* Conrad

CYPRIPEDIUM L.

1 Lip yellow, 2-3 cm long, strongly pouched. Stems 2-6 dm tall,
   with 3-5 broad, ovate leaves. Flower usually solitary, with a
   single, leaf-like bract. Lateral petals and sepals green-
   yellow to mottled purple, wavy-margined, 2-4 cm long. Frequent in low moist woods and valleys in the Black Hills and
   in the northeast part. Represented in South Dakota by the
   two varieties *parviflorum* (Salisb.) Fern. and *pubescens*
   (Willd.) Correll. May-June. Yellow Lady Slipper . . . . . .
   . . . . *Cypripedium calceolus* L.

1' Lip white with pale lavender veins, 2-2.5 cm long, deeply
   pouched. Stems 1-4 dm tall, with several broadly lanceolate
   leaves, the upper one as high as the solitary flower. Lateral
   petals and sepals lanceolate, 2-4 cm long, greenish, often
   with pale pink or lavender veins, twisted. Previously collected in wet meadows in the extreme eastern part in Clay,
   Brookings and Minnehaha counties. It now may be extinct.
   May-June. White Lady Slipper . . . *Cypripedium candidum* Muhl.

EPIPACTIS Siv.

Epipactis gigantea Dougl. Perennial, single stemmed plant, 3-6
dm tall, from creeping rhizomes. Leaves alternate, their bases
sheathing the stem, with blades broadly lanceolate, 7-12 cm long.
Flowers appearing solitary in the axils of upper leaves but the
inflorescence is a raceme with large, leaf-like bracts. Petals
and sepals distinguishable, the former pale pink or rose colored,
the latter green to purple. Lip 1-2 cm long, yellowish with red
or purple. Rare in the western part but locally abundant in valleys near streams in Fall River County. June-July. Helleborine.

GOODYERA R. Br.

1 Upper sepal and lateral petal fused to form a galea that is
   4-5 mm long and arches over the lip. Lip sac-like below.
   Leaves not with a white midvein. Plants 1-2 dm tall, with
   solitary stems. Leaves dark green, ovate, 2-3 cm long, borne
   near the base. Flowers several to many in a one-sided raceme
   about 4 cm long, greenish-white to pale pink. Lip about 3 mm
   long, with a deep, saccate base. Frequent in rich woods of
   the Black Hills. July-Aug . . . . *Goodyera repens* (L.) R. Br.

1' Upper sepal and lateral petal, the galea, 8 mm or more long.
   Lip not saccate. Leaves with whitened veins, especially the
   midvein. Plants 2-4 dm tall, from a rhizome. Leaves mostly
   basal, broadly ovate, 4-6 cm long. Flowering raceme 6-10 cm
   long, bracted, with several to many flowers tending to be

one-sided on the raceme. Flowers green to white, the lip 5-8
mm long, not deeply saccate. Rare at higher altitudes in rich
woods of the Black Hills. {*G. decipiens* of authors} June-
July. Rattlesnake Plantain . . . . *Goodyera oblongifolia* Raf.

HABENARIA Willd.

1 Principal leaves restricted to the basal part of the stem.
Leaves upward on the stem reduced to scales or bracts.

2 Spur 2-5 mm long, cylindrical. Flowering pedicels very
short or lacking, the flowers appearing sessile. Basal
leaves lanceolate, beginning to wither at flowering time.
Plants 2-5 dm tall, the inflorescence a spike-like raceme.
Flowers small, greenish-white, numerous, spirally arranged.
Lip 2-4 mm long, the sepals about 3 mm long. Rare in dry
woods of the Black Hills. July-Aug . . . .
Habenaria unalascensis (Spreng.) Wats.

2' Spur 15-25 mm long, curved and dilated at the distal end.
Flowering pedicels 5-10 mm long. Basal leaves 2, rounded
to obtuse in outline, not beginning to wither at flowering
time. Plants 3-5 dm tall, the inflorescence loosely 5-20
flowered. Petals white to greenish. Lip lance-linear,
1.5-2.0 cm long, arching out and downward. Rare in moist
woods of Lawrence County in the Black Hills. July-Aug.
Round-leaved Orchid . . . *Habenaria orbiculata* (Pursh) Torr.

1' Principal leaves not restricted to the basal part of the stem.

3 Lip 2-3 toothed or deeply lobed and fringed.

4 Flowers large and showy, whitish, the lip over 1 cm
long, with 3 deep lobes, these fringed. Plants of moist
meadows in the eastern part, the stems 4-8 dm tall.
Leaves lanceolate, 1-2 dm long, the lower ones larger.
Inflorescence a bracted spike 1-2 dm long, of many
flowers. Spur 2-5 mm long. Rare in moist meadows in
the extreme eastern part, now perhaps extinct. July.
Prairie Fringed Orchid . . . .
. . . . *Habenaria leucophaea* (Nutt.) Gray

4' Flowers greenish-white, the lip less than 1 cm long, not
deeply tri-lobed and fringed. Plants of the western
part.

5 Inflorescence with leaf-like bracts. Lip irregularly
2-3 toothed at the distal part. Spur 2-3 mm long,
pouched. Plants 2-5 dm tall, the stem thick and
leafy. Principal leaves on the lower part, lanceolate, 4-7 cm long. Flowers in loose, bracted racemes,

greenish, the upper sepal about 5 mm long. Lip 7-10
mm long, pendant. Frequent in moist or shaded woods
in the Black Hills. Ours var. *bracteata* (Muhl.) Gray.
June-July. Bracted Orchid . . . .
. . . . *Habenaria viridis* (L.) R. Br.

5' Inflorescence without leaf-like bracts. Lip entire
to toothed with small, lateral lobes and a small,
erect tubercle on the upper surface near the base.
Spur 3-6 mm long, not pouched. Stems 3-6 dm tall.
Leaves lanceolate, 10-20 cm long, not as large upward on the stem. Flowers in loose to compact racemes, greenish-yellow. Lip about 5 mm long, with
an irregular distal margin. Rare in moist wooded
ravines in the western part. Aug. Tubercled Orchid.
. . . . *Habenaria flava* (L.) R. Br.

3' Lip entire or nearly so, not lobed or fringed.

6 Flowers white. Base of lip broadened and tapering to
the tip. Lip pendant, 5-8 mm long. Plants stout, 4-8
dm tall, leafy. Principal leaves lanceolate, 6-10 cm
long, becoming smaller upward. Flowers in dense, race-
most spikes 8-20 cm long, the bracts varying from shorter
to longer than the flowers. Upper sepal 3-nerved and 5
mm long. Rare in low woods in the Black Hills. June-
July. White Orchid . . . .
. . . . *Habenaria dilatata* (Pursh) Hook.

6' Flowers greenish. Lip narrow, not broadened at the base.

7 Inflorescence loosely flowered. Spur of the flower
enlarged near the base, saccate. Plants 4-8 dm tall,
glabrous, leafy. Leaves lance-ovate, 4-9 cm long,
the principal ones at the middle of the stem. Inflorescence racemose, open, with ascending bracts.
Flowers greenish, the upper sepal broadly ovate, 3-
nerved. Lip about 6 mm long, narrow, entire. Infrequent to rare in rich woods or moist places in woods
of the Black Hills. June-July . . . .
. . . . *Habenaria saccata* Greene

7' Inflorescence congested and densely flowered. Spur
of the flower narrow, not enlarged at the lower end.
Plants 2-8 dm tall, glabrous, leafy below. Leaves
lanceolate, 3-8 cm long. Inflorescence a racemose
spike about 10 cm long, the bracts at the lower part
of the spike exceeding the flowers. The upper sepal
fused to the upper petals and forming a small hood.
Lip narrow, 5-6 mm long, the spur usually shorter

than the lip. Infrequent in low, wet woods of the
eastern part and in the Black Hills. June-July . . .
. . . . *Habenaria hyperborea* (L.) R. Br.

LISTERA R. Br.

Listera convallarioides (Sw.) Nutt. Small perennial herb with a
single pair of leaves halfway up the stem, from fibrous roots.
Plants 1-3 dm tall. Leaves broadly ovate with acute apexes, 4-6
cm long, opposite or nearly so. Inflorescence a loose-flowered
raceme 3-6 cm long. Flowers green-yellow, the sepals and petals
narrow, 1-nerved. Lip broadly obovate, 8-11 mm long, with an
apical notch. Rare in moist, shaded woods in the Black Hills.
June-July. Twayblade.

SPIRANTHES Rich.

1 Flowers in one rank on the spike, appearing spiralled due to
   twisting of the main rachis. Plants perennial, the stems 1-5
   dm tall, from fibrous roots. Leaves narrow, usually less than
   1 cm wide, extending up the stem, but becoming shorter upwards
   on the stem. Inflorescence 6-12 cm long, the flowers greenish-
   white to cream-colored. Petals 8-10 mm long, ovate and slightly pubescent beneath. Infrequent in mesic prairie and swales
   in the southeast part. Aug-Sept. Twisted Ladies' Tresses . .
   . . . . *Spiranthes vernalis* Engelm. & Gray

1' Flowers in two or more ranks or spirals on the spike.

   2 Paired protuberances at the base of the lip 0.5 mm long,
      projecting backwards. Lip oblong in outline, erose at the
      end. Plants 6-9 dm tall, glabrous and erect. Leaves most-
      ly basal, narrow, 15-20 cm long. Inflorescence densely
      flowered, spicate, usually 3-ranked. Flowers white, the
      lip 7-12 mm long. Frequent in low places of meadows and
      on floodplains in the eastern part. Aug-Sept. Ladies'
      Tresses . . . . *Spiranthes cernua* (L.) Rich

   2' Paired protuberances at the base obscure. Lip ovate in
      outline, strongly decurved, slightly constricted near the
      middle. Plants stout, 1-4 dm tall, glabrous. Leaves nar-
      rowly spatulate, mostly basal, the blades 10-20 cm long.
      Inflorescence 4-8 cm long, densely flowered, 3-ranked.
      Flowers creamy-white, the lip 8-12 mm long. Upper sepal
      and petals forming a curved hood 7-10 mm long. Frequent
      in meadows at higher altitudes in the Black Hills. July-
      Aug. Hooded Ladies' Tresses . . . .
      . . . . *Spiranthes romanzoffiana* Cham.

GROUP IV. DICOTYLEDONEAE

DICOTYLEDONS--Flower parts in 4's or 5's or multiples thereof.
Embryo with 2 seed leaves, opposite. Leaves usually net-veined.

1 Plants woody. Trees, shrubs or woody vines.

    2 Shrubs or woody, climbing vines . . . .
    . . . . Section 1.   page 145

    2' Trees with single, woody stems at least 1-2 meters tall.
    . . . . Section 2.   page 150

1' Plants herbaceous. Stem bases may be woody but obvious woody
stems and winter buds not present.

    3 Aquatic plants with stems submerged and rooted to the bot-
    tom or floating. Occasionally rooted at the margins of
    dessicating ponds or streams . . . .
    . . . . Section 3.   page 152

3' Terrestrial herbaceous plants.

    4 Perianth lacking or of a single series, the parts alike,
    a differentiated corolla not present (Apetalae) . . . .
    . . . . Section 4.   page 153

    4' Perianth present and differentiated into an outer series
    of sepals and an inner series of petals.

        5 Petals separate or united only at the very base
        (Polypetalae).

            6 Pistils several to many, of simple separate
            carpels . . . . . . Section 5.   page 157

            6' Pistil one or several carpels fused into a com-
            pound ovary.

                7 Ovary completely superior.

                    8 Corolla radially symmetrical (regular) . . .
                    . . . . Section 6.   page 157

                    8' Corolla bilaterally symmetrical or somewhat
                    irregular . . . Section 7.   page 161

                7' Ovary inferior or partly so, the lower part
                fused to other tissues . . . .
                . . . . Section 8.   page 162

        5' Petals united above the base (Gamopetalae).

            9 Ovary completely superior.

Section 1. Dicotyledonous Shrubs or Woody, Climbing Vines

1 Flowers appearing before the leaves in the spring.

   2 Flowers, at least the staminate ones, in catkins, without petals.

      3 Fruit a syncarp. Plants monoecious or dioecious . . . .
. . . . Moraceae   page 178

      3' Fruit a nut or capsule.

         4 Plants monoecious. Fruit 1 seeded nuts . . . .
. . . . Betulaceae   page 174

         4' Plants dioecious. Fruit many seeded capsules . . . .
. . . . Salicaceae   page 168

   2' Flowers not in catkins, with or without petals.

      5 Branches with short, sharp spines. Flowers in small clusters. Plants dioecious . . . .
. . . . Rutaceae   page 298

      5' Branches without spines or prickles.

         6 Corolla bilaterally symmetrical, papilionaceous . . .
. . . . Fabaceae   page 268

         6' Corolla radially symmetrical.

            7 Flowers white, petals exceeding 5 mm . . . .
. . . . Rosaceae   page 254

            7' Flowers yellow, petals 2-3 mm long . . . .
. . . . Anacardiaceae   page 305

1' Flowers appearing with or after the leaves in the spring or summer.

   8 Leaves imbricate-overlapping, scale-like, 1 mm long or shorter. Flowers rose-pink, in dense panicles . . . . . .
. . . . Tamaricaceae   page 314

   8' Leaves not imbricate-overlapping, longer than 1 mm.

9   Leaves opposite.

   10   Leaf blades simple.

      11   Margins of leaves entire.

         12   Leaves covered with silvery scales. Flowers
            axillary . . . Elaeagnaceae   page 320

         12' Leaves lacking silvery scales.

            13   Ovary superior.

               14   Flowers showy, lavender or pink . .
                 . . . . Oleaceae   page 346

               14' Flowers not showy, green or white .
                 . . . . Rhamnaceae   page 308

            13' Ovary inferior.

               15   Petals separate. Stamens 4 . . . .
                 . . . . Cornaceae   page 340

               15' Petals united. Stamens 5 . . . . .
                 . . . . Caprifoliaceae   page 397

      11' Margins of leaves toothed or lobed.

         16   Fruit deeply 3-5 lobed. Young branches
            green . . . . . Celastraceae   page 306

         16' Fruit not deeply lobed. Branches variously
            colored but not dark green.

            17   Ovary superior. Fruit a drupe with 2-4
               nutlets . . Rhamnaceae   page 308

            17' Ovary inferior. Fruit a 1-seeded drupe
               or capsule . . . .
               . . . . . . Caprifoliaceae   page 397

   10' Leaf blades compound.

      18   Leaflets 3 . . . . Ranunculaceae   page 216

      18' Leaflets 5 or more.

         19   Corolla large, red-orange . . . .
            . . . . Bignoniaceae   page 392

         19' Corolla small, white . . . .
            . . . . Caprifoliaceae   page 397

9' Leaves alternate.

   20   Leaf blades simple.

21 Margins of leaves entire.

   22 Branches with spines.

      23 Flowers lavender. Leaves more than 3 cm
         long . . . Solanaceae   page 378

      23' Flowers yellow. Leaves 2-3 cm long . .
         . . . . . . Berberidaceae   page 227

   22' Branches without spines.

      24 Stems trailing. Leaves evergreen . . .
         . . . . . . Ericaceae   page 341

      24' Stems erect or twining.

         25 Main stems twining. Leaves deeply
            lobed. Berries red . . . .
            . . . . Solanaceae   page 378

         25' Main stems erect or nearly so.

            26 Fruit a drupe. Flowers axillary
              . . Rhamnaceae   page 308

            26' Fruit a capsule or several
              seeded berry.

               27 Plants dioecious. Fruit a
                  capsule . . . .
                  Salicaceae   page 168

               27' Plants not dioecious. Fruit
                  a capsule or berry . . . . .
                  Ericaceae   page 341

21' Margins of leaves lobed or toothed.

   28 Leaves lobed.

      29 Leaves palmately veined and lobed.

         30 Branches twining or climbing.

            31 Branches with tendrils. Plants
              monoecious . . . .
              . . Vitaceae   page 309

            31' Branches without tendrils.
              Plants dioecious . . . .
              . . Menispermaceae   page 228

        30' Branches not twining or climbing.

            32 Flowers imperfect. Pistillate
              flowers maturing into a multiple

fruit (mulberry) . . . .
. . . . Moraceae   page 178

32' Flowers perfect.

    33 Stamens many. Ovary usually
       superior. Branches not spiny.
       . . Rosaceae   page 254

    33' Stamens 5. Ovary inferior.
       Branches often spiny . . . . . .
       . . Saxifragaceae   page 249

29' Leaves pinnately veined or lobed. The lobes
toothed . . . . Rosaceae   page 254

28' Leaves toothed but not lobed.

    34 Stems spiny. Teeth of leaf margins with
       spinulose tips . . . .
       . . . . Berberidaceae   page 227

    34' Stems not spiny. Teeth without spinulose
       tips.

       35 Leaf base not symmetrical . . . . . .
          . . . . . Betulaceae   page 174

       35' Leaf bases symmetrical.

          36 Flowers, at least the staminate ones,
             in catkins.

             37 Fruit a capsule with comose
                seeds. Plants dioecious . . . .
                . . Salicaceae   page 168

             37' Fruit not a capsule. Plants
                monoecious.

                38 Fruit a 1 seeded nut . . . .
                   Betulaceae   page 174

                38' Fruit a juicy multiple
                  fruit . . . .
                  Moraceae   page 178

          36' Flowers not in catkins.

             39 Anthers opening with apical
                pores . . . .
                . . Ericaceae   page 341

             39' Anthers not opening with apical
                pores.

40 Stems twining or trailing .
    Celastraceae   page 306

40' Stems erect or spreading . .
    Rhamnaceae   page 308

20' Leaf blades compound.

   41 Leaflets 3.

      42 Fruit a drupe . . . .
      . . . . Anacardiaceae   page 305

      42' Fruits achenes, fleshy drupelets, or hips
      . . . . Rosaceae   page 254

   41' Leaflets 5 or more.

      43 Leaves palmately compound.

         44 Stamens 5. Tendrils present. Stems not
         prickly . . Vitaceae   page 309

         44' Stamens many. Tendrils absent. Stems
         prickly . . Rosaceae   page 254

      43' Leaves pinnately compound.

         45 Leaflets toothed or dentate.

            46 Teeth of leaflets spinose-tipped . .
            . . . . Berberidaceae   page 227

            46' Teeth of leaflets not spinose-tipped.

               47 Flowers greenish. Leaves with-
               out stipules . . . .
               . . Anacardiaceae   page 305

               47' Flowers not green. Leaves usu-
               ally stipulate . . . .
               . . Rosaceae   page 254

         45' Leaflets entire or remotely dentate.

            48 Surface of leaflet with resinous
            glands, aromatic. Branches with
            stout, stipular spines . . . .
            . . . . Rutaceae   page 298

            48' Surface of leaflets without resinous
            glands. Branches without stipular
            spines . . . .
            . . . . Fabaceae   page 268

Section 2. Trees with a Single or Dominant
Woody Stem at Least 1-2 Meters Tall

1 Leaves simple.

    2 Leaves opposite or whorled on the branch.

        3 Margins of leaves entire.

            4 Leaves heart-shaped, in whorls of 3 at a node. Leaf
               surface softly pubescent . . . .
           . . . . Bignoniaceae   page 392

            4' Leaves spatulate to ovate, opposite or nearly so.
               Leaf surface with a silvery-scaly pubescence . . . .
           . . . . Elaeagnaceae   page 320

        3' Margins of leaves palmately lobed and veined . . . . . .
       . . . . Aceraceae   page 307

    2' Leaves alternate on the branch.

        5 Margins of leaves entire.

            6 Leaves with a silvery-scaly pubescence . . . .
           . . . . Elaeagnaceae   page 320

            6' Leaves glabrous to pubescent but not silvery-scaly.

               7 Twigs with sharp spines . . . .
             . . . . Moraceae   page 178

               7' Twigs lacking sharp spines.

                  8 Leaves palmately veined . . . .
               . . . . Fabaceae   page 268

                  8' Leaves pinnately veined . . . .
               . . . . Salicaceae   page 168

      5' Margins of leaves toothed or lobed.

        9 Leaves toothed but not lobed.

           10 Flowers in catkins.

              11 Fruit a capsule with hairy seeds. Leaves
                 narrow or lanceolate . . . .
             . . . . Salicaceae   page 168

              11' Fruit not a capsule. Leaves broader, ovate.

                 12 Plants monoecious. Fruit a 1-seeded nut
             . . . . . . Betulaceae   page 174

                 12' Plants dioecious. Fruit multiple, juicy
             . . . . . . Moraceae   page 178

10' Flowers not in catkins.

    13 Leaf bases obliquely asymmetrical.

        14 Leaves heart-shaped with a cordate base. Fruit a drupe . . . .
. . . . . **Tiliaceae**   page 310

        14' Leaves ovate to elliptic, the bases tapering. Fruit a samara . . . .
. . . . . **Ulmaceae**   page 176

    13' Leaf bases symmetrical . . . .
. . . . **Rosaceae**   page 254

  9' Leaves lobed.

    15 Leaf blades pinnately veined.

        16 Flowers green, the staminate flowers in catkins. Plants monoecious . . . .
. . . . **Fagaceae**   page 176

        16' Flowers not green, mostly showy, perfect . .
. . . . **Rosaceae**   page 254

    15' Leaf blades palmately veined.

        17 Leaves densely hairy beneath . . . .
. . . . **Salicaceae**   page 168

        17' Leaves not densely hairy beneath . . . . . .
. . . . **Moraceae**   page 178

1' Leaves compound.

  18 Leaves opposite on the branch.

    19 Leaflets 3-7, toothed. Fruiting samaras in pairs. Stamens 4 or more . . . . **Aceraceae**   page 307

    19' Leaflets 5-11, remotely toothed. Fruiting samaras solitary . . . . **Oleaceae**   page 346

  18' Leaves alternate on the branch.

    20 Leaflets entire or nearly so. Ovary 1-celled, superior, maturing as a pod or legume . . . .
. . . . **Fabaceae**   page 268

    20' Leaflets toothed.

        21 Plants monoecious. Flowers greenish. Staminate flowers in catkins. Pistillate flowers maturing as nuts . . . . . . **Juglandaceae**   page 174

21' Plants bisexual, flowers perfect. Flowers not
greenish, usually showy. Ovary maturing as a
fleshy pome . . . . . Rosaceae   page 254

Section 3. Aquatic Plants with Stems Submerged
and Rooted on the Bottom or Floating

1   Flowers or flower clusters at or above the water level.

  2   Leaf blades ovate to orbicular, floating, with a deeply
  cordate base or entire. Flowers usually solitary . . . . .
  . . . . Nymphaeaceae   page 215

  2' Leaf blades not large or orbicular.

    3   Leaves divided into filiform segments or deeply lobed
    and compound.

      4   Submerged leaves with small bladders. Flowers yellow,
      strongly bilabiate, on immersed scapes . . . .
      . . . . Lentibulariaceae   page 393

      4' Submerged leaves without bladders. Flowers not
      strongly bilabiate.

        5   Flowers in spikes or composite heads.

          6   Leaves opposite. Flowers in heads, usually
          showy . . . . . . Asteraceae   page 405

          6' Leaves whorled or verticillate. Flowers on
          immersed spikes, minute, not showy . . . . . .
          . . . . Haloragaceae   page 329

        5' Flowers solitary or in umbels. Leaves alternate.

          7   Pistils several. Flowers more than 5 mm across,
          solitary, yellow or white . . . .
          . . . . Ranunculaceae   page 216

          7' Pistils one in each flower. Flowers less than
          5 mm across, in umbels, white . . . .
          . . . . Apiaceae   page 331

    3' Leaves simple, entire or toothed.

      8   Flowers closely set in erect spikes, racemes or
      panicles.

        9   Ovary maturing into a pod or capsule. Petals 4;
        stamens 6 . . . . . . Brassicaceae   page 230

9' Ovary maturing into achenes or nutlets. Petals typically 5 or 5-lobed.

    10 Leaf bases with sheathing stipules at the nodes. Leaves alternate. Ovary maturing into a 1 seeded achene . . . . Polygonaceae   page 181

    10' Leaf bases lacking sheathing stipules. Leaves opposite. Ovary maturing into 1-4 nutlets . . . . Verbenaceae   page 368

8' Flowers solitary or clustered in axillary positions.

    11 Leaves whorled, narrow, 4-10 at a node. Flowers lacking a perianth . . . . Haloragaceae   page 329

    11' Leaves opposite.

        12 Flowers yellow, the petals fused. Leaves obovate rounded . . . . Scrophulariaceae   page 382

        12' Flowers not yellow, the petals, if present, not fused. Leaves linear to lance-shaped . . . . Lythraceae   page 321

1' Flowers or flowering parts usually submerged and not showy.

  13 Leaves dissected into harsh, filiform segments, whorled. Flowers minute, unisexual. Fruit a 1-seeded achene . . . . Ceratophyllaceae   page 215

  13' Leaves simple and narrow, opposite. Floating leaves, if present, spatulate. Fruit of 4 nutlets . . . . Callitrichaceae   page 304

Section 4. Herbaceous, Apetalous Plants or Perianth with a Single Series

1 Ovary inferior, at least the lower one-half fused to other tissues.

  2 Flowers clustered in heads or terminal locations and subtended with involucral or leaf-like bracts.

    3 Flowers of one kind. Perianth colored or corolla-like. Subtending bracts leaf-like . . . . Nyctaginaceae   page 205

3' Flowers often of two kinds. Perianth not colored or
corolla-like and usually lacking. Involucral bracts
not leaf-like . . . . Asteraceae   page 405

2' Flowers not clustered in heads and not subtended by involucral bracts.

    4 Ovary 2-celled and fruit 2-seeded.

       5 Perianth parts separate. Leaves alternate. Flowers
in umbels . . . . Apiaceae   page 331

       5' Perianth parts fused. Leaves opposite or whorled.
Flowers not in umbels . . . .
. . . . Rubiaceae   page 396

    4' Ovary 1-celled and fruit 1-seeded or fruit a capsule
and with more than 2 seeds.

      6 Leaves alternate or basal.

        7 Fruit 1-seeded. Plants partly parasitic on other
plants, particularly their roots or rhizomes . . .
. . . . Santalaceae   page 181

        7' Fruit a capsule. Plants not parasitic.

         8 Styles 6. Leaves basal, heart-shaped and with
a cordate base . . Aristolochiaceae   page 181

         8' Styles 2. Leaves not heart-shaped . . . . . .
. . . . Saxifragaceae   page 249

      6' Leaves opposite.

        9 Style 1 or stigma sessile. Flowers clustered in
leaf axils . . . . . Lythraceae   page 321

        9' Styles 2. Flowers not clustered in leaf axils . .
. . . . Saxifragaceae   page 249

1' Ovary superior.

  10 Pistils of several separate carpels.

    11 Stamens inserted and arising from below the carpels,
or plants dioecious, with only stamens or only pistils
. . . . Ranunculaceae   page 216

    11' Stamens inserted and arising from a hypanthium which
is cup-like.

      12 Carpels 5, usually in a ring and partly fused
laterally . . . . . . Crassulaceae   page 249

12' Carpels 1-several, not fused in a ring or
partly fused laterally . . . .
. . . . Rosaceae   page 254

10' Pistil of 1 carpel or of several carpels fused to form a
compound ovary. The compound ovary may be 1-celled and
1-seeded, however.

13 Ovary 1-celled or with one cavity.

14 Fruit a capsule. Ovules several.

15 Leaves opposite. Ovules on a central axis
within the capsule . . . .
. . . . Caryophyllaceae   page 208

15' Leaves alternate. Ovary with ovules attached
parietally.

16 Leaves rounded or lobed. Styles 2 . . . .
. . . . Saxifragaceae   page 249

16' Leaves lanceolate to oblong, entire.
Style 1 . . . Cistaceae   page 314

14' Fruit a utricle or one-seeded achene.

17 Flowers appearing petaloid.

18 Fruit a 3-angled achene. Stipules sheathing the stem, or flowers in small, involu-
crate heads . . . .
. . . . Polygonaceae   page 181

18' Fruit with a husk-like covering, not 3-
angled. Leaves opposite and entire . . .
. . . . Nyctaginaceae   page 205

17' Flowers not appearing petaloid.

19 Stipules present and conspicuous but not
sheathing the stem. Leaves opposite.

20 Flowers unisexual. Plants monoecious
or dioecious.

21 Leaves palmately veined or compound . . . .
. . . Moraceae   page 178

21' Leaves simple, often with stinging
hairs . . . .
. . . Urticaceae   page 179

20' Flowers perfect, leaves simple, the
    stipules scarious . . . .
    . . . . . Caryophyllaceae   page 208

19' Stipules not present. Leaves alternate or
   opposite.

    22 Perianth and bracts below the perianth
        scarious and dry. Flowers in dense
        spikes . . . .
        . . . . . Amaranthaceae   page 202

    22' Perianth and bracts not scarious.

        23 Fruit an achene. Flowers in
            clusters in the axils of alter-
            nate leaves . . . .
            . . . Urticaceae   page 179

        23' Fruit a utricle. Perianth and
            bracts fleshy green, often white-
            mealy . . . .
            . . . Chenopodiaceae   page 192

13' Ovary 2-several celled (carpelled). Ovules usually
   more than 1.

    24 Flowers unisexual, grouped in a cup-like, often
        glandular or petaloid structure. Ovary 3-
        carpelled . . . . . . Euphorbiaceae   page 299

    24' Flowers perfect, bisexual.

        25 Leaves whorled or verticillate, entire.
            Plants with stems prostrate . . . .
            . . . . Aizoaceae   page 206

        25' Leaves alternate, plants with upright or
            ascending stems.

            26 Plants usually annual. Perianth of 4
                parts, green. Stamens 6 . . . .
                . . . . . . Brassicaceae   page 230

            26' Plants usually perennial. Flowers in
                spikes, brownish. Stamens 2 . . . .
                . . . . . . Scrophulariaceae   page 382

Section 5. Herbaceous Dicotyledons, Petals Present.
Pistils Several, of Simple, Separate Carpels

1 Plants of trailing vines with alternate, palmately lobed
   leaves. Flowers unisexual . . . Menispermaceae   page 228

1' Plants with erect stems, not trailing.

   2 Stems and leaves succulent-fleshy. Petals yellow, flowers
      in cymes. Carpels 3-5 . . . . Crassulaceae   page 249

   2' Stems and leaves not succulent.

      3 Stamens numerous, fused in a column around the style,
         monadelphous. Carpels 5 or more in a ring . . . .
         . . . . Malvaceae   page 311

      3' Stamens not monadelphous.

         4 Leaves alternate.

            5 Petals and stamens arising from the calyx tube, or
               hypanthium, the flowers usually perigynous . . . .
               . . . . Rosaceae   page 254

            5' Petals and stamens arising from the receptacle,
               the flowers hypogynous . . . .
               . . . . Ranunculaceae   page 216

         4' Leaves opposite and pinnately compound. Plants pros-
            trate, the fruits sharply spinose and irregularly
            shaped . . . . Zygophyllaceae   page 297

Section 6. Herbaceous Dicotyledons with Radially Symmetrical
Flowers. Petals Separate. Pistil One or of Several
Carpels Fused into a Compound Ovary, Completely Superior

1 Sepals 2, often deciduous early in anthesis.

   2 Stems and often the leaves succulent-fleshy. Leaves entire
      . . . . Portulacaceae   page 207

   2' Stems and leaves not succulent. Leaves mostly deeply lobed
      or compound . . . . Papaveraceae   page 228

1 Sepals more than 2.

   3 Stamens numerous or at least more than twice the number of
      petals.

      4 Flowers with stamens fused in a column around the style,
         monadelphous . . . . . . Malvaceae   page 311

4' Flowers without monadelphous stamens.

    5 Leaves pinnatifid-lobed with spinose lobe tips.
       Petals white . . . . . . Papaveraceae   page 228

    5' Leaves not pinnatifid or with spinose lobe tips.

        6 Leaves simple.

           7 Style 1, petals 3 or 5 . . . .
              . . . . Cistaceae   page 314

           7' Styles 2, petals 5, yellow. Leaves opposite .
              . . . . Hypericaceae   page 313

        6' Leaves compound.

           8 Leaves trifoliolate, alternate. Stamens 6 or
              more. Fruiting capsule on a stipe . . . . . .
              . . . . Capparidaceae   page 247

           8' Leaves twice or more compound with more than 3
              leaflets or segments.

              9 Flowers in racemes . . . .
                 . . . . Ranunculaceae   page 216

              9' Flowers in dense heads or spikes . . . . . .
                 . . . . Fabaceae   page 268

3' Stamens twice as many as the petals, or fewer than twice as
   many.

    10 Stamens twice the number of petals.

        11 Sepals and petals 6 or more . . . .
           . . . . .Lythraceae   page 321

        11' Sepals and petals 4 or 5.

           12 Leaves simple or shallowly lobed.

              13 Style 1. Anthers opening by pores. Plants
                often with evergreen leaves . . . .
                . . . . . . Ericaceae   page 341

              13' Styles 2 or more.

                14 Ovary lobed. Principal leaves paired,
                  basal . . Saxifragaceae   page 249

                14' Ovary not lobed. Leaves cauline, op-
                  posite.

                  15 Leaves serrate. Flowers small, in
                    the axils of leaves . . . .
                    . . . Elatinaceae   page 313

15' Leaves entire or nearly so. Flowers not restricted to the leaf
axils. Stems often swollen at the
nodes . . . .
· · · Caryophyllaceae   page 208

12' Leaves compound or deeply divided.

16 Leaves opposite.

17 Leaves pinnately compound. Flowers
yellow. Plants prostrate. Fruit
sharply spined . . . .
· · · · · Zygophyllaceae   page 297

17' Leaves palmately compound or divided.
Fruits with long beaks . . . .
· · · · · Geraniaceae   page 293

16' Leaves alternate.

18 Styles 5. Leaves with 3 heart-shaped
leaflets. Plants less than 1 dm tall
· · · · · Oxalidaceae   page 295

18' Style 1. Leaves once or twice pinnately compound. Plants usually taller
than 1 dm . . . .
· · · · · Fabaceae   page 268

10' Stamens less than twice the number of petals.

19 Stamens as many as the petals.

20 Petals, sepals, and stamens 6. Leaf usually
solitary, ternately compound . . . .
· · · · · · · · · Berberidaceae   page 227

20' Petals and sepals 4 or 5. Leaves not ternately
compound.

21 Leaves alternate or basal.

22 Leaves palmately lobed . . . .
· · · · · Saxifragaceae   page 249

22' Leaf blades not palmately lobed.

23 Styles 2 or more, flowers yellow or
blue. Leaves narrowly lanceolate,
entire . . . .
· · · Linaceae   page 296

23' Style one or lacking.

24 Petals and sepals 4 . . . .
. . .   Brassicaceae   page 230

24' Petals and sepals a number other
than 4.

25 Leaves entire or toothed, the
flowers with ovaries maturing
into globose to ovoid capsules
. . Primulaceae   page 344

25' Leaves pinnately lobed or pinnately compound.

26 Leaves pinnately lobed.
The ovary maturing with a
long beak . . . .
Geraniaceae   page 293

26' Leaves once or twice pinnately compound. The ovary
maturing into a pod-like
legume . . . .
Fabaceae   page 268

21' Leaves opposite.

27 Leaf blades palmately lobed. Fruits with a
long beak . .   Geraniaceae   page 293

27' Leaf blades not palmately lobed.

28 Leaves serrate. Flowers in the axils
of leaves . . . .
. . . . .   Elatinaceae   page 313

28 Leaves entire.

29 Ovary with 4 or 5 cross-partitions.
The flowers yellow or blue . . . .
. . .   Linaceae   page 296

29' Ovary with a central placenta, not
partitioned. Flowers white to
lavender . . . .
. . .   Caryophyllaceae   page 208

19' Stamens numbering more than the petals but less than
twice as many as the petals.

30 Leaves opposite. Styles 2 or more . . . .
. . . .   Caryophyllaceae   page 208

30' Leaves alternate.

31 Sepals and petals 4.

    32 Stamens 6, tetradynamous, 4 long and 2 shorter . . . Brassicaceae   page 230

    32' Stamens 6 or more but not tetradynamous, elongate . . . Capparidaceae   page 247

31' Sepals 5. Petals 3 to 5.

    33 Leaves simple. Stamens 6 to 9. Fruit a capsule . . . Cistaceae   page 314

    33' Leaves once or twice pinnately compound. Fruit a legume or loment . . . . Fabaceae   page 268

Section 7. Herbaceous Dicotyledons with Bilaterally Symmetrical Flowers. Petals Separate or Nearly So. Pistil One or of Several Carpels Fused into a Compound Ovary, Completely Superior

1 Flowers with one or more obvious spurs or sacs formed by the petals or sepals or both.

    2 Leaves simple, shallowly lobed or entire.

        3 Spur sac-like, on the lower petal. Two of the 5 stamens projecting into the sac. Petals bearded within . . . . Violaceae   page 315

        3' Spur on the upper 3 sepals, petal-like, orange or yellow. Plants with watery, hollow stems . . . . Balsaminaceae   page 307

    2' Leaves compound or deeply lobed.

        4 Leaf blades twice or more pinnately or palmately divided. Spurs on the petals. Sepals 2, reduced to scales. Vegetation glaucous-glabrous . . . . Fumariaceae   page 229

        4' Leaf blades palmately lobed or divided or trifoliolately compound.

            5 Plants stemless, the leaves at soil level. Two of the 5 stamens projecting downward into the sac-like spur . . . . Violaceae   page 315

            5' Plants stemmed, the leaves cauline. Stamens not projecting downward into the spur. Flowers variously spurred . . . . Ranunculaceae   page 216

1' Flowers without obvious spurs or sacs but otherwise irregular.

  6   Corolla strongly bilaterally symmetrical.

      7   Corolla with two petals fused to form a keel. Stamens 5
         or 10. Leaves often compound . . . .
         . . . . Fabaceae   page 268

      7' Corolla with 3 petals. Two of the 5 sepals larger and
         petal-like. Stamens 6 to 8, fused in 1 or 2 groups.
         Leaves simple . . . . . . Polygalaceae   page 298

  6' Corolla only slightly irregular in that the petals are of
    unequal size.

      8   Leaves palmately compound. Stamens 6 or more. Flower
         parts not in 5's . . . . . Capparidaceae   page 247

      8' Leaves pinnately compound. Stamens 5 or 10. Flower
         parts in 5's . . . . Fabaceae   page 268

Section 8. Herbaceous Dicotyledons. Petals Separate. Pistil One or of Several Carpels Fused into a Compound Ovary, Inferior

1 Stamens 20-many, often in several series.

  2   Plants fleshy and succulent with spines. Leaves reduced or
    lacking . . . . Cactaceae   page 319

  2' Plants not fleshy but with stiff hairs. Leaves alternate,
    lobed or pinnatifid . . . . . Loasaceae   page 318

1' Stamens usually not more than twice the number of petals.

  3   Plants climbing vines with tendrils. Flowers unisexual . .
    . . . . Cucurbitaceae   page 402

  3' Plants not climbing.

    4   Stamens fewer or as many as the petals.

      5   Petals 5.

        6   Inflorescence umbellate.

          7   Leaves in a single whorl or solitary, compound,
            from the base or immediately beneath the in-
            florescence . . . Araliaceae   page 330

          7' Leaves alternate or basal, either simple or
            compound. Styles 2 from a stylopodium . . . .
            . . . . Apiaceae   page 331

        6   Inflorescence not umbellate.

8 Leaves simple, petioled and palmately veined.
   Styles 2 . . . . . Saxifragaceae   page 249

8' Leaves compound. Styles more or less than 2
   . . . . Rosaceae   page 254

5' Petals 2-4.

9 Petals 4. Flowers in a small dense cluster subtended by white bracts . . . .
   . . . . Cornaceae   page 340

9' Petals 2. Flowers in racemes . . . .
   . . . . Onagraceae   page 322

4' Stamens from more numerous than the petals to twice as
many.

10 Leaves ternately compound. Flowers greenish in
    small clusters. Stems from tuberous rootstocks . .
    . . . . Adoxaceae   page 400

10' Leaves not ternately compound.

11 Plants with succulent leaves. Sepals reduced
    . . . . Portulacaceae   page 207

11' Plants without succulent leaves. Sepals 4 or 5.

12 Style 1. Petals commonly 4 with 8 stamens
    . . . . Onagraceae   page 322

12' Styles 2. Petals commonly 5 . . . .
    . . . . Saxifragaceae   page 249

Section 9. Herbaceous Dicotyledons with Petals United, Gamopetalous. Ovary Superior, Flowers Radially Symmetrical

1 Stamens partially or completely fused to the corolla or other
flower parts.

2 Ovaries 2. Plants usually with milky juice. Fruits of
follicles.

3 Stamens separate. Styles united. Pollen not in waxy
masses . . . . Apocynaceae   page 349

3' Stamens united to the stigmas, the pollen in waxy
masses, styles separate. Corolla crested or horned . .
. . . . Asclepiadaceae page 350

2' Ovary 1, simple or compound.

4   Corolla scarious, translucent, not colored like usual
    petals, of 4 lobes. Leaves basal; flowers in spikes or
    heads . . . . Plantaginaceae   page 394

4' Corolla variously colored, not scarious.

    5   Ovary deeply 4-lobed, with the style arising cen-
        trally and from the base, gynobasic.

        6   Stamens 5. Leaves usually alternate, often entire
            . . . . Boraginaceae   page 360

        6' Stamens 2, rarely 3 or 4. Leaves usually opposite,
            toothed . . . . . . Lamiaceae   page 369

    5' Ovary not deeply 4-lobed.

        7   Stamens opposite the corolla lobes, usually 5.
            Fruit a capsule . . . Primulaceae   page 344

        7   Stamens alternate with the corolla lobes.

            8   Ovary with 1 cavity and seeds on a central
                placenta. Fruit a capsule.

                9   Leaves simple and opposite or, if trifoli-
                    ate, the segments not divided. Corolla
                    often plaited or folded in bud . . . .
                    . . . . Gentianaceae   page 347

                9' Leaves compound or pinnately divided, alter-
                    nate or basal. Inflorescence often scorpoid
                    . . . . Hydrophyllaceae   page 359

            8' Ovary with 2 or more cavities. The seeds not
                on central placentae.

                10   Stamens 2-4.

                    11 Fruit a capsule . . . .
                        . . . . . Scrophulariaceae   page 382

                    11' Fruit 4 1-seeded nutlets. Style from a
                        4-lobed ovary . . . .
                        . . . . . Verbenaceae   page 368

                10' Stamens 5.

                    12 Flowers funnelform, usually solitary.
                        Mostly twining vines or achlorophyllous
                        vines parasitic on green plants . . . .
                        . . . . . Convolvulaceae   page 353

                    12' Flowers not funnelform. Stems not
                        twining.

13 Ovary 4-lobed. Fruit 4 1-seeded nutlets . . . Boraginaceae   page 360

13' Ovary not 4-lobed. Fruit a capsule.

14 Style and stigma 1. Leaves alternate.

15 Corolla open or saucer-shaped.
Flowers in spikes or racemes .
. . Scrophulariaceae   page 382

15' Corolla tubular or shallowly
funnelform. Flowers in cymes
or racemes . . . .
. . Solanaceae   page 378

14' Styles 2 or the simple style with 2
or more stigmas.

16 Ovary 1 celled or incompletely
2-celled. Leaves compound.
Inflorescence scorpoid . . . .
. . Hydrophyllaceae   page 359

16' Ovary 2-4 celled.

17 Stigmas 3 or upper part of
style 3-parted. Ovary 3-
celled . . . .
Polemoniaceae   page 355

17' Stigmas 2 or style 2-parted.
Flowers funnelform . . . . .
Convolvulaceae   page 353

1' Stamens free from the corolla, arising from the receptacle.
Stamens 5 or 10. Plants often evergreen . . . .
. . . . Ericaceae   page 341

Section 10. Herbaceous Dicotyledons with Petals United,
Gamopetalous. Ovary Superior,
Flowers Bilaterally Symmetrical

1 Petals fused to form a keel or spur, flowers not strongly
bilabiate.

2 Petals 5, the 2 lower fused to form a keel. Stamens 10.
. . . . Fabaceae   page 268

2 Petals 4, with 1 or 2 lower ones fused to form a spur.
Stamens 6 . . . . Fumariaceae   page 229

1 Petals bilabiate or otherwise irregular.

   3 Plants parasitic on the roots or rhizomes of other plants.
Chlorophyll lacking. Stamens 4 . . . .
. . . . Orobanchaceae   page 392

   3' Plants not parasitic on other plants.

      4 Anther bearing stamens 5.

         5 Ovary deeply 4-lobed around the base of the central
style . . . . Boraginaceae   page 360

         5' Ovary not deeply 4-lobed. Flowers in terminal spikes
or racemes . . . . Solanaceae   page 378

      4' Anther bearing stamens 2 to 4, not including the sterile
ones that may be present.

         6 Ovary with 1 cell or cavity.

            7 Corolla strongly bilabiate, about 1 cm long. In-
florescence a slender spike. Fruit an achene . .
. . . . Phrymaceae   page 394

            7' Corolla not strongly bilabiate, up to 10 cm long.
Inflorescence an irregular cluster. Fruit a long,
2-beaked capsule . . Martyniaceae   page 394

         6' Ovary 2-4 celled.

            8 Fruit a capsule. Stem not square . . . .
. . . . Scrophulariaceae   page 382

            8' Fruit splitting into 2 to 4 seeded nutlets. Stems
often square or 4-angled in cross section.

               9 Vegetation with a minty odor. Corolla strongly
bilabiate . . . . Lamiaceae   page 369

               9' Vegetation lacking a minty odor. Corolla
weakly bilabiate or otherwise irregular . . . .
. . . . Verbenaceae   page 368

Section 11. Herbaceous Dicotyledons with Petals
United, Gamopetalous. Ovary Inferior

1 Flowers in dense heads subtended by involucral bracts.

   2 Stamens 5, usually fused in a cylindrical ring around the
styles. Flowers often of two kinds . . . .
. . . . Asteraceae   page 405

2' Stamens 2 or 4, not fused together in a ring. Heads of one
   kind of flowers. Leaves opposite . . . .
   . . . . Dipsacaceae   page 401

1' Flowers not in dense heads subtended by involucral bracts.

   3 Plants with succulent stems that are spiny. Leaves essen-
     tially lacking. Stamens 20 or more . . . .
     . . . . Cactaceae   page 319

   3' Plants lacking succulent stems. Leaves present. Stamens
      fewer than 20.

      4 Stamens free from the corolla.

        5 Anthers 10, opening by pores . . . .
          . . . . Ericaceae   page 341

        5' Anthers 5, opening longitudinally.

           6 Flowers radially symmetrical . . . .
             . . . . Campanulaceae   page 403

           6' Flowers bilaterally symmetrical . . . .
              . . . . Lobeliaceae   page 404

      4' Stamens fused to the corolla.

         7 Plants twining, with tendrils. Flowers unisexual . .
           . . . . Cucurbitaceae   page 402

         7' Plants not twining or with tendrils. Flowers bi-
            sexual.

            8 Leaves ternately compound, basal. Flowers small,
              greenish. Stamens twice the number of corolla
              lobes . . . . Adoxaceae   page 400

            8' Leaves not ternately compound, opposite or whorled.

               9 Inflorescence in spike-like racemes or panicles.
                 Stamens 3 . . . . Valerianaceae   page 400

               9' Inflorescence not in spike-like racemes or
                  panicles. Stamens 4 or 5.

                  10 Leaves with stipules. Ovary maturing into
                     2 1-seeded fruits . . . .
                     . . . . . . Rubiaceae   page 396

                  10' Leaves without stipules. Ovary not sepa-
                      rating at maturity . . . .
                      . . . . . . Caprifoliaceae   page 397

Family Salicaceae

(Key to Genera)

1   Buds with more than 1 scale. Catkins usually pendulous, the
    flowers subtended by disks. Plants small to large trees.
    Leaves lanceolate to deltoid . . . . *POPULUS*

1' Buds with 1 scale. Catkins usually upright, the flowers sub-
    tended by glands. Plants shrubby or small trees. Leaves
    narrow to ovate . . . . *SALIX*

POPULUS L.*

1   Petioles immediately beneath the blade flattened 90° to the
    flat part of the blade.

    2   Leaves round-ovate, often with cordate bases, pale green,
        2-8 cm long and 2-7 cm wide with abruptly pointed tips.
        Trees small, 5-12 meters tall, the bark smooth, grayish
        green. Buds shiny, not resinous. Fruits 3-5 mm long, on
        short pedicels. Frequent in foot hills of the Black Hills
        and scarps east to Todd County and in the northeast part.
        Apr-May. Quaking Aspen . . . . *Populus tremuloides* Michx.

    2' Leaves deltoid with broadly truncate or sub-cordate bases,
        bright green, the tips acuminate. Blades almost as broad
        as long. Trees to 20 meters, the bark rough. Buds shiny,
        resinous. Fruits ovoid, to 1 cm long, in loose, pendulous
        catkins 8-15 cm long. Common in alluvium and along streams
        and lakes over the state. Represented in South Dakota by
        the two varieties *deltoides* and *occidentalis* Rydb. Apr-
        May. Cottonwood . . . . *Populus deltoides* Marsh.

1' Petioles immediately beneath the leaf blade not flattened.

    3   Leaf blades pale whitened on the under surface, ovate to
        broadly lanceolate with cuneate or rounded bases, 7-12 cm
        long. Trees tall, to 30 meters, with reddish-gray, fur-
        rowed bark. Fruits ovoid, 5-7 mm long, on short pedicels,
        the fruiting inflorescence forming a short, compact raceme.
        Rare along streams and in canyons in Custer and Lawrence
        counties of the Black Hills and in the northeast part.
        Apr-May. Balsam Poplar . . . . . . *Populus balsamifera* L.

*The common cultivated species, *P. nigra*, the lombardy poplar;
P. bolleana, both columnar in shape; and *P. alba*, the white
poplar, are not included here.

3' Leaf blades green on the lower surface, not noticeably pale whitened when compared with the upper surface.

  4 Leaves broadly lanceolate with long acuminate tips, the blades up to 2 times as long as wide. The petioles one-half as long as the blades. Trees up to 20 meters tall, with broad, ascending branches forming a dome-like top. Bark gray-brown with branches a lighter yellow-brown color. Fruiting aments 7-15 cm long. Fruits ovoid, 6-8 mm long. Infrequent in valleys and along stream banks in the Black Hills. {Believed to be of hybrid origin between *P. angustifolia* James and *P. deltoides* Marsh.} Apr-May. Smooth Barked Cottonwood . . . . *Populus acuminata* Rydb.

  4' Leaves narrowly lanceolate with gradually tapered tips, the blades more than twice as long as wide. Petioles not more than one-third as long as the blades. Trees narrow with ascending branches, to 20 meters tall. Bark greenish, the twigs yellow-green. Fruiting aments 2-6 cm long, compact with blunt, glabrous fruits. Infrequent to rare along streams in the Black Hills. Apr-May. Narrow Leaved Cottonwood . . . . *Populus angustifolia* James

SALIX L.

1 Base of leaf blade or upper part of petiole at mature size with glands. Leaf margins also more or less glandular-serrate.

  2 Leaves white-glaucous on the under surface. Blades elliptic with tapering tips and bases, 4-7 cm long and 1-2 cm wide, shiny green on the upper surface. Plants shrubby, 1-4 meters tall, the branches dull green or brown. Rare in wet meadows in the Black Hills. June-July. Autumn Willow . . . . *Salix serissima* (Bailey) Fern.

  2' Leaves green on both surfaces; however, the lower surface may be lighter green but not white-glaucous.

    3 Leaf blades tapering to elongated acuminate tips. Twigs yellowish. Plants native to the Black Hills. Shrubby trees to 6 meters tall. Leaves lance-ovate, 5-10 cm long and 2-4 cm wide. Petioles 5-10 mm long, densely glandular at the distal region. Catkins 2-6 cm long, the fruiting capsules 4-6 mm long. Rare but locally frequent in moist soil in the Black Hills and in the northeast part. Apr-May. Shining Willow . . . . *Salix lucida* Muhl.

3' Leaf blades tapering to acute tips, with an ovate-
lanceolate shape. Twigs yellow-green. Plants introduced, not native. Small trees 2-8 meters tall. Leaves
dark green, firm and shiny, 4-10 cm long and 2-4 cm wide.
Bark gray on older part of trunk. Fruiting catkins 2-6
cm long, the fruiting capsules glabrous, about 5 mm long.
Infrequent as plantings over the state and not persisting. May. Laurel Willow . . . . . . *Salix pentandra* L.

1' Base of leaf blade and upper part of the petiole at maturity
without glands. Leaf margins entire to variously dentate but
not glandular.

4 Leaves linear or narrowly lanceolate to oblanceolate, up to
8 times longer than wide.

5 Leaf margins regularly and sharply serrate, the under-
side light green and somewhat glaucous, at maturity
glabrous. Leaf blades acuminate at both ends, 5-10 cm
long. Plants 1-3 meters tall, shrubby, with many clustered stems. Twigs slender, yellow-brown. Fruiting
catkins 1-2 cm long, the capsules hairy. Rare to infrequent in marshy or boggy meadows and along streams in
the northeast, the sandhills area in the south, and in
the Black Hills. May. Meadow Willow . . . .
. . . . *Salix petiolaris* Smith

5' Leaf margins entire or remotely denticulate. Surfaces
of leaves tomentose to glabrous.

6 Leaf margins remotely denticulate, not revolute
(rolled under). Blades linear to narrowly oblanceolate, 2-8 cm long. Plants shrubby or shrubby trees
2-5 meters tall. Fruiting catkins 2-5 cm long, the
capsules 2-5 mm long, becoming glabrate at maturity.
Widely distributed and common over the state on
shores and in low places. Represented in South
Dakota by the two subspecies *exigua* and *interior*
(Rowlee) Cronquist. May-June. Sandbar or Coyote
Willow . . . . *Salix exigua* Nutt.

6' Leaf margins entire and revolute, the blades permanently tomentose on both surfaces. Plants rarely
over 1 meter tall.

7 Catkins produced before the leaves are fully
formed. Low spreading shrubs of sandy prairie,
usually not more than 5 dm tall. Leaves linear
to oblanceolate, 2-5 cm long, crowded towards the
ends of branches. Catkins short, few-flowered,

the fruiting ones 1-2 cm long. Extremely rare on
prairie hillsides in the eastern part. Possibly
extinct. Apr-May. Dwarf Willow . . . .
. . . . *Salix humilis* Marsh.

7' Catkins produced after the leaves are fully formed.
Upright shrub of bogs or marshes, the stems 2-10
dm tall. Leaves oblong or linear-lanceolate, 4-8
cm long, mostly acute at both ends. Catkins on
leafy peduncles 1-5 cm long, the capsules white
tomentose. Rare in boggy places of the Black
Hills and in Marshall and Roberts counties of the
northeast part. May. Bog Willow . . . .
. . . . *Salix candida* Flugge

4' Leaves lanceolate to ovate, not more than 6 times as long
as wide at the widest part.

8 Leaf blades with long acuminate tips and glaucous undersurfaces. Usually trees up to 12 meters or more tall.

9 Leaves light green, thin, glabrous, the margins
finely serrate, white glaucous on the under surface.
Trees native, many times younger ones mistaken as
shrubs because 3-5 stems may develop from a base.
Leaves lance-ovate, 3-12 cm long, often with large
stipules. Twigs yellow to reddish brown; bark rough,
scaly. Very common on flood plains and along water
over the state. Apr-May. Peach-leaved Willow . . .
. . . . *Salix amygdaloides* Anderss.

9' Leaves dark green, shiny, firm, the margins serrate.
Trees introduced, often used for plantings.

10 Younger twigs brittle. Leaves lanceolate, 7-12
cm long, the petioles glandular-viscid when young.
Leaf margins with 4-6 teeth per cm. Trees up to
20 meters, with spreading branches and rough gray
bark. Catkins appearing with the leaves, 4-8 cm
long, the capsules slender, pedicelled. An oft-
planted tree that persists and occasionally escapes. Apr-May. Crack Willow . . . .
. . . . *Salix fragilis* L.

10' Younger twigs not brittle. Leaves lanceolate to
oblanceolate, 4-8 cm long. Leaf margins with 7-
10 teeth per cm. Trees to 20 meters or more, the
twigs golden yellow. Catkins appearing with the
leaves, 3-6 cm long, the capsules sessile, ovate.
Widely planted over the state and occasionally

escaping. Apr-May. White Willow . . . . . .
. . . . *Salix alba* L.

8' Leaf blades with acute-acuminate to obtuse tips. Shrubs
or at most, shrubby trees.

    11 Margins of leaves entire or nearly so. {Occasion-
ally *S. discolor* and *S. bebbiana* may have some
leaves almost entire. These can be found in the
next section.}

        12 Leaves oblanceolate to obovate, the apex rounded
to obtuse, 3-5 cm long. Leaves clustered at the
ends of twigs. Under surfaces of leaves often
pubescent. Shrubs commonly 3-4 meters tall, upright, the stems slender, with gray-green bark.
Fruiting catkins 3-7 cm long, the capsules pubescent, sessile. Infrequent in drier areas of
higher elevations in the Black Hills. Apr-May.
Western Pussy Willow . . *Salix scouleriana* Barr.

        12' Leaves oblong to elliptical, tapering acutely
at both ends of the blade, 2-7 cm long. Leaf
blades deep green above, glaucous below. Twigs
reddish brown, shining. A much-branched shrub
1-3 meters tall, the many separate stems in
clusters. Fruiting catkins 2-4 cm long, with
pubescent, sessile capsules. Infrequent in
moist meadows and other low wet, open places
at higher elevations in the Black Hills. Ours
var. *planifolia* (Pursh) Hutonen. June. Plane-
leaf Willow . . . . . . *Salix phylicifolia* L.

    11' Margins of leaves serrate or crenate-serrate, some
leaves only slightly so.

        13 Leaf surfaces more or less pubescent towards
maturity, sometimes whitened hairy below. Leaf
margins crenate-serrate to only slightly undulate. Blades tapered or wedge-shaped to the
petiole. Leaves 2-6 cm long. Plants shrubby,
2-5 meters tall, stems few to several, often
forming dense thickets. Bark gray and furrowed,
young twigs yellow-brown to reddish-brown.
Fruiting catkins 2-4 cm long, the capsules tapered, hairy, on pedicels 2-3 mm long. A frequent and variable species in the Black Hills
and the northeast lakes area represented by the
two varieties *bebbiana* and *perrostrata* (Rydb.)

Sarg. Apr-May. Bebb Willow . . . .
. . . . *Salix bebbiana* Sarg.

13' Leaf surfaces glabrous or glaucous towards maturity. Young leaves may be pubescent-hairy or
lanate.

14 Bases of leaf blades tapered to the petiole.
Leaf margins irregularly to remotely crenate-
dentate. Blades 5-8 cm long, elliptic to
oblanceolate. Upper surface dark green, the
lower one pale-glaucous. Large shrubs to 6
meters, the many stems forming a clumpy,
rounded shape. Catkins forming early, before the leaves appear, 2-7 cm long, thick
and hairy. Sporadic distribution in low
places and at the edges of thickets in the
Black Hills and east of the Missouri River.
Apr-May. Pussy Willow . . . .
. . . . *Salix discolor* Muhl.

14' Bases of leaves rounded or cordate. Leaf
margins closely serrate.

15 Leaves ovate to obovate, only 2-2.5
times longer than wide. Leaf tips acute
to obtuse. Leaves 3-6 cm long and 2-3
cm wide, dull glabrous above and pale
glaucous beneath. Shrub 2-3 meters tall,
spreading to upright. Young twigs dull
red, becoming gray-green. Stipules
prominent, persisting. Fruiting catkins
3-4 cm long, the capsules glabrous. Infrequent in swampy meadows and along
streams at higher altitudes in the Black
Hills. {incl. *S. padophylla* Rydb.} May-
June. Serviceberry Willow . . . .
. . . . *Salix monticola* Bebb. ex. Coult.

15' Leaves lanceolate to lance-ovate, with
tapering tips, the blades over 3 times
longer than wide, 4-6 cm long. Plants
shrubby, 2-3 meters tall. In our area
stem buds often infected by midges which
result in an ovoid, scaly growth up to 2
cm across called "pine cone gall." Stipules prominent, rounded, up to 1 cm
across. Frequent on shores and sandbars
over the state. {incl. *S. missouriensis*

Bebb., and *S. cordata* Muhl.} Represented in South Dakota by the two varieties *rigida* and *watsonii* (Bebb.) Cronq.
May. Diamond Willow . . . .
. . . . *Salix rigida* Muhl.

Family Juglandaceae

JUGLANS L.

Juglans nigra L. Tree 25-35 meters tall, the bark dark gray to
black, furrowed. Branches irregular, stout, the older ones becoming gnarled. Leaves pinnately compound, with 11-21 leaflets,
the leaflets lanceolate, soft pubescent, aromatic. Staminate
catkins 3-8 cm long, pendulous. Pistillate flowers usually solitary or paired in the leaf axils, petals 4, minute or lacking.
Pistils with 2 stigmas, maturing into a globose nut 3-5 cm across.
Infrequent in alluvial woods along the Big Sioux River in the
southeast part. Apr-May. Black Walnut.

Family Betulaceae

(Key to Genera)

1 Pistillate flowers few. Fruit a nut up to 1 cm across, enclosed or subtended by a leafy involucre . . . *CORYLUS*

1' Pistillate flowers many, arranged in catkins that are ovoid
or short-cylindric.

  2 Fruit a nutlet about 5 mm long, enclosed in a sac-like
  involucre . . . . *OSTRYA*

  2' Fruit a winged samara, lacking an involucre. Scales of the
  pistillate catkins 3-lobed . . . . *BETULA*

BETULA L.

1 Bark white to gray, separating into layers on maturing trees.
Trees small, usually less than 20 meters tall. Leaves ovate
to nearly rounded in outline, the distal part coarsely once
or twice-pinnate. Leaf blades 4-7 cm long. Pistillate catkins borne singly, maturing to 3 cm or more long in fruit.
Staminate catkins 1-4, clustered, on drooping pedicels. Frequent in the Black Hills and in the forest preserves of Harding County. Rare in the northeast. May. Canoe Birch . . .
. . . . *Betula papyrifera* Marsh.

1' Bark brown or gray-brown, not layered. Leaves less than 4 cm
   long.

   2 Plants shrubby, from less than 1 meter to not more than 3
   meters tall. Young leaves and branches densely puberulent
   and warty-glandular. Leaves usually less than 2 cm long,
   oval to elliptic, the apex rounded, finely serrate. Pistillate catkins 1-2 cm long and less than 5 mm across.
   Staminate catkins solitary, about 1 cm long. Infrequent
   in lowland valleys and along creeks in Lawrence and Pennington counties in the Black Hills. Ours var. *glanduli-
   fera* (Regel) Fl. May. Bog Birch . . . .
   . . . . *Betula glandulosa* Michx.

   2' Plants shrubby trees in clumps, the stems up to 4 meters
   tall. Young leaves and twigs glabrous to pubescent, with
   crystalline glands on the surface. Leaves 2-4 cm long,
   rounded to pointed at the apex, the margins serrate, sometimes irregularly so. Fruiting catkins 2-4 cm long. Staminate catkins several, 5-7 cm long. Frequent in forested
   ravines and along creeks and boggy places in the Black
   Hills and Harding County. {*Betula fontinalis* Sarg.} May.
   Mountain Birch . . . . *Betula occidentalis* Hook.

CORYLUS L.

1 Involucre surrounding nut of 2 broad, pubescent bracts that
  are wider than long. Trees shrubby, not over 4 meters tall,
  the young twigs pubescent and glandular. Leaves broadly
  ovate, their margins doubly serrate. Leaf blades 4-8 cm
  long, pubescent beneath. Pistillate flowers few in catkins,
  each subtended by a lacinate margined involucre. Nuts about
  1 cm in diameter. Infrequent in rich upland woods of the
  eastern part. Apr-May. American Hazelnut . . . .
  . . . . *Corylus americana* Walt.

1' Involucre surrounding nut elongate to a connate beak up to 3
   cm or more above the nut. Trees shrubby, 1-3 meters tall.
   Twigs and petioles glabrous to sparsely pubescent. Leaves
   ovate, 5-10 cm long, irregularly serrate with secondary serrations. Pistillate catkins of a few clustered flowers, the
   maturing nuts 1.5 cm in diameter. Frequent in rich woods of
   Marshall and Roberts counties in the northeast and in the
   Black Hills. Apr-May. Beaked Hazelnut . . . .
   . . . . *Corylus cornuta* Marsh.

OSTRYA Scop.

Ostrya virginiana (Mill.) K. Koch. Small tree up to 15 meters
tall, the bark of the main trunk rough and scaly. Leaves oblong
to ovate, with sharp double serrations, 7-12 cm long. Staminate
catkins dense, clustered and drooping, 3-5 cm long. Pistillate
catkins about 1 cm long, becoming 4-6 cm long in fruit, the nut
5-8 mm long, surrounded by an inflated, enclosing bract. Frequent in ravines and rich woods in the eastern, southern and
western parts of the state. Apr-May. Hop-hornbeam.

Family Fagaceae

QUERCUS L.

Quercus macrocarpa Michx. Large tree in the eastern part, a
small tree in the western part at the margins of forests at
lower altitudes in the Black Hills. Leaves 8-20 cm long, deeply
and coarsely pinnately lobed, the lobes rounded. The acorn 1-2
cm across, with a fringed involucre characteristic of the species.
The smaller western form has been referred to *Quercus mandanensis*
Rydb.; however, until there is evidence that it is a distinct
species, it is included here. Frequent on hillsides, especially
rich soils. May-June. Bur Oak.

Family Ulmaceae

(Key to Genera)

1   Leaf blades with more than one principal vein from the base.
    Flowers appearing with the leaves, partly staminate. Fruit a
    drupe with a hard stone . . . . *CELTIS*

1' Leaf blades with but a single principal vein from the base.
    Flowers appearing before the leaves on last year's wood, per-
    fect. Fruit a flattened samara . . . . *ULMUS*

CELTIS L.

Celtis occidentalis L. Tree to 30 meters or more. Bark becoming
warty-ridged on older trees. Leaves alternate, petioled, the
bases of the blades oblique. Leaves with sharply serrate margins, the blade ovate to lance-ovate, 3-8 cm long. Flowers in
small clusters, perfect or staminate, greenish, the perianth 4-6
parted, 3-5 mm long. Fruit globose, 7-9 mm in diameter, purple
to black. Frequent in upland and lowland woods and ravines over
the state. Apr-May. Hackberry.

ULMUS L.

1   Leaves glabrous on the upper surface, not scabrous to the
    touch.

   2   Branches with corky wings. Fruit pubescent, not cleft at
     the apex. Trees up to 30 meters. Leaves ovate or broadly
     ovate, doubly serrate with acuminate tips, 5-10 cm long.
     Flowers and fruits on slender pedicels, racemose. Samara
     ovate, without an apical notch, 12-15 mm long. Rare in
     rich, north-facing woods along the Big Sioux and Missouri
     Rivers in the extreme southeast part. Apr-May. Rock Elm.
     . . . . *Ulmus thomasi* Sarg.

   2' Branches without corky wings. Fruits glabrous, deeply
     cleft at the apex.

     3   Leaves mostly 5-9 cm long, oval-elliptical, the margins
      usually simply serrate. Fruiting pedicels erect. A
      small tree, usually not more than 20 meters tall. Flow-
      ers with brown to purple stamens, the fruits obovate,
      8-11 mm long. Commonly planted and escaping from culti-
      vation over the state. {Incorrectly referred to as
      Chinese Elm.} Apr-May. Siberian Elm . . . .
      . . . . *Ulmus pumila* L.

     3' Leaves mostly 8-14 cm long, ovate with acuminate tips,
      the margins doubly serrate. Fruiting pedicels pendulous.
      A large, graceful tree to 35 meters tall. Flowers in
      fascicles, on pedicels that are unequally elongate.
      Fruiting samara ovate with a cleft beak, 10-12 mm long.
      Common in alluvial and low, rich woods over the state.
      Commonly planted as a shade tree. Apr-May. American
      Elm . . . . *Ulmus americana* L.

1' Leaves not glabrous on the upper surface, very rough to the
    touch, scabrous. Trees to 25 meters, the bark rough, the
    inner surface mucilaginous. Leaves ovate to oblong, pubes-
    cent beneath, the blades doubly serrate, 10-20 cm long. Buds
    covered with dense, red hairs. Flowers in dense fascicles.
    Fruiting samaras almost orbicular, not cleft at the apex.
    Frequent in alluvial woods and low, loamy woods of ravines
    in the eastern part and the Black Hills. Apr-May. Red Elm,
    Slippery Elm . . . . *Ulmus rubra* Muhl.

Family Moraceae

(Key to Genera)

1 Plants trees or shrubs, often with milky juice. Fruits multiple.

   2 Stems thorny. Leaves entire, pinnately veined . . . . *MACLURA*

   2' Stems not thorny. Leaves palmately veined or lobed, the margins toothed . . . . *MORUS*

1' Plants herbaceous without milky juice. Fruits remaining separate.

   3 Leaves palmately compound. Plants erect herbs . . . . *CANNABIS*

   3' Leaves palmately lobed. Plants of twining stems . . . . *HUMULUS*

CANNABIS L.

Cannabis sativa L. Annual herbs, erect, the stems branched, to
1.5 meters tall. Plants dioecious. Leaves alternate or opposite,
5-9 lobed, the leaflets serrate, 4-9 cm long. Staminate flowers
pale green, without a perianth. Pistillate flowers in clusters
at the upper axils, also without a perianth. Fruits with a close
fitting bract, the achenes 2-4 mm long, ovoid. Frequent as a
weed in waste places over the state, more common in the eastern
part. July-Aug. Marijuana, Hemp.

HUMULUS L.

Humulus lupulus L. Plants dioecious, the stems twining from a
perennial base, 1-6 meters or more long. Leaves opposite, palmately lobed with 3-5 segments, the blades 3-8 cm long. Staminate flowers green, in panicles. Pistillate flowers in drooping
spikes, each with a persisting bract that becomes yellowish in
fruit. The fruiting cluster of bracted achenes becoming conspicuously yellow-green towards maturity. Infrequent in thickets
and brushy fencerows over the state. July-Aug. Hops.

MACLURA Nutt.

Maclura pomifera (Raf.) C. K. Schneid. Small trees usually not
over 6 meters tall, dioecious. Leaves ovate with acute to acuminate tips, the blades 6-20 cm long, alternate, with axillary
thorns. Staminate flowers in short spikes, not showy, green.
Pistillate flowers in globose clusters, maturing into a globose

syncarp of many achenes, the fruit 4-10 cm in diameter. Occasionally persisting from cultivation and possibly reproducing sparingly at roadsides and waste places of the extreme southern part. May-June. Osage Orange.

MORUS L.

Morus alba L. Dioecious or more commonly monoecious trees up to 8 meters tall. Leaves alternate, the margins serrate. The leaf-blades 3-8 cm long, often ovate to deeply lobed on the same tree. Staminate spikes 1-3 cm long, the flowers greenish, apetalous. Pistillate spikes 1-3 cm long, the flowers small, apetalous, green, the calyx of 2-6 parts. The many achenes becoming coalescent towards maturity. Fruit multiple, juicy, blackish red. Frequent to common in waste places. Originally planted and becoming widely naturalized in the southeast part. May-June. Mulberry.

Family Urticaceae

(Key to Genera)

1 Principal leaves alternate.

   2 Leaves ovate, broad, 4-8 cm across. Plants with stinging hairs . . . . *LAPORTEA*

   2' Leaves lanceolate, smaller, less than 3 cm across. Plants without stinging hairs . . . . *PARIETARIA*

1' Principal leaves opposite.

   3 Leaves lanceolate, with stinging hairs . . . *URTICA*

   3' Leaves ovate, without stinging hairs.

      4 Flowers in axillary spikes. Plants perennial, pubescent . . . . *BOEHMERIA*

      4' Flowers in axillary panicles. Plants annual, glabrous, the stems watery translucent . . . . . *PILEA*

BOEHMERIA Jacq.

Boehmeria cylindrica (L.) Sw. Perennial herb with erect stems 3-6 dm tall. Plants monoecious or dioecious. Leaves ovate, the blades coarsely serrate, 6-10 cm long, on well developed petioles. Flowers small, green, in dense axillary spikes, the staminate ones on the monoecious plants at the upper part of the spike. Stamens 4, the calyx 4-parted. Fruit an achene, enclosed in the

persisting calyx. Infrequent to rare in alluvial woods of the
southeast part. July-Aug.

LAPORTEA Gaud.

Laportea canadensis (L.) Wedd. Perennial herb 4-8 dm tall, the
stems mostly erect. Vegetation covered with stinging hairs that
give a severe burning sensation. Leaves ovate, the blade acuminate, 7-12 cm long. Plants monoecious, the staminate flowers in
small cymes in axils beneath the pistillate ones. Flowers small,
green, lacking petals. Stamens 5, sepals 5. Pistillate flowers
with 2-4 sepals. Fruit a small flattened achene 3-4 mm long.
Frequent to common in alluvial woods of the eastern part. July-
Aug. Wood Nettle.

PARIETARIA L.

Parietaria pennsylvanica Muhl. Small pubescent annual 1-4 dm
tall, the stems simple, mostly erect. Plants monoecious or
polygamous. Leaves lanceolate, 2-6 cm long, mostly entire.
Flowers in axils of the upper two thirds of the stem, mostly
monoecious. Flowers brown to green, lacking petals, in small
clusters. Staminate flowers of 4 stamens and 4 sepals. Pistillate flowers with a 4-parted calyx. Fruit a small achene usually about 1 mm across. Frequent in shady places of woods and
thickets over the state. July-Aug. Pellitory.

PILEA Lindl.

Pilea pumila (L.) Gray. Soft-stemmed annual 1-4 dm tall, the
stem smooth, pale green, watery. Plants monoecious or dioecious.
Leaves long petioled, the leaves shiny, glabrous. Blades ovate,
3-8 cm long, with serrate or crenate margins. Flowers green, in
small axillary cymes or panicles. Staminate flowers of 4 stamens
and 4-parted calyx. Pistillate flowers with a 3-parted calyx.
Fruit a small pale green achene up to 2 mm long, often with
purple spots. Rare in alluvial bottoms of the eastern part of
the state. July-Aug. Clearweed.

URTICA L.

Urtica dioica L. Perennial herbs with simple, erect stems which
may reach 1.5 meters in rich soil. Mostly dioecious; however,
some may be monoecious. Leaves opposite, 5-15 cm long, lanceolate with coarse teeth. Stinging hairs arm all the vegetative
parts. Flowers in axillary spikes or panicles, the pistillate
uppermost on monoecious forms. Staminate flowers of 4 stamens
and 4 subequal sepals. Pistillate flowers of 4 unequally shaped
sepals. Fruit a flattened achene about 1.5 mm long. Common over

the state in low alluvial woods and open rich soil that has
access to moisture. Ours ssp. *gracilis* (Ait.) Seland. July-
Sept. Nettle.

Family Santalaceae

COMANDRA Nutt.

Comandra umbellata (L.) Nutt. Plants rhizomatous perennials,
hemi-parasites on the roots of various prairie and woodland
species. Plants 5-20 cm tall, the stems usually clustered.
Leaves alternate or scattered, linear to elliptic, more or less
glaucous, 1-4 cm long, pale green. Flowers apetalous, white,
perfect, in terminal clusters, the sepals petaloid, 2-3 mm long.
Ovaries partly inferior. Fruits 4-8 mm long, ovoid, drupaceous.
Frequent to common in prairie and open woods over the state.
Represented in South Dakota by the two subspecies *pallida* (A.
DC.) Piehl. and *umbellata*. May-June. Bastard Toadflax.

Family Aristolochiaceae

ASARUM L.

Asarum canadense L. Perennial herbs from a slender rhizome,
acaulescent, usually with two cordate shaped leaves on petioles
6-10 cm long. One or two flowers on short pedicels at the soil
level. Calyx hairy, deep purple on their inner faces, three-
lobed. Stamens 12. Petals lacking. Fruiting capsule 1-3 cm
long, with several large, ovoid seeds. Rare in deep rich, north-
facing maple-basswood forest of Sieche Hollow in Marshall and
Roberts counties in the northeast part. May. Wild Ginger.

Family Polygonaceae

(Key to Genera)

1 Bases of leaves without sheathing stipules (ocreae). Flower
   clusters subtended by involucral bracts . . . . *ERIOGONUM*

1' Bases of leaves with sheathing stipules (ocreae). Flowers
   without subtending involucral bracts.

   2 Flowers in terminal corymbiform clusters. Leaves triangu-
      larly hastate, petioled . . . . *FAGOPYRUM*

   2' Flowers in spike-like racemes, either terminal or axillary.
      Leaves various.

3   Sepals 5, in one series, often petaloid and similar . .
. . . . *POLYGONUM*

3' Sepals 6, in two series, not petaloid. The inner series
often enlarged in fruit to form valves . . . .
. . . . *RUMEX*

ERIOGONUM Michx.

1   Plants annual or biennial, taprooted, without a woody rootstock.

2   Cauline leaves of the same shape but becoming reduced in
size upward. Principal leaves lanceolate, entire, 3-5 cm
long, densely tomentose on both surfaces. Flowering stems
3-6 dm tall, solitary or rarely a few, tomentose. Inflorescence open-branched, the flowers on dichotomously
branched rays. Flowers white to pink, the perianth 1-2
mm long. Common on dry plains and sandy hillsides from
the central part westward. July-Aug . . . .
. . . . *Eriogonum annuum* Nutt.

2' Cauline leaves much reduced or lacking. Basal leaves ovate
to reniform in outline.

3   Basal leaves glabrous to sparingly hairy on the lower
surface. The blade 1-2.5 cm long, on a petiole 1-3 cm
long. Cauline leaves elliptic, 0.5-1.5 cm long, similar
to the basal leaves. Flowers with hispid hairs on the
outer surfaces. Flowering stems 2-4 dm tall, the inflorescences open. Flowers white to yellow or pink, 1-2 mm
long, the involucres 1-2 mm long. Infrequent on dry
plains and hillsides of the western one-half. July-Aug.
. . . . *Eriogonum visheri* A. Nels.

3' Basal leaves densely pilose to tomentose below and less
so above. The blade about 2 cm long, on well developed
petioles. Cauline leaves much reduced or lacking.
Flowers glabrous. Flowering stems 2-5 dm tall, branched
above. Inflorescence much branched, the ultimate clusters several flowered and crowded on reflexed peduncles.
Perianth white to deep pink, 1-2 mm long, the involucres
1-2 mm long. Infrequent to rare on dry plains and hillsides in the western part. July-Aug . . . .
. . . . *Eriogonum cernuum* Nutt.

1' Plants perennial, with woody rootstocks.

4   Basal leaves usually over 5 cm long, spatulate to oblong.
Bracts below the inflorescence several, leaf-like.

Perianth bright yellow. Plants cespitose, the flowering
stems several, 1-2 dm tall, naked except for the leafy
bracts below the clustered inflorescence. Vegetative
surfaces white-tomentose. Inflorescence umbellate, the
perianth 4-6 mm long, showy. Frequent to common in prairies in the western one-half. July-Aug . . . .
. . . . *Eriogonum flavum* Nutt.

4' Basal leaves usually less than 5 cm long. Bracts below the
   inflorescence scale-like, less than 5. Perianth white to
   rose-colored.

   5 Flowers at the ends of dichotomously open branched
      stems. Plants shrubby or sub-shrubs 2-5 dm tall.
      Leaves 2-5 cm long, linear to oblanceolate, tomentose
      at least below. Inflorescence corymbose, much branched,
      up to 8 cm long. Flowers white to pink, the perianth
      2-4 mm long. Rare in the southwest part of the state
      but common westward in the plains. Ours var. *effusum*.
      June-July . . . . *Eriogonum effusum* Nutt.

   5' Flowers in dense heads or umbellate clusters. Plants
      densely cespitose or forming mats up to 2 dm across.
      Stems shrubby, the flowering stems up to 15 cm tall.
      Leaves essentially basal at the crown of the caudex,
      fewer upwards on the stem. Leaf blades linear-lanceo-
      late, 2-5 cm long, their margins revolute. Flowers in
      a contracted head-like umbel up to 1 cm across. Peri-
      anth white to rose colored, 3-4 mm long, glabrous to
      hairy on the external surface. Frequent in dry, sandy
      prairie in the western part. {incl. *E. multiceps* Nees.}
      Ours var. *pauciflorum*. July-Aug . . . .
      . . . . *Eriogonum pauciflorum* Pursh

FAGOPYRUM Mill.

Fagopyrum esculentum Moench. Annual herbs with simple or
branched stems 2-6 dm tall. Leaves alternate, triangularly
hastate, glabrous, the blades 2-7 cm long. Flowers perfect, in
corymbs or fascicled from upper axils of short bracts. Perianth
white or pink, 2-3 mm long. Fruiting achenes smooth and shiny,
about 7 mm long. Escaped from cultivation but not persisting.
July-Aug. Buckwheat.

POLYGONUM L.

1 Stems trailing or twining. The leaves broadly ovate to cordate. The outer sepals winged or keeled at maturity.

2   Outer sepals becoming conspicuously winged and much exceed-
    ing the achenes. Plants perennial, the stems 1-3 meters
    long. Stem sharply angled with roughened, minute teeth.
    Leaves ovate to broadly cordate with acuminate tips. In-
    florescence in racemes from upper axils, the perianth
    white. Achenes black, glossy, 3-4 mm long. Frequent in
    thickets and moist alluvial places over the state. July-
    Aug. False Buckwheat . . . . *Polygonum scandens* L.

2' Outer sepals keeled but not conspicuously winged, not much
    exceeding the achenes. Plants annual, twining, the stems
    up to 1 meter long, not sharply angled. Leaves triangu-
    larly cordate, their bases often sagittate. Inflorescence
    in interrupted racemes from upper axils, 2-6 cm long.
    Perianth green, 1.5-2 mm long. Achenes dull, 3-4 mm long.
    Frequent to common on roadside fences and in thickets over
    the state. June-Aug. Black Bindweed . . . .
    . . . . *Polygonum convolvulus* L.

1' Stems erect or spreading, not trailing or twining.

3   Flowers only in axillary locations; solitary or in short
    racemes.

    4   Flowers and fruits on deflexed pedicels or sessile and
        enclosed in the ocreae.

        5   Flowers hidden in the ocreae, sessile. Plants an-
            nual, branched, the stems glabrous, up to 3 dm tall.
            Leaves linear-lanceolate, the upper ones reduced to
            less than 1 cm long. Ocreae brown or reddish, becom-
            ing lacerate with age. Perianth green with white
            margins. Achenes dull, lanceolate, about 3 mm long.
            Rare to infrequent in sandy prairie over the state.
            Sept . . . . *Polygonum leptocarpum* Robs.

        5' Flowers and fruits deflexed on pedicels that are re-
            duced. Plants slender, annual, 2-4 dm tall. Leaves
            linear to narrowly lanceolate, 2-4 cm long, with sub-
            ulate tips. Flowers on pedicels 2-3 mm long, the
            perianth 3-4 mm long, with white to reddish distal
            margins. Achenes black and shiny, 3-4 mm long. Rare
            in rocky soil of middle altitudes in the Black Hills.
            July-Aug . . . . *Polygonum douglasii* Greene

    4' Flowers and fruits on erect pedicels.

        6   Principal leaves with two longitudinal folds near the
            midrib, the leaves linear, 1-3 cm long. Upper leaves
            reduced in size. Plants annual, slender, the branched
            stems 1-3 dm tall. Flowers few, scattered in upper

leaf axils. Achenes 2.0-2.5 mm long, sharply angular. Rare on Sioux quartzite in Minnehaha County in
the eastern part and in the Black Hills. July-Aug.
Slender Knotweed . . . . *Polygonum tenue* Michx.

6' Principal leaves flat, not longitudinally folded.
Upper leaves not appreciably reduced.

   7 Stems erect or ascending. Distal margins of
   sepals yellow-green, the outer 3 cucullate or
   prow-shaped.

      8 Leaves oval or oblong, not more than 3 times
      longer than wide. Stems ascending, much-
      branched, 1-5 dm tall. Leaf blades blue-
      green, tapered to the petiole. Ocreae white,
      persistent. Sepals unequal, the outer 3 longer
      and hiding the two inner ones. Achenes dull to
      shiny, dark olive, about 2.5 mm long. Common
      in waste places over the state. Aug-Sept . . .
      . . . . *Polygonum achoreum* Blake

      8' Leaves linear to narrowly ovate, usually 4
      times longer than wide. Stems erect, branched.

         9 Stems usually 5-20 cm tall, sharply angled,
         scabrous, branched from the base. Upper
         bracts sharply pointed. Flowers 1-4 in
         axils along the upper three fourths of the
         stem axis. Leaves lanceolate to oblong,
         usually less than 2.5 cm long. Achenes
         smooth and shiny-black. Rare to infrequent
         in lower altitudes of the Black Hills in
         the western part. July-Aug . . . .
         . . . . *Polygonum sawatchense* Small

         9' Stems usually 20-80 cm tall, not sharply
         angled or scabrous, with branches occurring
         along the length of the main stems. Upper
         bracts linear, not sharply pointed. Flowers
         1-3 in upper axils, many times enclosed by
         the fibrous, lacerate stipules. Leaves
         linear to lanceolate, the principal ones
         usually exceeding 3 cm in length. Achenes
         dull, with a granular surface. A highly
         variable and common species of dry to moist
         areas of prairie and waste land over the
         state. July-Sept . . . .
         . . . . . . *Polygonum ramosissimum* Michx.

7' Stems prostrate, spreading. The distal margins of
the perianth pink to rose-colored, flattened, not
cucullate or prow-shaped. Plants annual, up to 7
dm across. Leaves many, lance-oblong to oblanceolate, 1-3 cm long and up to 7 mm wide. Flowers
1-3 in the axils, the nodes many and telescoped.
Achenes brown, 2-3 mm long. A very common weed
of waste places, lawns and yards. {*P. aviculare*
L.} July-Sept. Common Knotweed . . . .
. . . . *Polygonum arenastrum* Jord.

3' Flowers in terminal spike-like racemes only or in axillary
locations and in terminal spike-like racemes.

10 Stems armed with downcurved prickles or spines. Stems
4-angled, twining, annual. Leaves sagittate, lanceolate to elliptical. Inflorescence short, seldom over
1 cm long, on long peduncles. Flowers dense, pink to
red. Achenes 2-3 mm long. Rare to infrequent in
marshy areas of Shannon and Todd counties in the sandhills of the southern part. July-Aug . . . .
. . . . *Polygonum sagittatum* L.

10' Stems not armed with prickles or spines.

11 Plants with simple stems, erect from fleshy rootstocks, the stems 1-3 dm tall. Flowering spike
solitary, terminal, the lower flowers replaced by
bulblets. Inflorescence 2-5 cm long and 4-8 mm
thick. Leaves few, the basal ones oblanceolate,
the upper ones reduced in size. Flowers white to
pink with stamens exserted. Rare in cool, moist
ravines at upper altitudes in the Black Hills.
July-Aug . . . . *Polygonum viviparum* L.

11' Plants with branched stems. Flowering spikes 1-
several, the lower flowers not replaced by bulblets.

12 Inflorescence terminal, rose-colored, usually
solitary. Plants perennial in marshy or wet
areas. Stamens usually 5.

13 Peduncle below the inflorescence glabrous.
Inflorescence short, 1-3 cm long and over
1 cm wide. Plants with long rhizomes, the
stems rooting at the nodes. Leaves elliptic, glabrous, 7-13 cm long. Achenes 2.5
mm long, dark brown, lenticular. Infrequent in potholes and marshes, principally

in the northern one half of the state.
{incl. *P. natans* Eat.} July-Aug . . . . .
. . . . *Polygonum amphibium* L.

13' Peduncle below the inflorescence pubescent.
Inflorescence 3-6 cm long and up to 1 cm
wide. Plants rhizomatous, the stems elongate, rooting at the nodes. Leaves ellip-
tic-lanceolate, pubescent. Flowers bright
pink to red. Achenes 2.5-3 mm long, lenticular, brown to black. Common in marshy
areas over the state, often forming a dense
growth in aquatic situations. July-Aug.
Water Smartweed . . . .
. . . . *Polygonum coccineum* Muhl.

12' Inflorescence axillary and terminal, white to
pink. Plants usually annual, of moist to drier
locations. Stamens usually more than 5.

14 Ocreae (sheaths) entire or split from the
distal margin but lacking bristles.

15 Inflorescences drooping or nodding.
Perianth white to pale pink, 2-3 mm
long. Plants 3 10 dm tall, with stout
stems. Leaves oblanceolate, 5-20 cm
long. The outer 3 perianth segments
becoming strongly 3-nerved in fruit.
Achenes 2 mm long, brown to shiny
black. Common over the state in low
places or areas that become occasionally inundated. July-Sept. Nodding
Willow Weed . . . .
. . . . . . *Polygonum lapathifolium* L.

15' Inflorescences erect or nearly so.
Perianth pink or lavender.

16 Styles or stamens exsert from the
perianth. Flowering racemes elongate, 2-8 cm long. Plants annual
to weakly perennial, the stems
mostly erect, 3-5 dm tall. Leaves
linear to lanceolate, the margins
ciliate, 3-8 cm long. Flowers dimorphic, some with styles exsert,
others with stamens exsert.
Achenes flattened, granular, black.
Infrequent in low places in the

southeast part. July-Sept . . . .
. . . . . . *Polygonum bicorne* Raf.

16' Styles and stamens included in the
   perianth. Flowering raceme 3-5 cm
   long. Plants annual, deep green,
   the stems 2-8 dm tall. Leaves
   lanceolate, 5-20 cm long. Flowers
   pink, 3-4 mm long, the outer 3
   perianth segments lacking prominent
   nerves when in fruit. Achenes 3-
   3.5 mm long, shiny brown or black.
   Common in rich, waste soil over the
   state. July-Sept . . . .
   . . . . *Polygonum pennsylvanicum* L.

14' Ocreae (sheaths) with bristles.

17 Outer surface of perianth glandular-
   dotted.

18 Inflorescences mostly erect in
   flower. Achenes black or nearly
   so, shining. Stamens mostly 8.
   Plants spreading to erect, the
   stems up to 8 dm tall, annual.
   Leaves numerous, 5-10 cm long,
   oblanceolate. Perianth with usually 5 segments. Achenes 3 mm
   long. Quite infrequent to rare
   in moist areas of the extreme
   eastern part. July-Aug. Water
   Smartweed . . . .
   . . . . . *Polygonum punctatum* Ell.

18' Inflorescences often drooping in
   flower. Achenes brown, dull, glandular. Stamens mostly 6. Plants
   erect or spreading, branched, to 6
   dm tall. Leaves lance-ovate. Perianth commonly 4-lobed, the flowers
   greenish white. Achenes 2.5-3.5 mm
   long. Infrequent in rich wet soil
   in the eastern and northeastern
   part. July-Sept . . . .
   . . . . . . *Polygonum hydropiper* L.

17' Outer surface of perianth not glandular-
   dotted.

19 Inflorescence slender, interrupted,
3-6 cm long. Flowers with 8 stamens. Plants perennial, stems to 1
meter long, rooting freely at the
nodes. Leaves lanceolate, numerous, 5-12 cm long. Achenes black
or dark brown, shiny, 2.5-3 mm
long. Rare in ponds and marshy
areas in the western part. July-
Aug. Water Pepper . . . .
. . *Polygonum hydropiperoides* Michx.

19' Inflorescence short, thick, continuous, seldom over 3 cm long. Flowers pink to rose, with 6 stamens.
Plants annual, the stems 2-7 dm
tall. Leaves with a conspicuously
dark blotch at the mid-section, the
blades lance-ovate. Achenes black,
smooth and shiny, 2.0-2.5 mm long.
Infrequent in rich soil of waste
areas over the state. July-Sept.
Lady's Thumb . . . .
. . . . . . *Polygonum persicaria* L.

RUMEX L.

1 Principal leaves with blades hastate or sagittate at the base.
Vegetation with a sour, acid taste. Plants dioecious, perennial from slender rhizomes, the stems 1-4 dm tall. Leaves
variable but the principal ones narrowly elliptic, 3-10 cm
long. Inflorescence narrow, paniculate, the flowers purple-
green. Fruits 1-2 mm long, tawny brown. Valves with entire
margins. Frequent in thin, sandy soil of waste places, more
common in the western one half of the state. May-June. Sheep
Sorrel . . . . *Rumex acetosella* L.

1' Principal leaves not hastate or sagittate.

2 Margins of valves of fruiting achenes with spinulose teeth.

3 Fruiting valves 4 mm or more long, coarsely toothed.
Plants perennial, the stems to 8 dm or more tall, sparingly branched. Leaves lanceolate, the blades 8-30 cm
long, flat to undulate-margined. Inflorescence paniculate with ascending branches, dense, leafy. Valves in
fruit usually with 3 tubercles. Introduced and occasional in moist swales and roadside ditches over the
state. May-June . . . . *Rumex stenophyllus* Ledeb.

3' Fruiting valves about 2 mm long, with elongate spinulose
   teeth. Plants annual, 2-6 dm tall. Principal leaves
   lanceolate, glabrous, the blades to 15 cm long. Inflorescence dense, leafy, in tight verticels, becoming reddish brown in fruit. Valves in fruit usually with 3
   small tubercles. Achenes 1 mm long. Frequent to common
   in sandy alluvium over the state. Ours var. *fueginus*
   (Phil.) Dusen. Aug-Sept. Golden Dock . . . .
   . . . . *Rumex maritimus*

2' Margins of the valves of fruiting achenes entire or undulate but not spiny toothed.

   4 Base of the tubercle or grain above the base of the
      valve. Tubercles 3. Pedicels lacking a visible joint.
      Plants perennial, stout, the stems up to 2 meters tall.
      Leaves large, the principal ones lanceolate with the
      blades rounded at their bases. Inflorescence of several
      ascending branches. Valves of the fruit rounded, 5-8 mm
      long and wide. Occasional in swamp and shallow water in
      the eastern part. July-Aug . . . *Rumex orbiculatus* Gray

   4' Base of the tubercle or grain even with the base of the
      valve. Pedicels with or without visible joints.

      5 Valves of fruiting achenes broadly rounded in out-
         line, the wings prominent, 7 mm or more across.
         Tubercle 1 or lacking.

         6 Mature valves up to 1.5 cm across, with or without
            a tubercle or grain. Plants perennial, from tap-
            roots, the stems up to a meter tall. Principal
            leaves basal, with long petioles, the blades 2-3
            dm long. Flowering panicles dense, many-flowered,
            with many ascending branches. Valves broadly
            rounded with a deeply cordate base. An occasional
            escape over the state. June-July. Patience Dock.
            . . . . *Rumex patientia* L.

         6' Mature valves 2-3 cm across, without a tubercle or
            grain. Plants perennial, from a spreading root-
            stock, the stems ascending, 2-6 dm tall. Leaves
            firm, leathery, 3-9 cm long, with ovate to oblong
            blades. Inflorescence paniculate, the perianths
            reddish. Maturing fruits with showy reddish
            valves, their margins wing-like, entire. A showy
            plant frequent on dry ridges and plains in the
            western part. May-June. Sour Greens . . . . . .
            . . . . *Rumex venosus* Pursh

5' Valves of fruiting achenes triangular to ovate in outline, the wings usually less than 7 mm across.

  7 Tubercles 1-3 on each fruit.

    8 Leaves crisp-margined and wavy, the blades of principal leaves oblong to lanceolate. Plants perennial from taproots, the stems rarely branched below the inflorescence, 2-8 dm tall. Maturing valves thin, ovate, about 5 mm long, with 3 grains, these often unequal in size. Frequent weed of roadsides and waste places, especially in the eastern part. May-June. Sour Dock . . . . *Rumex crispus* L.

    8' Leaves flat, the blades narrowly lanceolate to narrowly oblong, pale green, glabrous.

      9 Valves at maturity about 3 mm long; the leaves narrowly lanceolate or linear. Plants up to 1 meter tall, branched. Leaves numerous, varying in length, the blades willow-like. Inflorescence leafy, paniculate with ascending branches. Perianth segments greenish. Achenes 2 mm long. A common and variable species in a variety of habitats over the state. {incl. *R. triangulivalvis* Danser Rech. f. and *R. salicifolius* Weinm.} June-Aug. Willow-leaved Dock . . . . . . *Rumex mexicanus* Meissn.

      9' Valves at maturity about 5 mm long; the leaves lanceolate to oblong. Plants perennial, 4-8 dm tall, branched. Leaves pale green, flat. Inflorescence branched, loose, the flowers with jointed pedicels. Perianth green, about 2 mm long. Tubercles 1-3, often unequal in size. Frequent in moist ditches of roadsides and waste places over the state. May-July. Water Dock . . . . *Rumex altissimus* Wood

7' Tubercles lacking on the valves in fruit. Plants with a vertical taproot, perennial, the stem usually single, stout, up to 1 meter tall. Principal leaves long-petioled, the blades wavy-margined, oblong to ovate, 1-3 dm long. Inflorescence a large panicle with ascending branches. Valves 3-6 mm long, with reticulate veins. Infrequent in swales of prairie in the north and western part.

{incl. *R. fenestratus* Greene} June–July. Western
Dock . . . . *Rumex occidentalis* Wats.

Family Chenopodiaceae

(Key to Genera)

1 Stems fleshy, much jointed with thick internodes. Leaves reduced to scale-like structures . . . . *SALICORNIA*

1' Stems not jointed or fleshy. Leaves not scale-like.

  2 Leaves or leaf-like bracts narrowed to spiny or spiny-
tipped structures, sometimes shrubby.

    3 Flowers imperfect, the staminate ones in catkin-like
spikes. Plants perennial . . . . *SARCOBATUS*

    3' Flowers perfect, not in terminal spikes. Plants annual
. . . . *SALSOLA*

  2' Leaves not narrowed to spiny-tipped structures; however,
they may be linear in general shape.

    4 Flowers perfect or polygamous, perianth present on all
flowers.

      5 Fruiting perianth with obvious horizontal wings like
a halo.

        6 Flowers and fruits in the axils of leaves. Leaves
lanceolate, entire . . . . *KOCHIA*

        6' Flowers and fruits in spikes or branches thereof,
not axillary. Leaves toothed . . . *CYCLOLOMA*

      5' Fruiting perianth or fruits without obvious horizon-
tal wings; however, they may be vertically winged.

        7 Stamens more than 3. Fruits at maturity hidden
within the perianth.

          8 Leaves fleshy, linear, less than 2 mm wide,
entire . . . . *SUAEDA*

          8' Leaves not fleshy, more than 2 mm wide, often
toothed . . . . *CHENOPODIUM*

        7' Stamens 1–3, fruits at maturity visible, not hid-
den in the perianth.

          9 Fruits vertically winged, acute. Leaves en-
tire, linear . . . . *CORISPERMUM*

9' Fruits not winged. Leaves in ours hastate
lobed . . . . *MONOLEPIS*

4' Flowers imperfect, the pistillate ones lacking a perianth.

10 Vegetation with a scurfy or mealy surface, without
pubescence. Plants annual or perennial . . . . . .
. . . . *ATRIPLEX*

10' Vegetation with a stellate pubescence, neither
scurfy or mealy. Plants perennial with a stout
taproot . . . . *CERATOIDES*

ATRIPLEX L.

1 Plants annual, mostly monoecious. Leaves various, usually
toothed or lobed, not linear.

2 Fruiting bracts on many pistillate flowers appearing as
circular wings 8-12 mm across. Plants erect to spreading,
the stems branched, over 6 dm tall. Principal leaves about
7 cm long and 3-5 cm wide, greenish. Flowers in terminal
and axillary panicles, erect and spikelike. Fruits one-
seeded, flattened, but surrounded by basally connate bracts.
Infrequent as an escape in waste places. July-Aug. French
Spinach . . . . *Atriplex hortensis* L.

2' Fruiting bracts on pistillate flowers not circularly winged.

3 Foliage and young stems with a gray or white farinose,
mealy covering. Leaves mostly alternate.

4 Leaves conspicuously 3-nerved, entire. Fruiting
bracts of pistillate flowers not more than 4 mm long.
Plants 1-10 dm tall, branched from the base, erect.
Leaves 2-4 cm long, silvery white. Flowers in axillary clusters, the bracts on the pistillate ones oblong, entire. Rare on alkali flats west of the Missouri River. June-July . . . *Atriplex powellii* Wats.

4' Leaves not conspicuously 3-nerved. Fruiting bracts
of the pistillate flowers 4-10 mm long.

5 Margins of leaves dentate. Plants erect, 2-10 dm
tall, sparingly branched from the base. Vegetation white mealy. Leaf blades ovate to rhombic,
acute to obtuse at the base and apex. Flowers in
axillary clusters, the staminate and pistillate
mixed near the middle of the plant. Infrequent in
weedy and disturbed soil and alkali flats over the
state. July-Aug . . . . *Atriplex rosea* L.

5' Margins of leaves entire to undulate but not dentate. Plants rounded in outline due to branches
from the base, the stems 2-8 dm tall. Leaf blades
ovate to triangular, 2-4 cm long. Plants monoecious, the pistillate ones in axillary clusters.
Fruiting bracts 4-8 mm long, lacinate at their
distal ends. Common on clay soil, alkali spots
and eroded banks over the state. More common in
the western one-half. July-Aug . . . .
. . . . *Atriplex argentea* Nutt.

3' Foliage and young stems green, not covered with a gray-
white mealy substance. Leaves opposite or sub-opposite,
at least below.

6 Principal leaves less than 4 cm long, sessile. Staminate flowers clustered toward the tip of the flowering spike, the perianth with dorsally crested lobes.
Plants annual, with low, spreading branches. Stems
usually less than 3 dm tall. Leaves numerous, elliptic to ovate. Pistillate flowers solitary or few in
the axils of leaves. Staminate ones mostly in short,
terminal spikes. Seeds 1.5 mm long, subtended and
enclosed by a pair of bracts 2 mm long. Common on
clay soils and "badlands" in the western one-half.
July-Aug. Rillscale . . . .
. . . . *Atriplex dioica* (Nutt.) McBride

6' Principal leaves more than 4 cm long, petioled. Staminate flowers mixed with the pistillate ones in
axils of leaves and at the ends of branches. Plants
commonly over 3 dm tall, sparingly branched. Leaves
narrow to broadly ovate, the blade bases various.
Fruits with bracts deltoid to ovate, one-fourth to
one-half the length of the perianth. Seeds about
2.0 mm long. Common on alkali or saline soils over
the state. July-Aug. Spearscale . . . .
. . . . *Atriplex patula* L.

1' Plants perennial, dioecious. Leaves broadly linear with
rounded tips. Margins of leaf blades not toothed or lobed.

7 Pistillate plants with broad, elliptical bracts subtending
the fruits. Leaves tapering to a sessile base, scurfy
gray, 1-5 cm long. Plants perennial, woody, variously
branched, from 2-12 dm tall, with stems white to gray-
canescent. Flowers in sparse to congested spikes. Pistillate bracts completely surrounding the fruits. Frequent on
clay soil and alkali flats in the western one half of the

state. July-Aug. Shadscale . . . .
. . . . *Atriplex canescens* (Pursh) Nutt.

7' Pistillate plants lacking elliptical wings on the fruits.
Leaves tapering to a short petiole, the blades linear-
spatulate to oblanceolate, 2-5 cm long. Plants perennial,
the stems low, spreading, the main stem woody at the base,
usually less than 3 dm tall. Staminate plants with flowers
in leafy spikes, the pistillate plants with flowers in congested axillary spikes. Frequent on saline or alkaline
soils in the western part. July-Aug. Moundscale . . . . .
. . . . *Atriplex nuttallii* Wats.

CERATOIDES Gagnebin.

Ceratoides lanata (Pursh) J. T. Howell. Shrubby plants with
woody bases, the stems branched and spreading, up to 6 dm tall,
from a deep taproot. Leaves linear, up to 3 cm long, with a
stellate pubescence. Flowers in the axils of upper leaves, the
staminate ones uppermost. Pistillate fruiting bracts 4-8 mm
long, densely villous, becoming tawny brown at maturity. Frequent on dry hillsides of the western part. May be toxic to
grazing animals in winter time if eaten in large quantities.
{*Eurotia lanata* (Pursh) Moq.} May-June. Winterfat.

CHENOPODIUM (Tourn.) L.

1 Foliage with glands, distinctly aromatic. Plants variously
pubescent.

2 Perianth glabrous to lightly pubescent or slightly glandular. Plants annual or biennial, 3-9 dm tall. Leaves
fleshy, firm, sinuately-pinnatifid, 3-8 cm long. Flowers
sessile in bracted or bractless axillary spikes, numerous.
An ill-scented weed that is infrequent to rare in farmyards and waste places in the southeast part. Aug-Sept.
Wormseed . . . . *Chenopodium ambrosioides* L.

2' Perianth obviously glandular and pubescent. Plants annual,
the stems 1-3 dm tall, branched from below. Leaves up to
3 cm long, the margins pinnatifid or sinuately lobed. Inflorescence terminal, its length up to one-half the height
of the plant. Flowers small, greenish, numerous in axillary racemes. Rare to infrequent in dry to moist sandy
places in the western part. July-Aug. Jerusalem Oak . . .
. . . . *Chenopodium botrys* L.

1' Foliage without glands, not obviously aromatic. Plants not
pubescent.

3   Plants commonly less than 1.5 dm tall, occasionally up to
    2 dm, but usually low and spreading with many branches.
    Leaves thick, densely farinose, the blades up to 2 dm long
    with an irregular pair of basal lobes. Inflorescences
    short spikes of congested flowers. Perianth 5-parted.
    Pericarp separable from the seed. Seed horizontal (flat-
    tened dorso-ventrally), 0.9-1.1 mm across. Rare on dry
    clay banks of the Badlands. July-Aug . . . .
    . . . . *Chenopodium incanum* (S. Wats.) Aellen

3' Plants commonly more than 2 dm tall.

    4   Flowers and fruits in dense, globose glomerules over 5
        mm in diameter. Inflorescence reddish, spike-like in
        the upper axils of leaves. Plants annual, the stems
        simple to branched, 1-5 dm tall. Foliage not farinose.
        Principal leaves triangular, with sinuate margins and
        hastate bases, petioled. Perianth 3-parted, the fruits
        flattened laterally. Infrequent in dry, sandy soils of
        prairies and waste places in the western part. June-
        Aug. Strawberry Blite . . . .
        . . . . *Chenopodium capitatum* (L.) Asch.

    4' Flowers and fruits not in dense, globose glomerules.

        5   Perianth parts usually 3. Terminal flowers in the
            glomerules with horizontal fruits, the lateral ones
            with vertical fruits (flattened laterally). Plants
            of alkaline soils.

            6   Leaves densely white-farinose on the lower sur-
                faces, ovate in outline with 2-5 low, sinuate-
                dentate teeth on each margin. Blades mostly less
                than 1.5 cm wide and 2.5 cm long. Perianth parts
                glabrous. Plants decumbent to erect, 1-3 cm tall.
                Flowers sessile in axillary spikes. Frequent in
                alkaline swales over the state. {*C. salinum*
                Standl.} July-Aug . . . . *Chenopodium glaucum* L.

            6' Leaves glabrous, along with the perianth becoming
                tinged with red towards maturity. Leaves entire
                to few-toothed, more than 2 cm long, the principal
                leaves rhombic-ovate with 1 or more conspicuous
                teeth on each side. Perianth inconspicuously
                farinose. Plants usually low and spreading, pros-
                trate to erect, 1-4 dm tall. Flowers in axillary
                spikes that are shorter than the leaves. Frequent
                in brackish or alkaline areas over the state.
                June-Sept. Alkali Blite . . . .
                . . . . *Chenopodium rubrum* L.

5' Perianth parts usually 5. Seeds horizontal (flattened dorso-ventrally), only rarely a few vertical.

   7 Principal leaves linear or nearly so, 6 times or
more longer than broad, 1-nerved, the margins entire. Plants 2-5 dm tall, the main stem angled,
branched from the lower one-third. Inflorescence
of slenderly branched stems, the glomerules dis-
tantly spaced and overtopping the leaves. Pericarp readily separating from the seed. Seeds
black and shiny, 1.5 mm across. Infrequent in
sandy soils in the south and southwest part.
July-Aug . . . .
. . . . . *Chenopodium subglabrum* (Wats.) A. Nels.

   7' Principal leaves lance-ovate to rhombic or deltoid,
less than 6 times longer than broad, variously
toothed to entire.

      8 Flowers and fruits in various stages of matu-
rity in the same glomerules. Foliage green, of
a papery-thin texture. If leaves are farinose,
then only sparingly so on the lower surfaces.

         9 Principal leaf blades deltoid, the base with
a prominent upturned hastate lobe, hardly
longer than broad. Plants upright in habit,
over 2.5 dm tall, the stems slender. Flowers clustered on short, lateral branches,
giving the inflorescence a spike-like appearance. Pericarp maturing free from the seed
and exposed. Seeds 1.1-1.5 mm across. Infrequent in canyons and shady places over
the state. July-Aug . . . .
. . . . *Chenopodium fremontii* S. Wats.

         9' Principal leaf blades rounded or tapering to
the base, lance-ovate to rhombic-ovate in
outline, longer than broad.

           10 Leaves broadly ovate, 5-10 cm long, with
2-3 acute teeth on each side (maple-
leaved). Leaves bright green, thin,
glabrous except for the inflorescence.
Plants 3-8 dm tall, the stems mostly
simple. Inflorescence a pyramidal
paniculate spike, the flowers sessile.
Seeds 1.5-2.5 mm across. Frequent in
waste, shady places over the state.

{*C. gigantospermum* Aellen} July–Sept.
Maple-leaved Goosefoot . . . .
. . . . *Chenopodium hybridum* L.

10' Leaves lanceolate to ovate, up to 6 cm
long, entire, or the larger ones with
1-several short, ascending teeth. Upper leaves reduced to linear bracts
that subtend most of the inflorescence
branches. Plants up to 8 dm tall, deli-
cately branched, especially when growing in shady places. Flowers single to
few in a series of short, interrupted
spikes. Seeds black, 1.1–1.5 mm across.
Infrequent in dry woods, more common in
the eastern part. Aug–Sept . . . . . .
. . . . *Chenopodium standleyanum* Aellen

8' Flowers (fruits) in the glomerules at approximately the same stage of maturity. Foliage
dull green, membranous to coriaceous and
fleshy. Leaves and inflorescences variously
farinose.

11 Principal leaves narrowly lanceolate to
ovate, mostly less than 1.5 cm wide, 3-
several times longer than broad, densely
farinose. Plants low and diffuse to
upright-spreading, 1–8 dm tall. Inflorescence of glomerules in dense, paniculate spikes, farinose. Pericarp readily
separated from the seed, but the perianth
mostly covering the fruit at maturity.
Seed about 1 mm across, black, shiny. A
common species of dry places over the
state. June–Aug . . . .
. . . . . *Chenopodium dessicatum* A. Nels.

11' Principal leaves ovate to rhombic, entire
to variously toothed, 1–3 times longer
than broad.

12 Seeds smooth. Leaves membranous, the
principal ones variously toothed to
lobed, the upper ones reduced.

13 Sepals covering the fruit at maturity yellow-margined. Seeds 1.0–
1.5 mm across. Principal leaves
widest below the middle of the

blade, remotely triangular and irregularly toothed, the upper leaves
reduced but not appreciably different in shape. Plants relatively
upright, the stems simple to freely
branched, 2-8 dm tall. Inflorescence of dense, contiguous glomerules in paniculate spikes. Common
in waste places over the state.
June-Aug. Goosefoot . . . .
. . . . *Chenopodium album* L.

13' Sepals exposing the fruit at maturity, greenish-yellow at the margins. Seeds 0.9-1.2 mm across.
Principal leaves widest near the
middle or slightly below the middle, generally ovate-oblong in outline with shallow serrations. Upper leaves reduced in size and oblong in outline, entire. Plants
3-5 dm tall or taller in rich soil,
branched upward. Inflorescence of
flowers in distinct or close glomerules on axillary spikes. Robust
plants may have inflorescences of
loose, terminal panicles. Frequent
in a variety of habitats over the
state. July-Sept . . . .
. . . . *Chenopodium strictum* Roth.

12' Seeds minutely alveolate or reticulate.
Leaves thin-membranous to coriaceous.

14 Seeds 1.0-1.5 mm across. Leaves
thick, light green in color, narrow
to broadly ovate, with an obtuse
apex, irregularly toothed. Sepals
broadly keeled. Pericarp often re-
vealing a definitely yellowish
style base. Plants 3-9 dm tall,
much branched. Inflorescence
largely terminal, in slender paniculate spikes. Much like *C. album*
L. and often difficult to distin-
guish in immature plants. Common
in waste places over the state.
July-Sept . . . .
. . . *Chenopodium berlandieri* Moq.

14' Seeds 1.5-2.0 mm across, relatively
flat. Leaves thin-membranous, light
green in color, large, the principal
ones with sinuate-dentate margins,
others lanceolate. Sepals not prominently keeled. Plants with erect
stems up to a meter or more tall.
Inflorescence dense, much-branched,
becoming lead-gray in color towards
autumn, and plants becoming almost
completely leafless. Infrequent in
alluvial soils in the eastern part.
July-Aug . . . .
Chenopodium bushianum Aellen

CORISPERMUM L.

1 Spikes densely flowered, stout, usually more than 4 mm wide.
Fruits longer than 3 mm towards maturity, winged, narrower
than the subtending bract. Plants annual, with ascending to
erect branches, 1-5 dm tall. Leaves 2-6 cm long, linear, not
over 3 mm wide. Infrequent in sandy soil, especially on sand
bars, throughout the southern part of the state. Aug-Oct.
Bugseed . . . . *Corispermum hyssopifolium* L.

1' Spikes slender, not densely flowered, less than 4 mm wide.
Fruits less than 4 mm wide. Fruits less than 3 mm long,
winged, wider than the subtending bract. Plants annual,
branched above with lax flowering spikes. Leaves 2-5 cm
long, narrowly linear, less than 1 mm wide. Rare in dry,
sandy soil in the north and western part. Aug-Sept . . . . .
. . . . *Corispermum nitidum* Kit.

CYCLOLOMA Moq.

Cycloloma atriplicifolium (Spreng.) Coult. Annual plants with
many branches, becoming subglobose in shape, the stems up to 5 dm.
Leaves alternate, irregularly toothed or dentate, dropping when
the plant matures. Flowers in paniculate spikes, the fruits depressed, with a horizontal winged margin. Frequent in sandy soil
in the entire southern part. July-Aug. Winged Pigweed.

KOCHIA Roth.

Kochia scoparia (L.) Schrad. Annual with many branches, the stems
reddish or otherwise pigmented, often to 1.5 meters. Leaves various but mostly linear to lanceolate, up to 5 cm long. Flowers 1-
many, sessile in the axils of upper leaves, perfect or occasionally only pistillate. The color, texture and pubescence various.

Very abundant in waste or disturbed soil over the state. July-Aug. Summer Cypress, Fireweed.

MONOLEPIS Schrad.

Monolepis nuttalliana (Schultes) Greene. Annual herb with branched stems 1-3 dm tall. Leaves fleshy, hastate, on slender petioles, 1-3 cm long. Flowers sessile, clustered in the axils of upper leaves, often reddish. Sepals 1-3 mm long, the pericarp remaining attached to the fruit. Infrequent but locally abundant on alkaline clay or in swales in the western part. May-June. Poverty Weed.

SALICORNIA L.

Salicornia rubra A. Nels. Annual herb with fleshy branched and jointed stems, reddish, up to 10 cm tall. Leaves minute, opposite and scale-like around the stem. Flowers small, borne in clusters on fleshy spikes at the joints, the central one usually somewhat elevated. Infrequent at the margins of alkali lakes and ponds in the northern one-half. July-Aug. Saltwort.

SALSOLA L.

1 Mature calyx segments with thin, transverse wings. Upper leaves and bracts erect and spiny at the apex. Plants annual with many branches, the plants appearing globose, up to 1 meter tall. Leaves spiny, linear, 1-3 cm long. Flowers small, in the axils of upper bracts. Fruiting calyx 3-5 mm wide, enclosing the seed. A troublesome weed which became widely established in the drought of the 1930's. Frequent and locally abundant over the state. {*S. kali* var. *tenuifolia* Mey.} July. Russian Thistle . . *Salsola iberica* Sennen & Pau.

1' Mature calyx longitudinally keeled, not winged. Upper leaves and bracts erect to incurved, not rigidly spiny as above. Plants more upright, not as profusely branched, not globose in shape. Plants annual, mostly erect, 2-7 dm tall. Flowers small, in the axils of upper bracts. Fruiting calyx 3-5 mm wide, wingless, carinate on the back. A weed of recent introduction in the southern and northern part in disturbed places. July-Aug . . . . *Salsola collina* Pall.

SARCOBATUS Nees.

Sarcobatus vermiculatus (Hook.) Emory. Plants shrubby perennials, the stems irregularly branched, whitish, up to 1 meter tall. Leaves alternate, linear, 1-4 cm long, when deciduous the branches becoming spike-like. Flowers monoecious, the staminate ones in terminal spikes. Pistillate flowers in leafy bracts of lower axils, the perianth maturing into broad wings that surround the fruit. Frequent on alkali soils and clay banks of the southwest part. June-July. Greasewood.

SUAEDA Forsk. ex. Scop.

1 Plants annual, up to 4 dm tall. Perianth lobes unequal, some with minute, horn-like projections. Flowers crowded at the nodes of the stem. Vegetation glaucous. Leaves many, linear and rounded in cross section, 1-3 cm long. Upper leaves shorter and broad, lance-ovate. Perianth 5-lobed, the seed maturing horizontal or vertical, 1 mm across. Frequent in alkaline or saline soils over the state. July-Sept. Sea Blite . . . . *Suaeda depressa* (Pursh) Wats.

1' Plants perennial, up to 6 dm tall. Perianth lobes equal, erect, without horn-like projections. Flowers in the axils of upper leaves, on short, stout branches. Plants with woody bases, usually glabrous. Leaves numerous, dense, 1-3 cm long, narrowly linear. Upper leaves not much reduced. Seeds maturing horizontal, 0.8 mm across. Rare in alkaline soils in the northwest part. July-Aug . . . . . . *Suaeda intermedia* Wats.

Family Amaranthaceae

(Key to Genera)

1 Leaves alternate. Anthers with 4 cells but becoming as 2 cells when they dehisce . . . . *AMARANTHUS*

1' Leaves opposite. Anthers with 2 cells but becoming as 1 when they dehisce.

   2 Flowers in terminal spikes. Plants with a tomentose pubescence, especially on the under surfaces of leaves. Calyx with a dense woolly pubescence . . . . *FROELICHIA*

   2' Flowers axillary. Plants with a stellate pubescence. Calyx not woolly pubescent . . . . *TIDESTROMIA*

AMARANTHUS L.

1 Plants monoecious, the male and female flowers together on a spike or on separate spikes.

   2 Flowers primarily in axillary clusters. Leaf blades short and broadly spatulate, less than 3 cm long.

3   Stems prostrate and spreading radiately, forming mats
    2-8 dm across. Leaves pale green, obtusely rounded, on
    a well developed petiole. Flowers in dense clusters,
    the perianth parts 4 or 5. Bracts only slightly longer
    than the perianth. Fruits 2-2.5 mm long, lenticular and
    smooth. Common as a weed of waste places, lawns or cul-
    tivated areas over the state. July-Sept. Prostrate
    Pigweed . . . . *Amaranthus graecizans* L.

3' Stems ascending and bushy branched. Leaves elliptic to
    spatulate, the early vegetative ones longer than those
    on later flowering branches. Flowers in short axillary
    clusters with rigid bracts, greenish. Perianth of fe-
    male flowers usually 3-parted, the bracts longer than
    the perianth parts. Fruits lenticular, 1-2 mm long,
    with circumscissile dehiscence. Common weed of fields
    and waste places in the eastern part. Less common west-
    ward. July-Sept. Tumbleweed . . . *Amaranthus albus* L.

2' Flowers primarily in terminal or narrow paniculate spikes.
Leaf blades lanceolate or broader, usually exceeding 5 cm.
Plants normally not branching from the base.

4   Spikes slender, about 1 cm wide. Bracts subtending the
    pistillate flowers not exceeding 4 mm, thin, only as
    long as or slightly exceeding the sepals. Plants stout,
    the stems erect, 1 meter or more tall. Leaves 8-13 cm
    long, lanceolate to ovate. The inflorescence a compound
    terminal panicle, monoecious. Fruits 1.5-2 mm long, be-
    coming reddish at maturity. An occasional weed which is
    becoming locally abundant in waste places over the state.
    July-Sept. Prince's Feather . . *Amaranthus hybridus* L.

4' Spikes thick, 1.5-2 cm wide. Bracts subtending the
    pistillate flowers 6-8 mm long, rigid, conspicuously ex-
    ceeding the sepals. Plants with stout stems, usually
    branched, 1-1.5 meters tall. Leaves ovate, 6-10 cm
    long, with well-developed petioles. Inflorescence
    densely crowded, terminal and in upper axils, the spikes
    5-25 cm long. Fruits 3-4 mm long, the seeds reddish
    brown. Our most common pigweed of cultivated and waste
    places. Widespread and common over the state. July-
    Sept. Pigweed . . . . *Amaranthus retroflexus* L.

1' Plants dioecious, the male and female flowers on different
plants.

5   Perianth segments of pistillate plants present and readily
    distinguishable.

6 Sepals on pistillate flowers 5, their apexes obtuse. Male flowers with 5 unequal sepals, pointed but not spiny. Stems 3-9 dm tall, erect, mostly glabrous. Leaves lanceolate, 2-8 cm long. Inflorescence of slender terminal spikes, these branched to form a panicle. Bracts spine-tipped, shorter than the perianth. Fruit rounded, the seed about 1 mm across. Infrequent in dry, sandy soil in the south and western part. July-Aug . . . *Amaranthus arenicola* I. M. Johnst.

6' Sepals on pistillate flowers 1 or 2. Male flowers with 5 unequal sepals, the outer ones with the midvein ending in a rigid spine. Plants with stout stems, to 1.5 meters tall, often times with many branches. Petioles slender, the leaves 2-12 cm long, often reddish. Staminate flowers in a single terminal spike or having several lateral ones as well. Fruits of pistillate plants 1.5 mm across, the seed 1 mm. Infrequent in moist sandy soil in the south and east. Aug-Sept. Water Hemp . . . . *Amaranthus tamariscinus* (Nutt.) Wood

5' Perianth segments of pistillate plants usually lacking; occasionally 1 or 2 rudimentary sepals may be recognized. Outer sepals of staminate flowers of 5 unequal segments, not spiny. Plants prostrate to upright, up to 1.5 meters tall, the pistillate plants much-branched. Staminate plants smaller, the flowering spike slender. Leaves lanceolate to ovate, highly variable in length. Infrequent at the margins of ponds and streams in the eastern part. Aug-Sept. {*Acnida altissima* Ridd.} . . . . *Amaranthus tuberculatus* (Moq.) Sauer

FROELICHIA Moench.

1 Principal leaves 1-2 cm wide. Plants stout, erect, with erect branches 2-5 dm tall. Leaves short-petioled, 1-5 cm long, mostly towards the base of the stem. Inflorescence of spikes 1-8 cm long, the flowers sessile, each subtended by a scarious bract. Calyx densely woolly, flask-shaped, about 6 mm long, with wing-like crests. Rare to infrequent on sandy soils and roadsides in the sandhills area of the southern part. Ours var. *campestris* (Small.) Fern. July-Aug. Snake Cotton . . . . *Froelichia floridana* (Nutt.) Moq.

1' Principal leaves less than 1 cm wide. Plants slender, with several basal branches, 3-7 dm tall. Leaves linear to narrowly lanceolate, the upper ones much reduced. Inflorescence spike-like, 1-3 cm long. Calyx woolly, conic in shape, about 4 mm long, with 2 rows of lateral spines. Rare in sandy soil

in the southern part. Cottonweed. July-Aug . . . .
. . . . *Froelichia gracilis* (Hook.) Moq.

TIDESTROMIA Standl.

Tidestromia lanuginosa (Nutt.) Standl. Plants annual herbs with
branched prostrate or weakly ascending stems 1-6 dm long. Leaves
petioled, the blades 5-25 mm long, oval to rhombic. Upper and
lower surfaces of leaves with a stellate pubescence. Flowers in
small glomerules in the axils of leaves, the perianth 1-3 mm long,
not showy. Fruit small, rounded, 1-2 mm across. Rare on dry clay
and sandy soils in the southwestern part. July-Sept.

Family Nyctaginaceae

(Key to Genera)

1 Flowers sessile in heads, sometimes dense. The bracts beneath
   the flowering heads separate and distinct.

   2 Plants annual. Fruits with conspicuous membranous wings
      completely encircling . . . . *TRIPTEROCALYX*

   2' Plants perennial. Fruits with 3-5 thick wings on the
      angles, not completely encircling . . . . . *ABRONIA*

1' Flowers in clusters but pediceled. The bracts beneath the
   clusters united . . . . *MIRABILIS*

ABRONIA Juss.

Abronia fragrans Nutt. Perennial herbs with several stems lax or
ascending, 2-5 dm tall. Leaves opposite, glabrous, the blades
oblanceolate, 2-7 cm long. Flowers white, several sessile in
heads, the corolla tubes 1.5-2.5 cm long. Fruits less than 1 cm
long, with leathery wings. Infrequent in dry, sandy soil in the
southwestern part. May-July. Sand Verbena.

MIRABILIS L.

1 Leaves ovate to broadly cordate, petioled. Bracts subtending
   the inflorescence glabrous or essentially so. Stems several
   from a perennial rootstock, 4-9 dm tall, glabrous or sparsely
   pubescent. Leaf blades 3-8 cm long, cordate to deltoid.
   Flowers pale pink to red, about 1 cm long. Frequent in disturbed soils in a variety of habitats over the state. June-
   July. Wild Four-o'clock . . . .
   . . . . *Mirabilis nyctaginea* (Michx.) MacMill.

1' Leaves lanceolate to linear. Bracts subtending the inflorescence glandular-pubescent.

    2 Stem white-glaucous, essentially glabrous the entire length.
   Leaves linear, mostly less than 1 cm wide and 3-8 cm long.
   Stems to 7 dm tall, 1-5 from a stout root. Flowers 1-4 in
   open terminal clusters with pink-purple perianths 8-12 mm
   long. Frequent on dry prairie and exposed or eroded areas
   over the state. June-July . . . .
   . . . . *Mirabilis linearis* (Pursh) Hiemerl.

    2' Stem hirsute along most of the length but especially below.
   Leaves lanceolate, the principal ones usually wider than 1
   cm. Stems 2-10 dm tall, one or more from a perennial base.
   Leaf blades 2-8 cm long. Inflorescence terminal and axillary, cymose, commonly 3-flowered. Perianth pink to purple, 8-10 mm long. Frequent to common in dry or sandy
   prairie or disturbed places over the state. June-July . .
   . . . . *Mirabilis hirsuta* (Pursh) MacMill.

TRIPTEROCALYX Hook.

Tripterocalyx micranthus (Torr.) Hook. Annual herbs with branched
stems 1-5 dm tall, tending to be succulent. Leaves opposite, the
blades elliptic, 2-4 cm long, entire, the bases oblique. Flowers
in heads, greenish-white, sessile. Perianth petal-like, salverform, about 1 cm long. Stamens 5. Fruits 1.5-3.0 cm long, with
2-3 prominent encircling wings. Rare in sandy soil of the western part. May-June. Sand Puffs.

Family Aizoaceae

MOLLUGO L.

Mollugo verticillata L. Plants herbaceous annuals with radiating
prostrate stems 1-3 dm long. Leaves verticillate on the stems,
linear to spatulate, 1-3 cm long. Flowers 2-5 in axillary clusters, on short pedicels, the perianth white-margined, petaloid,
2-3 mm long. Stamens 3 or 5. Fruiting capsule oblong, 4-5 mm
long, 3-5 valved, with many seeds. A naturalized weed from
tropical New World areas now infrequent in moist areas in the
southern part of the state. July-Aug. Carpet Weed.

Family Portulacaceae

(Key to Genera)

1 Plants prostrate with obovate, fleshy leaves. Petals yellow
. . . . *PORTULACA*

1' Plants upright. Leaves other than obovate. Petals pink or white.

    2 Stem leaves linear, fleshy-succulent. Plants perennial. Inflorescence a cyme with naked peduncles . . . .
. . . . *TALINUM*

    2' Stem leaves perfoliate, a single pair fused completely around the stem to form a disk below the inflorescence. Inflorescence a terminal raceme, the flowers pale pink or white . . . . *MONTIA*

MONTIA L.

Montia perfoliata (Donn) Howell. Annual herb with soft stems 5-15 cm tall, from fibrous or tap roots. Basal leaves spatulate, entire, often as long as the stem. One pair of upper stem leaves present, opposite, fused to form a disk below the inflorescence, variable in size, 1-4 cm across. Inflorescence racemose, 1-8 cm long. Flowers pale pink or white, 2-4 mm long, the 2 sepals persistent. Stamens 5. Petals 5. Fruit a 3-valved capsule. Rare in moist places and springy soil in the Black Hills. May-June. Miner's Lettuce.

PORTULACA L.

Portulaca oleracea L. Prostrate, succulent-leaved annual with branches 1-6 dm long. Leaves obovate with obtuse tips, 1-2.5 cm long. Flowers solitary or few in leaf axils, the 4-6 petals yellow, 5 mm long. Stamens 8 or more. Ovary partly inferior. Fruiting capsule up to 4 mm long, dehiscing transversely, with many black seeds. Frequent in gardens, naturalized from the old world. July-Aug. Purslane.

TALINUM Adans.

Talinum parviflorum Nutt. Tufted perennial, cespitose from an elongate, branched taproot. Leaves basal, linear cylindric, succulent, 2-6 cm long. Flowering scapes 3-12 cm tall, branched above, the inflorescence cymose. The 5 or more petals pink, ephemeral, 3-6 mm long. Stamens 5, ovary superior. Fruiting capsules 3-4 mm long, flattened. Frequent to locally common on sandstone outcrops and Sioux quartzite over the state. June-Aug. Fameflower.

Family Caryophyllaceae

(Key to Genera)

1 Flowers lacking petals. Fruit one-seeded. Leaves with conspicuous membranous stipules . . . . *PARONYCHIA*

1' Flowers with petals. Fruit several to many seeded. If leaves
have stipules, they are not conspicuously membranous.

   2 Sepals forming a tubular or inflated calyx. Petals with a
narrowed base (clawed).

      3 Calyx with 1-3 pairs of bracts below . . . *DIANTHUS*

      3' Calyx without bracts below.

         4 Styles 2 (rarely 3).

            5 Flowers less than 1 cm long . . . . *GYPSOPHILA*

            5' Flowers 2 cm long or longer.

               6 Calyx ovoid, with angular wings. Petals with-
out appendages . . . . *VACCARIA*

               6' Calyx tubular, not winged. Petals with ap-
pendages . . . . *SAPONARIA*

         4' Styles 3-5 (rarely 4).

            7 Flowers perfect. Capsules opening with less than
10 teeth.

               8 Styles 3 (rarely 4). Capsule opening with 6
teeth. Calyx teeth much shorter than the tube
. . . . *SILENE*

               8' Styles 5. Capsules opening with 4-5 teeth.
Calyx teeth much longer than the tube . . . . .
. . . . *AGROSTEMMA*

            7' Flowers imperfect or occasionally perfect. Cap-
sules opening with 10 teeth . . . . *LYCHNIS*

   2' Sepals separate almost to the base. Petals lacking a
clawed lower portion.

      9 Petals more or less deeply incised to form 2 lobes.

         10 Fruits cylindrical, curved, opening with a row of
10 apical teeth . . . . *CERASTIUM*

         10' Fruits oval, splitting into 6 valves . . . .
. . . . *STELLARIA*

      9' Petals lacking a deep incision. Not obviously lobed.

11 Styles 4 or 5, as many as the sepals . . . .
. . . . *SAGINA*

11' Styles 3, sepals usually 5 . . . . . *ARENARIA*

AGROSTEMMA L.

Agrostemma githago L. Plants annual, herbaceous, with pubescent
stems to 8 dm tall, freely branched. Leaves opposite, linear-
lanceolate, almost erect and appressed to the stem, 5-10 cm long,
their surfaces white-hirsute. Flowers mostly solitary at the
ends of branches, red, the petals 2-3 cm long. Fruiting capsules
opening with 5 sutures but 1-celled. Seeds black, 2-3 mm long.
Rare in old fields and wasteland along roadsides. June-July.
Corn Cockle.

ARENARIA L.

1 Plants with rigid, subulate-pointed leaves.

2 Perennial plants forming a dense, cushiony branched plant
up to 15 cm across, from a stout, elongate taproot. Flowering stems erect.

3 Fruiting capsule dehiscing by 3 valves. Plants glandular-
pubescent. Stems several to many, from a branched crown;
the spreading plant up to 1 dm across. Flowering stems
up to 1 dm long, simple or sparsely branched. Inflorescence a leafy-bracted cyme, open. Sepals and petals 3-4
mm long, the former 3-nerved. Capsule as long as the
sepals or longer, with 3 entire valves. Rare in meadows
at higher altitudes in the Black Hills. July-Aug . . . .
. . . . *Arenaria rubella* (Wahlenb.) J. E. Smith

3' Fruiting capsule dehiscing by 6 valves. Plants puberulent to glandular. Stems many, forming a dense cushion
up to 1.5 dm across, from a stout, elongate taproot.
Flowering stems several, erect, 2-5 cm tall, the cauline
leaves longer than the basal leaves. Inflorescence a
dense, cymose cluster. Sepals 6-8 mm long, the petals
slightly longer. Capsule opening with 6 valves. Rare
on eroded clay or sandy slopes of the southwest part.
June-July . . . . *Arenaria hookeri* Nutt.

2' Annual plants or weak perennials with stems weakly ascending to prostrate, up to 2 dm tall, glabrous. Leaves linear-
subulate, 5-15 mm long, often with fascicles of secondary
leaves in their axils. Flowers in open cymes, the petals
broadly obovate, 4-6 mm long. Sepals 3-ribbed, acutely

pointed. Rare in rocky soil of ravines and low places in
the Black Hills. May-June . . . . *Arenaria stricta* Michx.

1' Plants with ovate to lanceolate leaves, their apexes acutely
pointed or obtuse.

   4 Leaves 3-8 mm long, ovate, with pointed apexes. Plants annual, with wiry stems 5-15 cm tall, commonly branched from
the base. Flowers several in open, terminal cymes. Petals
shorter than the sepals, the latter about 3 mm long. Fruiting capsules ovoid, with many gray to black seeds, each
about 0.5 mm long. Rare in sandy or rocky areas along
creeks and ponds in the Black Hills. May-June. Thyme-
leaved Sandwort . . . . *Arenaria serpyllifolia* L.

   4' Leaves about 3 cm long, lance-ovate, with obtuse apexes.
Plants usually perennial with soft stems weakly ascending.
Stems branched, puberulent. Flowers terminal in cymes or
solitary in the axils of upper leaves. Petals white, from
shorter than to longer than the sepals, the latter 3-6 mm
long. Frequent in rich, loamy upland woods in the eastern
part and in the Black Hills. {*Moehringia lateriflora* (L.)
Fenzl.} May-June . . . . *Arenaria lateriflora* L.

CERASTIUM L.

1 Plants annual. Fruiting capsules exceeding twice the length
of the calyx.

   2 Pedicels sharply deflexed at fruiting time and over twice
the length of the calyx. Stems glandular-pubescent, spreading, somewhat branched, 5-15 cm tall. Leaves narrowly
lanceolate, opposite. Inflorescence of bracteate cymes,
with slender, deflexed pedicels. Petals white or nearly
so, 3-6 mm long. Frequent in shaded, low thickets and alluvial areas over the state. Apr-May . . . .
. . . . *Cerastium nutans* Raf.

   2' Pedicels straight or slightly curved at fruiting time and
not over twice the length of the calyx. Stems often
branched at the base, viscid-hairy, 1-3 dm tall. Leaves
oblanceolate to oblong, opposite. Inflorescence an open
cyme, the flowers white or nearly so. Infrequent in dry,
sandy soils of the southern part and probably widespread
in the state. Similar to the preceding and previously considered a var. of that species. Apr-May . . . .
. . . . *Cerastium brachypodium* (Engelm.) Robins.

1' Plants biennial or perennial. Fruiting capsules not more than
twice the length of the calyx.

3 Petals almost twice as long as the sepals. Leaves linear
   to narrowly lanceolate, less than 6 mm wide and usually
   less than 2 cm long. Plants cespitose, glandular pubescent
   to glabrous, the stems 1-4 dm tall. Inflorescence of compact cymes at the ends of leafy branches. Petals 5-15 mm
   long. Common in prairie, at the edges of woods, and in
   waste places over the state. May-July. Mouse-ear Chick-
   weed . . . . *Cerastium arvense* L.

3' Petals shorter than to almost as long as the sepals.
   Leaves oblanceolate to spatulate, up to 8 mm wide and up
   to 2.5 cm long. Plants sprawling, the stems decumbent,
   branched, rooting at the nodes. Flowers several in open,
   dichotomously branched cymes, the petals 4-7 mm long, white
   to pink. Fruiting capsule cylindrical, up to 2 times the
   length of the sepals. An infrequent weed of waste places
   over the state. May-July . . . . . . *Cerastium vulgatum* L.

DIANTHUS L.

Dianthus armeria L. Plants annual or biennial herbs 2-5 dm tall,
mostly unbranched, erect, glabrous. Leaves narrowly lanceolate,
3-8 cm long, numerous in pairs. Inflorescence a congested terminal cyme, with several bracts. Stamens 10. Petals white to pink
or rose, 2-2.5 cm long, toothed at their distal ends. Calyx cylindrical, 1.5 cm long. An occasional escape in abandoned yards
and waste places and persisting as weeds. June-Aug. Deptford
Pink.

GYPSOPHILA L.

1 Plants annual, herbs with diffuse, slender stems 5-10 cm tall.
   Leaves linear-lanceolate, 1-3 cm long. Flowers on axillary
   pedicels, the petals pink to lavender, oblanceolate, 5-8 mm
   long. Fruiting capsule 3-4 mm long, with tuberculate seeds.
   Occasionally persisting as a weed in waste places after escaping from cultivation. In the western part. June-Aug.
   Baby's Breath . . . . *Gypsophila muralis* L.

1' Plants perennial, the stems freely branched and up to 1 meter
   tall. Vegetation glabrous or glaucous, the leaves opposite,
   linear to lanceolate, with connate bases. Flowers small, in
   diffuse, paniculate cymes, the petals 2-4 mm long. Fruiting
   capsule ovoid, dehiscing with 4-6 (5) valves, the seeds black,
   rugose. An occasional escape from cultivation and persisting
   as a weed in waste places and along roadsides. June-Aug . . .
   . . . . *Gypsophila paniculata* L.

LYCHNIS

1   Flowers perfect, bisexual. Leaves linear to lanceolate.
    Flowering calyx less than 12 mm long, not inflated in fruit.
    Plants 2-4 dm tall, stems 1-several, usually simple. Leaves
    mostly basal, the cauline leaves remote and reduced. Flowers
    on erect peduncles, arranged in loose cymes. Petals white to
    pink, usually not longer than the calyx. Infrequent on dry
    slopes and on roadsides in the western part. July-Aug . . . .
    . . . . *Lychnis drummondii* (Hook.) Wats.

1' Flowers imperfect, unisexual. Plants dioecious or polygamous.
    Leaves ovate, 3-10 cm long. Flowering calyx exceeding 12 mm,
    becoming inflated in fruit. Stems 4-10 dm tall, branched,
    several from a well-developed root. Inflorescence branched,
    irregular, the flowers white, odorous, night-blooming. Petals
    2-4 cm long, clawed, the lobes 3-5 mm long. Frequent over the
    state in waste places, especially on roadsides. June-Aug.
    White Campion . . . . *Lychnis alba* Mill.

PARONYCHIA Adans.

1   Plants densely cespitose, forming cushiony mats. Stems 4-8 cm
    long, the older ones covered with dead leaves. Leaves less
    than 6 mm long, linear with subulate tips, their stipules al-
    most as long as the leaves. Flowers solitary or in pairs, the
    petals lacking or rudimentary. Sepals tapering to an arched
    tip with an awn 1 mm long. Infrequent in dry, gravelly prai-
    rie or in rocky openings in pine forests in the western part.
    June-July . . . . *Paronychia sessiliflora* Nutt.

1' Plants cespitose with crowded, spreading branches. Stems
    seldom over 6 cm long, their internodes usually less than 5
    mm. Leaves more than 7 mm long, much exceeding their stipules.
    Flowers clustered at the ends of stems, the corolla lacking.
    Sepals 2-3 mm long, abruptly tapering to a very short awn.
    Bracts of the flowers as long as or longer than the sepals.
    Frequent in dry soils of hillsides in the southwestern part.
    June-July . . . . *Paronychia depressa* Nutt.

SAGINA L.

Sagina saginoides (L.) Britt. Biennial or perennial herb with
many stems from a crown forming a matted growth, the stems
slender, glabrous, usually not more than 5 cm long. Leaves many
in a basal rosette, linear and pointed, 5-10 mm long. Flowers
solitary at the ends of stems on slender pedicels. Sepals 2-3
mm long, erect, the 4 or 5 petals white, about one-half as long.

A circumpolar species that is rare in rock crevices at high altitudes in the Black Hills. May-Aug. Pearlwort.

SAPONARIA L.

Saponaria officinalis L. Perennial herb with many stems from
spreading rhizomes. Plants glabrous, 4-8 dm tall. Leaves opposite, lanceolate to ovate, with entire margins, 4-9 cm long.
Flowers white to pink, clustered in contracted terminal cymes
and in the axils of upper leaves. Petals 5, clawed. Stamens 10.
Fruiting capsule oblong, dehiscing by 4 teeth. Common on roadsides and persisting from abandoned farmyards in waste places
over the state. July-Sept. Bouncing Bet.

SILENE L.

1 Plants annual.

2 Stems 2-6 dm tall, with dark, glutinous areas between the
   nodes, otherwise glabrous. Leaves glabrous or sparsely
   pubescent, 3-5 cm long, linear or narrowly lanceolate.
   Flowers solitary or rarely few at the ends of slender
   pedicels. Petals white or pink, with a narrow claw, 6-9
   mm long. Fruiting capsule ovoid, 4-6 mm long. Frequent
   in dry sandy soil or on dry, eroded prairie knolls over
   the state. June-July. Sleepy Catchfly . . . .
   . . . . *Silene antirrhina* L.

2' Stems 2-5 dm tall, lacking sticky glutinous areas between
   the nodes, densely pubescent. Leaves ovate-lanceolate with
   a tapering petiole, 4-10 cm long, pubescent. Flowers in a
   loose cymose inflorescence, the petals white to pink,
   clawed, the clawed portion up to 2 cm long, the bilobed
   blade 6-8 mm long. Fruiting capsules inflated, 1.5 cm long.
   Rare as a weed of roadsides and waste places. {This plant
   often confused with *Lychnis alba* L., which is dioecious and
   larger.} July-Aug. Night-blooming Catchfly . . . .
   . . . . *Silene noctiflora* L.

1' Plants perennial.

3 Principal leaves whorled in groups of 4. Distal ends of
   the petals fimbriate into 3-6 lobes, the flowers appearing
   star-like. Plants perennial, the stems 1-several, 3-7 dm
   tall. Leaves lanceolate or narrower, 4-8 cm long. Flowers
   in a loose, terminal panicle. Frequent in rich woods of
   the southeast part. June-July. Starry Campion . . . . . .
   . . . . *Silene stellata* (L.) Ait. f.

3' Principal leaves opposite. Petals bilobed or shallowly
   lobed.

4   Vegetation, especially the stems, glaucous. Inflores-
    cence an open panicle. Plants perennial, the stems 2-7
    dm tall, often sprawling. Leaves lance-ovate, 3-8 cm
    long, the principal ones with clasping bases. Flowers
    white, the petal blades 3-6 mm long, bilobed. Calyx
    inflated in fruit, to 2 cm long. Infrequent as a weed
    on roadsides and railroad embankments in the eastern
    one-half. June-July. Bladder Campion . . . .
    . . . . *Silene cucubalus* Wibel.

4' Vegetation not glaucous; however, it may be glabrous to
    puberulent. Flowers few, usually solitary in the upper
    axils. Perennial from rhizomes, the stems 2-3 dm tall.
    Leaves lance-elliptic, mostly sessile, 5-10 cm long.
    Petals white, the lobes mostly entire. Calyx slightly
    inflated but mostly tubular in fruit, 1.5 cm long. Rare
    in rich woods of the Black Hills. June-July. White
    Campion . . . . *Silene nivea* (Nutt.) Otth.

STELLARIA L.

1 Leaves lance-ovate to ovate in outline, spreading.

  2 Stems and petioles with longitudinal lines of pubescence.
     Leaves distinctly ovate, 1-1.5 cm long, the lower ones
     petiolate. Stems weak and spreading, 1-3 dm long. Flow-
     ers solitary in axils or few in terminal leafy cymes.
     Petals 2-4 mm long, shorter than the sepals. Infrequent
     weed of lawns and gardens in the eastern part. April-June.
     Chickweed . . . . *Stellaria media* (L.) Cyrill.

  2' Stems and petioles glabrous. Leaves lanceolate, sessile,
     1.5-2.0 cm long. A matted perennial with many stems, 1-2
     dm long. Flowers solitary in axils, the petals 3-5 mm
     long, exceeding the sepals. Fruiting capsule ovoid, about
     as long as the sepals. A circumboreal species that is rare
     in rich soil of upland woods in the northeast part. July-
     Aug . . . . *Stellaria crassifolia* Ehrh.

1' Leaves linear to lance-shaped, ascending.

  3 Flowering pedicels spreading or curved downward. Leaves
     linear, less than 2 mm wide and over 3 cm long, occasion-
     ally spreading, but more commonly ascending. Inflorescence
     cymose, the flowers less than 4 mm across. Stems weak,
     spreading to ascending, up to 4 dm long. Frequent in moist
     woods and on stream banks or alluvial thickets over the
     state. May-July . . . . *Stellaria longifolia* Muhl.

3' Flowering pedicels ascending or erect. Leaves lanceolate, sharply ascending, less than 2 cm long. Inflorescence a few-flowered cyme or flowers axillary in upper leaves, the corolla 5 mm or more across. Stems usually erect but occasionally matted, 1-3 dm tall. Rare in moist woods in the Black Hills. June-July . . . . . *Stellaria longipes* Goldie

VACCARIA Medic.

Vaccaria segetalis (Neck.) Garcke. Herbaceous, taprooted annual herb up to 7 dm tall. Leaves opposite, lanceolate with auriculate bases, sessile or short petiolate, 4-8 cm long. Inflorescence an open, paniculate cyme, the flowers rose to light red. Petals slightly bilobed, the blade 5-7 mm long. Stamens 10. Fruiting capsule becoming 6-8 mm long, opening by 4 tooth-like valves. Occasionally occurring as a weed in waste places in the western part. June-Aug. Cowherb.

Family Ceratophyllaceae

CERATOPHYLLUM L.

Ceratophyllum demersum L. Aquatic plants with freely branched stems, submerged. Leaves 1-3 cm long, in whorls, dissected with minute teeth. Flowers minute, unisexual, lacking a perianth, solitary in axils of leaves. Fruit developing into a single achene 4-6 mm long, with basal spines. Infrequent in quiet water or slow streams in the eastern part of the state. June-July. Hornwort.

Family Nymphaeaceae

(Key to Genera)

1 Petals white, conspicuous. Sepals usually 4, green. Leaves peltate at the middle . . . . *NYMPHAEA*

1' Petals small, inconspicuous. Sepals 5-several, greenish-yellow to reddish. Leaves broadly rotund, not peltate . . .
. . . . *NUPHAR*

NUPHAR Sm.

Nuphar luteum (L.) Sibth. & Sm. Aquatic perennial from a creeping rhizome, the leaves erect above the water or the blades floating. Petioles up to a meter or more long. Leaf blades ovate to rounded, 2-4 dm across. Flowers on erect peduncles,

raised above the water, the perianth 3-5 cm across. Petals many,
small, scale-like. Sepals usually 6, the inner ones yellow or
reddish. Stamens many, yellow. Fruit ovoid, 4 cm or more long.
Infrequent in ponds, at lake margins, and in slow-moving streams
in the eastern part. Ours ssp. *variegatum* (Engelm.) Beal. July-
Sept. Yellow Water-Lily.

NYMPHAEA L.

Nymphaea tuberosa Paine. Aquatic perennial plants from stout,
tuberous rootstocks. Flowers and leaf blades floating on the
surface from long peduncles and pedicels. Petals white or pink,
numerous. Sepals 4, green. Stamens many, attached to the ovary.
Flowers 10-20 cm across. Ovary maturing into a capsule with several to many seeds. Rare in the eastern part. June-July. White
Water-Lily.

Family Ranunculaceae

(Key to Genera)

1 Flowers radially symmetrical and not possessing spurs.

   2 Perianth parts yellow.

      3 Sepals green; petals present. Fruits 1-seeded achenes
. . . . *RANUNCULUS*

      3' Sepals yellow; petals absent. Fruits several-seeded
follicles . . . . *CALTHA*

   2' Perianth parts principally colored other than yellow.

      4 Fruits 1-seeded achenes. Leaves simple to compound.

         5 Plants aquatic, with white petals . . . *RANUNCULUS*

         5' Plants terrestrial. Petals lacking.

            6 Leaves linear. Flowers solitary . . *MYOSURUS*

            6' Leaves compound.

               7 Flowers often imperfect. Sepals small. Leaves
alternate . . . . *THALICTRUM*

               7' Flowers perfect. Sepals well-developed.

                  8 Sepals usually 4, valvate in bud . . . . . .
. . . . *CLEMATIS*

                  8' Sepals usually 5, imbricate in bud . . . . .
. . . . *ANEMONE*

    4' Fruits berries or follicles. Leaves compound.

        9 Flowers in racemes. Fruits red or white berries.
        Leaves much dissected . . . . *ACTAEA*

        9' Flowers in corymbs or solitary. Fruits several-
        seeded follicles. Leaves ternately compound . . . .
        . . . . *ISOPYRUM*

1' Flowers bilaterally symmetrical or with one or more prominent
spurs.

    10 Petals 2, the upper sepal forming a hood; flowers not
    spurred . . . . *ACONITUM*

    10' Petals 4 or 5, the flowers spurred but not with a hood.

        11 Flowers bilaterally symmetrical, the upper sepal prom-
        inently spurred. Petals 4 . . . . *DELPHINIUM*

        11' Flowers radially symmetrical, the 5 petals prominently
        spurred . . . . *AQUILEGIA*

ACONITUM L.

Aconitum columbianum Nutt. Herbaceous perennial from a tuberous
crown with several thickened fibrous roots, the several stems 3-8
dm tall. Leaves alternate, petioled, the blades 3-5 lobed, sometimes deeply lacinate. Inflorescence a loose, terminal raceme,
many times with bracts. Flowers 3-5 cm high, lavender to deep
blue, irregular. The upper 2 petals raised on claws and under
the hood or helmet, the lower 3 petals vestigial or lacking.
Pistils 3-5, the fruit a follicle. Frequent in woods of the
Black Hills. July-Sept. Monkshood.

ACTAEA L.

Actaea rubra (Ait.) Willd. Herbaceous perennial with several
stems 4-8 dm tall, branched. Leaves large, few but compound with
several segments, the ultimate ones lobed and toothed. Flowers
somewhat congested, in terminal or axillary racemes. Petals
white, 2-3 mm long, not much exceeding the sepals. Fruits of
red or less commonly white berries 7-10 mm in diameter. The
white fruited form, *A. rubra* forma *neglecta* (Gillman) Robins,
occurs in rich woods of Marshall and Roberts counties in the
northeast part. Frequent in rich woods of the eastern and western part. May-June. Baneberry.

ANEMONE L.

1 Sepals 10 or more. Stems 6-15 cm tall, from small underground
tubers, the plants perennial. Basal leaves 3-lobed, deeply

divided.   Leaves subtending the peduncles sessile, more or
less divided.   Flowers white to purple, the 10-20 sepals petaloid on the inside but pubescent outside, each 1-2 cm long.
Fruiting achenes with dense hairs, the fruiting head ovoid to
cylindrical.   Frequent in dry, sandy prairie and plains over
the southern part of the state.   May-June . . . .
. . . . *Anemone carolinana* Walt.

1' Sepals 5-7.   Stems from taproots or rhizomatous roots.

2 Styles becoming elongate and plumose after flowering.
   Sepals 2-4 cm long, white to blue-purple.   Plants scapose,
   the leaves with linear segments.   Bracts beneath the flowers leaf-like.   Flowers mostly solitary on scapes 1-4 dm
   tall.   Achenes 3-4 mm long, with long, flexuous styles 2-4
   cm long.   Common in dry prairie over the state.   The state
   flower of South Dakota.   April.   Pasque . . . .
   . . . . *Anemone patens* L.

2' Styles less than 2 cm long after flowering, not plumose.
   Sepals less than 2 cm long.

   3 Achenes densely white hairy, forming a white cottony
      head towards maturity.

      4 Heads of fruits cylindrical, 2-4 cm long.   Styles
         pointed, persistent.   Plants 3-6 dm tall, with numer-
         ous basal leaves and several involucral leaves.
         Leaves ternate to dissected with linear to lanceolate
         divisions.   Flowers 1-several, umbellate, the sepals
         greenish-white, 8-10 mm long.   Frequent in open woods
         and prairie over the state.   June-July . . . . . .
         . . . . *Anemone cylindrica* A. Gray

      4' Heads of fruits globose to ellipsoid, not more than
         twice as long as wide.

         5 Involucral leaves short-petioled or sessile.
            Fruiting heads globose or nearly so.   The segments
            of principal leaves linear or narrowly lanceolate.
            Plants 1-5 dm tall, the vegetation white-strigose.
            Stems with 1-3 flowering peduncles.   Sepals green-
            ish white to pink, 6-10 mm long.   Frequent in open
            woods and on limestone ridges of the Black Hills.
            {*A. globosa* Nutt.}   June-July . . . .
            . . . . *Anemone multifida* Poir.

         5' Involucral leaves long-petioled.   Fruiting heads
            oval to ellipsoid.   The segments of principal
            leaves lance-ovate, 3-parted, with coarse teeth.

Plants 5-9 dm tall, hairy, but not grayish-
whitened. Inflorescence cymose, 1 to several
flowered. Sepals greenish white, to 1 cm long.
Rare in loamy woods of the eastern part and in
the western part. June-July. Tall Anemone . . .
. . . . *Anemone virginiana* L.

3' Achenes on the fruiting head not white hairy or forming
a cottony mass.

6 Stems branched, several-flowered, 2-6 dm tall.
Plants from spreading rootstocks. Principal leaves
3-5 parted. Involucral leaves mostly sessile. Flowers 1-3 on peduncles, the sepals white, 1-1.5 cm long.
Fruiting heads globose, the achenes pubescent. Common in meadows, openings of thickets and roadside
ditches over the state. May-June. Meadow Anemone
. . . . *Anemone canadensis* L.

6' Stems simple, unbranched, with one flower, 1-3 dm
tall. Plants from slender rhizomes. Involucral
leaves 3, each 2-3 parted. A single, long-petioled
leaf arising basally after flowering. Sepals white,
6-10 mm long. Achenes tapered with a hooked beak.
Rare but locally abundant in rich woods of Roberts
and Marshall counties of the extreme northeast part.
Apr-May. Woods Anemone . . . *Anemone quinquefolia* L.

AQUILEGIA L.

1 Flowers red to yellow, usually 3-4 cm long. Spurs much longer
than the blades of the petals. Plants with stems 3-9 dm tall,
with compound leaves, the ultimate segments with rounded,
shallow, crenate teeth. Flowers few, large and showy, solitary at the ends of branches. Fruits of 5 follicles, loosely
connivent at maturity. Infrequent on loamy, wooded hillsides
over the state. Apr-May. Columbine . . . .
. . . . *Aquilegia canadensis* L.

1' Flowers lavender to blue, about 2 cm long. Spurs slightly
longer than the blades of the petals. Plants 2-8 dm tall,
from branching root stocks. Leaves ternately compound, the
ultimate leaflets 3-lobed or coarsely toothed. Flowers few,
usually solitary. Spurs of the petals somewhat hooked at
their bases. Fruits of 5 follicles, 2-2.5 cm long. Infrequent on wooded hillsides at higher altitudes of the Black
Hills and in Harding County. May-June. Western Columbine . .
. . . . *Aquilegia brevistyla* Hook.

{*Aquilegia vulgaris* L., with flowers blue to pink or yellow and possessing strongly incurved spurs, is cultivated throughout our region and occasionally escapes. It resembles *A. canadensis* in habit.}

CALTHA L.

Caltha palustris L. Herbaceous perennials from creeping rootstocks, forming dense growths in moist or swampy places. Stems 3-7 dm tall, sparingly branched. Leaves cordate, reniform, with crenate margins, the blades 5-15 cm across, variously petioled. Flowers solitary or few at the ends of branches, the sepals yellow, 2-5 cm long, waxy-shining. Fruits of 5-12 flattened follicles 1-2 cm long. Infrequent, but locally abundant, in swampy areas of the northeast part. May-June. Marsh Marigold.

CLEMATIS L.

1 Flowers (sepals) white or cream-colored, several in axillary panicles.

    2 Principal leaves 5-7 pinnately foliolate, the leaflets lanceolate to narrowly ovate, with coarse teeth. Plants 2-4 meters, the stems climbing. Flowers in erect, bracteate panicles, the sepals cream-colored. Plants dioecious. Infrequent in low areas of woods in the Black Hills. July-Aug. Western Virgin's Bower . . . .
. . . . *Clematis ligusticifolia* Nutt.

    2' Principal leaves 3-foliolate, the leaflets ovate with coarse teeth or lobed. Plants 1-3 meters tall, the stems climbing on other vegetation. Plants polygamodioecious. The sepals white and petaloid, in axillary panicles. In fruit the inflorescence of many achenes, each one with a feathery beak 3-5 cm long. Frequent in alluvial and rich loamy woods of the eastern part. July-Aug. Virgin's Bower . . . . *Clematis virginiana* L.

1' Flowers (sepals) blue or lavender, solitary, on erect peduncles.

    3 Stems climbing or trailing. Sepals and stamens spreading. Sepals thin, not connivent.

        4 Leaves biternate, with secondary divisions lobed or toothed, 1.5-3 cm long. Stems climbing, 1-4 dm long. Flowers solitary on naked peduncles 5-12 cm long. Sepals 3-5 cm long, tapered acuminate, blue to purple. Infrequent or possibly rare in woods at lower altitudes in the Black Hills. June-July. Purple Virgin's Bower. . . . . *Clematis pseudoalpina* (Kuntze) Nels.

4' Leaves triternate, the tertiary divisions more finely
divided. Stems short, the plants low, matted together
and trailing but usually not climbing, 1-2 dm tall.
Flowers purple, solitary on peduncles, much like the
preceding. Fruiting achenes short pubescent, the
styles 3-5 cm long. Frequent in rocky soils of woods
of middle and upper altitudes of the Black Hills. June-
July . . . . . *Clematis tenuiloba* (Gray) C. L. Hitch.

3' Stems erect, not climbing. Sepals and stamens erect.
Sepals thick, connivent. Plants perennial, the stems 2-7
dm tall, mostly unbranched. Leaves pinnately decompound,
the ultimate divisions broadly lanceolate to ovate in outline. Flowers solitary and terminal, the sepals deep purple. Rare on hillsides in the extreme southwest part.
June-July . . . . *Clematis hirsutissima* Pursh

DELPHINIUM L.

1 Sepals white or essentially so, occasionally with blue spots.
Plants 2-8 dm tall, stems erect, from a deep-seated root system. Leaves palmately segmented into linear divisions, the
blade rotund in outline. Flowers in compact racemes, almost
spike-like. Sepals 10-12 mm long, almost straight. Fruiting
follicles 1.5-2 cm long, the seeds rough-surfaced. Frequent
in prairies over the state. May-June. Prairie Larkspur . .
. . . . *Delphinium virescens* Nutt.

1' Sepals blue, the inner or lower petals occasionally light
colored. Plants 2-5 dm tall, the stems 1-3 from a fibrous
to tuberous rootstock. Leaves mostly basal, rounded in outline, with palmately linear divisions. Inflorescence few to
several flowered, racemose. Sepals flared, deep blue, the
spur 10-13 mm long. Follicles 1.5-2 cm long, the seeds not
rough-surfaced. Frequent in dry woods and openings in the
Black Hills. May-June. Blue Larkspur . . . .
. . . . *Delphinium nuttallianum* Pritz.

ISOPYRUM L.

Isopyrum biternatum (Raf.) T. & G. Herbaceous perennial from
tuberously-thickened fibrous roots, the stems slender, 1-3 dm
tall. Leaves 2 or 3 ternate, the basal ones long petioled.
Leaflets 3-lobed. Flowers (sepals) white, 1-2 cm across.
Petals lacking. Flowers solitary or few at ends of branches
from axillary positions. Fruits divergent, 2-4, the follicles
with curved and tapered beaks. Rare in rich woods of the eastern part and in the Black Hills. Apr-May. False Rue Anemone.

MYOSURUS L.

Myosurus minimus L. Herbaceous annual with basal, linear leaves.
The flowering scapes usually less than 1 dm tall. Leaves linear
to narrowly spatulate, 2-4 cm long, tapering to a petiole. Flowers small, greenish-yellow, the pistils producing a spike of
achenes which may elongate to more than 5 cm at maturity. Rare
at the margins of dessicating ponds in the east-central part
where a number of artesian wells occur. May-June. Mouse Tail.

RANUNCULUS L.

1 Plants usually growing in water. At least some of the submersed leaves finely dissected.

   2 Petals yellow. Lower leaves dissected into decompound
      linear segments 1-2 mm wide. Emersed leaves broadly 3-
      parted. Stems erect to prostrate, 2-8 dm long, rooted in
      mud of aquatic areas. Flowers single to few on stout pe-
      duncles. Achene margins thickened around their bases.
      The stylar beak 1-2 mm long. Infrequent in lakes and
      ponds of the northeast part. May-June. Yellow Water-
      crowfoot . . . . *Ranunculus flabellaris* Raf.

   2' Petals white, sometimes yellowish at the base. Leaves all
      submersed and dissected into linear or filiform segments.

      3 Leaves petioled. Beak of maturing achenes less than 0.5
         mm long. Leaves collapsing when removed from the water.
         Flowering pedicels not recurved. Stems elongate,
         branched, 3-8 dm long. Flowers 1-1.5 cm across, at the
         water surface. Infrequent in ponds and slow streams
         over the state. Ours var. *capillaceus* (Thuill.) DC.
         May-July. White Water-crowfoot . . . .
         . . . . *Ranunculus aquatilis* L.

      3' Leaves sessile, not collapsing when removed from the
         water. Stipules present.

         4 Achenes with a beaklike style 0.5-1.0 mm long.
            Plants perennial, the stems lax and rooting from the
            lower nodes. Leaves 2-3 times dichotomously dis-
            sected into linear divisions, subrotund in outline.
            Flowering.pedicels curved, stout, 1-4 cm long. Flow-
            ers solitary at the ends of pedicels, 1-2 cm across,
            the petals white. Achenes obovate, with a persistent
            beak. Frequent in ponds and streams over the state.
            June-July . . . . . . *Ranunculus longirostris* Godr.

         4' Achenes beakless or the beak less than 0.5 mm long.
            Plants perennial, the stems lax and rooting at the

lower nodes. Leaves finely dissected into linear
divisions, borne on a stipular base. Flowers solitary, the petals white, 5-9 mm long, yellow at their
bases. Achenes transversely wrinkled, with an obovate shape. Similar to the preceding and perhaps
intergrading with it. Infrequent in ponds and
streams over the state. June-Aug . . . .
. . . . *Ranunculus subrigidus* W. Drew

1' Plants terrestrial or at the pond margins. At least the lower
leaves not finely dissected.

5 Stems usually less than 20 cm tall. Basal leaves entire to
deeply lobed.

6 Leaves simple, all basal. Blades glabrous, crenately
lobed. Fruiting head becoming elongate after flowering.
Plants tending to spread with stolons. Stems usually
less than 5 cm tall. Flowers small, 4-7 mm across.
Common in moist, sandy areas and on flood plains over
the state. June-Sept. Seaside Crowfoot . . . .
. . . . *Ranunculus cymbalaria* Pursh

6' Leaves above the base compound or deeply parted. Basal
leaves may be simple or compound. Fruiting heads globose to elongate.

7 Vegetation glabrous. Plants perennial, 4-10 cm tall.
Roots coarse, fleshy, lacking a main taproot. Lower
leaves lanceolate-ovate, mostly entire, tapering to a
petiole. Flowers solitary or few at main stem terminals. Petals 8-12 mm long. Infrequent in meadows
and marshy soil over the state. Apr-May . . . . . .
. . . . *Ranunculus glaberrimus* Hook.

7' Vegetation pubescent-hairy throughout.

8 Plants perennial, 8-20 cm tall. Lower leaves
ovate to rounded-crenate, long-petioled. Stem
leaves sessile, deeply lacinate. Flowers few to
several at branches or main stem terminals, the
petals oblong-elliptical, 5-8 mm long, exceeding
the sepals. Fruiting heads obovoid with achenes
about 1.5 mm long. Infrequent in meadows and
stream beds over the state. {incl. *R. ovalis*
Raf.} Apr-June. Prairie Buttercup . . . .
. . . . *Ranunculus rhomboideus* Goldie

8' Plants annual, scapose, usually less than 8 cm
tall. All leaves basal, the blades ternate to
biternate with linear divisions. Flowering and

fruiting peduncles leafless, up to 8 cm tall.
Flowers white to yellow, the petals 5-7 mm long.
Fruiting achenes in an elongate cluster up to 1.5
cm long. Achenes with a pointed beak 3-4 mm long.
The base of the achene with 2 lateral vesicles and
a single seed. Rare in waste places in the west-
ern part. Apr-May . . . .
. . . . *Ranunculus testiculatus* Crantz

5' Stems usually exceeding 20 cm and basal leaves lobed or
crenate.

9 Flowers less than 1 cm in diameter.

10 Basal leaves different in shape than the stem
leaves. Plants perennial.

11 Basal leaves cordate at their bases. Plants
glabrous or nearly so. Stems erect, 1-5 dm
tall, frequently branched above. Cauline
leaves sessile or nearly so, deeply 3 to 5
parted. Flowers with petals oblong, 2-3 mm
long, the fruiting head becoming ovoid. Common in rich woods and alluvial areas over the
state. Apr-May. Small Flowered Buttercup . . .
. . . . *Ranunculus abortivus* L.

11' Basal leaves tapered at their bases. Plants
usually hirsute to villous, occasionally only
glabrate.

12 Petals shorter than the sepals. Fruiting
head 2-4 mm long and less than 4 mm across.
Plants perennial, the stems 1-4 dm tall,
long-hairy to villous. Basal leaves 1-4
cm wide, some of them with 3 divisions.
Petals small, oblong. Achenes many in a
short ovoid head. Rare in rich woods of
the Black Hills. Apr-May . . . .
. . . . *Ranunculus micranthus* Nutt.

12' Petals as long as or longer than the sepals.
Fruiting heads cylindrical, 4-10 mm long and
over 4 mm in diameter. Plants perennial,
the stems 1-3 dm tall, glabrate to hirsute.
Principal basal leaves with rounded and
deeply divided blades. Stem leaves sessile
with linear divisions. Petals 3-8 mm long,
narrowly elliptic. Achenes glabrous to
pubescent, with recurved beaks. Rare in

valleys in the Black Hills. May-June . . .
. . . . *Ranunculus inamoenus* Greene

10' Basal and stem leaves alike in shape with 3-several
deeply divided lobes. Plants annual. Stems erect,
3-6 dm high, fleshy, freely branching. Flowers on
stout pedicels 1-3 cm long, from terminal or axillary positions. Petals 2-5 mm long. Achenes small,
numerous on a short-cylindric head. Common in
marshy soil, on flood plains and at pond margins
over the state. May-Aug. Cursed Crowfoot . . . . .
. . . . *Ranunculus sceleratus* L.

9' Flowers 1 cm or more in diameter. Plants usually with
spreading hairs.

13 Petals 2-5 mm long. Anthers less than 1 mm long.

14 Sepals distinctly longer than the petals. Fruiting head of achenes ovoid cylindric, 1-1.5 cm
long. Plants annual, or sometimes perennial,
branched, 3-6 dm tall. Leaves deeply 3-lobed.
Petals obovate, 2-4 mm long. Frequent in
marshes and moist meadows and woods over the
state. June-July . . . .
. . . . *Ranunculus pensylvanicus* L.

14' Sepals as long as to shorter than the petals.
Fruiting head of achenes globose-rounded, less
than 1 cm high. Plants perennial, 2-6 dm tall,
leafy branched. Leaves ternately divided, the
segments petiolate. Petals about as long as
the sepals, 4-5 mm long. Rare in marshy or
moist meadows and open woods of the Black Hills.
June-July . . . . . . *Ranunculus macounii* Britt.

13' Petals 7 mm or more long. Anthers exceeding 1 mm.

15 Basal leaves with blades reniform or cordate,
2-4 cm wide, on long petioles. Fruiting heads
ovoid to cylindric. Stem leaves linearly divided. Plants 1-4 dm tall, perennial. Flowers
showy, few to several at the ends of branches.
Infrequent in wet meadows and along streams in
the Black Hills. June-July . . . .
. . . . *Ranunculus cardiophyllus* Hook.

15' Basal leaves 3-parted or otherwise deeply divided. Fruiting heads rounded to sub-globose.

16 Stems slender, erect, up to 1 meter tall.
    Principal leaves palmately divided into 3-5
    segments, long-petioled, the blades cordate
    in outline. Flowers solitary at the ends of
    terminal or axillary pedicels, the petals
    10-12 mm long. Occasional along roadsides
    or at abandoned yards. Escaped or persist-
    ing after cultivation. May-July. Tall But-
    tercup . . . . *Ranunculus acris* L.

16' Stems thick, soft or stout, less than 6 dm
    tall. Principal leaves 3-parted.

    17 Plants stoloniferous, of wet soils.
        Stem soft and thick, becoming decumbent
        and rooting at the nodes. Achenes 3-4.5
        mm long, the beak straight. Plants per-
        ennial, up to 6 dm tall, the vegetation
        mostly glabrous. Principal leaf blades
        variously divided or incised. Flowers
        pedicellate, solitary or few on branches,
        the petals obovate, 9-15 mm long. Heads
        globose, 1 cm or more across. Frequent
        in low prairie swales, thickets and
        moist woods in the eastern part. May-
        June . . *Ranunculus septentrionales* Poir.

    17' Plants not stoloniferous, of drier soils.
        Stem stout, 3-4 dm tall, not rooting at
        the nodes. Achenes 2-3.3 mm long, the
        beak straight, subulate-tipped. Plants
        perennial, the stems with spreading,
        hispid hairs. Principal leaves 3-parted,
        irregularly incised. Flowers few on
        branches, the petals oblong. Fruiting
        heads subglobose, 6-10 mm across. Rare
        in dry woods in the eastern part. Apr-
        May . . . . . *Ranunculus hispidus* Michx.

THALICTRUM L.

1 Leaflets with more than 4 teeth, in outline with usually 3
   recognizable lobes, these 2-3 toothed. Plants usually not
   branched above.

   2 Compound leaf immediately below the inflorescence long-
      petioled. Plants 4-6 dm tall, flowering in May or early
      June. Dioecious. Male flowers with greenish yellow sta-
      mens. Fruiting achenes on female plants mostly sessile,

with ribs, about 4 mm long. Infrequent in rich woods of the Black Hills and in the northeast part. May-June. Dioecious Meadow-Rue . . . . *Thalictrum dioicum* L.

2' Compound leaf immediately below the inflorescence sessile. Plants 3-5 dm tall, usually smaller than the preceding, also, flowering later in June. Dioecious. Flowering panicles slender, erect. Male flowers with sepals 2-5 mm long, greenish. Pistillate flowers with sepals 1-3 mm long. Fruiting achenes 4-6 mm long, on short stipes. Infrequent in open woods and meadows in the northeast and in the Black Hills. June . . . . *Thalictrum venulosum* Trel.

1' Leaflets of usually 3 definite lobes which are mostly entire. Stems 1 meter or more tall, commonly branched above. Leaflets leathery, especially if growing in exposed areas; strongly veined below. Plants dioecious or polygamodioecious, the functionally male plants with large paniculate inflorescences. Anthers creamy-white. Achenes 3-5 mm long, with corky ribs, on short stipes. Common in meadows and woods over the state. June-July. Meadow Rue . . . . *Thalictrum dasycarpum* Fisch. and Avé-Lall.

Family Berberidaceae

(Key to Genera)

1 Plants shrubby. Stems armed with spines or leaves with spinulose tips or both . . . . *BERBERIS*

1' Plants herbaceous, the stem smooth, erect, bearing 1 large ternately compound leaf . . . . *CAULOPHYLLUM*

BERBERIS L.

1 Leaves smooth, entire, spatulate shaped, 2-3 cm long. Stems with simple spines. Flowers solitary or in clusters of 2 to 4. Plants branched, shrubby, to 2 meters. Flowers pale yellow, 6-8 mm across. Fruiting berries reddish, ellipsoid, about 1 cm long. Occasionally escaped from cultivation and persisting along fencerows and roadsides. May-June. Japanese Barberry . . . . *Berberis thunbergii* DC.

1' Leaves serrulate or dentate, the serrations with spinulose tips. Flowers in short racemes.

2 Stems spiny. Leaves simple, deciduous, 3-5 cm long, with prominent veins beneath. Plants shrubby, up to 2 meters

tall, freely branched. Flowering racemes of 10-15 flowers, the petals yellowish, entire, 4-7 mm long. Fruits about 1 cm long, ellipsoid. Occasionally escaped from cultivation. For years it has been eradicated whenever it occurs, it being the alternate host of the black stem rust of wheat. May. European Barberry . . . . . . *Berberis vulgaris* L.

2' Stems without spines. Leaves pinnately compound, the leaflets 3-7, with 7-18 spinose teeth, evergreen. Plants shrubby, the stems trailing, usually not over 5 dm tall. Flowers in racemes 3-6 cm long, petals 6, greenish-yellow. Fruiting berries ellipsoid to rounded, becoming dark blue at maturity. Frequent in ravines and on hillsides in the western part. Rare in the northeast part. {*Mahonia aquifolium* (Pursh) Nutt.} May-June. Oregon Grape . . . . *Berberis repens* Lindl.

CAULOPHYLLUM Michx.

Caulophyllum thalictroides (L.) Michx. Perennial herbs up to 7 dm tall. The stems and leaves glaucous. Leaves triternately compound, the leaflets oblong with 2-5 distal lobes. Flowers in a terminal panicle, perfect, the petals yellow, with glands. Fruiting capsules with several dark blue seeds, each 5-8 mm long. Infrequent in rich, loamy upland woods of Newton Hills State Park in Lincoln County and in the Sieche Hollow areas of Marshall and Roberts counties in the northeast. Apr-May. Blue Cohosh.

Family Menispermaceae

MENISPERMUM L.

Menispermum canadense L. Perennial twining or trailing vines usually trailing and suspended on other vegetation, the stems 1-4 meters long. Leaves alternate, palmately lobed, the blades 4-8 cm across. Plants dioecious, the flowers in cymose inflorescences that are axillary. Petals greenish-white, not showy. Fruits drupe-like, blue-black, up to 1 cm across. Infrequent in woods and thickets in the eastern one half of the state. June-July. Moonseed.

Family Papaveraceae

(Key to Genera)

1 Plants with spiny or prickly foliage. Flowers with fewer than 8 petals . . . . *ARGEMONE*

1' Plants lacking spiny foliage. Flowers with 8 or more petals
. . . . *SANGUINARIA*

ARGEMONE L.

Argemone polyanthemos (Fedde) GBO. Annual to weakly perennial
herbs with stems 3-8 dm tall, branched in the upper part. Leaves
bluish, glaucous, with irregular incised margins that are spiny.
Leaf blades broadly lance-ovate, sessile, 1-2.5 dm long. Flowers
solitary on short peduncles in upper axils, petals white or pink,
3-4 cm long. Fruiting capsules oblong-ovoid, up to 4 cm long,
prickly. Frequent in sandy or dry areas in prairies in the
southwestern part. June-July. Prickly Poppy.

SANGUINARIA L.

Sanguinaria canadensis L. Perennial herbs without aboveground
stems growing from a horizontal, creeping rootstock that exudes
a reddish juice when broken. Leaves basal, petioled, broadly
cordate with irregular lobes, the blades 6-20 cm across. Flowers on naked scapes, 5-15 cm tall, the petals 2-3 cm long, white
or pink. Stamens many, the 2 sepals falling early. Fruiting
capsule elongate, fusiform, 3-4 cm long. Frequent in rich woods
of the eastern part and in the Black Hills. Apr-May. Bloodroot.

Family Fumariaceae

(Key to Genera)

1 Flowers bilaterally symmetrical, the two outer petals saccate
   at their bases, white or whitish . . . . *DICENTRA*

1' Flowers asymmetrical, corolla with only one petal spurred.

   2 Petals yellow or yellow-orange, capsules linear, several
      seeded . . . . *CORYDALIS*

   2' Petals purple or red-purple, capsules rounded, 1-seeded
      . . . . *FUMARIA*

CORYDALIS Medic.

1 Flowers 5-9 mm long, the outer petal with a crest or wing on
   the back near the summit. Fruiting capsules erect or at least
   ascending, 1-2 cm long. Plants annual, at first erect, up to
   2 dm tall, then becoming lax to prostrate. Leaves glaucous
   green, with finer dissections than the following species.
   Flowers commonly cleistogamous. Infrequent in sandy soil of

roadsides and railroad embankments, especially in the eastern
part. May-June . . . . . . *Corydalis micrantha* (Engelm.) Gray

1' Flowers 12-15 mm long, the outer petal lacking a crest or wing
at the summit. Fruiting capsules spreading or pendant, 1-2.5
cm long. Plants annual or biennial, glabrous, up to 4 dm
high, branched and spreading. Flowers often not cleistogamous.
Frequent on rocky or sandy soil over the state. Ours represented by the two varieties *aurea* and *occidentalis* Engelm.
{incl. *C. montana* Engelm.} May-July . . . .
. . . . *Corydalis aurea* Willd.

DICENTRA Bernh.

Dicentra cucullaria (L.) Bernh. Perennial herb from many, small,
bulb-like corms. Leaves much dissected, basal, glabrous. Flowering scapes 1-3 dm tall. Flowers white or pale pink, the corolla with two prominent divergent spurs, the petals 12-16 mm long.
Sepals 2, small. Stamens 6, in 2 sets. Fruiting capsule 1-3 cm
long, tapered at both ends. Rare but locally abundant where it
occurs in rich woods of the eastern part. Apr-May. Dutchman's
Breeches.

FUMARIA L.

Fumaria officinalis L. Small herbaceous annuals diffusely
branched, the stems mostly less than 3 dm tall. Leaves many
parted, chiefly basal. Inflorescence racemose, the flowers 7-10
mm long, deep red or purple. Sepals 2; stamens 6. Fruits subglobose, one-seeded, about 3 mm across. Established as a weed
in waste places in the southeast part and perhaps elsewhere.
May-June. Fumitory.

Family Brassicaceae

(Key to Genera)

(In order to key out members of this family it is necessary to
have material in flower *and* in fruit.)

1 Petals yellow or yellowish.

2 Pods separated into two flattened halves (didymous), with
a constriction between the rounded portions . . . .
. . . . *PHYSARIA*

2' Pods not separated into two halves.

3 Fruits with a long stipe up to 1 cm long . . . .
. . . . *STANLEYA*

3' Fruits lacking a long stipe.

    4 Pods elliptic or oval to rounded in outline, not more than 2 times longer than broad (silicle).

        5 Silicle much flattened with a notched apex . . . . *ALYSSUM*

        5' Silicle rounded, the apex beaked.

            6 Fruits widest above the middle; leaves clasping the stem . . . . *CAMELINA*

            6' Fruits not widest above the middle.

                7 Leaves pinnatifid . . . . . . *RORIPPA*

                7' Leaves entire or toothed. Fruits globose.

                    8 Pods with a reticulate, pitted surface . . . . *NESLIA*

                    8' Pods smooth, not pitted . . *LESQUERELLA*

4' Pods several times longer than broad (silique).

    9 Plants glabrous or only sparsely pubescent.

        10 Leaves oblong, entire, the bases clasping the stem . . . . *CONRINGIA*

        10' Leaves dentate or pinnatifid.

            11 Fruits with seeds in 1 row in each cell . . . . *ERUCASTRUM*

            11' Fruits with seeds in 2 rows in each cell . . . . *DIPLOTAXIS*

9' Plants with hairs on leaves and stems or rarely without hairs.

    13 Hairs on vegetation simple, not branched, or occasionally glabrous.

        14 Fruit with a stout beak . . *BRASSICA*

        14' Fruit lacking a stout beak.

            15 Pods 4-angled; seeds in 1 row in each cell . . . . *BARBAREA*

            15' Pods rounded in cross-section; seeds in 1 or 2 rows in each cell.

                16 Fruits 1 cm or less long; seeds in 2 rows in each cell . . . . *RORIPPA*

16' Fruits 1.5 cm or more long; seeds
in 1 row in each cell . . . . . .
. . . . *SISYMBRIUM*

13' Hairs on the vegetation branched or forked.

17 Leaves divided into pinnatifid segments
. . . . *DESCURAINIA*

17' Leaves entire or toothed, not divided into
pinnatifid segments.

18 Fruits linear-elongate, mostly less
than 2 mm in diameter . . . .
. . . . *ERYSIMUM*

18' Fruits not linear-elongate, 2-4 mm in
diameter . . . . *DRABA*

1' Petals white, pink, or lavender, or sometimes lacking petals.

19 Fruits flattened at right angles to the septum, the suture
therefore on the face and at the middle of the flattened
surface.

20 Pods several-seeded.

21 Vegetation with simple hairs. Fruits orbicular or
rounded . . . . *THLASPI*

21' Vegetation with forked hairs. Fruits triangular,
the widest part above the middle . . *CAPSELLA*

20' Pods 2-seeded. Occasionally an extra seed may be
found in each cell.

22 Leaves grayish-green, densely pubescent, the margins dentate with sagittate bases clasping the
stem . . . . *CARDARIA*

22' Leaves green but not grayish pubescent. Leaves,
if their bases clasp the stem, with margins not
dentate . . . . *LEPIDIUM*

19' Fruits not flattened as above; almost twice as long as
wide or longer.

23 Flowers purple or lavender.

24 Plants less than 5 dm tall. Fruits not more than
2.5 cm long . . . . *CHORISPORA*

24' Plants usually exceeding 5 dm tall. Fruits 4-6 cm
long at maturity . . . . *HESPERIS*

23' Flowers white or tinged with pink.

25 Fruit on a stipe at least 1 mm long . . . .
. . . . *THELYPODIUM*

25' Fruit sessile, not on a stipe.

26 Plants small, annual, usually not exceeding
10 cm high . . . . *DRABA*

26' Plants larger, usually exceeding 1 dm high.

27 Pods three or more times longer than broad.

28 Leaves palmately divided . . . .
. . . . *DENTARIA*

28' Leaves pinnately toothed and divided
or entire.

29 Pods more or less flattened.

30 Hairs on vegetation branched.

31 Pods strongly flattened,
less than 1.5 cm long . .
. . . . . . *DRABA*

31' Pods not strongly flattened, more than 1.5 cm
long . . . . *ARABIS*

30' Hairs on vegetation simple or
lacking . . . . *CARDAMINE*

29' Pods rounded in cross-section.

32 Plants aquatic, with lax stems
. . . . *NASTURTIUM*

32' Plants terrestrial, with
slender stems . . . .
. . . . *ARABIDOPSIS*

27' Pods less than three times longer than
broad.

33 Vegetation glabrous. Plants with
stout stems, up to 1 meter tall . . .
. . . . *ARMORACIA*

33' Vegetation pubescent.

34 Pubescence on vegetation simple
. . . . *RORIPPA*

34' Pubescence on vegetation with
stellate or forked hairs . . . .
. . . . *BERTEROA*

ALYSSUM L.

1 Fruiting silicles stellate-pubescent; styles 0.5 mm long or
   less. Sepals persisting for a time after flowering. Plants
   annual or biennial, 1-3 dm tall, with branches from the main
   stem. Vegetative parts with a grayish, appressed pubescence.
   Leaves lanceolate, sessile, entire to serrate. Flowers cream-
   white, the petals 3-4 mm long. Fruit ovate to oval, 3-4 mm
   long, the septum at the rim. An occasional weed of roadsides
   and waste places. May-June . . . . . . *Alyssum alyssoides* L.

1' Fruiting silicles glabrous; styles 0.5-1 mm long. Sepals
   deciduous at flowering time. Plants annual, 1-3 dm tall,
   usually branched. Leaves oblanceolate, sessile, with a
   grayish-white stellate pubescence. Flowers smaller than in
   the preceding species. Fruits oval, 3-4 mm long. An occasional weed in dry soil of roadsides or waste places in the
   western part. May-June . . . . . . *Alyssum desertorum* Stapf.

ARABIDOPSIS Heynh.

Arabidopsis thaliana (L.) Heyn. Annual or weakly perennial plant
1-4 dm tall. Stems thin, erect, usually simple but occasionally
sparingly branched. Leaves mostly in a basal rosette, the cauline ones few, narrowly oblong. Vegetation with forked hairs.
Flowers small, white to pinkish, the petals 2-4 mm long. Fruits
cylindric, ascending, the siliques 0.5 to 1.5 cm long. Rare in
moist, wooded ravines in the Black Hills. May-June. Mouse-ear
Cress.

ARABIS L.

1 Fruiting pods erect or appressed towards or at maturity.

   2 Seeds with wings nearly 1 mm broad. Fruiting pods 2-3 mm
      wide, compressed. Plants biennial or perennial, stems 1-
      few from a branched caudex, 3-6 dm tall. Leaves spatulate,
      the lower ones rosulate, the upper ones many and overlap-
      ping, more lanceolate. Flowers white to pale pink, in 5-
      many flowered racemes. Infrequent to rare in the western
      part. June-Aug . . . . *Arabis drummondii* Gray

   2' Seeds wingless or wings less than 0.5 mm broad. Fruiting
      pods less than 2 mm wide.

      3 Fruits flat, plants at maturity usually not over 4 dm
         tall. Seeds in one row in each cell. Pods 5 cm long
         or less. Basal leaves in a rosette, the cauline leaves
         slender, erect, with clasping, auriculate bases. Petals
         white to gray-brown, 4-6 mm long. Frequent on calcareous

and gravelly soils over the state. Ours var. *pycnocarpa*
(Hopkins) Uline. May-June . . *Arabis hirsuta* (L.) Scop.

3' Fruits rounded in cross section. Plants at maturity
usually over 4 dm tall. Seeds in 2 rows in each cell
of the ovary. Pods commonly over 5 cm long. Biennial
with stems smooth in the upper part. Basal leaves
coarsely lanceolate, crowded. Cauline leaves with
auriculate-clasping bases. Flowers creamy-white, the
petals 3-6 mm long. Infrequent in dry soils of the
western part. Not as common as the preceding species.
May-June. Tower Mustard . . . *Arabis glabra* (L.) Bernh.

1' Fruiting pods spreading or pendant, not erect towards or at
maturity.

4 Cauline leaves with their bases auriculate or clasping the
stem.

5 Upper leaves glabrous or nearly so. Lower leaves pubescent with stellate hairs on both surfaces. Plants with
erect stems up to a meter tall. Petals white to pink,
4-7 mm long. Fruits at first erect but becoming spreading towards maturity. Frequent in sandy or rocky soil.
More common in the Black Hills region. June-July . . .
. . . . *Arabis divaricata* A. Nels.

5' Upper leaves distinctly pubescent. Lower leaves variously pubescent.

6 Fruits pendant towards maturity, the pedicels abruptly deflexed. Leaves lance-linear, usually not over
1 cm wide, more or less appressed to the stem. Stems
erect from a caudex, perennial, from 3 to 7 dm tall.
Flowers white to pink, the petals 6-8 mm long. Infrequent on dry, calcareous or rocky soil of the
western part. Ours var. *pinetorum* (Tidest.) Roll.
June-July . . . . *Arabis holboellii* Hornem.

6' Fruits widely spreading, usually less than 4 cm long
at maturity. Leaves thin, oblanceolate, over 1 cm
wide, their margins toothed. Stems 2-4 dm tall,
somewhat branched at the base, perennial. Petals
whitened, short, only slightly longer than the
sepals. Frequent in rich woods and on hillsides of
Clay, Lincoln and Union counties in the southeastern
part. May-June . . . *Arabis shortii* (Fern.) Gleason

4' Cauline leaves without auricled or clasping bases. Plants
up to 1 meter tall, biennial. Petals creamy-white, 3-5 mm
long. Fruits usually over 5 cm long, sickle-shaped.

Leaves lanceolate with slightly dentate margins, only
sparsely hairy. Rare in rich, loamy woods of Lincoln,
Marshall and Roberts counties in the east part. May-July.
Sicklepod . . . . *Arabis canadensis* L.

ARMORACIA Gaertn.

Armoracia rusticana Gaertn. Tall perennials up to 1 m with
thick, aromatic taproots. Stems mostly unbranched, with oblong
leaves 1-3 dm long. Lower leaves larger, long-petioled, the upper leaves smaller. Inflorescence of several racemes, terminal
and in upper axils. Flowers with petals 4-8 mm long, showy.
Fruiting pods 2 celled, ellipsoid. Escaping occasionally from
cultivation or persisting around old or abandoned farmsteads.
July-Sept. Horseradish.

BARBAREA R. Br.

1 Style tip beak-like, over 2 mm long. Principal basal leaves
   with several pairs of lateral lobes. Plants biennial, 2-8 dm
   tall, branched above, the upper leaves lobed. Flowers yellow,
   in crowded racemes that elongate in fruit. Fruiting siliques
   somewhat angled, erect or ascending, 1-3 cm long, with a distinct beak. Infrequent to rare as a weed in moist places or
   low fields in the eastern part. June-July . . . .
   . . . . *Barbarea vulgaris* R. Br.

1' Style tip short, the beak less than 1 mm long. Principal
   basal leaves with 1-2 pairs of lateral lobes. Plants biennial, 2-6 dm tall, the stems stiff, sometimes branched. Upper
   leaves pinnatifid. Flowering racemes terminal, compound, the
   petals yellow. Fruiting siliques 2-5 cm long, slightly compressed, strongly ascending. Infrequent in swamps and along
   streams in the Black Hills. June-July. Wintercress . . . . .
   . . . . *Barbarea orthoceras* Ledeb.

BERTEROA DC.

Berteroa incana (L.) DC. Plants annual to weakly perennial, the
stems branched from the base, 3-5 dm tall. Leaves mostly entire,
lanceolate; the surfaces covered with short, forked hairs. Flowers small, less than 5 mm high, with white petals. Pods less
than 1 cm long, oblong. Occasional weed that persists at old
farmsteads and in other waste places. June-July. Hoary Alyssum.

BRASSICA L.

1 Beaks of maturing fruits flattened and possessing 3 longitudinal nerves.

2 Fruits glabrous and about 2 mm wide. Beak not more than
   two-thirds as long as the body of the fruit. Plants annual, 3-9 dm tall, usually branched. Leaves simple to
   variously pinnatifid. Flowering racemes usually compound,
   up to 3 dm long. An introduced weed that occasionally becomes locally abundant. May-June. Charlock . . . . . .
   . . . . *Brassica kaber* (DC.) Wheeler

2' Fruits bristly or at least pubescent and from 3 to 5 mm
   wide. Beak shorter to somewhat longer than the body of
   the fruit. Plants annual, 2-8 dm tall, branched above.
   Leaves variously pinnate to pinnatifid, with 3 or more
   segments. Flowering racemes compound, spreading, up to
   3 dm long. An introduced weed of waste places over the
   state. May-June . . . . *Brassica hirta* Moench.

1' Beaks of maturing fruits 4-angled or rounded in cross-section
   and possessing 1 longitudinal nerve.

   3 Cauline leaves with bases clasping the stem, sessile. An-
      nual or sometimes biennial plants 3-9 dm tall. Lower
      leaves pinnatifidly dissected. Flowers yellow, the petals
      6-8 mm long. Fruits erect or ascending, rounded, 3-4 cm
      long. Occasionally escaped and persisting in the western
      part. May-July. Field Mustard . . *Brassica campestris* L.

   3' Cauline leaves not clasping the stem.

      4 Fruits mostly less than 2 cm long, erect and appressed
         towards maturity. Plants annual with simple branches,
         the stems up to 1 m tall. Leaves petioled, variously
         toothed or pinnatifid. Flowers 8-10 mm wide, the petals
         yellow, 7-8 mm long. Middle nerve of fruit as wide as
         the sutures. Frequent to common in old fields and waste
         places, especially in the eastern part. May-Sept.
         Black Mustard . . . . *Brassica nigra* (L.) Koch.

      4' Fruits 2-4 cm long, ascending but not closely appressed
         towards maturity. Plants annual, 5-10 dm tall, the stem
         branched. Leaves variously dentate to pinnatifid. Flow-
         ers pale yellow, the petals 6-8 mm long. Middle nerve
         of the fruit narrower than the sutures. An infrequent
         weed of waste places and occasionally escaped. May-July.
         Brown Mustard . . . . *Brassica juncea* (L.) Coss.

                        CAMELINA Crantz.

1 Stem hairy on the lower one-half. Petals 3-4 mm long. Mature
   pods about 5 mm long, their pedicels averaging about 1 cm.
   Plants slender, annual, 3-6 dm tall, sparingly branched.

Fruits obovoid, 6-7 mm long, with a persistent style. Flowers light yellow to greenish. Occasionally along roadsides and in weedy places over the state. May-June. False Flax . . . . *Camelina microcarpa* Andrz.

1' Stem glabrous on the lower one-half. Petals mostly 5-6 mm long, the flowers pale yellow. Mature pods about 7 mm long, obovoid or pyriform as in the preceding. Pedicels averaging about 1.5 cm long. Plants 3-6 dm high, with a greater tendency for branching than the preceding species. An annual that is seen occasionally in waste places or on roadsides over the state. June-July . . . . *Camelina sativa* (L.) Crantz.

CAPSELLA Medic.

Capsella bursa-pastoris (L.) Medic. Annual plants 1-6 dm tall, the stems solitary or variously branched. Leaves pinnatifid or pinnately lobed, 5 cm or more long. The inflorescence racemiform, flowers with white petals, about 2 mm wide. Fruits triangular to obcordate, the distal ends widest, 4-8 mm long. Very common in lawns, gardens and waste places, especially in the eastern part. Apr-June. Shepherd's Purse.

CARDAMINE L.

1 Stem leaves pinnately dissected, the terminal segment obovate. Stems about 5 dm tall, mostly erect and sparingly branched. Basal leaves few or lacking. Flowers creamy-white, about 4 mm wide. Fruits up to 3 cm long, ascending on short pedicels. Rare in moist or swampy places in the Black Hills and in the northeast part. Apr-June . . . . *Cardamine pensylvanica* Muhl.

1' Stem leaves simple, lacking dissections or lobes. Plants growing from a deep-seated tuberous bulb. Stems erect, up to 6 dm tall. Basal leaves ovate, on long petioles. Cauline leaves lance to oval-shaped, short petioled to sessile. Flowers white, the petals 7-11 mm long. Infrequent in swampy places in the northeast counties of Day, Grant, Marshall and Roberts. Apr-June . . . . *Cardamine bulbosa* (Schreb.) BSP

CARDARIA Desv.

Cardaria draba (L.) Desv. A perennial weed growing to 5 dm tall, the stems erect, with little or no branching. Leaves lance-ovate, 3-5 cm long, irregularly toothed, the bases sessile and auriculate. Flowers white, borne in elongate racemes, the pedicels up to 1.5 cm long. Fruits about 3 mm long, beaked, up to 5 mm wide, flattened with a cordate base. Rare in waste places over the state. May-June. Hoary Cress.

CHORISPORA R. Br.

Chorispora tenella DC. Annual or weak perennial 2-5 dm tall, branched at the lower part. Leaves coarsely toothed or with an irregular outline, lance-shaped, 3-8 cm long. Inflorescence terminal, racemiform, the flowers purple, up to 1 cm long. Sepals erect, 7-8 mm long, with sac-like bases. Fruits 2-3 cm long, erect or spreading, the beak prominent, 1 cm long or longer. Infrequent in waste places over the state. May-June.

CONRINGIA (Heist.) Link.

Conringia orientalis (L.) Dum. Plants annual, herbaceous, 3-6 dm tall; stems erect with little or no branching. Vegetation glabrous. Leaves oval to oblong, strongly clasping the stem. Inflorescence elongate racemes. Flowers pale yellow, the petals less than 1 cm long. Pods ascending, linear, 4-angled, 8 or more cm long at maturity. The seeds granular-roughened, in one row in each cell. Infrequent in waste places and along roads. May-July. Hare's Ear.

DENTARIA L.

Dentaria laciniata Muhl. Plants small herbaceous perennials 2-3 dm tall from rhizomes. Leaves 3-5 parted, the segments lanceolate with laciniate margins, in a whorl of three at the middle of the stem at flowering time. Flowers white to pink, racemosely arranged at the apex on the main stem. Petals 1-1.5 cm long. Fruit a linear pod about 3 cm long. Infrequent in rich woods of Roberts and Marshall counties in the northeast part and in the east in Lincoln County. Apr-May. Toothwort.

DESCURAINIA Webb. & Benth.

1 Leaves bi-pinnate or more dissected. Fruits more than 15 mm long. Leaf surfaces canescent with a stellate pubescence. Stems mostly erect, 3-7 dm tall. Flowers cream-colored, the petals 2-2.5 mm long. Plants annual or biennial. Frequent to common in overgrazed pastures, along roads or in waste places over the state. June-July. Herb Sophia . . . . *Descurainia sophia* (L.) Webb.

1' Leaves once-pinnate or deeply pinnatifid. Occasionally the larger leaves bi-pinnate. Fruits less than 15 mm long.

2 Fruits widest at the top (clavate) and tapering to the base. Seeds in the pod in two rows. Plants variously branched, the stems up to 6 dm tall. Lower leaves sometimes twice pinnate. Flowers pale yellow, the petals up to 3 mm long. Pedicels of the maturing fruits widely

spreading, the fruiting pods up to 12 mm long. Frequent to
common in waste places or dry sandy soil over the state.
Represented in South Dakota by the varieties *brachycarpa*
(Rydb.) C. L. Hitchc. and *intermedia* (Rydb.) C. L. Hitchc.
May-June. Tansy Mustard . . . .
. . . . *Descurainia pinnata* (Walt) Britt.

2' Fruits linear, up to 12 mm long, the seeds in one row in
the pod. Plants with erect stems up to 1 m tall. Leaves
narrowly lanceolate in outline with deeply incised pinnatifid segments. Vegetation variously pubescent to canescent.
Flowers pale yellow, the petals 2.5-3 mm long. Pedicels
narrowly ascending towards the maturity of the fruits, the
raceme dense. Infrequent in moist waste ground and on
river banks. Scattered over the state. June-July . . . .
. . . . *Descurainia richardsonii* (Sweet) O. E. Schulz

DIPLOTAXIS DC.

Diplotaxis muralis (L.) DC. Annual plant with branching stems
2-6 dm tall, the vegetation glabrous or sparsely pubescent.
Leaves pinnatifid or toothed, mostly basal. Inflorescence racemose, the pedicels up to 1.5 cm long. Flowers yellow, clawed,
4-7 mm long. Fruiting racemes elongate, the siliques ascending,
2-3 cm long, terete, with a short, distinct beak. Occasional
weed of waste, sandy places. June-Sept. Sand Rocket.

DRABA L.

1 Plants annual or at most winter annuals, flowering in spring
or early summer.

2 Vegetation, including the pedicels and fruits, pubescent to
rough-hairy. Plants 1-2 dm tall, the stems simple or sparingly branched at the base. Leaves 1-2 cm long, oblong to
obovate, the lower ones hairy. Inflorescence racemose at
maturity, 5-8 cm long. Flowers with or without petals.
When present, the corolla white. Fruiting pods widely
spreading. Apparently rare. Our only report from the
eastern part of the state. Apr-May . . . .
. . . . *Draba cuneifolia* Nutt.

2' Vegetation glabrous or nearly so, especially the upper
parts of the stem and fruiting pedicels.

3 Petals white, sometimes minute or lacking. Stems 2-10
cm tall, sparingly branched from the base. Leaves few,
basal, elliptical to ovate with a stellate pubescence.
Flowers 2-3 mm long, the petals often lacking. Fruits
linear, about 1 cm long, ascending. Frequent on sterile

soil and in dry and sandy exposed places over the state.
Apr-June . . . . *Draba reptans* (Lam.) Fern.

3' Petals pale yellow, becoming whitened towards senescence.
Stems up to 3 dm tall, usually not branched. Leaves
lance-ovate, 1-3 cm long, mostly basal. Flowers in a
loose raceme. Fruits 1 cm long, becoming ascending on
pedicels up to 3 cm long. Frequent in dry soil of open
woods and prairies of the western part, including the
Black Hills, and in the eastern part. Apr-June . . . .
. . . . *Draba nemorosa* L.

1' Plants biennial or perennial, flowering in summer.

4 Petals yellow.

5 Style present and distinguishable. Petals 4-6 mm long.
Sepals over 2 mm long. Stems up to 4 dm tall, slender
and branched from a simple or branched caudex. Leaves
mostly basal, lance-ovate, those on the stem sessile.
Inflorescence racemose, becoming elongate in fruit.
Fruits erect on ascending pedicels, 1-1.5 cm long. In-
frequent on wooded slopes and meadows in the Black Hills.
June-July . . . . *Draba aurea* Vahl.

5' Style very short or lacking. Petals 2-4 mm long. Sepals
less than 2 mm long. Stems 0.5-3.0 dm tall, simple or
with occasional branches. Leaves up to 4 cm long, mostly
basal. The cauline leaves ovate-lanceolate with irregular margins. Racemes 10-20 flowered, the pedicels exceeding the fruit length. Fruits erect, 8-11 mm long.
Rare in meadows or on dry slopes in the Black Hills.
June-July . . . . *Draba stenoloba* Ledeb.

4' Petals white, 3-5 mm long. Stems up to 2.5 dm tall,
branched from a caudex. Leaves mostly basal, numerous in
a rosette, spatulate to oblanceolate. Cauline leaves lance-
ovate. Inflorescence a dense raceme, elongating in fruit,
the lower pedicels axillary in upper leaves. Fruits erect,
with stellate pubescence, 5-12 mm long. Rare to infrequent
in rocky woods of upper altitudes in the Black Hills. June-
July. {*Draba cana* Rydb.} . . . . . *Draba lanceolata* Ledeb.

ERUCASTRUM Presl.

Erucastrum gallicum (Willd.) Schulz. Annual or sometimes biennial
plants 3-6 dm tall, sparingly branched above. Leaves deeply pinnatifid, the terminal segment the largest, blades to 12 cm long.
Inflorescence racemose, the flowers yellow. Stamens long, with a
basal gland. Fruit cylindrical, 2-2.5 cm long, with a conspicuous

beak. Occasional in waste places, appearing similar to species
of *Brassica*. May-July.

ERYSIMUM L.

1 Petals more than 1 cm long. Fruits toward maturity exceeding
5 cm. Biennial plants with simple erect stems or sparsely
branched below, 2-8 dm tall. Vegetation grayish due to dense
pubescence. Leaves mostly linear-lanceolate. Flowers bright
yellow in a dense raceme which elongates in fruit. Pods
widely spreading to ascending, up to 8 cm or more long. Common in open prairies over the state, especially westward.
May-June. Western Wallflower . . *Erysimum asperum* (Nutt.) DC.

1' Petals less than 1 cm long.

  2 Fruiting pedicel almost as thick as the fruit. Fruits
  spreading, over 5 cm long. Plants annual, 2-4 dm tall,
  mostly simple-stemmed. Basal as well as cauline leaves
  lanceolate, ashy-colored. Inflorescence of light yellow
  flowers, the petals 6-9 mm long. Fruits 6-8 cm long, almost 4-sided in cross section. Rare in dry prairies in
  the southern part. May-June . . . . . *Erysimum repandum* L.

  2' Fruiting pedicel about one-half as thick as the fruit.
  Fruits ascending, less than 5 cm long.

    3 Plants annual. Pedicels slender, in fruit up to 12 mm
    long. Petals not over 6 mm long. Stems erect, spar-
    ingly branched, up to 8 dm tall. Leaves lanceolate,
    commonly erect and somewhat appressed to the stem.
    Flowers bright yellow. Sepals 2-4 mm long. Fruits
    erect or ascending, up to 2.5 cm long. Frequent in
    moist soil and alluvial woods over the state. July-
    Aug. Wormseed Mustard . . . *Erysimum cheiranthoides* L.

    3' Plants perennial or biennial. Pedicels stout, in fruit
    usually not over 6 mm. Petals 6-10 mm long. Stems
    erect, 3-6 dm tall, usually simple, the rootstock not
    enlarged. Leaves linear, erect, remotely dentate, with
    a canescent surface. Flowers pale yellow, in a terminal
    raceme which elongates in fruit. Fruits up to 4 cm long.
    Infrequent in thin or sterile soil of open places. Also
    in open woods in the eastern part. June-July . . . . .
    . . . . *Erysimum inconspicuum* (Wats.) MacMill.

HESPERIS L.

Hesperis matronalis L. Plants perennial with erect stems up to
8 dm tall. Leaves lanceolate, margins dentate-sinuate, the

blades 5-15 cm. Flowers purple to occasionally white, in a
loose, terminal, racemose inflorescence. Petals up to 2 cm long,
showy, with an exsert claw. Fruits elongate, 4-6 cm long, with
a persistent style. Commonly escaping from cultivation and persisting, especially in the eastern part. May-June. Dame's
Rocket.

LEPIDIUM L.

1 Leaves on the stem ovate in outline, their bases clasping the
   stem. Lower leaves pinnatifid with linear segments. Petals
   narrow, over 1 mm long, yellow. Plants annual, the stems 2-5
   dm tall, branched above. Fruits elliptical, about 4 mm long.
   A weed of waste places that is becoming widely established.
   May-June . . . . *Lepidium perfoliatum* L.

1' Leaves on the stem not clasping or ovate.

   2 Petals usually absent or, if present, rudimentary. Fruits
      oblong to obovate, widest above the middle. Plants annual,
      2-5 dm tall, irregularly branched upward. Leaves on the
      upper part of the stem linear. The lower ones pinnatifid.
      Inflorescence a raceme which has flowers less than 4 mm
      across. In fruit the racemes are erect. Fruits 2-3.5 mm
      long, the upper part notched. Common as a weed in dry
      waste soil or on roadsides and in pastures. May-June.
      Peppergrass . . . . *Lepidium densiflorum* Schrader

   2' Petals white, conspicuous, about twice as long as the
      sepals. Fruits elliptic to ovate, widest at about the
      middle. Plants annual or biennial, the stems simple or
      branched, up to 4 dm tall. Basal leaves pinnatifid to
      twice pinnate, somewhat hirsute. Inflorescence racemose,
      the racemes many-flowered. Widespread as a weed in dry or
      moist soil over the state, especially in the east. May-
      June . . . . *Lepidium virginicum* L.

LESQUERELLA S. Wats.

1 Fruiting pedicels recurved and the fruits subglobose to ellipsoid, pendant. If the pedicels are straight, then the siliques
   more than 3.5 mm long.

   2 Plants annual to short-lived perennials, the stems 1-2 dm
      tall, spreading from the base. Inner basal leaves flat,
      with angular or dentate margins. Interior of fruiting
      valves usually glabrous. Principal leaves 2-5 cm long,
      oblanceolate and narrowing to a slender petiole. Inflo-
      rescence dense, in fruit becoming elongate, the pedicels
      usually secund. Similar to the following but usually of

smaller stature and less pubescent. Ours represented by
the two varieties *arenosa* and *argillosa* Rollins and Shaw.
Frequent in dry places in the western one-half. May-June.
· · · · · · · · · · · · *Lesquerella arenosa* (Rich.) Rydb.

2' Plants perennial, the stems 1-4 dm tall, widely spreading
to erect. Vegetation densely pubescent. Inner basal
leaves entire and involute when young. Interior of fruiting valves usually pubescent. Principal leaves oblanceolate and flat, 2-6 cm long. Inflorescence compact and
densely flowered, the pedicels becoming recurved after
flowering into an open fruiting cluster. Fruits 4-6 mm
long, usually subglobose. Frequent to common in the
western two thirds of the state. May-June. Bladderpod.
· · · · · · · · · · · *Lesquerella ludoviciana* (Nutt.) Wats.

1' Fruiting pedicels sigmoid or curved, the fruits erect, ellipsoid. Plants perennial.

3 Fruits 3-6 mm long, compressed at their apexes. Principal
leaves linear-oblanceolate, gradually narrowed to the petiole. Stems up to 25 cm tall, usually simple, densely pubescent. Inflorescence dense, not becoming conspicuously elongate in fruit. Frequent on dry plains and open areas in
the western part, including the Black Hills. May-June · ·
· · · · · · · · · · · · *Lesquerella alpina* (Nutt.) Wats.

3' Fruits 7-12 mm long, not compressed at their apexes. Principal leaves sub-orbicular or elliptical, gradually or
abruptly narrowing to the petiole, 2-5 cm long. Stems 5-20
cm tall, several, prostrate to ascending, from a woody caudex. Inflorescence densely flowered, becoming elongate and
loose in fruit. Pedicels sigmoid, the fruits erect. Rare
in dry, open places in the extreme southwest part. May-
June · · · · · · · · · · *Lesquerella montana* (Nutt.) Wats.

NASTURTIUM R. Br.

Nasturtium officinale R. Br. Plants perennial, rooted in mud
bottoms of quiet water. Stems lax and partly floating or submersed, rooting at the nodes, 1-5 dm long, usually only the top
erect. Leaves glabrous, pinnately compound, up to 5 cm long.
Inflorescence racemose with white flowers, the petals 3-4 mm
long. Fruiting pods 15 mm long, spreading or ascending. Infrequent in cool streams or ponds of the eastern part, especially
in the northeast counties. May-Aug. Watercress.

NESLIA Desv.

Neslia paniculata (L.) Desv.   Plants annual with erect stems 3-6
dm tall.   Leaves on the stem lanceolate with sagittate-clasping
bases.   Surface of the vegetation harshly hairy.   Flowers yellow,
the petals 2 mm or less long.   Inflorescence racemose, becoming
elongate in fruit.   Pods subglobose with a conspicuous surface
reticulation, about 2 mm in diameter.   Occasional in waste places
and old fields in the eastern part.   July-Aug.   Ball Mustard.

PHYSARIA (Nutt.) Gray

Physaria didymocarpa (Hood.) Gray.   Plants perennial with a stout
taproot or branched caudex.   Stems several from the base, spreading to erect, up to 1.5 dm tall.   Leaves ovate, mostly basal, arranged in a rosette-like pattern.   Leaf surfaces whitened with
dense, stellate hairs.   Flowers yellow, racemose, the petals up
to 1.4 cm long.   Fruits doubled (didymous), the pods obcompressed,
that is, the suture running around the periphery of the flattened
edge.   Infrequent in dry or sandy soil of the plains in the western part.   {incl. *P. brassicoides* Rydb.}   May-June.   Double Bladderpod.

RORIPPA Scop.

1   Plants perennial from a creeping rhizome, the stems lax-
spreading to ascending, up to 3 dm tall.   Leaves sessile,
partly clasping the stem, shallow to deeply sinuate, pinnatifid nearly to the midrib.   Flowers yellow, in racemes, the
petals 3.5-5.5 mm long.   Fruiting pedicels divergent and recurved, about as long as the fruits.   Infrequent in moist alluvial or sandy soil over the state.   Apr-July . . . .
. . . . *Rorippa sinuata* (Nutt.) Hitchc.

1' Plants annual or biennial, from a taproot.   Leaves entire to
deeply pinnatifid.

2   Stems less than 3 dm long, sprawling decumbent.   Leaves
usually fewer than 10.   Valves of fruiting pods papillose.

3   Fruiting siliques roughened with small papillae, tapering to the apex, not constricted at the center.   Plants
glabrous, much branched from the base.   Leaves lyrate-
divided, 2-5 cm long.   Flowering racemes terminal and
axillary.   Petals less than 1 mm long, shorter than the
sepals.   Fruits 3.5-7 mm long, about twice as long as
the pedicels.   Apparently rare in the western part.
June-Sept . . . . *Rorippa tenerrima* Greene

3' Fruiting siliques glabrous, only slightly tapering toward the apex if at all, constricted at the center.

Plants glabrous, much branched, decumbent to prostrate.
Principal leaves pinnately divided to the midrib. Flowering racemes lateral, developing in the axils of leaves.
Petals about 1 mm long, shorter than the sepals. Fruits
3-5 mm long, longer than the pedicels. Infrequent to
rare in alluvial soil along drainage pathways. July-
Sept . . . . *Rorippa truncata* (Jeps.) Stuckey

2' Stems more than 3 dm tall. Leaves usually more than 10.
Valves of fruiting pods glabrous.

    4 Principal cauline leaves unlobed, the margin entire to
irregularly serrate. Plants prostrate to erect, 3-5 dm
tall. Flowering racemes terminal and axillary, the
fruiting pedicels 2-5 mm long, divergent. Infrequent
to rare in moist areas. July-Sept . . . .
. . . . *Rorippa curvipes* Greene

    4' Principal cauline leaves pinnately divided to the mid-
rib, the margin angularly toothed. Plants mostly erect,
3-10 dm tall, much branched from the base. Flowering
racemes terminal and axillary, the fruiting pedicels
often over 7 mm long. Our most common species, occurring in moist areas over the state. May-Aug . . . . . .
. . . . *Rorippa palustris* (L.) Bess.

SISYMBRIUM L.

1 Pedicels 2-4 mm long, erect, causing the fruit to be appressed
to the stem. Pods tapered from base to tip, broader at the
base. Stem erect or sparingly branched above, 3-7 dm tall.
Lower leaves pinnatifid, the cauline and upper ones irregularly lobed and asymetrical. Flowers yellow, the petals 3 mm
long. Fruits 1-1.5 cm long on short, erect pedicels. Occasional as a weed in waste places in the eastern part. Apr-
July. Hedge Mustard . . . . *Sisymbrium officinale* (L.) Scop.

1' Pedicels more than 5 mm long, spreading in fruit, the pods
linear and not tapered.

    2 Fruits 5-9 cm long or longer. Pedicels stout, almost as
thick as the fruits. Stems tall, freely branching, up to
1 meter. Leaves deeply pinnatifid, up to 10 cm long.
Flowers pale yellow, the petals 6-7 mm long. Fruits rigid,
rounded in cross-section. Frequent as a weed of waste
places, roadsides and old fields over the state. June-
July. Tumbling Mustard . . . . . *Sisymbrium altissimùm* L.

    2' Fruits less than 4 cm long. Pedicels slender, much thinner
than the fruits. Plants annual, the stems branched, up to

1 meter tall. Principal leaves pinnatifid, the segments
with obvious dentation. Vegetation with long hairs. Flowers in dense racemes which elongate in fruit. Petals
bright yellow, 6-8 mm long. Becoming a common weed in the
eastern part along roadsides and in waste places. June-
July . . . . *Sisymbrium loeselii* L.

STANLEYA Nutt.

Stanleya pinnata (Pursh) Britt. Plants perennial from deep taproots, the stems simple to branched, 5-15 dm tall. Vegetation
somewhat glaucous. Leaves on lower part of stem pinnatifid, the
upper ones becoming irregularly crenate but not lobed. Flowers
bright yellow in dense racemes, the petals 1-1.5 cm long. Fruits
on prominent stipes that are ascending. Frequent on dry soils of
the plains in the western part. May-Aug. Prince's Plume.

THELYPODIUM Endl.

Thelypodium integrifolium (Nutt.) Endl. Plant biennial, 4-18 dm
tall. Leaves mostly glabrous, 4-10 cm long, the basal ones oblanceolate, toothed or lobed. Flowers in a dense, terminal,
paniculate inflorescence. Petals white to light lavender, 6-8
mm long, spreading. Fruits straight, spreading, 2-3 cm long.
Rare in moist open soils in lower altitudes of the Black Hills.
June-July.

THLASPI L.

Thlaspi arvense L. Plants annual, erect, 2-5 dm tall, often
branched from the middle upwards. Leaves oblong, 3-5 cm long,
clasping at the base, glabrous. Flowers in racemes that elongate
in fruit, the petals white, 2-4 mm long. Fruits flattened, oblong, 1-1.5 cm, with broad wings on the margins. Very common
weed of roadsides and disturbed places in the eastern part.
Apr-May. Pennycress.

Family Capparidaceae

(Key to Genera)

1 Flowering receptacle with a conspicuous, tubular, petaloid
   appendage which is pale pink to lavender in color. Petals
   unequal in length, their distal margins lacinate . . . .
   . . . . *CRISATELLA*

1' Flowering receptacle with a solid gland. Petals nearly equal
   in length, not lacinate.

2 Stamens usually 6. Fruits borne on a stipe more than 6 mm
   long. Plants glabrous. Petals lavender or pink, rarely
   white . . . . *CLEOME*

2' Stamens usually more than 6. Fruits sessile or with a very
   short stipe. Plants with glandular hairs. Petals white or
   pale pink . . . . *POLANISIA*

CLEOME L.

Cleome serrulata Pursh. Erect herbaceous annuals 3-8 dm tall,
the stems branched above, glabrous. Leaves alternate, 3-folio-
late, petioled, the leaflets lanceolate, 3-7 cm long. Upper
leaves short-petioled or sessile. Inflorescence a terminal
raceme, elongating in fruit. Flowers on spreading to recurved
pedicels. Petals about 1 cm long, rose-pink to rarely white.
Fruiting capsule 3-5 cm long, on a stipe 1-2 cm long. Common
in sandy areas and on floodplains over the state. June-July.
Bee Plant.

CRISTATELLA Nutt.

Cristatella jamesii T. & G. Slender annual plants with erect,
usually unbranched stems 1-4 dm tall. Surface of vegetation with
a viscid, glandular pubescence. Leaves trifoliolate, petioled,
the leaflets 1-3 cm long, linear to lance-shaped. Inflorescence
terminal, racemose, compact at first but elongating in fruit.
Petals white or cream colored, 3-4 mm long. Fruiting capsule
oblong, 2 cm long. Rare in dry, sandy soil of Bennett and Todd
counties in the southern part. June-July.

POLANISIA Raf.

Polanisia dodecandra (L.) DC. Annual herbs with glandular-viscid
vegetation, the stems much-branched from below, 1-6 dm tall.
Leaves trifoliolate, leaflets oblong, 1-3 cm long. Flowers in
terminal racemes, the petals white or pale pink. Petals about 7
mm long, the lobed or blade portion longer than the clawed portion. Stamens 1 cm long, slightly exsert from the corolla.
Fruiting capsules lanceolate, 3-4 cm long. Infrequent in sandy
soils and on flood plains over the state. Ours ssp. *trachysperma*
(T. & H.) Iltis. {*Polanisia graveolens* Raf.} June-July. Clammy
Weed.

Family Crassulaceae

(Key to Genera)

1 Petals present, conspicuously yellow. Plants succulent . . .
. . . . *SEDUM*

1' Petals lacking or if present, very inconspicuous. Plants not
succulent . . . . *PENTHORUM*

PENTHORUM L.

Penthorum sedoides L. Perennial herbaceous plants of moist
places, growing from a branched, underground rhizome. Stems 2-6
dm tall, sparingly branched above. Leaves lanceolate, about 7 cm
long, alternate. Flowers with 5 sepals and usually 10 stamens.
Fruits a radiately arranged group of follicles, becoming flattened at maturity. Frequent in low open places and in alluvium
in the eastern part. July-Aug. Ditch-stonecrop.

SEDUM L.

1 Leaves flattened, short, not over 4 mm long; densely overlapping one another on the stem. Petals bright yellow. Stems
densely tufted, 3-6 cm tall. Flowers in cymes. Clones containing a number of sterile prostrate branches. Quite rare
in moist places that are rocky or springy in the Black Hills.
June-July. Mossy Stone-crop . . . . *Sedum acre* L.

1' Leaves rounded, about 1 cm long; not densely overlapping each
other on the stem. Petals pale yellow. Plants tufted from
slender rootstocks. Stems 5-15 cm tall. Flowers in a forked
cyme. Lower leaves soon deciduous. Common on rocky slopes
and limestone outcrops in the western counties, including the
Black Hills. June-July. Stonecrop . . . .
. . . . *Sedum lanceolatum* Torr.

Family Saxifragaceae

(Key to Genera)

1 Plants woody shrubs. Flowers with inferior ovaries . . . . .
. . . . *RIBES*

1' Plants herbaceous. Flowers with ovaries superior to inferior.

  2 Leaves entire. Flowers solitary on leafless stems.
  Ovaries 4-carpelled . . . . *PARNASSIA*

  2' Leaves toothed or lobed. Flowers not solitary.

3 Ovary 1-celled.

    4 Plants from bulbs. Petals white or pink. Leaves deeply dissected . . . . *LITHOPHRAGMA*

    4' Plants from scaly rhizomes. Petals greenish. Leaves palmately lobed, the blade reniform . . *HEUCHERA*

3' Ovary 2-several celled.

    5 Petals white. Styles free above the ovary . . . . *SAXIFRAGA*

    5' Petals pink-red. Styles partially fused above the ovary . . . . *TELESONIX*

HEUCHERA L.

Heuchera richardsonii R. Br. Herbaceous perennial from a scaly, branched caudex, the stems short with leaves mostly basal. Leaf blades reniform, 3-6 cm across, with 7-9 crenate-undulate lobes. Surfaces hispid but becoming puberulent at maturity. Flower stems leafless, 3-6 dm tall, the flowers in a narrow thryse-like panicle. Petals 3-5 mm long, not showy, greenish, irregular. Fruiting capsule ellipsoid, with 2 beaks. Frequent to common in thickets, at edges of woods, thickets, and prairie swales over the state. June-July. Alum-root.

LITHOPHRAGMA Nutt.

1 Stem leaves with small bulblets in the axils. Hypanthium campanulate, hemispheric at the base. Seeds with spinose surfaces. Plants 5-15 cm tall, the stems simple. Leaves ternately divided, the blades 1-2 cm wide. Flowers white to pink, the petals 3-5 parted, 4-7 mm long. Rare in woods and openings at higher altitudes in the Black Hills. May-June. Woodland Star . . . . *Lithophragma bulbifera* Rydb.

1' Stem leaves without small bulblets in the axils. Hypanthium obconic, tapering to the base. Seeds wrinkled, without a spinose surface. Plants 1-3 dm tall, the stems mostly simple, glandular-pubescent. Leaves 3-5 cleft, the divisions deep, almost to the base. Blades 1-3 cm across, upper leaves sessile, the lower basal ones petioled. Flowers white-pink, the petals slightly unequal, 5-10 mm long, 3-5 lobed. Infrequent on open hillsides at middle altitudes in the Black Hills. May-June. Prairie Star . . . . *Lithophragma parviflora* (Hook.) Nutt.

PARNASSIA L.

Parnassia parviflora DC. Plants perennial, herbaceous, from a
short rootstock, the stems erect, simple to several from the
base. Leaves chiefly basal, thin, the blades 1-2 cm long, ovate
to elliptic, tapering to the petiole. Flowering stems essentially naked, 1-3 dm tall, the flowers solitary. Flowers typically pentamerous, the petals white-yellow, 4-7 mm long. Rare
to infrequent in moist, marshy or boggy places in the Black Hills.
July-Aug. Grass-of-Parnassus.

RIBES L.

1 Stems with bristles and spines or both. Flowers borne solitary or few in a cluster.

  2 Ovary and fruit with quite obvious hairs or prickles.

    3 Ovary with stiff prickles. Stamens at maturity about
      equaling the petals. Style not divided. Internodes
      variously thorny to glabrous. Plants erect, widely
      branched, stems 6-14 dm tall. Leaves 2-5 cm wide,
      variously 3-5 lobed. Flowers 1-3, the petals 2-2.5 mm
      long. Fruits 8-12 mm across, wine-red when mature.
      Infrequent in rich woods of Marshall, Minnehaha, Grant
      and Roberts counties in the east. May-June. Dogberry.
      . . . . *Ribes cynosbati* L.

    3' Ovary and fruit with glandular hairs. Stems spiny.
       Petioles of leaves glandular and hairy. Style divided
       nearly to the middle. Plants shrubby, 1-2 meters tall.
       Leaves 2-5 cm long, with deeply incised lobes. Flowers
       loosely spreading and drooping, 4-10 in a cluster.
       Petals light green to purple. Fruit dark purple to
       black, 6-8 mm across, glandular-hispid. Rare in moist
       wooded valleys in the Black Hills. June-July. Swamp
       Currant . . . . *Ribes lacustre* (Pers.) Poir.

  2' Ovary and fruit smooth, sometimes only slightly hairy when
     immature.

    4 Stamens 10-12 mm long, in full flower much longer than
      the sepals. Plants shrubby, 1-2 meters tall, with red-
      brown spines. Leaves 2-5 cm wide, sharply 3-5 lobed.
      Flowers pale green to nearly white, 3-4 in a cluster,
      drooping. Fruits 8-12 mm wide, purple when mature.
      Our most common species in woods throughout the state.
      Apr-May. Gooseberry . . . . . *Ribes missouriense* Nutt.

    4' Stamens 5-8 mm long, in full flower shorter than to
       equaling the sepals.

5   Flowers 5-8 mm long, the hypanthium tube campanulate and spreading upwards.

    6   Stamens as long as the sepals but almost twice the length of the petals. Leaf bracts ciliate-hairy but not glandular. Plants shrubby, 6-12 dm tall. Leaves irregularly rounded in outline, glabrous to softly hairy. Flowers about 5 mm long, the petals 2-2.5 mm long, greenish white to pink. Fruit globose, purple, 8-10 mm across. Rare to infrequent in rocky woods in the Black Hills. May-June. . . . . *Ribes hirtellum* Michx.

    6' Stamens distinctly shorter than the sepals but equaling the length of the petals or slightly surpassing them. Leaf bracts glandular-ciliate. Plants shrubby and sprawling, 6-15 dm tall. Leaves rounded in outline, irregularly 5-lobed, the surfaces more or less hairy. Flowers about 7 mm long, the petals 2-3 mm long, greenish white. Fruit globose, deep blue, 1 cm in diameter. Rare to infrequent in woods in the Black Hills. May-June . . . . *Ribes oxyacanthoides* L.

5' Flowers 10-12 mm long, the hypanthium cylindric. Petals 2.5-3 mm long, greenish purple. Style not divided at the apex. Plants shrubby, 4-8 dm tall, the stems bristly. Leaf blades thin, 1-4 cm across, remotely 5-lobed, finely pubescent. Flowers 1-3 on short peduncles, the fruits reddish-purple, 8-12 mm in diameter. Infrequent in valleys and along streams in the Black Hills and in Harding County. May-June. Spiny Currant . . . . *Ribes setosum* Lindl.

1' Stems unarmed. Flowers in pairs or in short racemes.

  7   Lower leaf surfaces with resinous yellow to orange dots, with a soft pubescence. Flowers greenish white in racemes of 5-10 flowers. Plants shrubby, 1-1.5 meters tall. Leaves 3-8 cm across, with 5 lobes, these coarsely dentate, more or less pubescent. Fruit black, glabrous, 6-10 mm across. Common in woods and thickets over the state. May-June. American Black Currant . . . . *Ribes americanum* Mill.

  7' Lower leaf surfaces not dotted, only slightly pubescent if at all.

    8   Flowers bright yellow. Leaves with two prominent indentations, giving them a three-lobed appearance. Plants shrubby, branched, 1-2 meters tall. Racemes 4-8 flowered,

the petals 2-3 mm long. Fruit black, globose, 8 mm in
diameter. Widely cultivated and escaped but native to
and commonly occurring throughout the state. {incl. *R.
aureum* Pursh.} May. Buffalo Currant . . . .
. . . . *Ribes odoratum* Wendl.

8' Flowers whitish, tinged with pink or red. Leaves with
   many small marginal indentations, appearing almost
   rounded in general outline. Plants low, shrubby, 3-8
   dm tall. Leaves reniform-rounded, 1-3 cm wide. Racemes
   1-4 flowered, the clusters drooping. Fruit red, 6-8 mm
   in diameter. Frequent on hillsides, rocky slopes and in
   wooded areas of higher altitudes throughout the Black
   Hills. {incl. *R. inebrians* Lindl.} May-June. Western
   Red Currant . . . . *Ribes cereum* Dougl.

SAXIFRAGA L.

1 Leaf blades reniform in outline, 1-1.5 cm across, but with 5-9
   shallow lobes. Flowers mostly fewer than 10, some of these
   replaced by bulbils. Plants perennial, 1-1.5 dm tall, from a
   fibrous rootstock, the stems mostly simple. Leaves and stems
   glandular-pubescent, grayish. Lower leaves petioled, the upper ones subsessile. Inflorescence racemose, with small,
   white flowers. Rare in moist rock crevices in higher altitudes in the Black Hills. June-July. Saxifrage . . . . . .
   . . . . *Saxifraga cernua* L.

1' Leaf blades lance-ovate, tapering to the petiole, up to 6 cm
   long. Flowers in a cymose-paniculate inflorescence, usually
   with more than 10 flowers. Plants perennial from short rhizomes, the stems clustered, 1-3 dm tall. Stems and young
   leaves pubescent with a reddish tomentum. Flowers white or
   tinged with pink, the petals 1-3 mm long, often with purple
   spots near their bases. Rare in moist, rocky crevices in the
   Black Hills. May-June . . . . . *Saxifraga occidentalis* Wats.

TELESONIX Raf.

Telesonix jamesii (Torr.) Raf. Herbaceous perennial from a scaly
rootstock, the stems several, glandular-pubescent, 5-15 cm tall.
Leaves mostly basal, the cauline ones much smaller. Blades reniform, with cordate bases, the margins crenate-dentate, 2-5 cm
across. Leaf surfaces pubescent. Flowers red-purple, in compact
terminal panicles. Petals up to 3 mm long, only slightly exceeding the calyx. Ovary half-inferior, with a well-developed hypanthium. Rare in moist rock crevices and talus slopes at higher
altitudes in the Black Hills. {incl. *T. heucheriformis* Rydb.}
July-Aug.

Family Rosaceae

(Key to Genera)

1 Plants woody, trees or shrubs.

    2 Leaves simple.

        3 Ovary inferior.

            4 Branches with thorny spines . . . . . . *CRATEGUS*

            4' Branches without thorny spines.

                5 Plants shrubby. Flowers in raceme-like inflorescences. Petals less than 2 cm long . . . . *AMELANCHIER*

                5' Plants tree-like. Flowers in umbel-like inflorescences. Petals exceeding 2 cm . . . *PYRUS*

        3' Ovary superior.

            6 Flowers with 1 ovary.

                7 Fruit an achene with a persistent, plumose style . . . . *CERCOCARPUS*

                7' Fruit a drupe (Plum or Cherry) . . . *PRUNUS*

            6' Flowers with 2-several ovaries, each becoming separate follicles.

                8 Leaves stipulate or stipular scars present. Leaves lobed, sometimes remotely . . *PHYSOCARPUS*

                8' Leaves without stipules and not lobed.

                    9 Plants low, forming cushion-like mats; leaves entire . . . . *PETROPHYTUM*

                    9' Plants with erect or ascending stems; leaves serrate . . . . *SPIRAEA*

    2' Leaves compound.

        10 Branches without prickles or thorns.

            11 Leaflets more than 9 . . . . *SORBUS*

            11' Leaflets less than 9.

                12 Fruit fleshy. Leaves 3 to 5 foliolate . . . . *RUBUS*

                12' Fruits dry achenes with hairs. Leaves 5 to 7 pinnate . . . . *POTENTILLA*

10' Branches with prickles or thorns.

    13 Carpels maturing as an aggregate fruit, fleshy, not enclosed in an hypanthium . . . . . . *RUBUS*

    13' Carpels maturing as small apples (hips), enclosed in the hypanthium . . . . *ROSA*

1' Plants herbaceous, not woody.

  14 Leaves trifoliolate or palmately compound.

    15 Petals white or tinged with pink.

      16 Plants stemless, leaves basal . . . *FRAGARIA*

      16' Plants with leafy stems.

        17 Styles elongate, persistent. Fruits of separate achenes . . . . *GEUM*

        17' Styles short, deciduous. Fruits of aggregate red ovaries, fleshy . . . . . . *RUBUS*

    15' Petals yellow or cream colored . . . . . *POTENTILLA*

  14' Leaves pinnately or ternately compound.

    18 Ultimate leaf segments linear. Leaves ternately compound. Petals white. Plants small, less than 3 dm tall . . . . *CHAMAERHODOS*

    18' Ultimate leaf segments not linear. Principal leaves pinnately compound.

      19 Flowers with hypanthiums of hooked bristles . . . . *AGRIMONIA*

      19' Flowers with a flat hypanthium, not armed with hooked bristles.

        20 Styles persistent, with a hooked joint or plumose . . . . *GEUM*

        20' Styles deciduous or short, not hooked or plumose . . . . *POTENTILLA*

AGRIMONIA L.

1 Main stem axis in the inflorescence glandular.

  2 Principal leaves with 3-9 leaflets. Plants perennial from a thick, fibrous rootstock, the stems 4-8 dm tall. The leaves coarsely serrate, sparsely hairy below. Flowers in terminal, spike-like racemes, the petals yellow, small. The hypanthium with at least 4 series of hooked bristles,

the maturing sepals also hooked. Rare to infrequent in
open woods in the south and western parts. July-Aug . . .
. . . . *Agrimonia gryposepala* Wallr.

2' Principal leaves with 11-23 leaflets. Plants perennial
    from fibrous roots, the stems densely hairy, 6-12 dm tall.
    Leaves lanceolate in outline, the segments sharply serrate,
    glandular and sparingly pubescent below. Flowers in spike-
    like racemes, the pedicels short. Hypanthium turbinate, 3
    mm across, the outer bristles much shorter than the inner
    ones. Rare in moist places in the sandhills in the south-
    ern part. July-Aug . . . . . . *Agrimonia parviflora* Ait.

1' Main stem axis in the inflorescence without glands.

3 Leaflets conspicuously glandular on the lower surfaces.
    Plants from fibrous roots, perennial, the stems 5-10 dm
    tall. Principal leaves with 7-11 leaflets. Inflorescence
    with dense hairs on the main axis. Petals yellow, 3 mm
    long, the fruiting hypanthium up to 5 mm, with 2-4 series
    of hooked bristles. Frequent in thickets and openings of
    woods in the Black Hills and in the eastern part. July-
    Aug . . . . *Agrimonia striata* Michx.

3' Leaflets finely pubescent but not glandular on the lower
    surfaces. Plants from tuberous thickened roots, the stems
    stout, up to 1 meter tall. Principal leaves with 5-13 leaf-
    lets, coarsely serrate. Axis of the inflorescence with a
    fine pubescence but lacking glands. Petals yellow, small,
    the mature hypanthium 2.3-3 mm high. Rare in dry woods in
    the southeast part. July-Aug . . *Agrimonia pubescens* Wallr.

AMELANCHIER Medic.

1 Leaf blades usually broadly truncate at the apex. Sepals tri-
    angular, 2.5-3 mm long, nearly as broad as long. Plants shrub-
    by with stoloniferous roots. Stems up to 3 meters high.
    Leaves 3-4 cm long, folded early in their development, becom-
    ing broad and glabrous at maturity. Fruits bluish at maturity,
    up to 1 cm in diameter. Frequent in open ravines, along shores,
    and at the edges of thickets over the state. Apr-May . . . . .
    . . . . *Amelanchier alnifolia* Nutt.

1' Leaf blades usually rounded or acute at the apex. Sepals 2.5-
    5 mm long. Plants shrubby with many stems up to 2 or more
    meters tall. In patches forming dense colonies. Subterranean
    stolons many. Leaves white tomentose at flowering time, be-
    coming glabrescent. Fruit globose, blackened at maturity.
    Common in open woods and ravines over the state, especially

in the southern part. Apr-May. {*A. humilis* Wieg.} . . . . .
. . . . *Amelanchier sanguinea* (Pursh) DC.

CERCOCARPUS HBK.

Cercocarpus montanus Raf. Woody shrubs up to 2 meters tall,
branched, the stems dark gray-brown. Leaves 2-4 cm long, oval
to ovate, serrate at the distal part of the blade, white hairy
on the under surface. Flowers perfect, 1-several on short pedicels in leaf axils or on short spurs. Petals absent, sepals 1-2
mm long, on a long calyx tube. Ovary maturing into a hairy-
styled achene 2-4 cm long. Infrequent on dry open hillsides of
the southern Black Hills in Fall River and Custer counties. Apr-
June. Mountain Mahogany.

CHAMAERHODOS Bunge.

Chamaerhodos erecta (L.) Bunge. Perennial herbs from short,
scaly tap roots. Stems 1-3 dm tall, erect, branched above.
Leaves 2-3 ternately divided, the basal ones in a rosette. Leaflets linear or slightly oblong, about 1 cm long. Flowers in flat-
topped bracteate cymes, the petals white. Stamens 5. Ovaries 5-
10, with lateral styles, the separate achenes about 1.5 mm long.
Frequent on dry, exposed plains and rocky outcrops of ridges in
woods and openings in Harding County. Ours var. *parviflora*
(Nutt.) C. L. Hitch. June-July.

CRATEGUS L.

1 Leaves broadest near the base, lobed or otherwise incised.
   Vegetation usually strongly pubescent, including the inflorescence branches, even towards maturity. Spines 3-5 cm long.
   Trees up to 12 meters tall. Flowers 2-3 cm across, white to
   rose-colored, in corymbs. Fruit scarlet, subglobose, 1.2-2.0
   cm wide, each with 4 or 5 nutlets. Infrequent in loamy woods
   and alluvial thickets in the eastern part of the state in the
   Big Sioux and Vermillion River watershed. May . . . .
   . . . . *Crategus mollis* (T. & G.) Scheele

1' Leaves broadest towards the middle, with an ovate shape, tapering towards the base. Leaf surfaces becoming glabrate towards
   maturity.

   2 Petioles with red raised glands. Teeth of leaves red-
      tipped. Nutlets of the fruit not pitted or concave on the
      ventral surface. Flowers 1.5 cm across. Stamens about 10.
      Fruit golden to red orange, 1 cm long. Spines 3 cm or more
      long. A round topped tree 2-7 meters tall. Banks and hill-
      sides of the western and eastern part. May . . . .
      . . . . *Crategus chrysocarpa* Ashe

2' Petioles without red raised glands. Teeth somewhat glandular but not red-tipped. Nutlets of the fruit pitted or
concave on the ventral surface. Flowers many, in dense
corymbs. Stamens about 20. Fruit deep red or scarlet at
maturity, subglobose, about 1 cm across. Spines 3-10 cm
long. A small tree up to 8 meters tall with ascending
branches. Ravines and wooded banks over the state. Ours
var. *succulenta*. May . . . . . . *Crategus succulenta* Link.

FRAGARIA L.

1 The terminal tooth of the terminal leaflet longer than the
two lateral adjacent ones. Leaves light green, of thin texture. Inflorescence usually surpassing the leaves. In fruit,
the achenes on the surface of the receptacle. Plants perennial, somewhat pubescent. Leaves varying much in size. Flowers in irregular cymes, sometimes dense, the petals about 9 mm
long. Frequent in thickets and woods, more common in the eastern part. Ours var. *americana* Porter. Apr-May . . . .
. . . . *Fragaria vesca* L.

1' The terminal tooth of the terminal leaflet shorter than the
two adjacent ones. Leaves deep green, of thicker texture than
the preceding. Inflorescence usually not surpassing the
leaves. In fruit, the achenes in pits of the receptacle, not
superficial. Plants strongly stoloniferous, glabrous to
sparsely pubescent. Flowers in bracteate cymes, the petals
obovate, 7-12 mm long. Common in woods over the state. Represented in South Dakota by the two varieties *glauca* Wats. and
illinoensis (Prince) Gray. Apr-May. Wild Strawberry . . . .
. . . . *Fragaria virginiana* Duchne.

GEUM L.

1 Style of ovary or fruit strongly hooked and jointed near or
above the middle. Plants with well developed stems.

2 Sepals red-purple, spreading or ascending at flowering.
Flower heads more or less nodding. Plants 3-5 dm tall,
the principal leaves with 3-5 leaflets, the terminal leaflet lobed or serrate. Flowers on pedicels that nod toward
maturity. Petals yellow to white with a red tinge, spreading. Infrequent to rare in swampy areas of lake and pond
margins and mesic prairies in the north and west part.
May-June . . . . *Geum rivale* L.

2' Sepals green, recurved at flowering time.

3 Petals white, 5-6 mm long. Plants 3-6 dm tall, with a
sparse to dense pubescence. Basal leaves on long

petioles, mostly trifoliolate. Stem leaves on short
petioles to almost sessile. Fruiting heads spherical,
1-1.5 cm across, the many achenes with a dense, hairy
pubescence. Frequent to common in thickets and woods
in the eastern part. Less common westward. June-July.
White Avens . . . . *Geum canadense* Jacq.

3' Petals yellow.

4 Terminal segment of the principal basal leaves much
larger than the adjacent lateral leaflets or segments. Lower part of the style glandular-pubescent.
Plants perennial from a spreading root stock, the
stems 1-few, 3-6 dm tall, hairy. Principal leaves
with 7-20 leaflets, or appearing pinnatifid. Flowers in cymes, the yellow petals 4-6 mm long, as long
as or slightly longer than the reflexed sepals. Infrequent at pond margins and swampy thickets over
the state. May-June . . . . *Geum macrophyllum* Willd.

4' Terminal segment of the principal basal leaves somewhat larger than the adjacent leaflets but alike in
shape. Lower part of the style mostly glabrous.
Plants perennial from a rootstock, the stems 1-
several, up to 8 dm tall. Leaves with 5-9 segments,
irregularly pinnatifid. Flowers in cymes, spreading,
the petals 5-8 mm long, about equaling the sepals.
Frequent in wet woods or thickets and openings in
the east and west parts. Ours var. *strictum* (Ait.)
Fern. June-July. Yellow Avens . . . .
. . . . *Geum aleppicum* Jacq.

1' Style of ovary or fruit not hooked or jointed, becoming tail-
like and plumose. Plants with short stems or scapose. Leaves
pinnatifid, the blades 5-12 cm long, with many segments. Flowering stems 2-4 dm tall, the sepals erect, reddish purple.
Petals yellow-pink to purple. Fruits with elongate plumose
styles. Frequent in meadows and prairies over the state.
May-June. Purple Avens . . . . *Geum triflorum* Pursh

PETROPHYTUM (Nutt.) Rydb.

Petrophytum caespitosum (Nutt.) Rydb. Plants cespitose, flattened, forming cushions on granitic or limestone rocks. Leaves
obovate to spatulate, entire, the blades 5-11 mm long. Vegetative surfaces densely silky, the leaves not deciduous. Inflorescence racemose, compact, on almost leafless stalks 3-6 cm tall.
Petals white, 1.5 mm long. Fruits of 3-5 follicles, each 1-2 mm
long and dehiscing on 2 sutures. Rare on limestone at higher

altitudes of the northern Black Hills. {*Spiraea caespitosa*
Nutt.} June–July. Rockplant Spiraea.

PHYSOCARPUS Maxim.

1 Fruiting carpels usually more than 2, united at the base,
   glabrous to pubescent, mostly 5–8 mm long. Plants shrubby,
   5–20 dm tall. Leaves rounded in outline, 3-lobed, the blades
   mostly 2–4 cm across, tending to be somewhat larger than in
   the following species. Flowers in terminal corymbs, the
   petals white, 2–3 mm long. Frequent in rocky or rich, wooded
   slopes over the state. {incl. *P. intermedia* (Rydb.) Schneid.}
   June–July. Ninebark . . . .
   . . . . *Physocarpus opulifolius* (L.) Maxim.

1' Fruiting carpels usually 2, fused to almost the middle, pubescent, 3–6 mm long. Plants shrubby, 5–10 dm tall. Leaves
   lobed, sometimes deeply, the blades 1.5–3 cm across, tending
   to be somewhat smaller than those in the preceding. Flowers
   in terminal corymbs, the petals 2–3 mm long. Infrequent in
   rocky woods and on ledges at middle altitudes of the Black
   Hills. June–July . . . . *Physocarpus monogynus* (Torr.) Coult.

POTENTILLA L.

1 Plant shrubby, freely branching, 30–100 cm tall. Ovaries and
   achenes with long hairs. Young branches with silky-pilose
   hairs but soon becoming smooth with reddish bark. Older bark
   shredding. Leaves 5–7 pinnate, appearing almost digitate.
   Flowers cream-colored, to 1.5 cm in diameter. Achenes densely
   hairy, the style attached below the middle. Frequent on hillsides and in valleys of the Black Hills and Harding County.
   June–July. Shrubby Cinquefoil . . . . *Potentilla fruticosa* L.

1' Plants herbaceous. Ovaries and achenes lacking hairs.

   2 Annuals or biennials, without prominent rootstocks. Leaves
      lacking whitish appearance due to dense hairs on upper or
      lower surfaces.

      3 Lower leaves 5–9 foliolate-pinnate, the leaflets 1–3 cm
         long. Achenes with a thickening on the inner margin.
         Stems 4–7 dm tall. Stamens many, usually more than 20.
         In flower the inflorescence densely to diffusely cymose.
         Infrequent along shores of ponds and on edges of allu-
         vial woods in the eastern part. June–July . . . .
         . . . . *Potentilla paradoxa* Nutt.

      3' Lower leaves usually 3-foliolate, less commonly 5-
         foliolate. Achenes not thickened on the inner margin.

4   Stems softly pubescent below. Stamens 15 or fewer.
Plants not coarse. Petals about one-half the length
of the sepals.

    5   Lower leaves frequently 5-foliolate. Calyx with-
out glands. Leaflets usually less than 2 cm wide
at the widest part, their margins serrate. Teeth
apexes of the leaflets spaced about 3-4 mm.
Plants spreading to erect, 10-40 cm tall. Inflorescence a leafy cyme, the petals 2-3 mm long.
Frequent along shores and alluvial woods in the
eastern part. June-July . . . .
. . . . *Potentilla rivalis* Nutt.

    5' Lower leaves 3-foliolate. Calyx with mealy glands.
Leaflets pubescent, commonly 2 cm wide, their margins serrate to crenate. Teeth apexes of the leaflets usually spaced about 5 mm. Plants 3-7 dm
tall. Flowers in cymes. Rare in open woods and
on flood plains in the western part. June-July.
. . . . *Potentilla biennis* Greene

4' Stems with stiff spreading hairs on the lower parts.
Stamens 15-20. Plants coarse, irregularly shaped
towards maturity, 3-6 dm tall, branched. Leaves 3-
foliolate, rarely some 5-foliolate. Petals at least
three-fourths the length of the sepals. Achenes with
noticeable ridges when mature. Very common in waste
places, edges of fields, and woods over the state.
June-July . . . . *Potentilla norvegica* L.

2' Perennials from a persisting rootstock or woody caudex.

6   Leaves green on both surfaces, never with whitish or
silky gray appearance.

    7   Basal leaves with 5 leaflets palmately arranged.
Plant hirsute, erect, 3-8 dm tall, the stems 1-
several, branched above. Flowers many in cymes
with flattened tops, the flowers with many pistils.
Stamens about 30. Petals pale yellow, showy. Infrequent in waste places in the eastern part but
spreading westward in the state. June-July . . . . .
. . . . *Potentilla recta* L.

    7' Basal leaves with leaflets digitate or pinnately ar-
ranged.

        8   Leaflets digitate, deeply dissected, the segments
1-2 mm wide. Plants densely branched from a
branching caudex. Stems spreading, less than 4 dm

tall. Flowers 1 cm across. A variable species
which is rare in higher altitudes in Custer,
Lawrence and Pennington counties of the Black
Hills. Ours var. *perdissecta* (Rydb.) C. L.
Hitchc. June-July . . . .
. . . . *Potentilla diversifolia* Lehm.

8' Leaflets pinnately arranged, these with serrated
margins but not deeply dissected. Leaflets 1 cm
or more wide at the widest part. Plants erect.

  9 Inflorescence of cymes open to diffuse and
spreading, the lateral branches not appressed.
Plants less than 4 dm tall.

    10 Leaflets 9-13, decreasing in size from apex
to base, many times with rudimentary leaflets adjacent and interspersed with regular
ones, these rounded in shape. Petals white
to cream-colored. Stems from a stoloniferous caudex, usually less than 3 dm tall.
Vegetation with brown hairs. Common on
hillsides and in rocky woods of Custer,
Lawrence and Pennington counties of the
Black Hills. June-July . . . .
. . . . *Potentilla fissa* Nutt.

    10' Leaflets 7-9, only slightly decreasing in
size from the apex, rounded with coarse
serrations. Petals bright yellow. Stems
several from a branching caudex, usually
3-4 dm tall. Easily confused with smaller
specimens of the following species *P.
arguta* Pursh. Infrequent on rocky hillsides of the Black Hills and on Sioux
Quartzite in the eastern part. June-July.
. . . . *Potentilla glandulosa* Lindl.

  9' Inflorescence of cymes narrowed and strictly
ascending, the lateral branches appressed and
ascending. Plants usually over 5 dm tall, the
stems villous. Flowers with sepals 6-8 mm
long, as prominent as the petals. Mature basal
leaves with 5-9 leaflets. Common in prairies
and plains over the state. June. Tall Cinquefoil . . . . *Potentilla arguta* Pursh

6' Leaves whitish on the lower leaf surface due to dense
hairs. Upper surfaces may also be whitened due to
dense hairs.

11 Basal leaves predominantly digitate-palmate, of 3-7
    leaflets.

    12 Leaves less than 3 cm at widest part, digitate.
        Leaflets pinnately dissected into lacinately
        lobed segments, the sinuses reaching over one-
        half the distance to the midvein.

        13 Flowers less than 1 cm across, borne in
            leafy bracted cymes from plants with ascend-
            ing leafy stems 1-3 dm tall. Leaves not
            woolly-hairy but with dense white pubescence
            below. Leaflets obovate, deeply incised,
            green above. Rare as an escape in waste
            places in the eastern part. June-July . . .
            . . . . *Potentilla argentea* L.

        13' Flowers usually over 1 cm across, borne in-
            conspicuously in bracteate cymes, solitary
            or few. Stems low, spreading, less than 1
            dm tall, from a stout taproot. Most leaves
            basal, woolly-hairy on both surfaces, digi-
            tately 5-foliolate, the leaflets deeply in-
            cised. Frequent in sandy prairie, rocky
            slopes and ridges of the Black Hills. Apr-
            May . . . . *Potentilla concinna* Rich.

    12' Leaves 4-10 cm at their widest part, the leaf-
        lets palmately arranged. Leaflets regularly
        toothed, the sinuses less than one-half the
        distance to the midvein. Upper surfaces of
        leaves greenish, less strigose than the lower
        surfaces. Plants 4-8 dm tall, from a heavy,
        branched caudex. Infrequent in meadows and
        valleys of the Black Hills. A highly variable
        species that is wideranging in the western
        United States. {Ours var. *glabrata* (Lehm.)
        Hitchc.} June-Aug . . . .
        . . . . *Potentilla gracilis* Dougl.

11' Basal leaves pinnate-pinnatifidly arranged, of 7-11
    leaflets or more.

    14 Flowers single on naked peduncles 4 cm or more
        long. Plants stoloniferous, almost strawberry-
        like. Petals yellow, 7-10 mm long. Upper leaf
        surfaces only sparsely hairy. Leaves 1-2 dm
        long. Mature achenes light brown with several
        longitudinal, corky wrinkles. Frequent in
        meadows, on prairies and at the edges of marshes

over the state, more common in the east. May-
July. Silverweed . . . . *Potentilla anserina* L.

14' Flowers on erect, leafy stems. Plants not
stoloniferous.

15 Leaflets lacinately lobed one half to three
fourths of the way to the midvein, the lobes
rounded. Stipules of leaves deeply cleft.
Leaves greenish above. Inflorescence
several-flowered, only narrowly branched.
Styles glandular-roughened at their bases.
Plants 4-8 dm tall. Petals about 6 cm long.
Common in prairies over the state. June-Aug
. . . . *Potentilla pensylvanica* L.

15' Leaflets toothed with obtusely pointed teeth.
Stipules entire. Leaves grayish-hirsute on
upper as well as lower surfaces. Inflorescence freely branched, many flowered.
Styles not glandular-roughened at their
bases. Plants 3-5 dm tall. Petals 6-8 mm
long. Frequent in open areas and prairies
in the Black Hills. {*P. effusa* Dougl.}
July-Aug . . . . . *Potentilla hippiana* Lehm.

PRUNUS L.

1 Flowers in clusters, the clusters sessile, from lateral buds.
Flowers and fruits usually less than 8 in each cluster.

2 Fruit somewhat flattened, with a groove on one side. The
stone flattened. {Plums}

3 Petals 8-10 mm long. Small trees, usually solitary in
habit, 3-4 meters tall. Fruits purple red, 2-3 cm
across, with a single stone. Leaves oblong-obovate,
6-10 cm long, often pubescent below but becoming glabrate. Flowers 2-5, the petals white. Rare along
roadsides and fencerows in the southeast part. Apr-
May . . . . *Prunus mexicana* Wats.

3' Petals 10-15 mm long. Small trees forming thickets,
the stems up to 5 meters tall. Stems becoming spiny or
thorned with age. Fruits red or yellow, the fleshy part
yellowish. Leaves lance-ovate, 4-10 cm long, their margins serrate. Common over the state. Forming thickets
along fencerows, water courses and edges of woods. Apr-
May. Wild Plum . . . . *Prunus americana* Marsh.

2' Fruit rounded, the stone round. {Cherries} Stems not becoming spiny.

4 Shrubs over a meter tall with ascending branches.
Leaves lance-ovate with tapering tips, their margins
serrate to the base of the blade. Flowers 3-10 in
corymbose clusters, the petals 5-7 mm long. Fruits
red, globose, 3-6 mm across. Infrequent in woods at
lower altitudes in the Black Hills and Harding County
in the western part. May-June. Pin Cherry . . . . . .
. . . . *Prunus pensylvanica* L.

4' Shrubs low, the branches rarely ascending to over a
meter. Leaves obovate, dark green, 3-6 cm long, becoming entire towards the base of the blade. Flowers appearing before or with the leaves, the petals white,
4-6 mm long. Fruits globose, reddish to black, 1-1.5
cm across. Common on hillsides and in wooded draws in
the plains over the state. Apr-May. Sand Cherry . . .
. . . . *Prunus besseyi* Bailey

1' Flowers in narrow racemes, many in a raceme. Fruits becoming
dark purple at maturity. Shrubs or small trees up to 10
meters tall, many times forming thickets in fencerows or at
the edges of woods. Leaves oblong-elliptic to lanceolate, 5-
10 cm long, nearly glabrous on the lower surface. Flowers
white, the petals 6-8 mm long. Fruits up to 1 cm across,
globose. Common over the state. {incl. *P. melanocarpa* (A.
Nels.) Rydb.} Apr-May. Choke Cherry . . . .
. . . . *Prunus virginiana* L.

PYRUS

Pyrus ioensis (Wood) Bailey. Plants shrubby or more usually
small trees up to 6 meters tall, usually thorny. Leaves broadly
elliptic, 6-9 cm across, with coarsely serrated lobes. Undersurfaces of leaves usually densely pubescent. Flowers white to
pink, the petals up to 2 cm long. Fruits almost globose, greenish at maturity, 2-2.5 cm in diameter. Rare in thickets and low
woods in the extreme eastern part along the Big Sioux River.
May-June. Wild Crabapple.

ROSA L.

1 Styles united into a column protruding from the calyx. Stems
trailing or climbing, flowers numerous, small, white. Occasionally planted in hedges and fencerows but probably not reproducing or persisting for long this far north. May-June . .
. . . . *Rosa multiflora* Thunb.

1' Styles distinct, scarcely exsert or not at all. Flowers solitary or corymbose.

   2   Stems 1-3 meters tall. Plants usually in thickets or
       wooded areas.

      3   Pairs of infrastipular thorns present. Teeth of the
         younger leaflets usually glandular on the long side.
         Leaflets 7-9, about 2 cm long. Petals usually not over
         2 cm long. Common in thickets and along fencerows over
         the state. {incl. *R. fendleri* Crepin} May-June . . . .
         . . . . *Rosa woodsii* Lindl.

      3' Pairs of infrastipular thorns lacking or not differenti-
         ated from other internodal prickles. Floral bracts and
         upper stipules usually stipitate-glandular. Sepals
         after anthesis erect and persistent. Leaflets 5-7,
         rarely 9, ranging from 1.5-4 cm in length. Quite com-
         mon in the western part, less so in the north and north-
         eastern part. Infrequent to common on wooded hillsides,
         ravines and in thickets. {incl. *R. engelmanii* S. Wats.}
         May-June . . . . *Rosa acicularis* Lindl.

  2' Stems usually less than 1 meter tall. Plants usually in
      open areas. Paired infrastipular thorns none or not dif-
      ferentiated from other internodal prickles. Stems many
      times unarmed.

      4   Leaflets ordinarily numbering 5 to 7, rarely 9, finely
         pubescent on the undersurfaces. Internodes commonly
         unarmed or with prickles on the lower part of the stem.
         Plant tending to be shrubby, not dying back to near the
         soil level. Stems 4-12 dm tall, irregularly branched.
         Flowers solitary or few (corymbose), borne on lateral
         branches of the previous year's growth. Common on road-
         sides, in prairies, and less common at the edges of
         thickets over the state. June. Wild Rose . . . .
         . . . . *Rosa blanda* Ait.

      4' Leaflets ordinarily numbering 9-11, glabrous or nearly
         so. Internodes prickly or thorny. Plants not shrubby,
         very low growing, tending to die back nearly to the soil
         level. Stems of the year erect, 2-4 dm tall. Inflores-
         cence corymbose, the flowers borne at the summit of the
         stems of the season as well as on lateral branches of
         the previous year. Common on prairies and plains over
         the state. {incl. *R. suffulta* Greene} May-June . . .
         . . . . *Rosa arkansana* Porter

RUBUS L.

1   Plants with thorns or bristles on the stems.

   2   Stems and pedicels with stout, hooked or curved spines, not
      glandular. Plants 1-2 meters tall. Leaves with 3-5 leaf-
      lets, the terminal one broadly ovate. Inflorescence in a
      dense cluster at the top of the stem, the petals white.
      Fruits black or dark purple. Infrequent in thickets and
      along streams over the state. May-June. Black Raspberry.
      . . . . *Rubus occidentalis* L.

   2' Stems and pedicels with straight and rather weak bristles,
      also minutely glandular. Plants to 2 meters tall, spread-
      ing. Leaves with 3-5 leaflets. Inflorescence in an open
      cluster at the top and in upper axils of the stem. Fruit
      red, never black. Common in thickets and open woods of
      the state. Ours ssp. *sachalinensis* (Levl.) Focke. June.
      Red Raspberry . . . . *Rubus idaenus* L.

1' Plants without thorns or bristles on the stems.

   3   Leaves compound with 3 leaflets, these 3-6 cm long at
      maturity. Stems herbaceous and weak with prostrate stolons.
      Plants seldom over 5 dm tall. Inflorescence 1-3 flowered,
      terminal. Flowers white, 1 cm wide. Fruit dark red to
      purple. Infrequent on moist slopes in woods of the Black
      Hills and on the east-facing slope at the western side of
      Big Stone Lake in Roberts County. June-July. Dwarf Black-
      berry . . . . *Rubus pubescens* Raf.

   3' Leaves simple but lobed, at maturity 15-30 cm wide. Stems
      commonly over a meter tall, becoming woody. Inflorescence
      in a long-peduncled cluster. Flowers 3 cm wide. Fruit
      red, to 2 cm in diameter. Common at high altitudes in
      Lawrence and Pennington counties in the Black Hills.
      {incl. *Rubacer parviflorum* (Nutt.) Rydb.} July-Aug.
      Thimble Berry . . . . *Rubus parviflorus* Nutt.

SORBUS L.

Sorbus scopulina Greene. Plants shrubby with several stems to 3
meters tall. Leaves pinnately compound, with 9-13 leaflets, the
leaflets 3-5 cm long, serrate. Inflorescence a compound cyme 2-4
cm across. Flowers white, the petals oval, 5-6 mm long. Fruit
yellow to red orange, globose and glossy, 1 cm in diameter. Rare
in deep ravines in the northern Black Hills. June. Western
Mountain Ash.

SPIRAEA L.

1   Inflorescence conical or elongate. Leaves lanceolate, 3-6 cm
    long, their margins with mostly regular serrations. Plants
    shrubby, 5-20 dm tall, the stems erect. Inflorescence pubes-
    cent on the branches. Flowers white, the petals about 2 mm
    long. Fruiting follicles glabrous and shiny. Infrequent in
    mesic swales or openings in woods in the Black Hills and the
    eastern part. July-Aug. Meadowsweet . . . .
    . . . . *Spiraea alba* Du Roi

1' Inflorescence paniculate, irregularly rounded or flat topped.
    Leaves broadly lanceolate to oval or elliptical, their margins
    with irregular serrations. Plants shrubby with ascending
    branches, 2-8 dm tall, from a creeping rootstock. Branches of
    the inflorescence sparsely hairy to glabrous. Flowers white,
    the petals about 2 mm long. Fruiting follicles glabrous. In-
    frequent in woods at middle altitudes of the Black Hills.
    Ours var. *lucida* (Dougl.) C. L. Hitchc. June-July. Mountain
    Meadowsweet . . . . *Spiraea betulifolia* Pall.

Family Fabaceae

(Key to Genera)

1   Plants woody shrubs or trees.

    2   Shrubs; rarely becoming tree-like.

        3   Flowers yellow, in clusters of 2-3; pod about 2 cm
            long . . . . *CARAGANA*

        3' Flowers purple, in racemes; pods less than 1 cm long
            . . . . *AMORPHA*

    2' Trees.

        4   Stems with spines or thorns.

            5   Trees with paired stipular spines up to 1 cm long
                . . . . *ROBINIA*

            5' Trees with branched thorns 5-15 cm long . . . .
                . . . . *GLEDITSIA*

        4' Stems not armed. Leaves 2-several compound . . . . . .
            . . . . *GYMNOCLADUS*

1' Plants herbaceous. Occasionally woody stem bases are apparent.

    6   Principal leaves simple.

7 Anthers of 2 forms, stamens monadelphous. Flowers
yellow . . . . *CROTALARIA*

7' Anthers uniform, stamens diadelphous. Flowers with
purple pigments . . . . *ASTRAGALUS*

6' Principal leaves compound.

    8 Leaves twice pinnate. Flowers in globose heads.

        9 Stems with curved prickles; flowers pink . . . . . .
        . . . . *SCHRANKIA*

        9' Stems without prickles; flowers whitish . . . . . .
        . . . . *DESMANTHUS*

    8' Leaves once-pinnate, trifoliolate, or leaflets sometimes
only two.

        10 Leaflets even-numbered.

            11 Terminal part of leaf ending in a tendril.

                12 Style of flowers rounded at end with hairs
covering the terminal 1 mm . . *VICIA*

                12' Style of flowers flattened at the end with
hairs only on the inner edge . . . .
        . . . . *LATHYRUS*

            11' Terminal part of leaf not ending in a tendril
        . . . . *CASSIA*

        10' Leaflets odd-numbered.

            13 Leaves with 3-5 leaflets.

                14 Vegetation with glandular dots on the sur-
face.

                    15 Petals 5, papilionaceous.

                        16 Leaves digitately foliolate . . . .
        . . . . *PSORALEA*

                        16' Leaves pinnately foliolate . . . .
        . . . . *DALEA*

                    15' Petals 1-5, not papilionaceous . . . . .
        . . . . *PETALOSTEMON*

                14' Vegetation lacking glandular dots.

                    17 Leaflets with serrations.

                        18 Flowers in racemes . . *MELILOTUS*

                        18' Flowers in capitate clusters.

19 Fruits straight . . . *TRIFOLIUM*

19' Fruits coiled or sickle-shaped . . .
. . . . *MEDICAGO*

17' Leaflets with entire margins.

20 Fruit a loment; breaking transversely
into indehiscent 1-seeded units or of
one unit.

21 Flowers pink-purple. Fruits 1-
several jointed . . . *DESMODIUM*

21' Flowers creamy-white. Fruits consisting of a 1-seeded unit and a
stalk . . . . *LESPEDEZA*

20' Fruit a legume; a longitudinally dehiscent pod with several seeds.

22 Flowers in clustered heads.

23 Leaflets 5 . . . . *LOTUS*

23' Leaflets 3 . . . . *TRIFOLIUM*

22' Flowers in racemes, in axils of
leaves, solitary, or few at the
ends of long peduncles.

24 Stamens distinct. Leaves with
large ovate stipules 1-2 cm wide
. . . . *THERMOPSIS*

24' Stamens diadelphous. Stipules,
if present, smaller.

25 Stems trailing or climbing.

26 Leaflets 5, rarely as
many as 7 . . . .
. . . . . *APIOS*

26' Leaflets usually 3.

27 Style glabrous.
Keel petal not
strongly curved
. . . *AMPHICARPAEA*

27' Style pubescent.
Keel petal strongly
curved . . . . . .
. . . *STROPHOSTYLES*

25' Stems not trailing or climbing.

28 Plants caespitose, flowers in racemes . . . . .
. . . . . *ASTRAGALUS*

28' Plants with upright
leafy stems, flowers
solitary to few, axil-
lary . . . *LOTUS*

13' Leaves with more than 5 leaflets.

29 Vegetation with glandular dots (distinctly visible with 10x magnification).

30 Corolla papilionaceous.

31 Fruit few-seeded, spiny . . *GLYCYRRHIZA*

31' Fruit 1-2 seeded, not spiny . . . .
. . . . *DALEA*

30' Corolla not papilionaceous.

32 Petals 5, stamens 5 . . . . *PETALOSTEMON*

32' Petal 1, stamens 10 . . . . *AMORPHA*

29' Vegetation lacking glandular dots.

33 Plants with stems trailing or climbing . . .
. . . . *APIOS*

33' Plants with stems not trailing.

34 Leaves palmately compound . . . .
. . . . *LUPINUS*

34' Leaves pinnately compound.

35 Flowers in umbellate heads . . . . .
. . . . *CORONILLA*

35' Flowers in racemes.

36 Stamens 10, filaments distinct
. . . . *SOPHORA*

36' Stamens 9 plus 1, diadelphous.

37 Fruit a legume with longitudinal dehiscence.

38 Plants without leafy
stems. Keel petal

abruptly beaked . . . . .
. . . . . . *OXYTROPIS*

38' Plants with leafy stems.
Keel petal not abruptly
beaked . . *ASTRAGALUS*

37' Fruit a loment with transverse dehiscence . . . . . .
. . . . *HEDYSARUM*

AMORPHA L.

1 Plants much-branched woody shrubs 1-3 meters tall. Leaflets
exceeding 2 cm in length, pubescent, 11-25 per leaf. Flowers
in racemes 6 cm or more in length, deep purple, the banner of
the corolla 6 mm long. Pods glandular, less than 1 cm long.
Frequent in brushy alluvial areas and along streams and ponds
over the state. More common in the eastern part. June.
False Indigo . . . . *Amorpha fruticosa* L.

1' Plants low woody shrubs less than a meter tall. Leaflets 5-15
mm long, rarely longer.

2 Leaflets mostly glabrous at maturity. Leaves with a distinct petiole. Stems usually not over 3 dm tall. Leaflets
elliptic, 5-10 mm long, from 15 to 30 per leaf. Flowers in
racemes at the end of branches, the racemes usually solitary, 3-5 cm long. Flowers deep lavender, the pods about
4 mm long. Infrequent in upland prairies as an undershrub
over the state. June-July . . . . . . *Amorpha nana* Nutt.

2' Leaflets densely white canescent at maturity. Leaves very
short petioled. Stems 4-9 dm tall, several from a heavy
root. Leaflets 8-15 mm long or rarely longer, 15-50 per
leaf. Flowers in clustered racemes, each raceme 4-10 cm
long. Petals deep purple. Pods hairy, about 4 mm long.
Very common in prairie and on plains over the state. June-
July. Lead Plant . . . . *Amorpha canescens* Pursh.

AMPHICARPAEA Elliott.

Amphicarpaea bracteata (L.) Fern. Perennial twining herb with
stems to 1 meter long. Leaves trifoliolate, the leaflets rhombic-
ovate, 2-5 cm long. Inflorescence few-flowered from long, axillary peduncles. Flowers white to lavender, the petals about 1.2
cm long. Fruiting pod flat, oblong and pointed at each end, 2-3
seeded. Infrequent in both upland and alluvial woods over the
state. July-Aug. Hog Peanut.

APIOS Fabr.

Apios americana Medic. Perennial twining herb of woods, the
stems 1 or more meters long, from tuberous roots. Leaves with
5-7 leaflets which are lance-ovate in shape, each 4-6 cm long.
Flowers in dense racemes at the ends of axillary peduncles.
Petals brown-purple, 1.0-1.5 cm long. Fruiting pods linear,
several seeded, becoming coiled after dehiscence. Infrequent
to rare in rich woods over the state. July-Aug. Ground Nut.

ASTRAGALUS L.

1 Leaves principally simple (unifoliolate). Occasionally the
   plant may contain some compound leaves.

   2 Pods inflated and mottled, 2.5-5 cm long and half as wide.
      Leaves long and filiform. Stems 1-4 dm, from a rhizome.
      Leaves mostly linear phyllodes. Flowers in racemes, 2-5
      cm long, the petals 7-12 mm long, mostly yellow with purple-
      tinged keels. Fruits thin-walled, inflated, 1-celled. Fre-
      quent in sandy areas of the south and western part. Ours
      var. *filifolius* (Gray) Herm. May-June . . . .
      . . . . *Astragalus ceramicus* Sheld.

   2' Pods narrowly oblong. The leaves spatulate, mostly basal,
      the plant acaulescent. Stems many times prostrate, from a
      branching caudex. Leaves 2-5 cm long, the surface strigose,
      their apex acute. Racemes 1-3 cm long, few-flowered.
      Petals brown-purple, 7-9 mm long. Fruits 1-1.5 cm long,
      compressed laterally. Frequent in dry soil of the western
      part of the state. May-June . . . .
      . . . . *Astragalus spatulatus* Sheld.

1' Leaves principally compound.

   3 Leaves trifoliolate. Stems short and branched from a
      heavy caudex, the vegetative parts forming a hemispheric
      cushiony arrangement close to ground level.

      4 Flowers white or off-white in color; however, some may
         have purple-tinged keels. Inflorescence sessile. Calyx
         cylindrical.

         5 Petals with hairs. Plants flowering in June and
            early July. Stems 1-3 cm long, several. Leaves
            palmately trifoliolate, covered with dense, silvery
            hairs. Flowers sessile in the axils of leaves,
            yellow-white, 1-1.5 cm long. Fruits up to 1 cm
            long, 1-celled. Although reported from Fall River
            County, I have not seen a specimen. It is in Wyo-
            ming and Colorado so should be looked for in extreme

southwest South Dakota. June–July . . . .
. . . . *Astragalus hyalinus* M. E. Jones

5' Petals glabrous. Plants flowering in late May or
   early June. Stems 1–4 cm long, several from the
   branched caudex. Leaves 3–7 cm long, palmately
   trifoliolate, covered with silvery hairs. Flowers
   with purple-tinged keels 2–3 cm long, borne in axillary clusters. Fruits 7–9 mm long with a short beak;
   one-celled. Much like the preceding species, but
   having larger flowers. Frequent in dry prairies in
   the western one half of the state. May–June . . . .
   . . . . *Astragalus gilviflorus* Sheld.

4' Flowers pink-purple. Inflorescence with peduncles.
Calyx campanulate.

6 Flowers 9–13 mm long. Stipules on leaf bases glabrous. Leaves 1–4 cm long, covered with silvery
   hairs. Flowers 1–3 in racemes, the peduncles 7–16
   mm long. Plants mat-forming with short stems.
   Fruits 4–7 mm long, 1-celled, sessile. Rare on dry,
   exposed calcareous soils in extreme southwest counties. Apr–May . . . . . *Astragalus barrii* Barneby

6' Flowers 6–8 mm long. Stipules on leaf bases villous.
   Leaves trifoliolate but occasionally 5-foliolate, 1–2
   cm long, covered with dense, villous hairs. Flowers
   6–8 mm long, purple, in 2–4 flowered racemes. Stems
   short, the plants forming low mats. Fruits 5–6 mm
   long, covered with pilose hairs. Although I have
   not seen specimens from South Dakota, it is in eastern Wyoming and northwest Nebraska. It should occur
   in the extreme southwest part. May–June . . . . . .
   . . . . *Astragalus sericoleucus* A. Gray

3' Leaves with 5-many leaflets. Plants various in habit.

7 Leaflets ending in bristly-tipped spines. Stems 5–20 cm
   long, prostrate from a branching caudex. Leaves 1–2 cm
   long with 5–7 spinulose tipped leaflets. Flowers 5–6 mm
   long, purple-tinged or white, 1–3 in a group. Fruits
   5–7 mm long, 1-celled, sessile. Rare to infrequent in
   open dry areas, on bluffs, or dry places in the western
   one-half. June–July . . . *Astragalus kentrophyta* Gray.

7' Leaflets not spine-tipped. Plants more or less upright.

8 Vegetative surfaces with malpighian hairs, i.e.,
   attached at the middle with two points 180° from
   each other.

9   Plants low growing, the stems poorly developed or
    mostly acaulescent.

   10   Flowers yellow or cream colored (occasionally
        purple-tinged ones are found). Leaves green-
        ish with long, villous hairs. Calyx campanu-
        late. Leaves 6-14 cm long, leaflets 9-19,
        each slightly over 1 cm long. Flowers 6-10
        mm long, the later ones cleistogamous and
        short-peduncled. Fruits 2 cm long, lunate,
        7-8 mm wide. Common over the state in sandy
        or dry prairie. Apr-June . . . .
        . . . . *Astragalus lotiflorus* Hook.

   10' Flowers pink-purple. Leaves silvery gray due
        to appressed hairs. Calyx cylindrical, the
        tube 7-8 mm long, with teeth 2-2.5 mm long.
        Leaves 5-11 cm long, the leaflets 5-15 mm,
        elliptic. Flowers 12-20 mm long, purple.
        Fruits oblong, drying dark, 15-25 mm long
        and 7-8 mm broad. Quite common in prairie
        over the state. Apr-June . . . .
        . . . . *Astragalus missouriensis* Nutt.

9' Plants with stems well-developed, the branches
    decumbent to erect.

   11   Ovaries or pods with a distinct stipe from the
        receptacle. Vegetation possessing a distinc-
        tive odor characteristic of seleniferous
        plants. Stems 2-6 dm long, several from a
        branched caudex. Leaves 6-14 cm long, 13-27
        leaflets. Racemes 3-15 cm, several to many
        flowered. Petals yellowish. Pods 2.5-3 cm
        long, on a stipe 4-8 mm long. Common in low
        prairies or moist sandy places over the state;
        more common westward. July-Aug . . . .
        . . . . *Astragalus racemosus* Pursh.

   11' Ovaries or pods sessile on the receptacle.
        Vegetation lacking distinctive odor.

      12   Flowers principally yellow. Fruits with
           glabrous surfaces. Plants blooming in
           July and August. Stems 2-6 dm tall, sev-
           eral from a branched caudex. Leaves 10-
           13 cm long, with 15-30 leaflets. Flower-
           ing racemes 6-15 cm long, many-flowered.
           Petals 1.0-1.5 cm long. Frequent to com-
           mon in a variety of open habitats over

the state. July-Aug . . . .
. . . . *Astragalus canadensis* L.

12' Flowers principally lavender. Fruits with
a strigose surface. Plants blooming in
May and June. Stems 1.5-4 dm tall, several from a caudex, weakly erect to decumbent. Leaves 6-12 cm long. Flowers 13-16
mm long. Pods 7-12 mm long, 2-celled,
erect. Common in dry prairie and in exposed places on knolls or edges of bluffs
over the state. {incl. *A. striatus* Nutt.}
May-June . . . *Astragalus adsurgens* Pallas

8' Vegetative surfaces with hairs basally attached.

13 Calyx tube cylindric.

14 Plants cushiony-caespitose. Vegetation
densely covered with woolly hairs. Stems
approximately 6 cm long, not well-developed,
several from a caudex. Leaves 3-9 cm long
with 7-15 leaflets. Racemes about 3 cm long,
few to several flowered, the petals yellowish
with a purple-tipped keel, 20-25 mm long.
Pods 10-20 cm long, oblong, also densely villous. Rare in the western part. May-June.
. . . . *Astragalus purshii* Dougl.

14' Plants not cushiony-caespitose. Vegetation
variously hairy, but not woolly.

15 Ovaries and pods with a distinct stipe
within the calyx. Fruiting pods usually
reflexed towards maturity.

16 Flowers principally pink to purple.
Upper surface of fruiting pods with
two furrows running longitudinally
(bisulcate). Stems 15-70 cm, several
from a branched caudex. Leaves 6-10
cm long with 17-29 leaflets, oblong
to elliptic. Racemes 4-19 cm long,
many-flowered; the petals pink-purple.
Pods linear-oblong, up to 2 cm long.
Frequent in the western part. June-
July . . . *Astragalus bisulcatus* Gray

16' Flowers not pink-purple; mostly white
to yellow. Pods three-angled in
cross-section. Vegetation with loose,

long hairs. Stems 2-5 dm tall, several
from a woody caudex. Leaves 6-12 cm long
with 20-30 leaflets, elliptic to oblanceolate. Racemes 2-14 cm long, several to
many-flowered, the petals 1.5-2.0 cm long,
whitish with a purple-tinged keel. Pods
mostly glabrous, partially 2-celled. Infrequent in the western part, including
the Black Hills. May-June . . . .
. . . . . . *Astragalus drummondii* Dougl.

15' Ovaries and pods usually sessile within the
calyx. Fruiting pods usually not reflexed
towards maturity.

17 Leaflets elliptic to ovate or oblanceolate. Flowers principally pink-purple.

18 Vegetation with long-villous hairs.
Fruiting pods curved, glabrous.
Plants 1-3 dm tall with several stems
from a caudex. Leaves 10-20 cm with
fine spreading hairs. Racemes several to many-flowered, the petals
1.5-2.0 cm long, greenish-purple.
Pods oblong, 13-20 mm, sessile.
Very rare in sandy prairies of the
southwestern part. May-June . . . .
. . . . *Astragalus mollissimus* Torr.

18' Vegetation strigose but not with long
villous hairs. Fruiting pods not
curved towards maturity.

19 Ovaries and pods pubescent.
Plants from rhizomes. Calyx
teeth half as long as the tube.

20 Flowers erect. Fruiting pods
oblong, covered with long,
villous hairs. Stems 1-4 dm
long, few to several, decumbent. Leaves 4-7 cm long.
Leaflets 13-19, narrowly elliptic. Racemes 1-4 cm long,
few to several flowered, the
petals purplish, 1.5-2.5 cm
long, erect. Pods 8-10 mm
long, erect, apparently 2-

celled. In dry prairie, they may produce only 1 stem. Common in prairies over the state. May-June . . . . *Astragalus agrestis* Dougl.

20' Flowers spreading. Fruiting pods ovoid, with fine strigose hairs. Stems 1-3 dm long, from a creeping rhizome. Leaves 5-9 cm long with 13-30 leaflets. Racemes 1-3 cm, few-flowered, petals 14-18 mm long, spreading, purplish. Pods 1-2 cm long, ovoid with a stout beak. Frequent and widespread in the state in sandy swales or prairie flats. May-June . . . . *Astragalus plattensis* Nutt.

19' Ovaries and pods glabrous. Plants from a stout caudex. Calyx teeth less than half as long as the tube. Stems 5-50 cm long, several decumbent or prostrate, radiating from a woody caudex. Leaves 4-11 cm long, leaflets 15-27, narrowly lanceolate. Racemes 2-4 cm, few to several flowered, petals 14-20 mm long, pink-purple, but fading to yellow. Pods ovoid-globose, inflated, 16-27 mm long, frequently reddish. Common in prairie over the state. Apr-June . . . . *Astragalus crassicarpus* Nutt.

17' Leaflets linear. Flowers yellowish with a purple-tipped keel. Plants from a caudex. Stems 2-5 dm long, several, ascending to erect. Leaves 4-9 cm, leaflets 7-17, linear, their bases decurrent with the rachis. Racemes 3-10 cm long, the flowers tightly clustered, but elongated in fruit. Petals 18-20 mm long, yellowish. Fruiting pods oblong to ovoid, 11-23 mm long, cross-ribbed, 1-celled. Very rare in sandy prairie in the extreme southwest part. It possesses the characteristic odor of a seleniferous plant. June-July . . *Astragalus pectinatus* (Hook) G. Don

13' Calyx tube campanulate.

21 Stipules reflexed, leaf-like or foliaceous. Stems
    3-7 dm tall, few to several from a woody caudex.
    Leaves 9-16 cm long, with 9-15 leaflets, oblong to
    obtusely lanceolate, 2-5 cm long. Racemes 3-7 cm
    long, several flowered, the petals 8-10 mm, whitish
    with a purple-tipped keel. Pods glabrous, somewhat
    inflated, 2-3.5 cm long. Ours plants restricted to
    the Black Hills. June-July . . . .
    . . . . . *Astragalus americanus* (Hook) M. E. Jones

21' Stipules erect or spreading, not foliaceous.

    22 Ovaries and pods with a stipe at least 1 mm
        long within the calyx.

        23 Plants from a creeping rhizome. Ours re-
            stricted to the Black Hills at middle alti-
            tudes in meadows or in pine woods. Fruit-
            ing pods with black hairs toward maturity.
            Stems 1-4 dm, several, decumbent-spreading
            from rhizomes. Leaves 7-10 cm long, leaf-
            lets 19-25, ovate to elliptic. Racemes 2-
            6 cm, the flowers several, crowded at the
            ends of the peduncles. Flowers 7-12 mm,
            purple, fading to yellow. Pods 12 20 mm,
            straight to slightly curved, strigose with
            black hairs. Frequent in the Black Hills.
            June-July . . . . . . *Astragalus alpinus* L.

        23' Plants from a caudex. Fruiting pods gla-
            brate towards maturity. Not restricted to
            pine woods of the Black Hills.

            24 Leaflets narrowly elliptic with acute
                apexes. Fruiting pods lunate to sickle
                shaped, the stipe 5-10 mm long. Stems
                1.5-4 dm, few to several from a caudex.
                Leaves 3-8 cm long, leaflets 9-19, 1-2
                cm long. Racemes several flowered,
                clustered at end of peduncle. Flowers
                8-11 mm long, yellow-white, the keel
                purple-tipped. Pods papery, glabrous
                towards maturity, 2.5-3.5 cm long.
                Frequent in meadows and woods of the
                Black Hills and the western part. May-
                June . . . .
                . . . *Astragalus aboriginorum* Richards

            24' Leaflets elliptic to oblong with obtuse
                apexes. Pod not lunate, the stipe 5 mm
                long or shorter.

25 Flowers yellowish. Fruiting raceme
longer than the peduncle, the latter
1-4 cm. Stems 2-5 dm long, several
from a branching caudex. Leaves 5-8
cm long, several flowered. Flowers
7-9 mm long, the keel purple-tipped.
Pods 12-18 mm long, nearly flat, 1-
celled. Frequent in sandy or dry up-
lands in the western one-half. June-
July . . . . *Astragalus tenellus* Pursh

25' Flowers purplish. Fruiting racemes
usually longer than the single peduncle.
the latter 5-15 cm long. Stems 2-6 dm
long, several from a branching caudex.
Leaves 5-8 cm, leaflets 15-25, each one
6-10 mm long. Racemes 4-15 cm long,
several flowered, each flower 8-10 mm
long. Fruiting pods 13-20 mm long,
one-celled, round in cross section.
Frequent in dry or sandy soil and
prairies over the state. June-July . .
. . *Astragalus flexuosus* (Hook) G. Don

22' Ovaries and fruiting pods sessile within the calyx.

26 Fruiting pods 10-25 mm long, rounded in cross-
section. Stems several, slender, 1-4 dm tall,
several from a branched caudex. Leaves pinnately compound, with 13-25 leaflets, the leaflets lanceolate to oblong, 1-2 cm long. Inflorescence extending above the leaves, 6-12 flowered. Flowers white with purple tips and keels,
the calyx with dark, strigose hairs. Fruits
linear, sessile and pendant. Rare in open
woods in the Black Hills. Ours var. *hylophilus*
(Rydb.) Barneby. June-July . . . .
. . . . *Astragalus miser* Dougl.

26' Fruiting pods less than 10 mm long, laterally
compressed.

27 Calyx teeth less than 1 mm long. Fruiting
pods curved, stems 2-8 dm long, several
from a caudex. Leaves 4-9 cm, with 7-15
leaflets, narrowly linear to oblong. Racemes 4-12 cm long, several to many flowered. Flowers 4-8 mm long, purplish. Pods
6-8 mm long, ovate, boat-shaped, strigose,

and cross-ribbed. Frequent in dry and open
grasslands in the western one half of the
state. May-June . . . .
. . . . *Astragalus gracilis* Nutt.

27' Calyx teeth exceeding 1 mm. Fruiting pods
straight. Stems 1-4 cm long, several from
a branching, woody caudex. Leaves 2-5 cm
long, leaflets 9-13, each leaflet 8-12 mm
long, lanceolate with an acute tip. Racemes 1-3 cm, few to several flowered.
Flowers 5-9 mm, purple, sometimes whitish.
Pods 7-10 mm long, laterally compressed, 1-
celled. Rare in the Badlands and other
dry, sterile soil of the western part.
June-July . . . .
. . . . . . *Astragalus vexilliflexus* Sheld.

CARAGANA Lam.

Caragana arborescens Lam. Perennial woody shrubs or small trees
3-5 m tall. Stems several-branched. Leaves pinnate, often fascicled, leaflets 8-12, rounded at the apex, 1.5-2.5 cm long.
Flowers yellow, perfect, in fascicles of 2-4, the petals up to
2 cm long. Fruiting pods 2 cm long, sessile, several-seeded.
Frequently planted for windbreaks, hedgerows and persisting.
June-July. Pea Tree.

CASSIA L.

Cassia fasciculata Michx. Plants herbaceous annuals with sparingly branched stems 3-6 dm tall. Leaves pinnately compound, the
leaflets elliptic, about 1.5 cm long, each leaf possessing 6-12
pairs. Inflorescence of several flowers on axillary peduncles.
Flowers yellow with red stamens and staminodes. Petals subequal,
1-2 cm long. Fruiting pods linear, flat, not segmented. Locally
common in alluvial and sandy openings in the southeast part.
{*Chamaecrista fasciculata* (Michx.) Greene} July-Aug. Partridge
Pea.

CORONILLA L.

Coronilla varia L. Perennial herbs with ascending branches, the
stems 3-5 dm tall. Leaves odd-pinnate, the leaflets 11-21, each
one 1-2 cm long, obovate. Flowers in globose umbels, from axillary peduncles. Petals light purple to pink, 1-1.5 cm long, the
keel petal curved. Fruiting pods 2-4 cm long, jointed transversely. Occasionally escaped or persisting from cultivation
over the state. May-July. Crown Vetch.

CROTALARIA L.

Crotalaria sagittalis L. Plants annual, sparingly branched, 1-4
dm tall. Leaves simple, lanceolate, pubescent. Leaf blades 3-6
cm long, sessile or nearly so, the short petioles decurrent on
the stem. Flowers yellow, in few-flowered racemes terminating
the stem branches. Petals 8-10 mm long. Fruiting pods becoming
inflated, 2-3 cm long, 2-several seeded. Rare in open sandy or
waste areas in the extreme southeast counties of Clay, Lincoln,
Minnehaha and Union. July-Aug. Rattlebox.

DALEA Willd.

1 Stems and leaves pubescent. Flowering spikes about 1.5 cm in
   diameter. Plants perennial, few to several stems 3-6 dm tall
   from a woody base. Leaves 2-4 cm long, with 5 pinnately arranged leaflets. Flowers yellow, in dense, silky spikes 2-5
   cm long. Infrequent on dry, exposed prairie knolls throughout
   the southern one half of the state. July-Aug . . . .
   . . . . *Dalea aurea* Nutt.

1' Stems and leaves glabrous. Flowering spikes less than 1.5 cm
   in diameter.

   2 Plants annual. Flowering spikes dense, cylindric. Plants
      several branched, up to 5 dm tall. Leaves with 15-35 leaf-
      lets pinnately arranged, each leaflet about 4 mm long.
      Flowers white or rose-pink in dense, silky spikes 2-6 cm
      long. Petals 2.5-3 mm long. Frequent in sandy prairie in
      the eastern part. {*Dalea alopecuroides* (Willd.) Rydb.}
      July-Sept . . . . *Dalea leporina* (Ait.) Bullock

   2' Plants perennial. Flowering spikes lax and open, the flow-
      ers scattered. Stems branched above, 3-8 dm tall. Leaves
      with 5-11 leaflets, each leaflet about 7 mm long. Flowers
      white, the petals 12-15 mm long. Bracts with scarious mar-
      gins and black dotted surfaces. Calyx tube densely silky,
      turning bronze at maturity. Frequent in dry prairies and
      on plains over the state. June-July . . . .
      . . . . *Dalea enneandra* Nutt.

DESMANTHUS Willd.

Desmanthus illinoiensis (Michx.) MacMill. Plants perennial,
herbaceous to shrubby. Stems smooth, strongly angled, up to 1
meter tall, with irregular branches. Leaves bipinnate with 12-
25 pinnae, the leaflets numerous, linear, each one 2-3 mm long.
Flowers in compact, peduncled heads, the petals white or green,
short, the stamens long-exsert. Fruiting pods elongate, 1-2 cm
long, thin, becoming spirally twisted. Frequent in thickets of

ravines or on prairie hillsides over the state. May-June.
Prairie Mimosa.

DESMODIUM Desv.

1   Terminal leaflet almost as wide as long with an abruptly
    acuminate tip. Fruit with a long stipe. The dorsal margin
    of the fruit almost straight. Plants perennial, the stems
    1-4 dm tall. Leaves clustered towards the upper portions of
    the stems on long petioles. Flowers in terminal panicles, on
    peduncles 6-8 mm long. Joints of the fruit somewhat triangu-
    lar, glabrous. Rare to infrequent in rich woods of the east-
    ern part. July-Aug . . . . *Desmodium glutinosum* (Muhl.) Wood

1' Terminal leaflet longer than wide; not acuminately pointed.
    Fruit on short stipes, both margins of the fruit convex.

    2   Leaves pubescent with hooked hairs on the lower surface.
        Petiole much longer than the stalk of the terminal leaflet.
        Stems erect, up to a meter or more tall. Leaflets ovate-
        lanceolate, strongly reticulate with dense pubescence on
        both surfaces. Flowers 8-10 mm long, usually in a terminal
        raceme. Fruits with 3-6 joints, pubescent, rounded on both
        margins. Rare in alluvial, sandy thickets in extreme south-
        east counties. July-Aug . . . . *Desmodium illinoense* Gray

    2' Leaves glabrous to sparsely pubescent, without hooked hairs.
        Petiole only slightly longer than the stalk of the terminal
        leaflet. Plants 1-2 meters tall, the stems erect, only
        sparingly branched above. Leaflets oblong to lance-oblong,
        5-8 cm long. Inflorescence of several terminal racemes
        that are densely flowered. Flowers 10-13 mm long, purple.
        Fruits with 3-5 joints, each segment rhombic in outline.
        Common in woods and thickets over the state. July-Aug . .
        . . . . *Desmodium canadense* (L.) DC.

GLEDITSIA L.

Gleditsia triacanthos L. Trees up to 20 meters tall. Stems with
branched thorns 4-10 cm long. Leaves once or twice-pinnate, the
leaflets elliptic to lance-ovate, 1-2.5 cm long. Flowers polygamous or perfect, small, greenish. Staminate racemes densely
flowered, short-peduncled. Fruiting pods 10-30 cm long, flattened, becoming curved and twisted towards maturity. Infrequent
in rich woods and wooded coulees of the eastern part. It commonly is planted and persists elsewhere in the state. May-June.
Honey Locust.

GLYCYRRHIZA L.

Glycyrrhiza lepidota Pursh. Plants perennial with erect stems 4-8 dm tall, sparingly branched above. Leaves pinnately compound, with 11-19 oblong leaflets. Flowers white to pale yellow, 12-15 mm long, in erect racemes. Fruits 1.5 cm long, becoming dark brown, covered with hooked spines, with few to several seeds. Common in moist prairies and swales over the state. June-July. Licorice.

GYMNOCLADUS Lam.

Gymnocladus dioica (L.) Koch. Trees up to 30 meters tall. Leaves alternate, bipinnate to decompound with 5-9 pinnae. Leaflets ovate, 2-7 cm long, with acuminate tips. Flowers polygamous, in terminal racemes or panicles. Petals greenish. Fruiting pods flattened, oblong, 10-20 cm long and 3-4 cm wide, curved. Seeds few to several, embedded in pulpy tissue. Seed coats very hard and resistant to breaking. Rare to infrequent in rich wooded draws along the Big Sioux River in Lincoln and Union counties. May-June. Kentucky Coffee-tree.

HEDYSARUM L.

1 Mature fruits (loments) usually less than 6 mm wide at the widest point. Wing-margins of the fruit less than 1 mm wide. Flowers 13-15 mm long, purple to white. Leaflets tending to be more lanceolate than the following. Plants 3-7 dm tall, several stems arising from the crown. Frequent in moist soil of woods at higher altitudes in Custer, Lawrence and Pennington counties. Ours var. *philoscina* (Nels.) Rollins. June-July . . . . *Hedysarum alpinum* L.

1' Mature fruits (loments) usually exceeding 7 mm in width at the widest point. Wing-margins of the fruit 1-2 mm wide. Flowers 16-20 mm long, pink to purple. Leaflets with brownish glandular dots, also tending to be more ovate-elliptic in shape than the preceding. Plants 4-8 dm tall, stems several, branching, from a woody crown. Rare in Lawrence and Pennington counties at higher elevations. June-Aug . . . . *Hedysarum occidentale* Greene

LATHYRUS L.

1 Flowers white to cream-colored. Leaves with 3-4 pairs of leaflets, ovate. Stipules to 3 cm long. Stems up to 8 dm long, often trailing. Inflorescence a 5-8 flowered raceme, the petals about 1.5 cm long. Fruiting pods 4 cm long, glabrous. Frequent in thickets and woods of the Black Hills and

in rich woods of Marshall and Roberts counties of the northeast. May-July . . . . *Lathyrus ochroleucus* Hook.

1' Flowers pink to lavender.

  2 Stems winged, sometimes only slightly so.

    3 Leaflets 2, paired, each one 6-10 cm long. Petioles
      winged. Stems climbing or trailing, 1 meter or more
      long, with broad wings. Inflorescence peduncled, 10-
      20 cm long, the flowers in racemes. Petals 1.5-2.0 cm
      long. Occasionally escaped and persisting in the west-
      ern part. June-Aug. Everlasting Pea . . . .
      . . . . *Lathyrus latifolius* L.

    3' Leaflets 4-12, linear to elliptic. Petioles not winged.
      Stems slender, climbing, 3-6 cm long. Tendrils often
      branched. Inflorescence on peduncles about as long as
      the subtending leaf, the raceme of 2-6 flowers, red to
      purple. Fruiting pods linear, 4-5 cm. Infrequent in
      moist places and in woods of the northeast part. June-
      July . . . . *Lathyrus palustris* L.

  2' Stems wingless.

    4 Leaflets narrow, less than 5 mm wide, linear in outline.
      Stems 10-20 cm long. Leaves with 2-6 pairs of leaflets,
      these 2-3 cm long. Frequent on hillsides and in open
      places of the south and western parts. {*L. ornatus*
      Nutt.} May-June . . . . . . *Lathyrus polymorphus* Nutt.

    4' Leaflets wider, 1.5-2.0 cm, ovate in outline. Stems 6-
      10 dm tall, usually erect. Leaves with 4-7 pairs of
      leaflets, veiny. Peduncles 5-10 cm long. Racemes
      short, 10-14 flowered, the corolla purple or faintly
      lavender, 12-15 mm long. Fruiting pods 4-5 cm long,
      glabrous. Frequent in upland prairie swales and open-
      ings of woods in the northeast and in the Black Hills.
      June-July . . . . *Lathyrus venosus* Muhl.

LESPEDEZA Michx.

Lespedeza capitata Michx. Plants perennial with mostly simple
stems 5-12 dm tall. Plant surfaces villous. Leaves trifoliolate, the leaflets up to 4 cm long. Inflorescence forming a
dense capitate cluster of many flowers, each flower 8-12 mm
long. Petals cream-colored, subtended by leafy bracts which
become rust brown and showy. Fruit pubescent, shorter than the
petals and bracts. Rare in mesic prairie of Union and Clay counties of the southeast part. July-Aug. Bush-clover.

LOTUS L.

1   Plants perennial with trailing, prostrate stems rooting at
    nodes. Flowers in axillary peduncled umbels. Leaves with 5
    leaflets, the lower pair stipulate-like. Petals bright yel-
    low, the bases tinged with red. Flowers 8-12 mm long. Pod
    becoming 2-3.5 cm long with brownish-black seeds. Commonly
    planted as cover on roadsides and persisting, especially in
    the eastern part. June-Aug. Bird's-foot Trefoil . . . . . .
    . . . . *Lotus corniculatus* L.

1' Plants annual with stems mostly erect, 1-4 dm tall. Flowers
    1-3 in the axils of upper leaves on peduncles up to 3 cm long.
    Leaves generally with 3 leaflets; however, occasionally lower
    leaves may have 5 leaflets and upper leaves may be simple.
    Flowers 4-6 mm long, the petals cream-yellow, tinged with red.
    Fruiting pods 1-3 cm long, with 4-10 seeds. Frequent in dry
    prairie, openings in woods, or on rocky or sandy soil over
    the state. {incl. *L. americanus* Bisch.} May-July . . . . .
    . . . . *Lotus purshianus* Clem. & Clem.

LUPINUS L.

1   Plants annual, less than 20 cm tall. Stems with scurfy brown
    hairs. Branches spreading near the ground, the main stem
    short. Seeds usually 1 or 2 in each pod, each seed shaped
    like a hockey puck. Flowers about 1 cm long, varying in
    color from blue to white. Frequent in dry, sandy soil in
    the western half of the state. May-June . . . .
    . . . . *Lupinus pusillus* Pursh.

1' Plants perennial, over 30 cm tall. Stems from a woody base,
    not spreading; branched. Seeds more than two in each pod.
    Flowers 8-14 mm long, varying in color from violet to white.
    Our plants can be segregated into the following two varieties:

        Flowers mostly 5-9 mm long, leaflets 3-6 cm long . . . .
        . . . . *L. argenteus* var. *parviflorus* (Nutt.) Hitchc.

        Flowers mostly 8-12 mm long, leaflets 5-7 cm long . . .
        . . . . *L. argenteus* var. *argenteus* Pursh.

    Frequent in a variety of habitats in the western part of the
    state. May-June. Lupine . . . . . . *Lupinus argenteus* Pursh.

MEDICAGO L.

1   Plants perennial. Flowers exceeding 6 mm in length.

    2   Flowers blue or purple. Herbaceous stems 2-8 dm tall, from
        deep, perennial roots. Leaflets oblanceolate, serrate,

over 2 cm long. Flowers in loose heads on short peduncles.
Petals 8-11 mm long. Fruiting pods becoming coiled towards
maturity, pubescent. Widely cultivated and persisting for
a time in open areas and on roadsides. June-July. Alfalfa
. . . . *Medicago sativa* L.

2' Flowers yellow. Plants with sparingly branched stems decumbent to erect, up to 1 meter tall, from deep roots.
Leaflets oblanceolate, more linear than the preceding, up
to 2 cm long. Flowers 6-8 mm long, in short racemes, mostly yellow with occasionally some tinged with violet. Fruiting pods falcate or almost straight, 3-4 mm long. Occasionally escaped and becoming established in open places and
along roadsides. July-Aug. Yellow Lucerne . . . .
. . . . *Medicago falcata* L.

1' Plants annual. Stems prostrate or weakly ascending, spreading
to a half meter or more. Leaflets obovate, 1-2 cm long. Flowers yellow in globose heads, the petals less than 4 mm long.
Fruiting pods black, about 3 mm long, one-seeded. A common
weed of waste places over the state. May-Aug. Black Medic
. . . . *Medicago lupulina* L.

MELILOTUS Adans.

1 Flowers white. Stems 5-20 dm tall, erect, branched above,
from deep biennial roots. Leaves trifoliolate, the leaflets
serrate. Flowers in racemes 4-10 cm long, the petals about 5
mm long. Fruiting pods 2-4 mm long, glabrous. Commonly cultivated and frequently escaping. July-Sept. White Sweet
Clover . . . . *Melilotus albus* Desr.

1' Flowers yellow. Stems freely branched, up to 1.5 meters tall.
Plants annual or biennial. Leaflets elliptic, 1-2.5 cm long.
Flowers in racemes up to 10 cm long. Fruiting pods 3 mm long.
Common as an agricultural legume and escaping in waste places
and on roadsides over the state. May-Aug. Yellow Sweet
Clover . . . . *Melilotus officinalis* (L.) Lam.

OXYTROPIS DC.

1 Petals pink to purple. Pubescence on leaves appearing branched
and attached at the middle, with the branches appressed. Stems
very short, the leaves produced at soil level. Leaves with 7-
17 leaflets, the leaflets narrow to oblong. Inflorescence in
spikes 4-10 cm long, the petals 1.5-2 cm long, occasionally
white. A wide-ranging species over the state in a variety of
dry or prairie habitats. May-July. Purple Locoweed . . . . .
. . . . *Oxytropis lambertii* Pursh.

1' Petals white to cream-colored or yellowish, only rarely tinged
with purple. Pubescence on vegetation of simple hairs.

   2 Flowers mostly 12-15 mm long. Yellowish, the keel petals
   without blotches. Foliage green at maturity. Leaflets
   many, up to 31. Flowers in dense spikes, the petals occasionally tinged with purple. Fruiting pods becoming membranous at maturity. Infrequent on rocky or sandy places
   in Custer, Lawrence and Pennington counties of the Black
   Hills and in the southern part. May-July . . . .
   . . . . *Oxytropis campestris* (L.) DC.

   2' Flowers mostly 18-25 mm long, whitish and with keel tips
   faintly tinged with purple. Foliage silvery-gray and
   sericeous at maturity. Leaflets up to 21. Inflorescence
   in spike-like racemes. Fruiting pods thick-walled and
   hardened. Frequent in prairie openings and meadows at
   middle elevations in the western part, especially in the
   Black Hills. May-July. White Locoweed . . . .
   . . . . *Oxytropis sericea* Nutt.

PETALOSTEMON Michx.

1 Vegetative surfaces glabrous to sparsely hairy. Leaflets 3-9.

   2 Flowers white. Calyx glabrous or with a few hairs.

      3 Bracts beneath the calyx slender-tipped and conspicu-
      ously surpassing the calyx. Spikes dense and uninter-
      rupted. Leaflets commonly more than 1.5 cm long.
      Plants perennial, the stems 3-8 dm tall, glabrous.
      Leaflets 5-9, usually 7, slightly oblanceolate or ob-
      long. Inflorescence 2-5 cm long. Frequent in dry, up-
      land prairies over the state. June-July. White Prairie
      Clover . . . . *Petalostemon candidum* Michx.

      3' Bracts beneath the calyx as long as or slightly longer
      than the calyx, deciduous. Spikes loose and interrupted.
      Leaflets rarely exceeding 1.5 cm, usually about 1 cm.
      Plants perennial, 3-7 dm tall, the stems simple to
      branched. Leaflets 5-9, linear to lanceolate. Inflo-
      rescence 2-5 cm long, flexuous and slender. Common in
      dry prairie over the state. June-July. Western Prairie
      Clover . . . . . *Petalostemon occidentale* (Gray) Fern.

   2' Flowers pink-purple. Calyx sparsely to conspicuously hairy.
   Stems several from a heavy perennial root. Leaflets 3-7,
   linear to narrowly lanceolate, 1-2 cm long. Inflorescence
   in cylindric spikes 2-6 cm long, the flowers with petals
   pale pink to purple. Frequent in dry plains and prairies

over the state. June-July. Purple Prairie Clover . . . .
. . . . *Petalostemon purpureum* (Vent.) Rydb.

1' Vegetative surfaces densely covered with short hairs (villous)
that turn rust colored towards maturity. Leaflets 9-17, linear to oblong, 7-10 mm long. Plants with one-several stems
2-6 dm tall from a woody base. Flowering spikes 2-5 cm long,
the flowers light pink to purple. Calyx and bracts densely
villous. Infrequent on sandy, exposed prairie knolls and on
rocky openings of woods of the western part. July-Aug . . .
. . . . *Petalostemon villosum* Nutt.

PSORALEA L.

1 Flowers exceeding 10 mm in length. Plants from enlarged,
tuberous roots. Leaves with usually 5 leaflets.

2 Stems densely hairy, usually not over 3 dm tall. Leaves
principally at lower part of the stem, their leaflets gray-
strigose on the undersurfaces. Flowers in dense cylindric
spikes 2-8 cm long. Fruiting pods becoming 2 cm long,
ovoid and beaked, with strigose hairs. Common in native
prairies over the state. May-June. Tipsin, Prairie Turnip
. . . . *Psoralea esculenta* Pursh.

2' Stems sparingly hairy, mostly glabrous below, commonly to 5
dm or more tall, branching above. Leaves distributed along
the stem. Leaflets 5, each 2-4 cm long, slightly hairy beneath. Flowers in dense heads or spikes 2-5 cm long. Corolla blue. Fruiting pods ovoid, about 8 mm long. Frequent in sandy or dry prairies in the south and western
parts of the state. Apr-June . . *Psoralea cuspidata* Pursh.

1' Flowers less than 10 mm long. Plants from roots not tuberous-
ly enlarged. Leaves usually with 3 leaflets.

3 Vegetative surfaces conspicuously whitened due to dense,
silvery hairs.

4 Leaflets principally 5, linear. Plants with stems 3-5
dm tall from creeping rootstocks. Stems branched above.
Principal leaves with 5 leaflets, each leaflet 2-3 cm
long. Inflorescence a lax, interrupted raceme. Flowers
7-10 mm long, rose to purple. Fruits ovoid, flat, 6-8
mm long. Apparently rare in sandy places of Pennington
and Washabaugh counties in the southwest part. June-
July . . . . *Psoralea digitata* Nutt.

4' Leaflets usually 3, ovate. Perennial from rhizomes.
Stems 2-5 dm tall, irregularly branched. Leaves occasionally 5-foliolate, the leaflets 2-5 cm long. Flowers

almost sessile in interrupted spikes. Petals deep blue
or purple, 7-10 mm long. Fruiting pod ovoid, beaked,
almost 1 cm long. Common in prairie over the state.
June-July . . . . *Psoralea argophylla* Pursh.

3' Vegetative surfaces glabrous to sparsely hairy but not
silvery white.

   5 Corolla principally white with keels tinged with purple.
Fruiting pods rounded, 4-5 mm long, glandular. Plants
1-4 dm tall, from long, creeping rhizomes, the stems
strongly glandular. Leaflets greenish, 2-3 cm long,
elliptic to narrowly obovate. Flowering racemes with
10-40 flowers, almost spike-like. Frequent on sandy
soils and sand bars over the state. May-Aug . . . . .
. . . . *Psoralea lanceolata* Pursh.

  5' Corolla principally blue or purple. Fruiting pods ovoid.

     6 Leaflets oblanceolate to obovate. Fruits 7-9 mm,
with short beaks. Plants with erect stems, branched,
2-6 dm tall. Leaflets obtuse with abrupt points.
Flowers in slender racemes 1-4 cm long. Petals 5 mm
long, light blue. Frequent in dry prairies and sandy
plains in the western part. Ours var. *tenuiflora*.
July-Aug . . . . *Psoralea tenuiflora* Pursh.

    6' Leaflets narrowly linear, 2-3 mm wide. Fruiting pods
8 mm long, with conspicuous beaks. Plants 4-10 dm
tall, the stems widely branched. Inflorescence lax,
the flowers scattered. Petals 7-8 mm long, blue.
Rare in sandy soil in the south and western part.
July-Aug . . . . . . *Psoralea linearifolia* T. & G.

ROBINIA L.

Robinia pseudoacacia L. Erect tree up to 20 meters or more, with
erect branching. Leaves pinnately compound, with 7-19 leaflets,
the stipules modified into a pair of spines up to 1 cm long.
Leaflets 2-5 cm long, oval to elliptic. Flowers in axillary,
dropping racemes, numerous, the petals white and fragrant. Corolla 2-3 cm long. Fruit a flattened legume 5-10 cm long. Infrequently planted in the southern part and persisting. May-
June. Black Locust.

SCHRANKIA Willd.

Schrankia nuttallii (DC) Standl. Perennial shrub with prickly
stems that branch and spread irregularly to a meter or more in
height. Leaves bipinnate with 6-12 pairs of pinnae, the leaflets

numerous, each 4-6 mm long. Flowers in pink to red globose heads,
on axillary peduncles that are prickly. Fruiting pods 3-8 cm
long, sessile, 4-angled, with prickly spines. Frequent in dry
soils and exposed areas in the central part of the state. July-
Aug. Sensitive Brier.

SOPHORA L.

Sophora nuttalliana Turner. Plants herbaceous, perennial from a
creeping rootstock. Stems up to 30 cm tall, sparingly branched.
Leaves odd-pinnate, the leaflets 13-21, elliptic. Vegetative
surfaces silky-canescent. Flowers in axillary or terminal racemes, the petals white, pale yellow, or violet. Flowers about
1.5 cm long. Fruiting pods 3-4 mm long, rounded in cross section. Infrequent in dry soils of the south and western part of
the state. {incl. *S. sericea* Nutt.} Apr-June.

STROPHOSTYLES Ell.

1 Flowers exceeding 8 mm in length. Leaflets oval to remotely
  3-lobed, not more than twice as long as wide. Plants twining
  annuals, the stems up to a meter long. Leaves pinnately 3-
  foliolate. Flowers in capitate clusters on long peduncles,
  the petals rose-purple to green. Fruiting pods linear, 4-6
  cm long. Infrequent in sandy or alluvial thickets in the
  central and eastern part. Aug-Sept. Wild Bean . . . .
  . . . . *Strophostyles helvola* (L.) Ell.

1' Flowers less than 8 mm long. Leaflets linear to lanceolate,
   at least twice as long as wide. Plants decumbent twining to
   erect, the stems 1-4 dm tall. Leaves pinnately 3-foliolate,
   the terminal leaflet on a longer pedicel. Flowers 1-several
   on axillary peduncles, the petals pink to purple. Fruiting
   pods 2-3 cm long. Frequent on dry prairies over the state.
   July-Aug . . . . . *Strophostyles leiosperma* (T. & G.) Piper

THERMOPSIS R. Br.

Thermopsis rhombifolia Nutt. Plants perennial with simple stems
10-35 cm tall. Leaves trifoliolate, the rhombic leaflets up to
3 cm long, glabrous on the upper surface. Ovate stipules at the
base of the petioles large and leaf-like. Flowers in racemose
inflorescences, yellow, the petals up to 2 cm long. Fruiting
pods becoming curved to a half circle at maturity, 8-12 seeded.
Frequent on hillsides and open places of the western part.
{incl. *T. arenosa* A. Nels.} May-June. Golden Pea.

TRIFOLIUM L.

1   Calyx tube densely pubescent.

  2   Flowering heads globose or nearly so, 2-3.5 cm in diameter, on short peduncles or sessile. Leaflets ovate, broadest near the middle. Plants perennial with a short taproot, the stems 3-8 dm tall. Inflorescence 50-many flowered, the petals up to 20 mm long, deep red. Commonly cultivated and escaped throughout our range. June-Aug. Red Clover . . . . *Trifolium pratense* L.

  2' Flowering heads ovoid to cylindric, on conspicuous peduncles. Leaflets obovate, broadest above the middle. Plants annual, the stems 3-7 dm tall. Vegetation with coarse appressed to spreading hairs. Inflorescence commonly 3-5 cm long, many flowered, the petals 12-15 mm long, crimson. Infrequently planted and persisting in waste places in the eastern part. May-July. Crimson Clover . . . . *Trifolium incarnatum* L.

1' Calyx tube glabrous to sparsely hairy.

  3   Flowers 12-17 mm long. Flowering heads terminal and usually solitary. Plants perennial from a taproot, stems glabrous, from 2 to 4 dm tall. Leaflets narrowly elliptic. Inflorescence 2-3.5 cm long and almost as wide. Petals pale red to purple. Rare in meadows of the eastern part. June-July . . . . *Trifolium beckwithii* Brew.

  3' Flowers less than 10 mm long. Flowering heads axillary on conspicuous peduncles.

    4   Stems creeping, stoloniferous. Petals white or with pink edges. Vegetation glabrous or sparsely pubescent. Stems up to 5 dm long, usually creeping but occasionally erect. Flowers in long-peduncled heads, the petals 5-9 mm long. Very common in lawns, roadsides and in waste places. May-Aug. White Clover . . *Trifolium repens* L.

    4' Stems decumbent to erect, not stoloniferous. Petals pink to red but occasionally white, turning brown after pollination. Plants perennial, the stems 3-7 dm tall. Heads globose, on peduncles 2-6 cm long. Frequently planted and escaping or persisting in old fields and along roadsides. May-Aug. Alsike Clover . . . . *Trifolium hybridum* L.

VICIA L.

1   Flowers few, usually in pairs, borne sessile in the axils of upper leaves. Plants annual, the stems up to 1 meter,

ascending or climbing. Tendrils at the upper part of stem.
Stipules large, sharply serrate. Flowers white to lavender,
2-3 cm long. Fruiting pod flattened, 4-8 cm long. Occasionally escaped from cultivation over the state. July-Aug.
Spring Vetch . . . . *Vicia sativa* L.

1' Flowers 6 or more in racemes, on axillary peduncles.

    2 Base of flowers gibbous, or bulged to one side, the pedicel
      therefore appearing lateral. Racemes one-sided, dense.

        3 Plants densely hirsute. Flowers mostly over 15 mm long,
          purplish. Plants annual or biennial, the stems up to 1
          meter long, spreading. Fruiting pods 2-3 cm long, sev-
          eral seeded. Escaped and persisting along roadsides and
          railroad embankments in the western part. June-July.
          Winter or Woolly Vetch . . . . *Vicia villosa* Roth

        3' Plants glabrous to pubescent. Flowers 10-15 mm long,
          purple. Plants perennial, the stems climbing, up to 1
          meter long. Leaflets mostly 12-18, linear. Stipules
          entire. Fruiting pods 1.5-2.0 cm long, several-seeded.
          Naturalized in the northern Black Hills. May-July.
          Tufted or Bird Vetch . . . . *Vicia cracca* L.

    2' Base of flowers not gibbous, the pedicels basal. Plants
      perennial, the stems twining or climbing, usually 5 dm or
      more tall. Stipules sharply serrate. Leaflets 8-12, vary-
      ing much in shape and size. Racemes loose, few-flowered,
      usually 4-10. Flowers red to purple, 12-20 mm long. Our
      common species, occurring in thickets and meadows over the
      state. Ours represented by the two varieties *americana* and
      minor Hook. June-July. American Vetch . . . .
      . . . . *Vicia americana* Muhl.

Family Geraniaceae

(Key to Genera)

1 Leaf blades pinnately veined and divided. Styles in fruit
  spirally twisted . . . . *ERODIUM*

1' Leaf blades palmately veined and divided. Styles in fruit
  coiled back . . . . *GERANIUM*

ERODIUM L.

Erodium cicutarium (L.) L'Her. Plants annual, herbaceous, becoming much branched from the base, up to 3 dm tall. Leaves

oblanceolate, the principal ones pinnately divided, 3-8 cm long.
Flowers in peduncled cymes, 2-several, the petals pink or rose-
colored. Flowers 1 cm or more wide. Fruits spindle-shaped, the
body 4-5 mm long. The style becoming spirally twisted when dry.
Occasionally escaped as a weed in waste places. May-July.
Stork's Bill.

GERANIUM L.

1 Plants annual or biennial. Petals less than 10 mm long.

   2 Sepals less than 4 mm long, not bristle-tipped. Fertile
      stamens 5. Stems to 5 dm tall, diffusely branched. Leaves
      with 7-9 divisions, the basal ones long-petioled. Inflo-
      rescence dense, the petals red-purple, 4-6 mm long. Fruits
      9-11 mm long. Rarely established as a weed in waste places
      in the western part. May-Aug . . . . *Geranium pusillum* L.

   2' Sepals 4-8 mm long, ending in bristly tips. Fertile sta-
      mens 10.

      3 Beak of the style column, including the stigma, 4-5 mm
         long. In fruit the pedicel much longer than the calyx.
         Plants weakly stemmed, 1-5 dm tall. Principal leaves
         cordate-rounded, deeply divided into 5 segments. Flow-
         ers usually borne in pairs on dichotomously branched
         peduncles. Petals pinkish, about as long as the sepals.
         Rare in upland woods of the Black Hills. June-July . .
         . . . . *Geranium bicknellii* Britt.

      3' Beak of the style column, including the stigma, 2-3 mm
         long. In fruit, the pedicel about equaling the calyx.
         Plants annual, the stems simple to branched, 1-5 dm
         tall. Vegetation hirsute to pilose. Leaves cordate-
         rounded, deeply divided into 5 segments, varied in size.
         Flowers in congested inflorescences, divaricately spread-
         ing in fruit. Petals pink to light rose. Frequent as a
         weed in thin, sandy soils over the state. May-Aug . . .
         . . . . *Geranium carolinianum* L.

1' Plants perennial. Petals exceeding 10 mm in length.

   4 Petals white to rose-colored or white with purple veins.
      Stems 4-8 dm tall, mostly erect, with spreading hairs or
      glabrate. Hairs in the inflorescence with a purplish cast.
      Leaves palmately 5-parted, the indentations deep. Flowers
      few, pediceled. Petals 10-18 mm long. Fruits glandular-
      hairy, 4-7 mm long. Frequent in valleys and moist soil of
      the Black Hills. [Variously pubescent forms with lavender
      petals are found in the Black Hills. These are believed to

be the result of hybridization with *G. viscosissimum* F. &
M.} June-July . . . . *Geranium richardsonii* F. & T.

4' Petals rose to purple. Stems pubescent to densely hirsute.

  5 Inner surfaces of petals glabrous. Styles with a fine
    pubescence. Stems up to 5 dm tall, from a stout root-
    stock. Leaves mostly long-petioled, basal, except for
    a single pair at mid-height. Leaf blades in laciniate
    segments. Flowers 2-4 mm across, the petals up to 2 cm
    long, lavender. Infrequent in rich woods of Marshall
    and Roberts counties in the northeast part. Apr-May.
    . . . . *Geranium maculatum* L.

  5' Inner surfaces of petals with white, glandular hairs.
    In South Dakota only in the Black Hills.

    6 Stems glandular-puberulent above and hirsute below.
      Lower petioles hirsute. Plants perennial, the stems
      up to 6 dm tall. Leaves 5-parted, cordate-rounded,
      the segments separated by deep incisions. Flowers
      pink to purple, the petals up to 2 cm long. Infre-
      quent but locally abundant in open woods and alluvial
      areas in the Black Hills. May-July . . . .
      . . . . *Geranium viscosissimum* F. & M.

    6' Stems glabrous to spreading-puberulent above, not
      glandular. Lower petioles glabrous. Plants peren-
      nial, the stems 2-5 dm tall. Leaves 5-parted, much
      like the preceding. Flowers pink to purple, the
      petals up to 2 cm long. Infrequent to rare in open
      woods and alluvial areas in the Black Hills. Occupy-
      ing the same area as the preceding species; some con-
      sider it a variety of it. May-July . . . .
      . . . . *Geranium nervosum* Rydb.

Family Oxalidaceae

OXALIS L.

1 Flowers rose to lavender. Plants without above ground stems,
growing from a scaly bulb. Leaves 1-2 cm across, 3-foliolate,
the leaflets obcordate. Flowers in small clusters on peduncles overtopping the leaves. Petals 1-1.5 cm long. Frequent
in upland prairies and openings in woods over the state. Apr-
May . . . . *Oxalis violacea* L.

1' Flowers yellow. Plants with leafy stems above ground.

2   Plants from rhizomes. Stipules at leaf bases usually
    lacking.

    3   Fruiting pedicels erect, the fruits erect, glabrous or
    nearly so. Flowers 1-8 in cymose umbels. Plants peren-
    nial, the stems prostrate to ascending, 1-4 dm tall.
    Leaflets green or green-purple, 1-2 cm across. Corolla
    yellow, the petals 4-8 mm long. Fruiting capsules 12-
    15 mm long, erect or spreading. Frequent to common as
    a weed in alluvial places over the state. {*O. europa*
    Jord.} May-Sept. Yellow Wood Sorrel . . . .
    . . . . *Oxalis stricta* L.

    3' Fruiting pedicels arched or deflexed in fruit, the
    fruits erect, pubescent. Flowers 2-5 in umbels. Plants
    perennial, gray-green, the stems erect but becoming de-
    cumbent and lax, 1-5 dm tall. Leaflets 1.5-2.0 mm wide.
    Flowers pale yellow, the corolla 4-10 mm long, the in-
    florescence mostly overtopping the leaves. Fruits 1.5-
    3.0 cm long, abruptly pointed. Common in woods and
    waste places over the state. May-Sept . . . .
    . . . . *Oxalis dillenii* Jacq.

2' Plants lacking rhizomes but with creeping stems. Leaf
    bases usually with stipules. Hairs on the vegetation
    lacking septations. Plants with trailing stems, only the
    upper portion erect, 2-6 dm long. Leaflets 1-3 cm across.
    Flowers on peduncles with 2-5 in a cluster. Fruiting cap-
    sules 1.5-2.0 cm long, grayish, erect. Infrequent weed of
    disturbed areas, especially yards and gardens. {incl. *O.*
    repens Thunb.} May-Sept . . . . . . *Oxalis corniculata* L.

Family Linaceae

LINUM L.

1 Petals yellow to orange-yellow. Sepals often with glands or
   glandular hairs on their margins.

    2 Petals about 8 mm long. The septum separating the seeds in
   the capsule not complete, partially exposing the seed. The
   margins of this incomplete septum ciliate. Capsules about
   3 mm long. Plants annual, 2-6 dm tall, branched, glabrous.
   Leaves 1-2 cm long, linear, deciduous shortly after maturation. Infrequent in moist to upland prairies in the southern and eastern tiers of counties. June-July . . . . . .
   . . . . *Linum sulcatum* Riddell

2' Petals up to 12 mm long. The septum separating the seeds complete, the seeds not exposed. Capsules about 5 mm long. Plants perennial from a taproot, the stems 1-4 dm tall, branched above. Leaves few, becoming deciduous after maturation. Petals yellow, oftentimes with a reddish tinge on the inside lower parts. Frequent to common in upland prairie over the state. In South Dakota represented by the two varieties *rigidum* and *compactum* (A. Nels.) Rogers. June-Aug. Yellow Flax . . . . *Linum rigidum* Pursh.

1' Petals blue to pale lavender, sometimes white. Sepals may be toothed but lack glands.

3 Plants perennial. Inner sepals entire. Stems 2-6 dm tall, only sparingly branched in the upper part. Leaves linear, ascending, up to 2 cm long. Flowers solitary or few at the ends of upper branches, the petals blue to off-white, 13-17 mm long. Frequent in prairies and thickets in the western part. Infrequent to rare in the eastern tiers of counties. Ours var. *lewisii* (Pursh.) Eat. & Wright. June-July. Blue Flax . . . . *Linum perenne* L.

3' Plants annual. Inner sepals toothed or at least ciliate. Stems 2-7 dm tall, erect. Leaves narrow, erect. Flowers in a terminal cymose inflorescence, the petals blue, 1-1.5 cm long. Stamens elongate, extending beyond the petals. Fruits about 7 mm long. A cultivated plant that escapes occasionally but does not persist. June-Aug. Common Flax . . . . *Linum usitatissimum* L.

Family Zygophyllaceae

TRIBULUS L.

Tribulus terrestris L. Plants prostrate, annual from a taproot, the stems radiating to a distance of 5 dm or more. Leaves pinnately compound with 4-8 pairs of leaflets each 6-10 mm long. Flowers 1-few on short axillary peduncles, the petals white or pale yellow. Fruits maturing into several 2-spined segments, each 4-8 mm long. The mature fruits can penetrate tires and cause very painful penetration of animal and human feet. Frequent in waste places over the state. June-Sept. Puncture Vine.

Family Rutaceae

ZANTHOXYLUM L.

Zanthoxylum americanum Mill. Perennial shrubby tree to 4 meters, much branched and with sharp stipular spines up to 1 cm long. Leaves pinnately compound with 5-11 leaflets, the leaflets 1-5 cm long. Vegetation aromatic, with glandular dots. Plants dioecious, the flowers appearing before the leaves in small axillary clusters. Petals 4-5, small, oval, green to yellow. Sepals lacking. Fruits follicular, 4-6 mm long, also glandular-aromatic. Frequent in draws, at the edges of thickets, and in woods in the eastern part of the state. Apr-May. Prickly Ash.

Family Polygalaceae

POLYGALA L.

1 Plants annual, the stems solitary from a taproot.

    2 Flowering racemes dense and head-like or short-cylindric, 1-3 cm long. Plants 1-4 dm tall, the stems simple or sparsely branched above. Leaves linear, alternate. Flowers white to rose-purple, the wings oval, 4 mm long, becoming longer in fruit. Rare in open woods in the Black Hills. July-Aug . . . . *Polygala sanguinea* L.

    2' Flowering racemes slender, mostly less than 1 cm long and 5 mm wide at the base, tapering to the apex. Plants slender, the stems single below but branched upward. Leaves linear, 1-3 cm long and less than 3 mm wide, in whorls at least on the lower part. Flowers white, 1-2 mm long. Fruiting capsules dropping as soon as mature. Frequent in sandy prairie and sterile soils over the state. Aug-Sept . . . . *Polygala verticillata* L.

1' Plants perennial from a thick, woody root. Stems several from the base.

    3 Principal leaves linear, less than 2 mm wide and 1-2.5 cm long. Stems 2-4 dm tall, mostly glabrous. Flowering racemes elongate, tapering, the flowers white-green, 2-4 mm long. Fruiting capsules elliptic, up to 3 mm long. Common on dry prairies and sandy slopes or openings in woods over the state. June-July. White Milkwort . . . . *Polygala alba* Nutt.

    3' Principal leaves lance-ovate, 1-2.5 cm wide and 2-5 cm long. Stems several, 2-4 dm tall, glabrous to puberulent.

Flowering racemes 2-4 cm long and 6-8 mm wide, slightly
tapering to the top. Flowers white, the petals 2-5 mm
long. Fruiting capsules rounded, 2-3.5 mm long and about
as wide. Infrequent to rare on sterile soil or rocky outcrops in the Black Hills of the western part. June-July.
Seneca Snake-root . . . . *Polygala senega* L.

Family Euphorbiaceae

(Key to Genera)

1 Flowers lacking a perianth, clustered. Juice milky. Male
   flowers reduced to a single stamen. Female flowers represented by a single 3-carpelled ovary. Stamens (male flowers)
   and the female flower in a cup-shaped involucre called a cyathium . . . . *EUPHORBIA*

1 Flowers with a perianth. Juice not milky.

   2 Pubescence of leaves simple. Leaf blades ovate-lanceolate
      with crenate margins . . . . *ACALYPHA*

   2' Pubescence of leaves, especially the lower surface, of
      branched (stellate) hairs. Leaves oblong-lanceolate,
      mostly entire . . . . *CROTON*

ACALYPHA L.

Acalypha virginica L. Annual plants with stems 2-5 dm tall.
Leaves alternate, the blades broadly lanceolate with petioles
one-half the length of the blades, the latter 2-6 cm long.
Plants monoecious, the pistillate flowers in the same axillary
spikes as the staminate ones. Female flowers are below the male
on the spike and subtended by lobed bracts 8-12 mm long. Infrequent as a weed in waste places in the southeast counties. Aug-
Sept. Three-seeded Mercury.

CROTON L.

Croton texensis (Klotzsch) Muell. Plants annual, branched, to
5-15 cm tall. Plants dioecious, the male smaller. Leaves and
stems densely canescent with a stellate pubescence giving the
vegetation a yellow-gray appearance. Male flowers in racemes
1-2 cm long, the calyx present but petals lacking. Female flowers few, apetalous, with branched styles. Frequent in dry sandy
soil and on roadsides and railroad right-of-ways in the southwestern part. June-July. Skunkweed.

EUPHORBIA L.

1 Glands of the involucre (cyathium) with white or colored petaloid appendages.

2 Leaves opposite, their bases not symmetrical. Stipules present.

3 Capsules completely glabrous.

4 Plants annual. The bases of leaves usually oblique.

5 Stipules united into a triangular, white scale-like structure. Plants rooting at nodes forming a dense mat, the stems 5-25 cm long. Leaves 2-6 mm long. Capsules to 1.3 mm long. Seeds smooth or nearly so. Infrequent in waste places or disturbed soil of the southern part of the state. July-Sept . . . . *Euphorbia serpens* HBK.

5' Stipules not united. Plants not rooting at the nodes.

6 Plants with stem tips and leaf bases pubescent. Stems decumbent to erect, 5-50 cm long. Leaves mottled reddish to solid red below. Capsules 1.6-2.0 mm long. Seeds with angles rather sharp with low transverse ridges, the angles lighter than the dark brown to black faces. Infrequent weed of the southeast part, especially in moist, sandy soil. July-Sept . . . . *Euphorbia nutans* Lag.

6' Plants glabrous throughout.

7 Leaves entire.

8 Mature capsules exceeding 2 mm in length. Plants decumbent to erect, up to 9 cm. Appendages of the gland more than 0.5 mm long. Leaves linear to elliptic, up to 3 cm long, and 2-5 mm wide. Capsules 2-3 mm long, seeds with low irregular ridges with coat white to mottled. Frequent in dry, sandy soils, particularly in the western part. Ours var. *intermedia* (Engelm.) Wheeler. July-Sept . . . . *Euphorbia missurica* Raf.

8' Mature capsules less than 2 mm long. Plants prostrate to slightly ascending. Appendages of glands less than 0.5 mm

long. Leaves oblong to ovate, 4-12 mm
long and 2-4 mm wide. Capsules 1.5-2.0
mm long, sharply 3-lobed. Seeds with low
rounded angles, white to light brown with
mottling. Rare in sandy soil of southern
part. July-Sept . . . .
. . . . *Euphorbia geyeri* Engelm.

7' Leaves toothed or otherwise serrate.

    9 Seeds with 3-7 prominent ridges running
transverse. Leaves not red-mottled, the
bases unequal. Plants semi-prostrate to
slightly ascending, 2-25 cm long. Mature
capsules 1.2-1.8 mm long. Very common
weed in a variety of habitats over the
state. June-Sept . . . .
. . . . . *Euphorbia glyptosperma* Engelm.

    9' Seeds with irregularly raised wrinkles,
not prominently transverse. Leaves red-
mottled above, especially along the midrib. Plants prostrate to decumbent,
stems 5-20 cm long. Mature capsules 1.5-
2.0 mm long; seeds sharply angled, 1.0-
1.6 mm long. Wide-ranging as a weed in
sandy or alluvial areas as well as in
disturbed places. June-Sept . . . . . .
. . . . . . *Euphorbia serpyllifolia* Pers.

4' Plants perennial. The bases of the leaves slightly
cordate. Stems 3-12 cm long, prostrate to semi-erect.
Leaves 3-7 mm long and 2-5 mm wide. Stipules to 1 mm
long, the segments narrowly linear. Mature capsule
1.5-2.5 mm long, only 1 to 2 seeds per capsule, with
several irregular ridges running transversely from
the raphe. Rare in the extreme southwest part of
the state. July-Sept . . *Euphorbia fendleri* T. & G.

3' Capsules pubescent, plants annual.

    10 Fruiting capsules pubescent primarily on the angles.
Seeds with narrow, sharp, transverse ridges. Plants
prostrate, spreading, up to 6 dm across, the stems
crisply hairy. Leaves irregularly oblong with unequal bases, 3-15 mm long. Cyathia with peduncles
1-2 mm long, the glands oval, reddish. Mature capsules 1.2-1.5 mm long, pubescent on the angles,
especially above. Seeds with transverse ridges extending from the longitudinal facets. Rare in sandy

soil in the southern part. July-Sept . . . . . .
. . . . *Euphorbia prostrata* Ait.

10' Fruiting capsules uniformly pubescent. Seeds lacking sharp, transverse ridges.

   11 Seeds with low, broad transverse ridges, the
   base of the seed rounded. Stems prostrate to
   ascending, 5-35 cm long, variously hairy.
   Leaves 4-15 mm long, 2-6 mm wide, typically
   red-mottled above along the midvein, green to
   reddish below. Capsules 1.2-2.0 mm long, uniformly pubescent. Seeds with 3-5 low transverse
   ridges passing through the angles. Frequent in
   waste or disturbed areas and along roadsides in
   the southeastern part. July-Sept. Wart Weed.
   . . . . *Euphorbia maculata* L.

   11' Seeds pitted or mottled, the base of the seed
   truncate or concave. Plants prostrate to weakly
   ascending, stems 5-35 cm long. Vegetation densely villous throughout. Leaves 3-10 mm long,
   ovate to oblong-linear, green. Capsules 1.4-2.2
   mm long, seeds 1-1.5 mm, typically tan-white
   with brown mottling. Frequent in sandy prairie
   soils of the southern part. July-Sept . . . . .
   . . . . *Euphorbia stictospora* Engelm.

2' Leaves alternate, opposite, or whorled, their bases symmetrical. Stipules lacking or gland-like.

   13 Plants annual. Leaf-tips pointed.

      14 Leaves with petioles opposite, the blades lanceo-
      late to linear. Stems 1-6 cm tall, erect, with
      numerous slender branches. Cyathia on short pedi-
      cels in upper axils. Glands greenish-yellow. Cap-
      sules 3-5 mm long, glabrous. Seeds 2.5-3.3 mm long,
      with roughened coat. Infrequent in sandy prairies
      in the southern part. July-Sept . . . .
      . . . . *Euphorbia hexagona* Nutt.

      14' Leaves sessile, oval to oblong, alternate on the
      lower part of the stem. Plants 3-9 dm tall, erect,
      with few or no branches. Inflorescence umbel-like,
      with floral leaves whorled, the margins showy white.
      Capsule 3-lobed, 4-6 mm long. Seeds 3-4 mm long,
      prominently tuberculate. Frequent to common in dry
      slopes, eroded areas and disturbed prairie, more
      common in the southern part. July-Oct. Snow-on-
      the-mountain . . . . . . *Euphorbia marginata* Pursh.

13' Plants perennial. Leaf-tips obtuse or rounded. Stems
up to 1 meter tall. Leaves alternate on the stem,
linear-oblong. The bracteal leaves beneath the inflorescence opposite. Capsules 3-4 mm long, becoming glabrous towards maturity. Seeds 2.2-3.0 mm long, with
shallow depressions over the surface. Rare in the
southeastern part on dry, sandy soil. July-Sept.
Flowering Spurge . . . . *Euphorbia corollata* L.

1' Glands of the involucre (cyathium) without white or colored
appendages.

15 Involucral glands numbering 1 or 2 per cyathial unit.

16 Principal stem leaves mostly opposite and uniform in
shape, pubescent above. Floral bracts green and mottled with red. Plants annual, erect, 1.5-5.0 dm tall.
Leaves ovate to lanceolate, with wide dentations on
the margins. Capsules 3-5 mm long, sharply tuberculate. Frequent in dry sandy soils, roadsides and
railroad embankments of the southern part. July-Sept
. . . . *Euphorbia dentata* Michx.

16' Principal stem leaves alternate; linear and entire,
the floral leaves lobed and wider. Floral bracts
green, with bases reddish. Annual, stems to 8 dm.
Leaves essentially glabrous. Capsules 3-4 mm long.
Seeds 2.5-3.0 mm long, sharply tuberculate. Rare in
alluvial woods of the southeast part. {*E. heterophylla* L.} July-Sept. Fire on the-mountain . . . . .
. . . . *Euphorbia cyathophora* Murr.

15' Involucral glands numbering 4 or 5 per cyathial unit.

17 Maturing ovary and capsule smooth to slightly roughened. Glands with horn-like projections.

18 Style and stigma of developing ovary 1.5-3.0 mm
long. Plants from deep roots, the colonies usually dense in areas where it occurs. Stems 3-7
dm tall, glabrous. Leaves lanceolate, scattered,
those beneath the flowering portion whorled.
Seeds mottled, caruncle large, flat. Considered
a noxious weed. Common in waste areas, old fields
and roadsides. {*E. esula* L.} June-Aug. Leafy
Spurge . . . . *Euphorbia podperae* Croiz.

18' Style and stigma of developing ovary up to 1.0 mm
long but not exceeding it.

19 Stem leaves narrow, usually not more than 3
mm wide. Perennials from creeping rhizomes.

Stems 1-3 dm tall. Leaves 1-3 cm long, linear-
lanceolate, numerous. Occasionally persisting
from cultivation, especially near old ceme-
teries or homesteads. Rare in the southeastern part. May-June. Cypress Spurge . . . . .
. . . . *Euphorbia cyparissias* L.

19' Stem leaves ovate, usually exceeding 4 mm in
width. Plants 1-3 dm tall, branched, the stem
leaves alternate, 1-1.5 cm long. Capsules 4-
4.5 mm long, glabrous. Seeds 2-3 mm long,
with shallow depressions. Rare in dry sandy
soil of extreme southwestern part. June-Aug.
. . . . . . *Euphorbia robusta* (Engelm.) Small

17' Maturing ovary and capsule with prominent papillae.
Glands without horn-like projections. Plants annual
or less frequently biennial, 2-4 dm tall. Leaves
alternate, oblanceolate to spatulate, glabrous. Capsules 3-3.5 mm long, the seeds with wrinkled reticulations. Frequent in moist to dry sandy soil or prairies, particularly in the western part. {*E. dictyo-
sperma* Fisch. and Mey. and *E. obtusata* of Amer. auth.
in part.} Apr-May . . . . . *Euphorbia spathulata* Lam.

Family Callitrichaceae

CALLITRICHE L.

1 Leaves, submersed and floating, linear, 1-3 mm wide. Fruits
1.5 mm wide and 1 mm long, nearly circular in outline. Stems
short and slender, usually not exceeding 6 cm. Leaves opposite, with one vein, entire, about 1 cm long. Flowers axillary, small, imperfect. A circumboreal species sporadically
distributed in shallow, wooded streams in the northern part.
{*C. autumnalis* L.} July-Aug. Water Starwort . . . .
. . . . *Callitriche hermaphroditica* L.

1' Leaves variable. The floating ones oblanceolate or spatulate,
only the submersed ones linear. Fruits rounded to slightly
longer than wide. Surfaces with sculptured markings.

2 Margins of the fruit winged on the upper part. Stems
slender, to 20 cm long. Floating leaves with 3 veins, the
submersed ones linear and with one vein. Sculpturing of
fruit surface in vertical lines, the margins slightly
winged. Fruits 1 mm long, not quite as wide. Occasional

in slow moving streams in the northeast part and in the
Black Hills. {*C. palustris* L.} July-Aug . . . .
. . . . *Callitriche verna* L.

2' Margins of the fruit not winged. Stems longer, to 4 dm.
Floating leaves obovate, to 1 cm wide, 3-nerved. Sculpturing of fruit surfaces irregular, not in vertical lines.
Fruits roughly heart-shaped, almost as wide as long, about
1 mm. Occasional in shallow ponds and waterholes in the
western and northwestern part. July-Aug . . . .
. . . . *Callitriche heterophylla* Pursh.

Family Anacardiaceae

(Key to Genera)

1 Leaflets three to many. If only 3, the terminal leaflet sessile and less than 4 cm long. Fruit surface pubescent . . . .
. . . . *RHUS*

1' Leaflets three, the terminal one petioled and exceeding 4 cm.
Fruit surface glabrous . . . . *TOXICODENDRON*

RHUS L.

1 Leaflets several to many.

  2 Leaves the twigs glabrous. Fruits bright red with short
hairs less than 1 mm long. Shrubs sparsely branched, up
to 3 meters tall. Leaflets 12-30, narrowly lanceolate
with many teeth, 5-8 cm long. Inflorescence upright and
paniculate, 10-15 cm long, the flowers greenish-white.
Fruit a drupe, 4-7 mm across, becoming bright red. Common
in upland areas and on slopes of draws and ravines over
the state. June-July. Smooth Sumac . . . . *Rhus glabra* L.

  2' Leaves and twigs softly and densely hairy. Fruits with
dense hairs 1-2 mm long. Tall shrubs to small trees up
to 5 meters tall. Leaflets 10-30, lanceolate with serrations on both margins. Inflorescence paniculate, upright, the flowers green. Fruit a drupe, becoming dull
red and beset with hairs. Occasionally escaped from cultivation and persisting. Probably not naturalized. June-
July. Staghorn Sumac . . . . *Rhus typhina* L.

1' Leaflets three, the terminal one sessile. Plants low, bushy
shrubs usually less than 1.5 meters tall. Flowers small,
greenish, appearing before the leaves in the spring. Fruits

red, pubescent, in short, spike-like clusters. Quite common
on hillsides and in draws from Yankton County west in the
south and over the state in the southwest one-half to the
northwest corner. {*R. trilobata* Nutt.} May. Skunk-bush
Sumac . . . . *Rhus aromatica* Ait.

TOXICODENDRON Mill.

Toxicodendron rydbergii (Small) Greene. Shrub or sub-shrub usually less than 1 meter tall but occasionally to 3 meters tall or
climbing vines up to 4 cm in diameter. Flowers appearing after
the leaves in late spring, greenish-white, monoecious, the inflorescences in the axils of upper leaves. Fruits berry-like
drupes, white or green, 6-9 mm in diameter. Frequent to common
in all habitats over the state but most common in alluvial or
low wooded areas. {*Rhus radicans* L.} June. Poison Ivy.

Family Celastraceae

(Key to Genera)

1 Leaves alternate. Stems twining on other shrubs or trees.
   Flowers functionally unisexual . . . . *CELASTRUS*

1' Leaves opposite. Stems erect from a shrubby base. Flowers
   perfect . . . . *EUONYMOUS*

CELASTRUS L.

Celastrus scandens L. Woody shrubs with twining stems to 4
meters tall. Stems irregularly branched. Leaves elliptic to
ovate, 5-8 cm long, with acuminate tips. Flowers in terminal
or axillary panicles, green to cream-colored. Functional female
flowers forming yellow-orange fruits 7-9 mm in diameter that
burst in the fall to expose a bright orange arillate structure
with 1-3 seeds. Infrequent in alluvial woods over the state,
more common eastward. May-June. Bittersweet.

EUONYMOUS L.

Euonymous atropurpureus Jacq. Perennial woody shrub with erect
stems to 4 meters. Leaves strictly opposite, broadly ovate, 6-
10 cm long, with finely serrated margins. Inflorescence a number
of few flowered cymes, the flowers dull red to brown, not showy.
Fruits 4-lobed, 1.5 cm across, maturing to expose a dull red aril
covering a single seed. Rare in rich woods in the southeast part.
June. Wahoo, Burning Bush.

Family Aceraceae

ACER L.

1 Leaves simple.

  2 Sinuses between the principal leaf lobes shallow, not ex-
    tending halfway to the base. Trees 15-20 meters tall.
    Leaves opposite, the blades pale green, firm, glabrous.
    Flowers long-pedicelled, in corymbs, petals lacking.
    Fruits with slightly spreading wings, 3-5 cm long. In-
    frequent in rich loamy upland woods of Marshall and Roberts
    counties in the extreme northeast part, and along the Big
    Sioux River. April. Sugar Maple . . *Acer saccharum* Marsh.

  2' Sinuses between the principal leaf lobes extending to below
    the middle of the leaf. Trees up to 30 meters tall. Leaves
    opposite, the blades pale green above and silvery white be-
    low, usually 5-lobed. Flowers in umbels, each cluster uni-
    sexual. Fruits with widely spreading wings, 5-6 cm long.
    Common in alluvial areas in the eastern part, rare west-
    ward. April. Silver Maple . . . . . . *Acer saccharinum* L.

1' Leaves compound, the 3-7 leaflets coarsely and irregularly
   serrate or with lobes. Trees with many branches, the main
   stem often 15 meters or more tall. Plants dioecious, the
   pistillate flowers racemose, the staminate ones in drooping
   clusters, both apetalous. Fruits 3-4.5 cm long, the wings
   only slightly spreading. Common in alluvial woods and along
   streams over the state. April. Box Elder . . *Acer negundo* L.

Family Balsaminaceae

IMPATIENS L.

1 Flowers yellow-orange with brown spots. The spur of the enlarged sepal curved sharply, the free area about 7 mm long.
  Stems 5-10 dm tall, glabrous, branched in the upper part.
  Leaves pale green, glaucous, 3-8 cm long, petioled. Flowers
  on drooping pedicels, 2-3.5 cm long. Fruiting capsules about
  2 cm long, dehiscing forcibly when touched. Infrequent in
  moist, alluvial woods and thickets over the state. June-Aug.
  Spotted Touch-me-not . . . . *Impatiens biflora* Walt.

1' Flowers pale yellow with brown spots. The spur of the enlarged sepal curved at a right angle to the sepal, about 4-5
   mm long. Stems 8-15 dm tall, branched, glabrous. Leaves as
   in the preceding but larger, to 6 cm across. Flowers 2.5-4
   cm long, in wide-spreading racemes. Fruiting capsules 2-2.5

cm long. Rare in alluvial woods of the eastern part. July-
Sept. Pale Touch-me-not . . . . *Impatiens pallida* Nutt.

Family Rhamnaceae

(Key to Genera)

1 Leaves principally 3-veined from the base of the blade.
Flowers white or essentially so. Fruits dry, capsule-like
. . . . *CEANOTHUS*

1' Leaves pinnately veined. Flowers greenish. Fruits fleshy,
drupe-like . . . . *RHAMNUS*

CEANOTHUS L.

1 Leaf margins entire, the lower leaf surface pubescent beneath.
Branches spiny. Shrubs 2-7 dm tall, branched above, the
younger branches densely hairy. Leaves oblong, 1-2.5 cm long.
Flowers clustered in terminal racemes, the petals 1-2.5 mm
long. Fruits rounded, about 4 mm across, somewhat lobed. In-
frequent on limestone ridges at upper altitudes of the Black
Hills and in western Custer and Lawrence counties. June-July.
. . . . *Ceanothus fendleri* A. Gray

1' Leaf margins toothed. Branches not spiny.

   2 Upper leaf surfaces glossy, the lower surface velvety-hairy.
   Leaves thick, evergreen, the blades 4-8 cm long, broadly
   elliptic to ovate, aromatic. Shrubs 1-2.5 meters tall,
   spreading in patches. Inflorescence paniculate, the petals
   2-2.5 mm long. Fruits lobed, irregularly spherical, 3-6 mm
   across. Rare in the Black Hills but locally abundant in
   the Terry Peak area of Lawrence County. June-July. Mountain Balm . . . . *Ceanothus velutinus* Dougl.

   2' Upper leaf surfaces not glossy. Leaves not thick, deciduous. The blades 2-6 cm long, lance-ovate to elliptical.
   Low woody shrubs 2-8 dm tall, from a woody rootstock.
   Flowers white, in corymbose clusters, the petals about 2
   mm long. Fruits globose, 5 mm across, persisting on
   branches the following year. Frequent to common on prairie hillsides and open thickets over the southern part of
   the state. Ours var. *pubescens* (T. & G.) Shinners. May-
   June . . . . *Ceanothus herbaceus* Raf.

RHAMNUS L.

1 Plants dioecious. Leaves opposite, with 3 principal lateral
   veins. Blades elliptic ovate, 3-5 cm long. Shrubs or small
   trees up to 4 meters tall, the branches thorny. Flowers with
   4 sepals, the petals erect, 1-1.5 mm long. Fruit a drupe with
   4 stones. Commonly planted and occasionally escaped and persisting in the eastern part. May-June. Buckthorn . . . . . .
   . . . . *Rhamnus catharticus* L.

1' Plants monoecious. Leaves sub-opposite or alternate. Branches
   not thorny.

   2 Sepals 5. Petals lacking. Fruits with 3 stones. Leaves
      with serrated margins, alternate. Plants low, shrubby,
      sometimes trailing, the stems 3-8 dm tall. Leaves mostly
      more than 5 cm long, lance-elliptic. Flowers in clusters
      of 1-3 at leaf bases, the sepals 1-2 mm long. Rare in
      moist woods of the Black Hills. May-June. Alder Buck-
      thorn . . . . *Rhamnus alnifolia* L'Her.

   2' Sepals 4, petals present, 4. Fruits with 2 stones. Leaf
      margins very finely serrulate. Plants shrubby, 5-20 dm
      tall. Leaves sub-opposite or alternate, the blades 3-8
      cm long, lanceolate. Male flowers mostly 2 or 3 in leaf
      axils, the female ones usually solitary. Infrequent in
      woods and thickets of ravines in the eastern one-half.
      May-June . . . . *Rhamnus lanceolata* Pursh.

Family Vitaceae

(Key to Genera)

1 Leaves compound . . . . *PARTHENOCISSUS*
1' Leaves simple . . . . *VITIS*

PARTHENOCISSUS Planch.

1 Inflorescence of cymes or terminal panicles, not dichotomously
   branched. Tendrils branched and having adhesive disks. Plants
   climbing, the stems 5 meters or more tall. Leaves palmately
   compound, the 5 leaflets dull green above, 6-10 cm long, with
   coarse serrations. Inflorescence a panicle longer than wide,
   a central axis usually apparent. Fruits dark, about 5-6 mm
   long, with 1-3 seeds. Infrequent in rich loamy woods and alluvial woods in the southeastern part. June-July. Virginia
   Creeper . . . . *Parthenocissus quinquefolia* (L.) Planch.

1' Inflorescence with dichotomous branches. Tendrils sparingly
branched, mostly lacking adhesive disks. Plants climbing or
trailing, 1 meter or more long. Leaves palmately compound,
5-foliolate, the leaflets glossy green on the upper surfaces.
Inflorescence dichotomously branched, usually as wide as or
wider than long, the two diverging branches also dichotomous.
Fruits 7-9 mm long, with 3-4 seeds. Common in woods and
thickets over the state. June-July. Woodbine . . . .
. . . . *Parthenocissus vitacea* (Knerr.) A. S. Hitchc.

VITIS L.

1 Transverse diaphragms in pith near the bases of flowering
branches less than 2 mm thick. Leaves usually deeply 3-5
lobed as well as conspicuously serrate. Maturing fruits
glaucous-whitened. Plants perennial with high-climbing
vines. Leaves 5-14 cm long, glabrous to sparsely pubescent.
Inflorescence a panicle, the flowers functionally unisexual.
Fruits 6-8 mm across, with 1-3 seeds. Frequent to common in
thickets and alluvial woods over the state. May-June. River-
bank Grape . . . . *Vitis riparia* Michx.

1' Transverse diaphragms in pith near the bases of flowering
branches usually exceeding 2 mm in width. Leaves coarsely
dentate but only shallowly 3-lobed. Maturing fruits tending
to be less glaucous than the preceding. Plants perennial with
high climbing vines. Leaves rounded in outline, 10-15 cm
across, pubescent below. Inflorescence loosely flowered, 10-
15 cm long. Fruits 5-10 mm in diameter. Infrequent in thickets and alluvial woods over the state. May-June. Frost Grape
. . . . *Vitis vulpina* L.

Family Tiliaceae

TILIA L.

Tilia americana L. Tree to 30 meters. Leaves broadly ovate to
sub-rotund, the blade bases oblique. Margins sharply serrate.
Leaf surfaces becoming glabrous towards maturity. Flowers in
cymes on a long peduncle from a leaf-like bract that is narrow
and elongate. Petals yellow green, with staminodes, the flowers
fragrant. Corolla 7-10 mm long. Fruit a nut-like berry 7-9 mm
in diameter. Frequent in remnants of rich woods, usually in
ravines and north-facing loamy slopes, along the eastern part of
the state from Union and Yankton counties in the south to Marshall
and Roberts counties in the northeast. June-July. Linden, Bass-
wood.

Family Malvaceae

(Key to Genera)

1 Fruit a 5-carpelled capsule with loculicidal dehiscence.
   Petals yellow with purple bases . . . . *HIBISCUS*

1' Fruit of 5 to many carpels in a ring and separating at
   maturity.

   2 Each carpel with 2 or more seeds . . . . . . *ABUTILON*

   2' Each carpel with only 1 seed.

      3 Stigmatic surface across the top of the abruptly widened
   style top. Flowers salmon red . . . . . . *SPHAERALCEA*

      3' Stigmatic surface along the sides of the upper part of
   the style.

         4 Flowers 7-10 cm across. Bractlets below the flower
   6-9, united at their bases . . . . . . *ALTHAEA*

         4' Flowers less than 6 cm across. Bractlets below the
   flower 3 or fewer.

            5 Petals obcordate with a notch at the apex. Carpels
   without beaks . . . . *MALVA*

            5' Petals truncate without a notch at the distal end.
   Carpels beaked . . . . *CALLIRHOE*

ADUTILON Mill.

Abutilon theophrasti Medic. Annual herbs 0.5-1.5 meters tall,
sparingly branched. Leaves alternate, petioled, the blades
rounded, with crenate margins, 10-25 cm across. Vegetation
covered with a soft pubescence. Flowers mostly axillary in the
upper part, the petals yellow, 1-1.5 cm long. Fruits with sharply tipped carpels. Frequent in weedy pastures and abandoned
fields in the eastern part. Aug-Sept. Velvet Leaf.

ALTHAEA L.

Althaea rosea L. Tall perennials with simple stems to 2 meters
or more, coarsely hairy. Leaves large, the blades cordate-
orbicular, 1-3 dm across. Flowers many in peduncled clusters
in the upper axils, white to purple, up to 10 cm across. Fruits
of 15 or more carpels around a central axis, separating at maturity. Commonly planted and persisting around homesteads and
waste places. July-Sept. Hollyhock.

CALLIRHOE Nutt.

Callirhoe alcaeoides (Michx.) Gray. Plants perennial from a
thick rootstock, the stems branched, to 4 dm tall. Leaves ovate
to triangular in outline, palmately veined, with several deep,
irregular incisions. Lower leaves long-petioled. Flowers in
loose corymbs, or solitary in upper axils, the petals pink or
rose. Corolla 1-1.5 cm long. Rare in dry soil in the southern
part. June-July. Poppy Mallow.

HIBISCUS L.

Hibiscus trionum L. Much branched annual spreading irregularly,
the stems 1-5 dm tall. Vegetation with coarse hairs. Leaves
mostly 3-5 parted, irregularly lobed, ovate to orbicular in outline. Flowers on axillary peduncles, the petals 3-4 cm long,
yellow, with deep purple bases. Fruiting capsules subglobose,
hairy. Frequent as a weed over the state. July-Aug. Flower
of an Hour.

MALVA L.

1   Petals white or rosy tinged, usually not much longer than 1 cm.

    2   Mature fruits rounded in outline, the individual carpels
        flattened at the distal surface and rugose-roughened, their
        lateral surfaces radially veined. Plants annual with
        spreading stems, the leaves orbicular to reniform in out-
        line, 3-5 cm across. Flowers clustered in the axils of
        leaves, the pedicels to 3 cm long. Fruits about 10-
        carpeled. A common weed of lawns and farmyards over the
        state. June-Sept. Common Mallow . . *Malva rotundifolia* L.

   2' Mature fruits crenately shaped in outline, the individual
        carpels rounded on the distal surface, not roughened, the
        lateral surfaces not veined. Plants annual, much like the
        preceding. Flowers fascicled in the axils of leaves, the
        petals obcordate and whitish. Fruits of 12-15 carpels.
        Frequent as a weed of farmyards, lawns and roadsides in
        the eastern part. June-Sept . . . . *Malva neglecta* Wallr.

1' Petals red to lavender, 1-2 cm long. Plants biennial, the
    stems ascending to erect, 3-8 dm tall. Leaves shallowly
    lobed, reniform, on petioles 2-8 cm long, the blades 4-10
    cm across. Flowers in clusters on upper axillary peduncles,
    the petals much longer than the calyx. Occasionally escaped
    from cultivation and persisting in disturbed or waste places.
    July-Aug. High Mallow . . . . *Malva sylvestris* L.

SPHAERALCEA St.-Hill.

Sphaeralcea coccinea (Pursh) Rydb. Herbaceous perennial from a
deep spreading root system, the stems mostly 0.5-2.0 dm tall,
simple to clustered. Leaves mostly 3-parted into irregular
lobes, the blades 2-6 cm long. Flowers in terminal clusters or
racemes, the petals salmon red, 1 cm or more long. Vegetation
with branched hairs. Frequent to common on prairies and plains
over the state. May-July. Cowboy's Delight, Scarlet Mallow.

Family Hypericaceae

HYPERICUM L.

1   Stamens numerous, in 3-5 clusters. Fruiting capsule 3-celled
    when seeds begin to mature. Plants perennial, usually in
    dense patches, the stems tough, 4-8 dm tall, branched above.
    Leaves linear-oblong, 2-4 cm long, their bases rounded but
    not meeting around the stem. Flowers in compound cymes, the
    petals up to 1 cm long, yellow with black dots. Fruiting
    capsule 6-9 mm long. Occasional as a weed over the state.
    June-July. St. John's Wort . . . . . *Hypericum perforatum* L.

1' Stamens usually not more than 12, not in clusters. Fruiting
    capsule 1-celled when seeds begin to mature. Plants annual,
    stems sparingly branched above, 3-6 dm tall. Leaves lanceo-
    late, 2-4 cm long, broadly obtuse or rounded at the base, the
    basal margins meeting around the stem. Flowers yellow, numer-
    ous in cymes, the petals 2.5-4 mm long. Fruiting capsule up
    to 7 mm long. Adventive as a weed in waste places. July-Aug
    . . . . *Hypericum majus* (Gray) Britt.

Family Elatinaceae

(Key to Genera)

1   Flowers with 5 petals and sepals. Vegetation sparsely puberu-
    lent . . . . *BERGIA*

1' Flowers with 3 petals and sepals. Vegetation glabrous . . . .
    . . . . *ELATINE*

BERGIA L.

Bergia texana (Hook.) Schubert. Plants much-branched, annual,
herbaceous with stems to 2 dm tall. Leaves opposite, thick,
ovate with tapering bases, 1-3 cm long. Flowers clustered in

axils of leaves, perfect. Perianth 5-parted, the petals white,
2-3 mm long. Capsules 5-celled, 2 mm or more long. Rare in
alluvial soil in the south central part. July-Aug.

ELATINE L.

Elatine triandra Schkuhr. Small aquatic annuals usually immersed
in water or creeping on mud, the stems flaccid, branched, 3-10 cm
long. Leaves opposite, oblong to oblanceolate, 3-5 mm long.
Flowers small, in the axils of leaves, 3-parted. Ovary 2-4
celled, the seeds with rows of areoles slightly curved. Infrequent in shallow water of ponds and lakes over the state. July-
Aug. Waterwort.

Family Tamaricaceae

TAMARIX L.

Tamarix ramosissima Ledeb. Shrubby trees up to 5 meters tall.
Stems branched, not stout. Leaves small, scale-like, overlapping, about 1 mm long. Inflorescence a dense panicle or series
of spikes of rose-colored flowers, perfect. Petals 1-1.5 mm
long. Fruiting capsules tapering to the tip, 2-4 mm long. Infrequent in low places along streams and on river banks, but
becoming established along streams in the eastern part, especially the Missouri River. {*T. gallica* of authors} July-Sept.
Salt Cedar.

Family Cistaceae

(Key to Genera)

1 Plants with 5 yellow petals or petals lacking . . . .
  . . . . *HELIANTHEMUM*
1' Plants with 3 red petals, these very small . . *LECHEA*

HELIANTHEMUM Mill.

Helianthemum bicknellii Fernald. Plants perennial from short rhizomes. Stems 2-3 dm tall, usually solitary. Leaves lanceolate-
elliptic in shape, 2-3 cm long and 5-8 mm wide. Flowers early in
the season with petals, borne in terminal racemes. Later flowers
borne on short axillary branches, these without petals. Capsules
of the petalous flowers 4-5 mm across, those of the later apetalous flowers only 2 mm across. Rare to infrequent on rock outcrops of open woods of the Black Hills. June-July. Frostweed.

LECHEA L.

Lechea intermedia Leggett.   Plants perennial, the stems clustered from rootstocks.   Branches erect, 2-3 dm tall.   Leaves linear, 1-2 cm long, only 2-3 mm wide.   Flowers with 3 small red petals, clustered in small panicles at the ends of short branches.   Sepals slightly longer than the petals.   Capsules spherical, 3-valved, with 2-6 seeds.   Apparently very rare in open pine woods in the Black Hills.   July-Aug.   Pinweed.

Family Violaceae

VIOLA L.

1 Plants with aboveground leafy stems.   Flowers on axillary peduncles.

  2 Flowers principally yellow.

    3 Leaves longer than broad, the blades oblong to lanceo-late, 2-6 dm long, the margins almost entire.   Stems short, the leaves borne near the base.   Vegetative sur-face glabrous to puberulent.   Flowers yellow with brown-purple lines, 5-12 mm long, the spur short.   Peduncles about as long as the leaves.   Frequent to common in prairies over the state.   Apr-May.   Yellow Prairie Violet . . . . *Viola nuttallii* Pursh

    3' Leaves broadly orbicular to ovate with cordate bases, the upper ones on well-developed petioles.   Leaf blades crenate to dentate, their surfaces mostly smooth.   Stems on mature plants 1-4 dm tall.   Flowers yellow, in axil-lary peduncles 2-6 cm long.   Fruits about 1 cm long, glabrous to pubescent.   Frequent in alluvial woods in the eastern part and in woods of the Black Hills and Harding County.   {*V. pensylvanica* of authors}   Apr-May.   Common Yellow Violet . . . . *Viola pubescens* Ait.

  2' Flowers principally purple or white with purple.

    4 Petals white with purple lines or margins.   Stipules mostly entire with scarious margins.   Plants mostly glabrous, with several stems 2-4 dm tall.   Leaves on well developed petioles, the blades 5-10 cm across, cordate at the base, their margins crenately toothed.   Petals about 1.5 cm long, with a short spur.   Frequent to common in woods over the state.   Ours var. *rugulosa* (Greene) Hitchc.   May-June.   Canada Violet . . . . *Viola canadensis* L.

4' Petals purple. Stipules with teeth at the distal end.
Plants glabrous to pubescent, essentially without stems
early in the spring but these develop soon. Leaves
mostly cordate-ovate, 1-3 cm wide, their margins crenately toothed. Flowers on elongate peduncles, the
spur 4-6 mm long. Fruit 4-5 mm long. Frequent in
rock crevices and woods of the Black Hills. May-July.
. . . . *Viola adunca* Smith

1' Plants lacking above ground stems, acaulescent. Flowers on
peduncles that arise from the crown near the soil level.

5   Leaves deeply lobed, usually 3-parted, the ultimate segments linear. Plants glabrous to sparingly pubescent.
Flowers bright violet, the lateral petals bearded on the
inside. Flowering peduncles mostly taller than the leaves.
Plants commonly producing apetalous (cleistogamous) flowers
on erect peduncles. {Hybridizes with *V. pratincola* Greene,
V. nephrophylla Greene, and *V. sororia* Willd., giving
intermediate forms.} Common on prairie over the state.
May-June. Prairie Violet . . . . . *Viola pedatifida* G. Don

5' Leaves not deeply lobed into linear segments.

6   Rootstocks very slender. Plants rare, restricted to
cool and shady habitats at higher altitudes in the
Black Hills.

7   Petal faces white with violet lines or marginal
coloration. Leaf blades reniform, the lower surface
mostly pubescent. Flowers on short peduncles, the
petals not bearded. Spur short, not more than 2 mm
long. Rare in cool, north facing woods of the Black
Hills at higher altitudes. May-June. Kidney-leaved
Violet . . . . *Viola renifolia* Gray

7' Petal faces purple. Leaves broadly ovate with cordate bases, their surfaces sparsely hirsute. Plants
small, the flowers on short peduncles. Petals beard-
less, the spur large, 5-7 mm long. Rare in shady
ravines in the Black Hills. May-June. Great Spurred
Violet . . . . *Viola selkirkii* Pursh

6' Rootstocks thick and short. Plants not restricted to
cool, shady forests.

8   Plants with some leaves cleft or lobed at their bases.
Plants glabrous, the leaf blades variously shaped.
Some leaves ovate-deltoid with crenate margins, others
with 3-7 lobes. Flowers deep purple, the petals narrow. Cleistogamous flowers on erect peduncles.

Infrequent in dry to moist open swales or on loamy
banks in the east part. Apr-May . . . .
. . . . *Viola viarum* Pollard

8' Plants with leaves not dissected or lobed, the blades
   entire to serrate with cordate bases.

   9 Leaves and petioles pubescent. Leaf blades ovate
    to almost reniform, 5-8 cm across. Flowers deep
    lavender to pale violet, on peduncles about as
    tall as the leaves. Lateral petals bearded, the
    keel petal whitened at the base. Infrequent in
    rich upland and alluvial woods in the eastern
    part. Hybridizes with *V. nephrophylla* Greene
    and *V. missouriensis* Greene. Apr-May. Downy
    Blue Violet . . . . *Viola sororia* Willd.

   9' Leaves and petioles mostly glabrous.

    10 Flowers pale violet. Leaves deltoid shaped,
     usually longer than wide, their bases cordate.
     Apetalous flowers on peduncles that are pros-
     trate. Flowers on short peduncles, the
     spurred petals not bearded within. Sepals
     with a narrow light-colored margin. Infre-
     quent in alluvial woods in the eastern part.
     Apr-May . . . . . . *Viola missouriensis* Greene

    10' Flowers deep violet. Leaves ovate-cordate,
     usually as wide as long.

     11 Spurred petal glabrous. Cleistogamous
      flowers on short, prostrate peduncles.
      Leaves with an acute apex. Plants 8-25
      cm tall, from a thick short rootstock.
      Petals 10-15 mm long, with a short spur.
      Common in meadows, thickets and waste
      places over the state. Apr-June. Meadow
      Violet . . . . . *Viola pratincola* Greene

     11' Spurred petal bearded. Cleistogamous
      flowers on erect peduncles. Leaves
      broadly cordate with a rounded apex.
      Plants 5-20 cm tall, from a moderately
      thick ascending rootstock. Petals 10-15
      mm long, the spur about 4 mm long. Infre-
      quent in marshy meadows and low wet places
      in woods in the eastern and western part.
      May-June . . . . *Viola nephrophylla* Greene

Family Loasaceae

MENTZELIA L.

1   Plants annual. Stems narrow, not exceeding 4 mm in diameter.
Petals less than 6 mm long.

   2   Upper leaves narrow, usually linear and entire to shallowly
dentate. Seeds with conspicuous tubercles on their surfaces. Plants with decumbent to erect stems 1-4 dm tall,
white with glabrous to sparsely pilose. Principal leaves
oblanceolate with dentate margins. Flowers with golden-
yellow petals 3-4 mm long. Fruiting capsules 1-1.5 cm
long. Rare in dry, sandy soil in the western part. June-
Aug . . . . *Mentzelia albicaulis* Dougl.

   2' Upper leaves ovate-lanceolate, entire, sessile. Seeds with
only minutely roughened surfaces. Stems 1-3 dm tall, whitish and usually pilose, branched above. Principal lower
leaves ovate to lanceolate with sinuate margins. Flowers
in loose, terminal cymes, the petals yellow, 3-6 mm long.
Fruits 1-2 cm long. Rare on dry, sandy soil in the western part. June-July . . . . . . *Mentzelia dispersa* Wats.

1' Plants perennial or biennial. Petals 8 mm long or longer.

   3   Flowers usually with only 5 petals, these not more than 15
mm long. Seeds 5 or less per capsule, the seeds not winged.
Plants roughly spinose-hairy. Stems branched, up to 6 dm
tall. Leaves ovate-lanceolate, sessile to short-petioled,
their margins irregularly toothed. Flowers yellow-orange,
with 15-25 stamens. Rare on eroded sandy or exposed rocky
soil in the western part. June-July. Stick-leaf . . . . .
. . . . *Mentzelia oligosperma* Nutt.

   3' Flowers usually with 10 petals, each 2 cm or more long.
Fruiting capsules with many seeds, winged or with a margin.

      4   Petals 5-7 cm long, usually closed during daylight
hours. Fruiting capsules 3-6 cm long. Plants 3-7 dm
tall, branched. Leaves 5-12 cm long, petioled, with
pinnatifid or sinuate margins. Flowers creamy-white,
with up to 50 stamens. Very common on eroded clay
banks and bluffs over the state. July-Aug. Sand Lily
. . . . *Mentzelia decapetala* (Pursh) Urb. & Gilg.

      4' Petals 3-4 cm long, usually open during the daylight
hours. Fruiting capsules 1.5-3 cm long, the seeds many,
winged. Stems up to 5 dm tall, erect, branched from the
base. Leaves lanceolate, pinnatifid or irregularly sinuate, up to 7 cm long. Flowers creamy-white, with 30 or

more stamens. Infrequent on plains and eroded hillsides in the western part. July-Aug . . . . *Mentzelia nuda* (Pursh) T. & G.

Family Cactaceae

(Key to Genera)

1 Plants not jointed, the stems more or less subglobose.

   2 Flowers produced at the apex of tubercles that are adjacent to the areoles . . . . *ECHINOCEREUS*

   2' Flowers produced at the base of the grooves or between the tubercles . . . . *CORYPHANTHA*

1' Plants jointed, the stems more or less flattened to cylindrical in cross section . . . . *OPUNTIA*

CORYPHANTHA (Engelm.) Lem.

1 Flowers white to greenish. Fruits becoming tinged with red towards maturity, usually less than 1 cm long. Stems usually solitary, 3-6 cm tall. Areoles with 1 main spine surrounded by many smaller spines. Flowers 2-3 cm long. Fruit subglobose. Frequent in dry prairie or sandy soil in the western part. {*Mammillaria missouriensis* Sweet} June . . . . *Coryphantha missouriensis* (Sweet) Britt. and Rose

1' Flowers red or lavender. Fruits remaining green, from 1-2 cm long at maturity. Stems solitary or in clusters, 3-8 cm tall, cylindric to subglobose. Areoles with 3-5 main spines and numerous smaller marginal spines. Flowers 2-3 cm long. Fruit globose to ellipsoid, juicy. Frequent to common in dry prairies and sandy soil over the state. {*Mammillaria vivipara* (Nutt.) Haw.} May-June. Pincushion Cactus . . . . *Coryphantha vivipara* (Nutt.) Britt. and Br.

ECHINOCEREUS Engelm.

Echinocereus viridiflorus Engelm. Perennial subglobose or cylindrical plants 2-6 cm tall. Stems solitary or few in a cluster. Ribs usually 13, acutely angled, the flowers produced near the areoles that are on the ribs, appearing lateral. Corolla yellow-green, 2-3 cm long. Fruits green, about 1 cm long, spiny. Seeds black. Rare in dry, sandy soil in the southwest part. May-June. Hedgehog Cactus.

OPUNTIA Hill.

1   Joints of the stems cylindrical and finger-like, easily de-
    tached. Stems prostrate or weakly ascending, up to 2 dm tall
    from a perennial base. Areoles white-woolly, with several
    spines 1-3 cm long. Flowers yellow, the corolla 3-5 cm long.
    Fruits 1-2 cm long, tubercled, dry and spiny. Frequent in
    dry prairies and sandy, exposed areas over the state. May-
    June . . . . *Opuntia fragilis* (Nutt.) Haw.

1' Joints of the stems flattened, not easily broken or detached.

    2   Areoles on mature joints separated by a distance of 1.5 cm
        or more. Fruits succulent, at least 4 cm long. Plants
        prostrate or weakly ascending, the joints 8-10 cm long.
        Areoles with 1-3 spines, each up to 3 cm or more long.
        Flowers yellow to pink, the corolla about 7 cm long.
        Fruits juicy and edible. Locally abundant on dry prairie
        and plains from the east central area to the western part.
        {*Opuntia humifusa* Raf.} June-July. Prickly Pear . . . . .
        . . . . *Opuntia compressa* (Salisb.) MacBr.

    2' Areoles on mature joints separated by a distance of less
        than 1.5 cm. Fruit dry, usually not more than 2 cm long.
        Stems prostrate, forming mats, the joints 7-10 cm long.
        Areoles small, with 1-2 principal spines, each elongate
        to 3 cm or more. Flowers yellow to orange, the corolla
        4-7 cm long. Fruit spiny and dry, oblong. Infrequent but
        may be locally abundant on prairies and plains of the west-
        ern one-half. June-July . . . . . *Opuntia polycantha* Haw.

Family Elaeagnaceae

(Key to Genera)

1   Leaves alternate. Plants with flowers perfect or unisexual
    . . . . *ELAEAGNUS*

1' Leaves opposite. Plants dioecious . . . . . . *SHEPHERDIA*

ELAEAGNUS L.

Elaeagnus angustifolia L. Shrub or small tree up to 5 meters
tall, much branched; at times with many spiny stems. Leaves
alternate, scurfy green or silvery, oval, 3-6 cm long. Flowers
clustered, small, inconspicuous, the perianth yellowish. Fruit
about 1 cm long, ellipsoid, silvery. Commonly escaped from cultivation and widely established along streams, fencerows and old
dwellings. May-June. Russian Olive.

SHEPHERDIA Nutt.

1 Leaves silvery brown below. Shrubs less than 2 meters tall,
not thorny. Branches scurfy brown when young. Leaves dark
green on the upper surface, ovate in outline, 3-5 cm long.
Flowers small, on the male plants clustered at nodes of previous year's growth, greenish-yellow to brown. Female flowers
few. The fruits about 5 mm long, yellow-red. Common in higher elevations of the Black Hills . . . .
. . . . *Shepherdia canadensis* (L.) Nutt.

1' Leaves green above and below. Shrubs often exceeding 2 meters,
often thorny. Branches whitish, turning green or brown.
Leaves 2-4 cm long, oblong, opposite. Flowers brown, clustered at the nodes. Fruits ovoid-round, about 5 mm long, becoming scarlet red when mature in July or August. Frequent
to common at the edges of woods, in ravines and along streams
and lakes over the state. Apr-May. Buffalo Berry . . . . . .
. . . . *Shepherdia argentea* Nutt.

Family Lythraceae

(Key to Genera)

1 Plants annual or biennial. Petals 4, not conspicuous and
sometimes lacking.

  2 Fruiting capsule of 3-4 valves, dehiscing along the septa
  . . . . *ROTALA*

  2' Fruiting capsule rounded with several locules, splitting
  irregularly at maturity . . . . *AMMANIA*

1' Plants erect perennials. Petals 5 or more, conspicuous . . .
. . . . *LYTHRUM*

AMMANIA L.

Ammania coccinea Rottb. Plants annual herbs with simple or
branching stems up to 5 dm tall. Leaves strictly opposite,
clasping at their bases, linear-oblong, from 2 to 4 cm long.
Flowers axillary, sessile or in short cymes, each with a subglobose hypanthium. Petals, when present, rose to purple, short.
Fruit a membranous capsule about 5 mm across. Frequent in low,
marshy areas and dessicating pond margins over the state. July-
Aug. Tooth Cup.

LYTHRUM L.

1 Plants mostly more than 7 dm tall, coarse, perennial from a
  rhizome. Principal leaves 5-8 cm long, lanceolate, with
  rounded bases. Flowers crowded along the upper branches,
  spike-like, 1-5 dm long. Petals red-purple, up to 1 cm long.
  Stamens 10-14, twice as many as the petals. Rarely escaped
  from plantings and becoming successfully established in moist,
  low areas in the eastern part. July-Aug. Purple Loosestrife.
  . . . . *Lythrum salicaria* L.

1' Plants usually less than 7 dm tall, not as coarse as the preceding, perennial from rhizomes. Principal leaves 2-5 cm
  long, lance-ovate, their bases tapering to rounded. Flowers
  solitary or few in the axils of upper leaves. Petals purple,
  about 5 mm long. Stamens as many as the petals. Infrequent
  over the state but locally abundant in marshy floodplains
  along the Missouri River, especially the eastern part. {*L.
  alatum* Pursh} July-Aug . . . . *Lythrum dacotanum* Nieuw.

ROTALA L.

Rotala ramosior (L.) Koehne. Plants annual or biennial, the
stems 4-angled, 5-30 cm tall. Leaves linear to oblong, 1-3 cm
long, sessile and opposite on the stem. Flowers usually solitary, rarely more than one, in the axils of upper leaves. Perianth 4-parted, the petals 3-5 mm long. Fruiting capsule 2-3 mm
across, subglobose. Rare in swampy or alluvial places in the
southeast part. July-Sept.

Family Onagraceae

(Key to Genera)

1 Flowers with 2 petals and 2 sepals. Fruits oval with bristly
  hooked hairs . . . . *CIRCAEA*

1' Flowers with 4 petals and 4 sepals. Fruits several times
  longer than wide.

  2 Fruits indehiscent, nut like.

    3 Ovary with 3 or 4 locules. Staminal filaments with
      scale-like appendages near their bases . . *GAURA*

    3' Ovary with 1 locule. Staminal filaments lacking scale-
      like appendages . . . . *STENOSIPHON*

  2' Fruits a dehiscent capsule, elongate.

    4 Seeds with a dense tuft of hairs . . . . . *EPILOBIUM*

4' Seeds without hairs.

    5 Ovary two celled. Leaves less than 3 mm wide . .
    . . . . *GAYOPHYTUM*

    5' Ovary four celled. Leaves wider than 3 mm and
    petals exceeding 3 mm in length.

        6 Petals less than 5 mm long. Flowers small,
        sessile in axils of leaves. Plants annual . .
        . . . . *BOISDUVALIA*

        6' Petals exceeding 5 mm. Flowers pedicellate in
        upper axils of leaves. Plants annual to peren-
        nial.

            7 Stigmas of flowers divided into 4 linear
            lobes . . . . *OENOTHERA*

            7' Stigmas of flowers enlarged at the tip.
            They may be toothed but not divided into
            linear lobes . . . . *CALYLOPHUS*

BOISDUVALIA Spach.

Boisduvalia glabella (Nutt.) Walpers. Plants small, annual,
branched from the base and spreading, the stems 1-3 dm long.
Leaves 1-1.5 cm long, alternate, lanceolate, sessile. Flowers
in the axils of upper leaves, the petals 2-4 mm long, pink,
deeply bi-lobed. Fruiting capsules less than 1 cm long, slightly
curved and tapered from the base. Infrequent in open alluvial
areas of the western part. June-July.

CALYLOPHUS Spach.

1 Plants cespitose, with stems shorter than 2 dm, or plants
acaulescent. Plants perennial from a woody caudex. Stems
and foliage gray with strigose hairs. Leaves linear, 1-3 cm
long. Flowers solitary, borne in the upper leaf axils.
Petals 1-2 cm long, the hypanthium tube 3-4 cm long. Rare
on limestone outcrops of western Custer and Fall River counties in the extreme western part. {*Oenothera lavandulaefolia*
T. & G.} Ours ssp. *lavandulifolius* (T. & G.) Towner and Raven.
May-June . . . . *Calylophus hartwegii* (Benth.) Raven

1' Plants 1-several from a perennial base, the stems 1-6 dm tall.
Leaves linear-oblong, 2-4 cm long, with serrated margins.
Flowers in upper axillary portions of the stem, the petals
8-12 mm long, becoming reddish on the inside. Hypanthium
tube 1-1.5 cm long. Fruiting capsules 1.5-2.0 cm long,

cylindrical. Very common in prairies and on open plains over
the state. June-July . . *Calylophus serrulatus* (Nutt.) Raven

CIRCAEA L.

1   Principal leaves ovate-oblong, their blades at least 6 cm
    long, twice as long as wide. Fruits 2-celled and 2-seeded.
    Plants 3-10 dm tall, mostly erect, only sparingly branched,
    with opposite leaves. Inflorescence a many-flowered raceme,
    the petals 2.0-3.5 mm long. Fruits rounded in outline, with
    hooked bristles. Infrequent in moist upland and alluvial
    woods of the Black Hills and in the extreme northeast part.
    {*C. quadrisulcata* (Maxim.) Franch. and Sav.} Ours ssp. *cana-
    densis* (L.) Asch. & Magnus. July-Aug. Enchanter's Night-
    shade . . . . *Circaea lutetiana* L.

1' Principal leaves ovate with cordate bases, their blades 2-5
    cm long, not more than twice as long as wide. Fruits 1-celled
    and mostly 1-seeded. Plants smaller, usually not over 3 dm
    tall, their stems weak, from a tuberous root. Inflorescence
    a short raceme, the petals about 3 mm long. Fruits ovoid, 2.0
    mm long, with hooked bristles. Rare in marshy or boggy places
    in woods at higher altitudes in the Black Hills. July-Aug . .
    . . . . *Circaea alpina* L.

EPILOBIUM L.

1   Plants annual, with a taproot. Stems freely branched, 4-8 dm
    tall. Leaves alternate, narrowly lanceolate to linear, 3-6 cm
    long. Flowers in terminal racemes. Petals light pink to rose-
    colored, 5-8 mm long. Fruiting capsule 2-3 cm long. Seeds
    with yellow-brown hairs. Frequent in sandy soil of woods,
    especially in the western part. July-Aug . . . .
    . . . . *Epilobium paniculatum* Nutt.

1' Plants perennial, not taprooted. Stems simple to variously
    branched.

    2   Petals 1 cm long or longer, purple. Stems simple, 1 meter
        tall or taller, from a spreading rhizome system. Leaves
        lanceolate, glabrous, 6-15 cm long. Inflorescence a con-
        spicuous terminal raceme with many flowers, the petals 1-2
        cm long. Fruiting capsules 4-7 cm long, their seeds with a
        tawny brown to off-white tuft of hairs. Frequent in dis-
        turbed areas in the Black Hills. July-Sept. Fireweed . .
        . . . . *Epilobium angustifolium* L.

    2' Petals less than 1 cm long, white or pink. Stems usually
        much less than 1 meter tall.

3 Leaves entire, their margins rolled under, revolute. Stems rounded, without decurrent lines running down from the bases of the leaves.

    4 Upper surfaces of leaves glabrous, the principal ones 4-8 mm wide. Before or at the beginning of flowering, the tips of the branches nodding. Plants perennial with thin stolons, the stems erect, 1-4 dm tall. Leaves lanceolate, opposite. Flowers few or solitary, the petals white to pink, 4-5 mm long. Fruiting capsule 3-7 cm long, the seeds with tawny-brown hairs. Rare in moist soil of the western part. July-Aug . . . . *Epilobium palustre* L.

    4' Upper surfaces of leaves evenly strigose, the principal ones less than 3 mm wide. Tips of branches not nodding before flowering. Plants perennial, from long, underground stolons, the stems simple to branched, 3-5 dm tall. Flowers many in terminal racemes, the petals 4-6 mm long, white to pink. Capsule slender, 3-5 cm long, the seeds with light, tawny hairs. Frequent in thickets and moist places over the state. July-Aug . . . . *Epilobium leptophyllum* Raf.

3' Leaves toothed and flat. Stems angled, with decurrent lines running down from the bases of leaves.

    5 Principal cauline leaves with distinct, narrow petioles without wings. Coma on seeds brown. Plants perennial, the stems 5-10 dm tall, branched above. Leaves lanceolate, 5-15 cm long, with many distinct teeth. In bud the flowers with projecting sepal tips. Flowers pink, the petals 3-5 mm long. Capsules 3-4.5 cm long. Infrequent in moist places over the state. July-Aug . . . . *Epilobium coloratum* Bieh.

    5' Principal cauline leaves nearly sessile or with winged petioles. Coma on seeds white. Plants perennial, the stems erect, branched above, 3-8 dm tall. Leaves opposite, the principal ones 3-6 cm long. Flowers in dense racemes, the petals white to pink, 4-8 mm long. Fruiting capsules 5-9 cm long. A common and variable species of moist or marshy places over the state. July-Aug. Willow Herb . . . . *Epilobium adenocaulon* Haussk.

GAURA L.

1 Plants annual with stems to 1.5 meters tall. Principal leaves
   6-12 cm long, lanceolate to lance-ovate, the margins undulate
   or remotely toothed. Inflorescence of terminal spikes at the
   ends of upper branches. Flowers numerous, the petals pink or
   white, 2-3 mm long. Fruits 5-8 mm long, nutlet-like. Frequent to common over the state in sandy, prairie areas that
   are disturbed. July-Aug . . . . *Gaura parviflora* Dougl.

1' Plants perennial with stems usually less than 5 dm tall.
   Principal leaves 2-4 cm long, narrowly lanceolate to oblanceolate with entire to shallowly toothed margins. Flowers many,
   in spikes 5-15 cm long. Petals rose-pink to occasionally
   white, 3-4 mm long. Fruits angled, 4-8 mm long, with persisting, subtended bracts. Common in upland prairie and other dry
   areas over the state. June-July . . . .
   . . . . *Gaura coccinea* (Nutt.) Pursh

GAYOPHYTUM A. Juss.

Gayophytum diffusum T. & G. Plants annual, diffusely branched,
the slender stems 2-5 dm tall. Leaves linear, less than 3 mm
wide, from 1-4 cm long. Flowers minute, the petals 1-3 mm long,
white to light pink. Fruiting capsules 5-8 mm long, the seeds
about 1 mm long. Occasional in sandy or rocky soil of disturbed
areas in the western part. Ours ssp. *parviflorum* Lewis & Szwey.
June-July. Baby's Breath.

OENOTHERA L.

1 Plants acaulescent, perennial, cespitose from a thick taproot.

   2 Petals pale yellow, 1-2 cm long. Fruiting capsules ovoid
      with winged angles. Leaves oblong to oblanceolate in out-
      line, the margins shallow to deeply pinnatifid. Petioles
      arising from the crown, shorter than the blades. Fruiting
      capsules 1-2 cm long. Apparently rare in dessicating flat
      areas in the northwest part. May-June . . . .
      . . . . *Oenothera flava* (Nels.) Garrett

   2' Petals white, aging pink, 2-4 cm long. Fruiting capsules
      cylindric to lance-shaped, without winged angles. Leaves
      and flowering pedicels from the soil level. Leaf blades
      4-10 cm long, the margins sinuately toothed to pinnatifid,
      the teeth triangular. Flowers showy white, mostly closed
      during the bright daylight. Fruiting capsules 3 cm long,
      ovoid. Frequent to common in dry prairies of the western
      two thirds of the state. Ours represented by the two

subspecies *caespitosa* and *montana* (Nutt.) Munz. May-July
. . . . *Oenothera caespitosa* Nutt.

1' Plants with well developed stems above ground, the stems annual to perennial.

   3 Flowers with yellow petals, becoming faded or occasionally red-orange with age.

      4 Ovaries and fruits tapering upward. Seeds in maturing ovaries horizontally oriented and with sharp angles.

         5 Leaves thick, appearing gray due to dense strigose or hirsute hairs. Gland-tipped hairs not evident in the inflorescence. Plants mostly biennial, up to 2 meters tall. Leaves spatulate to lanceolate, strigose to hirsute. Inflorescence leafy, the flowers yellow. Petals 8-15 mm long, the fruiting capsules 2-4 cm long. Frequent to common over the state in waste places. Ours represented by the subspecies *strigosa* (western part) and *canovirens* (Steel) Munz (widespread over state). June-Sept . . . .
. . . . . . *Oenothera strigosa* (Rydb.) Mack. & Bush.

         5' Leaves thinner, appearing green, not gray-hairy. Gland-tipped hairs evident in the inflorescence. Plants biennial from a taproot, mostly simple, up to 2 meters tall. Leaves lanceolate, 3-12 cm long, the margins entire to wavy. Flowers yellow, the petals 1-2.5 cm long. Fruiting capsules 1.5-4.0 cm long. Frequent to common in dry, disturbed soil over the state. Ours ssp. *centralis* Munz. June-Sept . . . . *Oenothera biennis* L.

      4' Ovaries and fruits cylindrical. Seeds in maturing ovaries vertical, not sharply angled.

         6 Flowers arranged in a terminal spike, the upper leaves reduced. Petals rhombic-ovate to obovate, 1.5-2.0 cm long. Plants biennial, the stems erect, 3-8 dm tall, the vegetation silky strigose. Principal leaves sinuate-pinnatifid to subentire, becoming smaller upward. Inflorescence a dense, terminal spike, the flowers yellow. Rare in sandy prairie in the southern part. June-July . . . .
. . . . *Oenothera rhombipetala* Nutt.

         6' Flowers in the upper leaf axils, the upper leaves not much reduced. Petals obovate, 5-18 mm long. Plants annual to biennial, 1-6 dm tall, several stems common.

Vegetation variously strigose to subglabrous. Leaves
dentate to sinuately pinnatifid or lacinate, 2-5 cm
long. Flowers yellow, turning pink as they wither.
Fruiting capsules 2-3 cm long. Infrequent in sandy
prairies in the western two-thirds. May-June . . . .
. . . . *Oenothera laciniata* Hill.

3' Flowers with white petals, becoming rose-colored with age.

7 Fruiting capsules oblong and gently tapering at each
end, fusiform. Flower buds nodding prior to opening.
Stem leaves usually deeply pinnatifid.

8 Plants annual or winter annuals. Petals 1.5-4.0 cm
long. Fruiting capsules over 2 cm long. Stems 3-5
dm tall, branched from the base. Leaves at the soil
level spatulate, forming a rosette. Stem leaves
pectinate-pinnatifid at their margins, with 4-8 narrow divisions. Flowers solitary in the upper axils,
showy. Frequent to common in dry plains and prairies
in the western two-thirds. June-July . . . .
. . . . *Oenothera albicaulis* Pursh.

8' Plants perennial from slender horizontal rootstocks.
Petals 7-12 mm long. Fruiting capsules less than 2
cm long. Stems erect, 1-few in a cluster, 1-2.5 dm
tall, the flowers in the upper leaf axils. Stem
leaves oblong to lanceolate, with deep, pinnatifid
divisions. Infrequent in dry places of plains and
prairies of the southwestern part, including the
southern Black Hills. June-July . . . .
. . . . *Oenothera coronopifolia* T. & G.

7' Fruiting capsules cylindric, slightly tapering to the
summit. Principal leaves entire to shallowly dentate,
not deeply pinnatifid.

9 Vegetation at upper part of the plant glabrous except
for glandular pubescence in the inflorescence. Leaves
glabrous. Plants perennial from creeping rootstocks,
the stems erect, 4-10 dm tall. Epidermis of stems
white, and exfoliating. Principal leaves 4-9 cm long,
only remotely toothed. Flowers with a disagreeable
odor, the petals 2-3 cm long, becoming pink with age.
Capsules 2-3 cm long, erect. Frequent to common in
dry, sandy prairie over the state. June-July . . . .
. . . . *Oenothera nuttallii* Sweet.

9' Vegetation in the upper part of the plant strigose-
canescent. Leaves cinerous. Plants perennial from

creeping rootstocks, stems 1-several, 1-5 dm tall.
Principal stem leaves 1-4 cm long, entire to shallowly dentate. Epidermis of stem exfoliating. Flowers
with a fragrant odor, the petals 1.5-2.5 cm long.
Capsules spreading, 1.5-4 cm long. Apparently rare
but should occur in sandy soil in the southern and
western part. Ours represented by the ssp. *latifolia*
(Rydb.) Munz. June-July . . *Oenothera pallida* Lindl.

STENOSIPHON Spach.

Stenosiphon linifolius (Nutt.) Heynh. Plants perennial, herbaceous from a woody base, the stems 0.5-2 meters tall. Plants
with few to several branches. Vegetation glabrous to glaucous.
Leaves lance-linear to lanceolate, mostly entire, 2-5 cm long,
sessile or nearly so. Inflorescence of dense, narrow spikes,
the spikes becoming 2-3 dm long in fruit. Flowers white, the
petals rhombic, 4-5 mm long. Fruits 2-3 mm long, ovoid. Rare
in disturbed places in the Black Hills. Aug-Sept.

Family Haloragaceae

(Key to Genera)

1 Leaves narrow, simple. Flowers perfect, axillary . . . . . .
. . . . *HIPPURIS*

1' Leaves pinnately dissected, compound. Flowers monoecious or
polygamous . . . . *MYRIOPHYLLUM*

HIPPURIS L.

Hippuris vulgaris L. Perennial aquatic herbs having simple stems
from rhizomes rooted in mud. Stems often times projecting above
the water line. Leaves whorled, narrow, 4-10 or more at a node,
each 1-3 cm long. Flowers axillary, perfect, sepals and petals
lacking. Stamens 1, pistil solitary. Fruit about 2 mm long,
1-4 seeded. Occasionally brought by wildfowl and persisting in
ponds or lakes over the state. June-Aug. Mare's-Tail.

MYRIOPHYLLUM L.

1 Carpels smooth on the dorsal surface. Stamens 8.

2 Leaves subtending the flowers (usually emersed), entire or
slightly dentate, shorter than the flowers. Submerged
leaves verticillate in 4's or 5's, each leaf pinnatifid
into thread-like divisions. Flowers of staminate

inflorescences deep purple. Frequent in ponds or lakes over the state. Ours var. *exalbescens* (Fern.) Jeps. July-Aug. Water Milfoil . . . . . *Myriophyllum spicatum* L.

2' Leaves subtending the flowers pinnatifid, much longer than the flowers. Submerged leaves verticillate in 3's or 4's, the ultimate divisions thread-like. Flowers of staminate inflorescences pink or greenish. Infrequent in ponds and lakes of the northern part. July-Aug . . . . *Myriophyllum verticillatum* L.

1' Carpels roughened and 2-keeled on the dorsal surface. Stamens 4.

3 Floral leaves lanceolate or wider with serrated edges, in whorls of 3-6. Foliage leaves whorled in 4's or 5's, each with over 5 pairs of filiform segments. Staminate flowers with faint pink or white petals. Infrequent to rare in ponds in the north and western parts. July-Aug . . . . *Myriophyllum heterophyllum* Michx.

3' Floral leaves linear with pectinate or serrate divisions, in whorls of 3-6. Foliage leaves alternate or scattered, pectinate with 3 or more pairs of filiform segments. Staminate flowers with purple petals. Infrequent in ponds and lake shorelines over the state. July-Aug . . . . *Myriophyllum pinnatum* (Walt.) BSP

Family Araliaceae

(Key to Genera)

1 Leaves alternate or basal, 2 or more times compound. One or more umbels of flowers on each plant . . . . . . *ARALIA*

1' Leaves once palmately compound and having one umbel of flowers on each plant . . . . *PANAX*

ARALIA L.

1 Leaves arising directly from the rhizome. The petiole erect, the segments ternately compound. Leaflets 5-8 cm long, finely serrate. Inflorescence of usually 2 or more umbels, the peduncle naked, from the rhizome. Plants usually 5 dm or more tall. Frequent on shady and rocky banks and in rich upland woods in the Black Hills and in the eastern part. May-June. Wild Sarsaparilla . . . . *Aralia nudicaulis* L.

1' Leaves widely spreading, pinnately compound, the leaflets large, ovate, to 15 cm. Inflorescence of several umbels in

in a large panicle, the flowers greenish, 2 mm wide. Plants
stout, 1 meter or more tall. Fruiting berries rounded, 5 mm
across, red-purple. Rare to infrequent in rich woods of the
eastern part. July. Spikenard . . . . . *Aralia racemosa* L.

PANAX L.

Panax quinquefolium L. Plants perennial, herbaceous, with fusiform branch roots. Stems simple, 2-4 dm tall. Leaves in a single whorl, usually 5-foliolate, the leaflets ovate or broader,
2-10 cm long. Inflorescence a simple umbel that is terminal.
Flowers yellow or green, the petals 1-1.5 mm long. Fruiting
berries 8-11 mm across, becoming red. Rare in rich woods in
the eastern part. July-Aug. Ginseng.

Family Apiaceae

(Key to Genera)

1 Ovary and fruit possessing scales or bristles.

  2 Principal leaves simple. Inflorescence globose, dense,
     without rays . . . . *ERYNGIUM*

  2' Principal leaves compound. Inflorescence umbellate, with
     rays.

    3 Fruit several times longer than wide with a tapering
       apex. Lower portion of the fruit with straight, down-
       ward projecting bristles . . . . *OSMORIIZA*

    3' Fruit ovoid to almost globose, not more than 3 times
       longer than wide.

      4 Leaves much dissected, the ultimate segments less
         than 1 cm wide. Fruits ribbed, with the bristles
         on the ribs . . . . *DAUCUS*

      4' Leaves 3-7 parted, the leaflets exceeding 1 cm.
         Fruits subglobose, covered with hooked bristles
         . . . . *SANICULA*

1' Ovary and fruit glabrous to pubescent, lacking scales or
   bristles.

  5 Leaves pinnately divided, the principal ones having more
     than 3 ultimate divisions or leaflets.

    6 Plants of aquatic or marshy habitats.

      7 Leaves once pinnately compound.

8 Leaflets regularly serrate. Stem solitary, stout,
up to a meter or more tall . . . . . *SIUM*

8' Leaflets irregularly incised. Stem sparsely
branched, rarely over 5 dm tall . . *BERULA*

7' Leaves twice or more pinnately compound.

9 Ultimate leaf segments serrated, exceeding 1 cm
in length, linear to broadly lanceolate . . . . .
. . . . *CICUTA*

9' Ultimate leaf segments not serrate, usually not
exceeding 1 cm in length, filiform . . . .
. . . . *PTILIMNIUM*

6' Plants of drier habitats.

10 Above ground stems well developed, exceeding 2 dm
in height.

11 Leaflets, ultimate divisions of leaves, filiform, less than 2 mm wide, flowers yellow . . .
. . . . *ANETHUM*

11' Leaflets, ultimate divisions of leaves, linear
or broader, at least exceeding 2 mm in width.

12 Flowers yellow . . . . *PASTINACA*

12' Flowers white or pink.

13 Stems purple spotted, usually exceeding
1 meter . . . . *CONIUM*

13' Stems not purple spotted or exceeding 1
meter.

14 Leaves once-pinnate. Plants with
tuberous, fascicled roots . . . . .
. . . . *PERIDERIDIA*

14' Leaves twice or more pinnate.
Plants taprooted . . . *CARUM*

10' Above ground stems poorly developed, plants mostly
acaulescent. If present, stems less than 2 dm tall.

15 Fruits rounded in cross section or compressed
at right angles to the commissure, i.e., the
suture on the side. Ribs of the fruit thickened . . . . *MUSINEON*

15' Fruits flattened, compressed parallel to the
commissure, i.e., the sutures on the edge. Ribs
of the fruit more or less winged.

16 The median or dorsal rib of each mericarp
    winged . . . . *CYMOPTERUS*

16' The median or dorsal rib of each mericarp
    filiform, not winged . . . . . *LOMATIUM*

5' Leaves palmately or ternately compound, or trifoliolate.

17 Flowers white.

18 Leaves ternate with very large leaflets, usually
    over 1 dm wide. Plants stout, erect, up to a
    meter or more tall . . . . *HERACLEUM*

18' Leaves trifoliolate with leaflets ovate, not over
    8 cm wide. Plants smaller, less than 8 dm tall
    . . . . *CRYPTOTAENIA*

17' Flowers yellow.

19 Leaflets entire . . . . *TAENIDIA*

19' Leaflets serrate . . . . *ZIZIA*

ANETHUM L.

Anethum graveolens L. Plants annual, herbaceous, the stems glabrous and branched above, 3-10 dm tall. Leaves pinnately dissected into numerous filiform divisions. Inflorescence of compound umbels that are terminal and in the upper axils. Flowers
yellow, strongly scented. Fruits 3-5 mm long, the ribs narrowly
winged. Occasionally escaped from cultivation but usually not
persisting. July-Aug. Dill.

BERULA Hoffm.

Berula erecta (Huds.) Cov. Plants slender, perennial in aquatic
habitats. Stems pale green, glabrous, 2-8 dm tall. Roots spreading, fascicled. Leaves pinnately decompound, the ultimate segments irregularly lobed or incised. Flowers white, in compound
umbels, with 6-15 rays, each 1-3 cm long. Fruits 1-2 mm long,
usually not maturing. Infrequent in marshy soil of seepage
slopes or in shallow water of springs or bogs over the state.
Ours var. *incisum* (Torr.) Cronq. July-Aug. Water Parsnip.

CARUM L.

Carum carvi L. Biennial or perennial herbs with taproots, the
stems glabrous, branched, up to 1 meter tall. Leaves pinnately
dissected, the ultimate divisions narrow. Principal leaves 1 dm
or more long and 3-8 cm wide. Inflorescences terminal and in
upper axillary locations, the umbels compound, with 8-15 rays.

Flowers white or occasionally pink. Fruit oblong, 3-4 mm long,
prominently ribbed. Cultivated and rarely escaping in the Black
Hills and in the eastern part. June-July. Caraway.

CICUTA L.

1   Leaflets narrow, less than 5 mm wide. Upper leaf axils often
    with small bulbs. Plants slender, the stems sparsely branched,
    3-8 dm tall. Leaves dissected, the ultimate segments toothed,
    sometimes only remotely so. Flowers in small umbels usually
    less than 5 cm across, the petals white. Fruits, when matur-
    ing, rounded in outline, about 2 mm long. Infrequent in
    marshy or swampy soils of the northcentral and northeast part
    of the state. Poisonous to humans. July-Aug . . . .
    . . . . Cicuta bulbifera L.

1' Leaflets lanceolate, more than 5 mm wide. Upper leaf axils
    lacking small bulbs. Plants stout, the stems branched, up to
    a meter or more tall. Leaves twice or more pinnate, the ulti-
    mate segments 5-7 cm long, sharply serrate. Flowers in dense
    umbels that are terminal and in upper axils. Umbels 5-12 cm
    across, the petals white. Fruits ovoid, 2-5 mm long, ribbed.
    Frequent to common in moist or swampy places over the state.
    Poisonous to humans. July-Aug. Water Hemlock . . . .
    . . . . Cicuta maculata L.

CONIUM L.

Conium maculatum L. Plants erect, freely branched, biennial, up
to 1-2 meters tall. Stems more or less purple-spotted, glabrous.
Leaves broadly triangular to ovate in outline, 2-4 times compound,
the ultimate segments incised, 4-6 mm long. Flowers in many
umbels, each umbel 4-7 cm across. Petals white. Fruits ovoid,
2-3 mm long. Infrequent over the state in waste places that are
moist. All parts of the plant are poisonous. July-Aug. Poison
Hemlock.

CRYPTOTAENIA DC.

Cryptotaenia canadensis (L.) DC. Plants perennial, herbaceous,
with branching stems 3-7 dm tall. Leaves alternate, petioled,
trifoliolate, the leaflets lanceolate to ovate, 4-10 cm long.
Margins of leaflets serrate to doubly serrate. Flowers in irregular umbels, the flowers few. Petals white, very small.
Fruits dark-colored towards maturity, 4-6 mm long, curved. Infrequent in rich, loamy woods or alluvial woods in the eastern
part. June-July. Honewort.

CYMOPTERUS Raf.

1 Leaves fleshy and whitened-glaucescent, the ultimate segments
   pinnately lobed, not very distinct due to crowding of the
   tips. Flowers white or more commonly tinged with purple.
   Fruits with conspicuous wings not narrowed towards the base.
   Plants 5-15 cm tall, from a fleshy taproot. Leaves ovate in
   outline, 2-7 cm long. Fertile rays of the umbels 3-6, the
   pedicels 2-4 mm long. Fruits 6-10 mm long. Infrequent on
   dry soil in the western one half. May-June . . . .
   . . . . Cymopterus montanus Nutt.

1' Leaves not fleshy or glaucescent, the ultimate segments pinnately lobed but distinct, up to 2 mm wide. Flowers white,
   in dense umbels. Fertile rays 3-5, the pedicels usually not
   more than 1 mm long. Fruits with wings narrowed towards the
   base. Plants 5-20 cm tall, from a fleshy taproot. Frequent
   in dry soil of prairies and on clay banks in the western two
   thirds of the state. May-June . . . .
   . . . . Cymopterus acaulis (Pursh) Raf.

DAUCUS L.

Daucus carota L. Biennial herb from a stout taproot, the stems
simple or branched, 4-8 dm tall. Leaves oblong in outline, once
to several pinnately compound, the ultimate segments toothed.
Flowers in compound umbels 4-8 cm across. Petals white, the
central flower of the umbel often times purple or otherwise dark.
Umbels subtended by well-developed pinnatifid bracts. Fruits oblong, flattened dorsally, 3-4 mm long. Frequent in waste soil
and on roadsides as a weed. July-Aug. Wild Carrot.

ERYNGIUM L.

Eryngium planum L. Plants perennial, from a woody taproot, the
stems stout, mostly unbranched, 5-8 dm tall. Lower leaves petioled, the blades ovate to oblanceolate, 8-12 cm long, almost
equaling the petioles. Upper leaves sessile, palmately or irregularly divided. Inflorescence cymose-branched, the heads
blue flowered, 1-1.5 cm high. Bracts beneath the heads 1-3 cm
long, lacinately divided. Fruits 4-6 mm long, ovoid, scaly.
Rare in sandy soil at Lake Madison in Lake County, apparently
adventive and brought in by migratory water fowl. July-Aug.

HERACLEUM L.

Heracleum sphondylium L. Plants perennial, 1-2 meters tall, the
stems pubescent. Leaves ternately compound, the leaflets stalked,
1-3 dm broad, incised. Petioles winged on the margins at the

lower one-half. Inflorescence of compound umbels, terminal and
axillary, the principal ones 1-2 dm across. Flowers white, the
outer ones of the umbel often larger and with petals irregular.
Fruits obovate, dorsally flattened, about 1 cm long, pubescent.
Frequent in marshy or rich, damp soil over the state. Ours ssp.
montanum (Schl.) Briq. July-Aug. Cow Parsnip.

LOMATIUM Raf.

1 Fruit oblong to orbicular, less than 2 times longer than wide.

2   Ovary and young fruit pubescent, becoming sparsely pubescent towards maturity. Leaves with ultimate divisions
many and crowded, 1-3 mm long, giving a lacy appearance.
Flowers yellow, the bractlets villous. Flowering scapes
curved and ascending, later becoming mostly prostrate, 6-
20 cm long. Fruits 5-10 mm long, the wings almost equaling the body width. Frequent in prairies over the state,
more common westward. Ours var. *foeniculacum*. Apr-June.
. . . . *Lomatium foeniculaceum* (Nutt.) Coult. & Rose

2' Ovary and young fruit glabrous. Leaves tri-pinnate to
ternate, to 2 dm long, the ultimate divisions clearly
discernible, 2-8 mm long. Flowers white to pink, their
bractlets scarious margined, glabrous. Flowering scapes
erect to ascending, 4-20 cm long. Fruits 3-10 mm long,
glabrous, with narrow wings. Common in prairies and on
dry plains over the state. May-June . . . .
. . . . *Lomatium orientale* Coult. & Rose

1' Fruit narrowly oblong, more than 2 times longer than wide.

3   Plants strictly glabrous throughout. Plants without a
main stem above ground but the crown with several to many
leaf sheaths of previous years. Leaves once or twice pinnate, the ultimate segments linear, the blades ascending
to erect. Flowers yellow, on scapes 20 cm or more. Fruits
8-10 mm long. Infrequent to rare in dry soil of the extreme western part. June-July . . . .
. . . . *Lomatium nuttallii* (Gray) McBride

3' Plants villous to tomentose, the ovary sparingly pubescent
but becoming glabrous. Plants with a short main stem above
ground, from an elongate taproot. Leaves 3 to 4 times compound, 3-12 cm long. The ultimate segments of the leaves
2-6 mm long, not strictly linear. Flowers yellow or white
to pink, the umbels on short stalks. Bractlets leaf-like,
not scarious margined. Fruit 1-2 cm long, with wings
nearly as wide as the body. Rare in dry soil of the

extreme western part. May-June . . . .
. . . . Lomatium macrocarpa (Nutt.) Coult. & Rose

MUSINEON Raf.

1' Plants with an above ground main stem, the branches erect or
   spreading. Leaves ovate to oblong in outline, up to 1 dm
   long, bluish glaucous, with the ultimate divisions pinnately
   lobed. Flowers yellow, the umbels 1-3 cm across at flowering
   time. The umbels with 10-25 branches. Fruits 4 mm long,
   ovate. Frequent in dry soils of prairie and eroded areas in
   the western two thirds of the state. May-June . . . .
   . . . . Musineon divaricatum (Pursh) Nutt.

1' Plants lacking a main stem above ground, the perennial caudex
   branched. Leaves numerous, once or twice pinnate, narrowly
   oblong in outline, the segments linear to filiform, less than
   1 mm wide. Flowers off-white to yellow, in dense umbels.
   Fruits oblong to ovoid, 3-6 mm long. Frequent to commonly
   widespread in prairies and rocky outcrops in woods in the
   western part. May-June . . . . . . Musineon tenuifolium Nutt.

OSMORHIZA Raf.

1 Mature styles curved outward but not more than 1 mm long.
   Involucels (small bracts immediately subtending the umbel-
   lets) lacking.

   2 Fruit obtuse or convexly curved to the summit below the
      style. Rays and fruiting pedicels widely spreading or
      reflexed. Plants perennial, 3-6 dm tall. Leaves 2 or 3
      times ternate, glabrous to pilose. Fruits about 1.5 cm
      long, the styles very short, curved. Rare in rich woods
      in the Black Hills. June-Aug . . . .
      . . . . Osmorhiza depauperata Nutt.

   2' Fruit linear, dished in or concavely curved to the summit
      below the style. Rays and fruiting pedicels spreading to
      ascending. Plants perennial from a taproot, the stems
      solitary or few, 3 dm or more tall. Leaves twice ternate,
      thin, with coarse teeth. Flowers in several umbels of 3-8
      rays. Fruits 12-18 mm long. Infrequent to rare in woods
      of the Black Hills. June-July . . . .
      . . . . Osmorhiza chilensis H. & A.

1' Mature styles straight or nearly so, 1-4 mm long. Involucels
   present, of several bractlets.

   3 Stems and leaves softly villous. Styles in fruit less than
      2 mm long. Bases of stems and roots not anise-scented.

Plants 3-8 dm tall, branched, the vegetation densely hairy.
Leaves ternately compound, the leaflets 4-8 cm long, coarsely toothed. Flowers in irregular umbels. Fruits 18-20 mm
long, with a slender, tapering base. Common in woods and
on shaded hillsides over the state. May-June . . . .
. . . . *Osmorhiza claytonii* (Michx.) Clarke

3' Stems and leaves glabrous or sparsely hirsute. Styles in
fruit 2-4 mm long. Bases of stems and roots anise-scented.
Plants 3-8 dm tall, stout, sparingly branched above.
Leaves 2-3 times ternately compound, the leaflets 2-7 cm
long, coarsely toothed. Flowers in compound umbels, not
dense, the petals white. Fruits 2-3 cm long, slender at
the base. Frequent to common in rich loamy and alluvial
woods over the state, more common eastward. Represented
in South Dakota by the two varieties *longistylis* (common)
and *villicaulis* Fern. (rare in the southeast). May-June
. . . . *Osmorhiza longistylis* (Torr.) DC.

PASTINACA L.

Pastinaca sativa L. Plants biennial, the stems 1-several from a
taproot, 6-12 dm tall. Leaves pinnately compound, up to 4 dm
long, the leaflets ovate with serrated or cleft margins, sessile
on the main rachis. Inflorescence of compound umbels, terminal
and axillary, each up to 10 cm across. Flowers yellow, the
fruits broadly elliptical, 5-6 mm long. Frequent to common on
roadsides and near abandoned homesteads over the state. June-
July. Parsnip.

PERIDERIDIA Reichb.

Perideridia gairdneri (H. & A.) Mathias. Plants perennial from
an enlarged taproot, the stems usually solitary, erect, 3-8 cm
tall. Leaves once-pinnate with long and narrow ultimate segments. Flowers in compound umbels, each 2-4 cm across. Rays
of each umbel up to 2-5 cm long. Fruits almost rounded, 2-3 mm
long and almost as wide. Rare in moist meadows in the western
part. Aug-Sept. Squaw-Root.

PTILIMNIUM Raf.

Ptilimnium capillaceum (Michx.) Raf. Plants annual, herbaceous,
with slender branched stems 3-5 dm tall. Leaves pinnately dissected, 4-10 cm long, the ultimate divisions filiform, usually
not exceeding 1 cm in length. Flowers white, in compound umbels
2-5 cm across. Fruits broadly ovoid, 2-3 mm long, the lateral
ribs forming a margin around the fruit. Rare in swampy places

and slow moving streams in the extreme southern part of the
state. July-Aug. Mock Bishop Weed.

SANICULA L.

1 Styles in fruit longer than the bristles. Plants perennial.

    2 Sepals linear with pointed tips. Petals green to white.
       Fruits up to 6 mm or more long. Plants perennial, the
       stems 4-8 dm tall, branched. Leaves 5-7 palmately divided,
       the segments oblanceolate, serrate. Lower leaves long-
       petioled, the upper ones almost sessile. Flowers in ir-
       regular compound umbels, perfect or staminate. Fruits ses-
       sile, about 6 mm long. Frequent in rich loamy woods in the
       eastern part and in the Black Hills. July-Aug . . . . . .
       . . . . *Sanicula marilandica* L.

    2' Sepals ovate, their tips acute. Petals yellowish. Fruits
       about 3 mm long. Plants perennial, the stems slender, 3-7
       dm tall, branched. Leaves 3-5 parted, the leaflets 5-7 cm
       long, oblanceolate. Margins of the leaflets serrate to ir-
       regularly incised. Lower leaves long petioled, the upper
       ones progressively shorter petioled. Flowers in small
       umbels, the fertile ones in pedicels to 1 mm long. Fruits
       broadly obovoid, on short stipes, the body about 3 mm long.
       Infrequent in moist to dry woods in the eastern part. June-
       July . . . . *Sanicula gregaria* Bickn.

1' Styles in fruit shorter than the bristles. Plants biennial,
   fibrous rooted, the stems branched, 3-9 dm tall. Leaves
   palmately 3-parted, the basal ones on long petioles. Leaflets broadly lanceolate in outline, their margins irregularly
   serrate with some deeper incisions. Flowers in small umbels,
   most of them fertile. Imperfect staminate flowers few.
   Fruits almost globose, 3-6 mm long, their styles short. Infrequent in woods in the south and eastern parts. July-Aug.
   . . . . *Sanicula canadensis* L.

SIUM L.

Sium suave Walt. Plants perennial, erect, the stem usually solitary, 4-9 dm tall, from a crowned taproot. Leaves pinnately compound, the petioles with winged margins. Leaflets 7-13, sessile,
narrowly lanceolate, their margins serrate. Flowers in compound
umbels, these 3-12 cm across. Petals white. Fruits ovate, 2-3
mm long and almost as wide. Frequent in swampy places over the
state. July-Aug. Water Parsnip.

TAENIDIA Drude.

Taenidia integerrima (L.) Drude. Plants perennial, herbaceous, the stems 4-7 dm tall. Stems glabrous or glaucous. Plants with fibrous roots. Leaves ternately compound, the leaflets mostly entire, ovate to oblong, 2-4 cm long. Flowers in loose umbels, the flowers white. Fruits elliptic or oblong, somewhat flat-tened, 3-4 mm long. Rare on open rocky wooded hillsides in the eastern part. May-June.

ZIZIA Koch.

1 Principal basal leaves simple, cordate at the base, 2-6 cm long, upper leaves simple to ternate. Plants perennial, the stems 3-6 dm tall. Flowers yellow, in compact compound umbels. Fruit ovate to oblong, 3-4 mm long. Frequent to common in meadows and moist swales over the state, more common eastward. June-July . . . . *Zizia aptera* (Gray) Fern.

1' Principal basal leaves ternately divided, the leaflets ovate to lanceolate. Leaflets varying much in length, from 2-8 cm, their margins sharply serrate. Flowers bright yellow, in dense compound umbels, the rays of each umbel 1-3 cm long. Fruits ellipsoid, 3-4 mm long. Common on roadsides in meadows and open woods in the eastern part and in the Black Hills. May-June. Golden Alexander . . . . . . *Zizia aurea* (L.) Koch.

Family Cornaceae

CORNUS L.

1 Plants small and herbaceous, less than 3 dm tall. Flowers in a dense head-like cluster subtended by 4 large, whitish peta-loid bracts. Fruits red. Leaves ovate with pinnate venation, petioled, 3-7 cm long. Frequent in rich woods of the western part, especially the Black Hills. May-July. Bunchberry . . .
. . . . *Cornus canadensis* L.

1' Plants small trees or shrubs. Flowers in cymes or paniculate. Bracts none or minute. Fruits blue or white.

  2 Pedicels or young twigs reddish, slightly hairy to glabrous. Leaves with short appressed hairs beneath. Fruit white.

    3 One and two year old twigs red; pith white to lead-colored. Leaves broadly ovate with 5-7 pairs of veins. Branches loosely spreading, stoloniferous. Our most common species forming dense thickets along roadsides

and in low places over the state. June-July. Red
Osier . . . . *Cornus stolonifera* Michx.

3' One and two year old twigs brown or gray; pedicels, especially those in fruit, reddish. Pith brown. Shrubs
2-3 meters tall. Leaves ovate to narrowly elliptic with
long-acuminate tips. Moist to dry open areas and in
open woods of flood plains, especially in the eastern
part. Ours ssp. *racemosa* (Lam.) J. S. Wils. May-June.
Gray Dogwood · · · · · · · · · · *Cornus foemina* Mill.

2' Pedicels and young twigs gray to brown, densely tomentose.
Fruit white or blue towards maturity.

4 Lower surface of leaf blade with loose, spreading pubes-
cence. Leaves ovate to elliptic, scabrous above.
Shrubs or small trees to 5 meters tall. Fruit white.
Frequent in moist alluvial areas in the eastern part.
May-June. Rough-leaved Dogwood . . . .
. . . . *Cornus drummondii* May.

4' Lower surface of leaf with pale appearance due to short
appressed hairs. Leaves with broadly rounded bases and
acuminate tips, nearly glabrous above. Upright shrub
up to 3 meters tall. Fruit blue or bluish-white. Infrequent in open moist soil or in valleys in the eastern
part. Ours ssp. *obliqua* (Raf.) J. S. Wils. June-July.
Pale Dogwood · · · · · · · · · · *Cornus amomum* Mill.

Family Ericaceae

(Key to Genera)

1 Plants with chlorophyll. Leaves not scale-like.

2 Ovary inferior. Shrubs up to 2 dm or more tall. Leaves
deciduous . . . . *VACCINIUM*

2' Ovary superior. Low shrubs or herbs less than 2 dm tall.
Leaves evergreen.

3 Petals united. Fruit fleshy. Prostrate shrubs with
bright, glabrous, green leaves . . . . . . *ARCTOSTAPHYLOS*

3' Petals distinct. Fruit not fleshy.

4 Leaves not restricted to basal part of stem. Flowers
corymbose . . . . *CHIMAPHILA*

4' Leaves basal on the stem. Inflorescence scapose, the
flowers in racemes or single . . . . . *PYROLA*

1' Plants lacking chlorophyll. Leaves scale-like. Petals united,
   the flowers in racemes . . . . *PTEROSPORA*

ARCTOSTAPHYLOS Adans.

Arctostaphylos uva-ursi (L.) Spreng. Plants shrubby, freely
branched and prostrate, the stems spreading to 2 meters square
or more. Leaves bright green, obtusely rounded, entire, 1-3 cm
long. Flowers white to rosy-tinged pink, the petals united to
form an urn about 5 mm long. Fruit a red berry up to 1 cm with
up to 5 nutlets. Common at higher elevations of the Black Hills
and Harding County. May-June. Bearberry.

CHIMAPHILA Pursh.

Chimaphila umbellata (L.) Bart. Plants perennial, shrub-like
herbs with stems up to 3 dm tall. Leaves oblanceolate, up to 6
cm long, sharply serrate, petioled. Flowers in small corymbs,
the petals distinct, 5-6 mm long, rose-pink. Fruit a capsule
about 5 mm in diameter. Rare in rich pine woods at higher elevations in the Black Hills. July. Prince's Pine, Pipsissewa.

PTEROSPORA Nutt.

Pterospora andromedea Nutt. Plants 3-8 dm tall with unbranched
stems from a densely matted root system. Plants are saprophytic
on pine humus and lack chlorophyll. Leaves scale-like, fleshy,
yellowish to red-brown. Flowers in elongate racemes, the pedicels recurved. Corolla urn-shaped, up to 1 cm long. Fruit
globose, a capsule 8-10 mm across. Old stems persist as dried
fibrous stalks for several years. Frequent in pine woods of the
Black Hills and Harding County. July-Aug. Pine-drops.

PYROLA L.

1 Flower single, recurved at the top of a few-bracted scape less
   than 10 cm tall. Leaf blades elliptic to obovate, 1-2 cm long,
   the petioles one-half to almost as long as the blade. Flower
   1.2-2.5 cm across, white, the style 2-3 mm long, straight.
   Fruit a capsule which is almost spherical, about 6 mm in diameter. Rare in deep canyons of rich woods in Lawrence and
   Pennington counties in the Black Hills. June-July. One-
   flowered Wintergreen . . . . *Pyrola uniflora* L.

1' Flowers more than one in racemes.

   2 Styles of flowers straight, the raceme one-sided (secund).
      Leaves separated on the stem by distinct internodes.
      Plants spreading by rhizomes, the stems 5-10 cm tall.

Leaves ovate to elliptic, 2-4 cm long. Flowers 5-6 mm
across, the petals white, in racemes of 10 flowers or less.
Capsule about 4 mm across. Infrequent in deep canyons of
Lawrence and Pennington counties in the Black Hills. June-
July . . . . *Pyrola secunda* L.

2' Styles curved downward or to one side. The racemes not 1-
sided. Leaves more or less crowded at the base of the
plant.

3 Upper surfaces of leaves with white or pale areas along
the veins. Leaves ovate to elliptic, 3-6 cm long, coriaceous, deep green except for the pale streaks or mottled areas. Flowers yellow to greenish-white, turning
purple, on scapes 10-15 cm tall, the petals 6-9 mm long.
Plants spreading by rhizomes. Rare in Deadwood Gulch in
Lawrence County. July-Aug . . . . . . *Pyrola picta* Sm.

3' Upper leaf surfaces not streaked or mottled.

4 Petals rose-pink to purple. Flowering stems 2-3 dm
tall. Leaves suborbicular to round with a slight
cordate base, 2.5-4 cm across, dark green above.
Flowers 10-15 mm across with petals 5 mm or longer.
Frequent in moist, shady woods of the Black Hills
and Harding County. July-Aug . . . .
. . . . *Pyrola asarifolia* Michx.

4' Petals pale yellow to greenish white.

5 Petals white, generally lacking greenish veins.
Leaves rotund or broadly elliptic, 2-2.5 cm wide,
the blade slightly decurrent on the petiole.
Flowering stems 1.5-2.5 dm tall, the racemes
about 10-flowered. Flowers with white petals 8-
10 mm long. Rare in shady moist woods of Custer
and Pennington counties in the Black Hills. Ours
var. *americana* (Sweet) Fern. July-Aug . . . . . .
. . . . *Pyrola rotundifolia* L.

5' Petals white with green tinged veins. Sepals as
wide as long or wider.

6 Principal leaves less than 3 cm long, excluding
the petiole. Flowering stems with less than 10
flowers, 1-2.5 dm tall. Leaf blades sub-rotund
to rounded. Flowers with petals 3-6 mm long.
Rare in woods of Lawrence and Pennington counties in the Black Hills. June-July . . . . . .
. . . . *Pyrola virens* Schweig.

6' Principal leaves exceeding 3 cm long, excluding
the petiole. Flowering stems usually with more
than 10 flowers; the stems 1.5-3 dm tall. Leaf
blades broadly elliptic, decurrently narrowing
to the petiole. Flowers with petals 5-7 mm
long. Infrequent in upland woods of the Black
Hills and in Harding County. July . . . . . .
. . . . *Pyrola elliptica* Nutt.

VACCINIUM L.

1 Leaves 2 cm or more long. Berries blue-purple, 7-9 mm long.
Plants usually 5 dm or more in height, the branches spreading.
Flowers 5-7 mm long, rose-yellow, solitary on short pedicels
in the axils. Very rare in rich woods of Deadwood Gulch near
Lead in Lawrence County. May-June. Mountain Huckleberry.
. . . . *Vaccinium membranaceum* Dougl.

1' Leaves 7-15 mm long. Berries bright red, 3-5 mm across.
Plants low with slender branches, usually not exceeding 2 dm
in height. Flowers about 4 mm long, the petals fused, pinkish, borne on short pedicels. Fruits sweet, rounded. Infrequent in woods at higher altitudes in Lawrence and Pennington
counties in the Black Hills. Apr-June. Grouseberry . . . . .
. . . . *Vaccinium scoparium* Leiberg

Family Primulaceae

(Key to Genera)

1 Plants leafy stemmed. Flowers axillary or in racemes . . . .
. . . . *LYSIMACHIA*

1' Plants without leafy stems. The leaves basal. Flowers on
scapes.

2 Corolla lobes sharply reflexed, showy. Leaves oblanceolate, 5-15 cm long . . . . *DODECATHEON*

2' Corolla lobes erect, small, not showy. Leaves obovate,
small, usually less than 1 cm long . . . . . *ANDROSACE*

ANDROSACE L.

1 Bracts below the flower cluster lance-ovate, 2-6 mm long.
Plants commonly less than 5 cm tall. Basal leaves sparingly
toothed or entire, usually less than 1 cm long, hairy. Flowers in an umbellate inflorescence, the corolla 2-3 mm wide,

white. Frequent on sandy prairies and sandy, disturbed areas
over the state. Apr-May. Rock Jasmine . . . .
. . . . *Androsace occidentalis* Pursh

1' Bracts below the flower cluster lance-subulate, about 2 mm
long. Plants up to 10 cm tall. Basal leaves 2 cm or more
long, with marginal dentations, glabrous to glandular-hairy.
Flowers umbellate, the corolla white to pink, 4 mm long.
Fruit globose, maturing within the persistent calyx. Rare
to infrequent on rocky soil of ravines in Lawrence County of
the Black Hills. June-July . . . *Androsace septentrionalis* L.

DODECATHEON L.

1 Principal leaves usually over 6 cm long, sinuate-margined.
Anthers about 6 mm long, over 3 times longer than the filament. Plants perennial, the scapes 2-4 dm tall, tending to
be larger in all respects than the following species. Inflorescence umbellate or flowers solitary, the petals purple to
rose, the lobes sharply reflexed. Frequent in wet meadows in
the Black Hills and Harding County. May-July. Shooting Star.
. . . . *Dodecatheon pauciflorum* (Durand) Greene

1' Principal leaves usually less than 6 cm long, with entire margins. Anthers about 5 mm long and less than 3 times the length
of the filament. Plants perennial, the scapes 1-4 dm tall and
averaging smaller than the preceding species. Inflorescence
of several flowers in an umbel, or the flowers solitary, the
petals rose-colored. Petals 10-18 mm long, the lobes sharply
reflexed. Frequent in meadows in the Black Hills. Similar to
the preceding and possibly intergrading with it. May-July . .
. . . . *Dodecatheon pulchellum* (Raf.) Merr.

LYSIMACHIA L.

1 Leaves with punctate dots. Flowers in short, axillary racemes
1-3 cm long. Staminodes absent. Plants perennial, the stems
mostly simple, 3-6 dm tall, from creeping rhizomes. Leaves
opposite, entire, the blades lanceolate, 4-8 cm long. Flowers
on peduncles 2-5 cm long, the petals 5-7, yellow. Infrequent
in marshy or swampy areas or on lakeshores in the northeast
part and in the Black Hills. June-July. Tufted Loosestrife.
. . . . *Lysimachia thrysiflora* L.

1' Leaves without punctate dots. Flowers solitary or few in the
axils of leaves. Staminodes alternating with the stamens.

2 Principal leaves ovate or at least lance-ovate, the petioles ciliate on each side. Flowers 2-3 cm across. Plants

erect, branched, 2-8 dm tall, from a rhizomatous base.
Leaves 4-12 cm long, mostly opposite. Flowers few on
axillary peduncles. Fruiting capsules maturing within
the persistent sepals. Frequent to common in alluvial
thickets and moist wooded areas over the state. June-
July. Fringed Loosestrife . . . . . *Lysimachia ciliata* L.

2' Principal leaves lanceolate to linear-lanceolate, the petioles with few hairs or none on their sides. Flowers 1.5-
2.5 cm across. Plants erect, the stems stout, little
branched, from short rhizomes. Leaves 3-10 cm long, opposite. Flowers on axillary peduncles, smaller than the preceding species. Infrequent in moist thickets and in rich
wooded ravines in the northeast counties. June-July . . .
. . . . *Lysimachia hybrida* Michx.

Family Oleaceae

(Key to Genera)

1 Leaves pinnately compound, opposite. Trees . . *FRAXINUS*
1' Leaves simple, opposite. Shrubs . . . . *SYRINGA*

FRAXINUS L.

Fraxinus pennsylvanica Marsh. Trees to 20 meters tall. Leaves
pinnately compound, opposite, the leaflets 5-7, lanceolate to elliptic. Sparingly pubescent on the under surfaces of leaves.
Flowers in panicles, inconspicuous, polygamo-dioecious. Fruiting
samaras 3-5 cm long, the wing spatulate. Frequent to common in
thickets and wooded ravines and along streams over the state.
Represented in South Dakota by the two var. *pennsylvanica* and
subintegerrima (Fern.) Vahl. Apr-May. Green Ash.

SYRINGA L.

Syringa vulgaris L. Shrubs with several stems, to 3 meters tall.
Leaves simple, opposite, their surfaces glabrous. Leaf blades 5-
7 cm long, ovate. Flowers lilac to white, the petals forming a
cylindric tube up to 1 cm long. Fruiting capsules 1-1.5 cm long,
with flat, winged seeds. Occasional escape from plantings but
more commonly persisting for long periods of time around old
dwellings. May. Lilac.

Family Gentianaceae

(Key to Genera)

1 Plants of very moist or boggy places. Leaves trifoliolate, alternate . . . . *MENYANTHES*

1' Plants of drier habitats. Leaves simple, whorled or opposite.

   2 Petals with conspicuous spurs or sacs projecting downward . . . . *HALENIA*

   2' Petals not spurred.

      3 Stem leaves whorled, 3 or more. Corolla lobes with a large, glandular spot . . . . *SWERTIA*

      3' Stem leaves opposite, sessile.

         4 Corolla with plaits or folds in the sinuses . . . . *GENTIANA*

         4' Corolla without plaits or folds in the sinuses.

            5 Styles filiform, deciduous. Petals 3-4 cm long, usually deep purple . . . . *EUSTOMA*

            5' Styles short and stout. Petals 1-1.5 cm long, white to greenish or violet . . . . *GENTIANELLA*

EUSTOMA Salisb.

Eustoma grandiflorum (Raf.) Shinners. Plants annual or weak perennial herbs with 1-several erect stems, 2-5 dm tall. Leaves lance-ovate, sessile, 2-4 cm long, glabrous. Inflorescence cymose-paniculate, the flowers 3-4 cm long. Petals lavender to almost white. Style stout, almost as long as the ovary. Rare in open meadows near Cascade Springs in Fall River County. July-Aug. Prairie Gentian.

GENTIANA L.

1 Flowers less than 3 cm long at anthesis. Principal leaves less than 3 cm long and 1.5 cm wide. Plants perennial, the stems 1-3 dm tall, often clustered at the base from a woody rootstock. Stem leaves lanceolate, widest near the middle. Flowers borne terminally in the upper axils of leaves, the petals blue, 2-3 cm long. Rare in moist meadows of the western part. Ours from Harding and Pennington counties in the Black Hills. July-Aug . . . . . . *Gentiana affinis* Griseb.

1' Flowers more than 3 cm long at anthesis. Principal leaves exceeding 3 cm in length, and 1.5 cm in width.

2   Corolla bottle shaped and almost closed at the top when in
    flower. Stems 3-6 dm tall, single or few from a perennial
    base. Leaves ovate-lanceolate, up to 10 cm long. Inflo-
    rescence a dense, sessile cluster of flowers terminal or
    in the upper 1 or 2 leaf axils. Extremely rare in wet
    meadows over the state. Aug-Sept. Bottle Gentian . . . .
    . . . . *Gentiana andrewsii* Griseb.

2' Corolla with lobes of petals erect or slightly spread at
    flowering time. The plaits or folds between the lobes
    fimbriate. Stems 2-4 dm tall, mostly simple, 1-several
    from a perennial rootstock. Leaves lanceolate, not more
    than 5 cm long, opposite. Inflorescence in a terminal
    cluster or the upper few axils. Petals deep blue. Rare
    in native prairie of eastern counties and in the Black
    Hills. {*G. puberula* Michx.} Aug-Sept. Downy Gentian . .
    . . . . *Gentiana puberulenta* Pringle

GENTIANELLA Moench.

Gentianella amarella (L.) Borner. Plants annual herbs with stems
1-4 dm tall, erect with branches from near the base and upwards.
Basal leaves spatulate, the principal stem leaves lanceolate and
2-4 cm long. Many flowers borne terminally and axillary on ascending peduncles in upper leaves. Corolla varying from white
to greenish or lavender, the petals 8-14 mm long. Frequent in
valleys and moist ravines of the Black Hills and Harding County.
{*Gentiana strictiflora* (Rydb.) Nels.} Aug-Sept.

HALENIA Borkh.

Halenia deflexa (Sm.) Griseb. Plants herbaceous annuals 2-4 dm
tall, branched above. Leaves opposite, ovate, the principal ones
sessile, 2-4 cm long. Inflorescence of terminal and upper axillary cymes, the flowers mostly greenish yellow. Petals spurred,
6-8 mm long. Flowers 4-parted. Infrequent in moist valleys or
ravines in the western part, principally the Black Hills. July-
Aug. Spurred Gentian.

MENYANTHES L.

Menyanthes trifoliata L. Plants perennial with a short stem from
a thick rootstock. Leaves 3-foliolate, on long petioles chiefly
basal. Leaflets ovate, glabrous, 5 cm or more long. Inflorescence racemose, on a scape which extends above the leaves. Flowers white to pink, 1.0-1.5 cm long. Apparently extremely rare in
the eastern part. Our only specimen from boggy marshland 2 miles
south of Elkton in Brookings County. May-June. Buckbean.

SWERTIA Adans.

Swertia radiata (Kell.) Ktze. Plants herbaceous perennials up to
a meter or more tall, the stems stout, erect. Leaves whorled on
the stem, 10-20 cm long, narrowly oblong to lanceolate, glabrous
to sparingly puberulent. Inflorescence many-flowered, terminal
and in the axils of upper leaves. Flowers greenish, with purple
dots. Petals 4 or 5 parted, with a pair of glandular areas on
the lobes. Frequent in moist areas of woods, valleys and ravines
in the Black Hills. May-June.

Family Apocynaceae

APOCYNUM L.

1 Flowers more than 5 mm long. Corolla 2-3 times the length of
   the calyx, pinkish. Leaves drooping and spreading, slightly
   to densely pubescent. Plants up to 1 meter tall, freely
   branching and lacking a principal stem in the upper part.
   Fruiting follicles slender and pointed, 1-2 dm long. Infrequent on the edges of woods and on shrubby hillsides over the
   state. {Hybrids resulting from crosses between this species
   and the following two species have been called *A. medium*
   Greene] June-July. Spreading Dogbane . . . .
   . . . . *Apocynum androsaemifolium* L.

1' Flowers less than 4 mm long, the corolla only slightly longer
   than the calyx. Flowers greenish-white. Leaves ascending.

   2 Leaves short-petioled and elliptic, rounded at the base.
      Mature follicles 12 cm long or longer. Coma of seeds 2 mm
      or longer. Plants to 1 meter tall. Leaves 5-11 cm long,
      lanceolate to broadly elliptic. Flowers white or occasion-
      ally tinged with pink, terminal in irregular clusters,
      erect. Frequent in open woods, along roadsides, and in
      moist open places over the state. June-July. Indian Hemp.
      . . . . *Apocynum cannabinum* L.

   2' Leaves sessile and obtusely rounded to cordate at the base.
      Mature follicles less than 12 cm long. Coma of seeds up to
      2 mm long. Plants up to 6 dm tall, somewhat branched above.
      Leaves oblong-lanceolate, 7-8 cm long, variously pubescent.
      Flowers greenish-white, in terminal clusters, erect. Fre-
      quent in moist to dry, sandy open places over the state.
      June-July . . . . *Apocynum sibiricum* Jacq.

Family Asclepiadaceae

ASCLEPIAS L.

1 Hoods of the crown without a small, incurved horn. Flowers small and greenish.

2 Inflorescence of one umbel of flowers at the terminal portion of the stem. Plants 2-3 dm tall, the stems covered with coarse hairs. Leaves lance-ovate in shape, 4-6 cm long, sparingly pubescent on both surfaces. Flowers on short peduncles, the corolla lobes 4-5 mm long. Fruiting follicles 5-8 cm long, softly pubescent. Infrequent on dry prairie hillsides in the eastern part. {*Acerates lanuginosa* (Nutt.) DC.} June-July . . . . *Asclepias lanuginosa* Nutt.

2' Inflorescence of 2-several umbels along the upper part of the stem or terminal. Plants glabrous to slightly pubescent.

3 Leaves linear. Umbels 2-several along the upper part of the stem. Each umbel with 10-15 flowers. Hoods of each flower slightly toothed. Plants from a thick rootstock, the stems 1-few, 2-5 dm tall. Leaves 5-10 cm long, the margins slightly rolled under. Fruiting follicles slender, about 6 cm long, erect. Rare in dry prairie in the south and western part. {*Acerates angustifolia* (Nutt.) Dec.} July . . . . . *Asclepias stenophylla* Gray

3' Leaves narrowly lanceolate to ovate. Umbels usually fewer than 4, with many flowers. Hoods of each flower entire. Plants perennial, the stems ascending or spreading, 2-5 dm tall. Leaves highly variable in shape, from narrow to broadly lanceolate or ovate, 4-6 cm long. Flowers with corolla lobes 4-6 mm long. Fruiting follicles pubescent, 5-8 cm long, erect on deflexed pedicels. Frequent to common in usually upland prairie or sandy soil over the state. {*Acerates viridiflora* (Raf.) Eaton} July. Green Milkweed . . . . *Asclepias viridiflora* Raf.

1' Hoods of the crown with a small, incurved horn. Flowers variously colored but principally lavender.

4 Principal leaves at the middle of the stem alternate. Flowers yellow to bright orange. Stem densely hairy, lacking milky juice. Plants 3-6 dm tall, stems several from a deep taproot. Leaves 5-10 cm long, linear to lanceolate. Inflorescence of 1-several umbels, the petals 7-10

mm long. Fruiting follicles 7-11 cm long, erect. Infre-
quent in mesic prairie and meadows in the eastern part.
Ours ssp. *interior* Woods. June-July. Butterfly Weed . .
. . . . *Asclepias tuberosa* L.

4' Principal leaves at the middle of the stem opposite or
   whorled. Flowers other than yellow to bright orange.
   Stems not densely hairy. Plants with milky juice.

   5 Leaves narrow and linear, less than 2 mm wide, whorled
     on the stem.

     6 Stems clustered from the base, usually less than 2
       dm tall. Leaves linear, less than 1 mm wide, aver-
       aging about 3 cm long, numerous. Inflorescence of
       small 2-several flowered umbels terminal and in up-
       per axils, with short peduncles. Flowers greenish-
       white, the corolla lobes 3-4 mm long. Fruiting fol-
       licles narrow, 3-6 cm long, erect. Frequent on dry
       prairie over the state. July-Aug. Dwarf Milkweed.
       . . . . *Asclepias pumila* (A. Gray) Vail

     6' Stems, if branched, only sparingly so, commonly over
        2 dm tall. Leaves 1-2 mm wide, 4-6 cm long, numerous
        in whorls of 3-5. Umbels several in the upper axils,
        the flowers greenish white, on short peduncles.
        Petal lobes 4-5 mm long. Fruiting follicles slender,
        4-5 cm long, mostly glabrous, erect. Common in prai-
        ries, along roadsides, and waste places over the
        state. June-Aug. Whorled Milkweed . . . .
        . . . . *Asclepias verticillata* L.

   5' Leaves lanceolate to ovate, oppositely arranged on the
      stem.

      7 Plants, including leaves, strictly glabrous.

        8 Leaf bases clasping the stem, sessile. Blades
          ovate to broadly lanceolate, the apexes obtuse.
          Corolla dull or pale lavender. Stems to 8 dm
          tall, stout, mostly simple. Leaves 10-15 cm long.
          Inflorescence 1-several umbels sub-terminal on the
          stem. Flowers with corolla lobes 11-13 mm long.
          Fruiting follicles glabrous, 8-10 cm long. Infre-
          quent in mesic prairie and roadside ditches in the
          southeast part. June-Aug. Smooth Milkweed . . .
          . . . . *Asclepias sullivantii* Engelm.

        8' Leaf bases not clasping the stem, the blades with
           distinct petioles, lanceolate to linear oblong, 4-
           15 cm long. Stems erect, stout, branched above,

frequently exceeding 1 meter. Several umbels in
the upper axils or terminal, the corolla lobes
pale red to pink, 6-7 mm long. Fruiting follicles
slender, glabrous, 4-5 cm long. Frequent to common in swampy or marshy places over the state.
June-Aug. Swamp Milkweed . . . .
. . . . *Asclepias incarnata* L.

7' Plants hirsute to densely pubescent.

    9 Stems 3-6 dm tall. Flowers greenish white, occa-
sionally with some light purple. Fruiting follicles pubescent without soft, filiform processes.

        10 Lobes of the corolla about 5 mm long. Leaves
ovate, acute at both ends of the blade, usually not over 3 cm at the widest part. Stems
usually simple, slender, erect. Inflorescence
a solitary and terminal umbel or with a few in
the upper axils. Follicles slender, mostly
erect on short pedicels. Infrequent in meadows and open woods in the eastern part and in
the Black Hills. June-July . . . .
. . . . *Asclepias ovalifolia* Decne.

        10' Lobes of the corolla 7-10 mm long. Leaves
broadly ovate, emarginate to retuse at both
ends of the blades, often to 5 cm at the
widest part. Stems spreading or ascending,
mostly simple, stout. Inflorescence of dense
umbels, lateral in the upper axils, on short
peduncles. Follicles erect, 5-10 cm long, on
deflexed pedicels. Rare in sandy prairie soil
of Bennett and Todd counties in the southern
part. July-Aug. Sand Milkweed . . . .
. . . . *Asclepias arenaria* Torr.

    9' Stems 7-15 dm tall. Flowers greenish purple.
Fruiting follicles with a dense pubescence and
soft filiform processes.

        11 Hoods of the flowers 11-15 mm long, narrow,
much longer than the adjacent stamens. Stems
stout, erect, 6-10 dm tall. Leaves ovate,
with short petioles, the blades broadly ovate.
Inflorescence of several umbels, each 5-7 cm
across. Corolla pale purple with a greenish
tinge, the corolla lobes about 1 cm long.
Fruiting follicles 7-10 cm long, on deflexed

pedicels. Common in dry to moist prairie over the state. July-Aug. Showy Milkweed . . . . *Asclepias speciosa* Torr.

11' Hoods of the flowers 6-8 mm long, ovate, only slightly exceeding the stamens. Stems simple, up to 1 meter or more tall. Leaves thick, broadly elliptical, with a soft pubescence. Umbels numerous, each with many flowers. Corolla pale purple or greenish, the lobes 8-10 mm long. Fruiting follicles broadly lanceolate, up to 1 dm long. Very common along roadsides, in waste places, and in prairie over the state. June-Aug. Common Milkweed . . . . *Asclepias syriaca* L.

Family Convolvulaceae

(Key to Genera)

1 Plants without chlorophyll, parasitic on herbs or shrubs, yellowish, no roots present, leaves reduced to scales . . . . *CUSCUTA*

1' Plants with chlorophyll, not parasitic.

2 Corolla not over 1 cm long. Styles 2, separate almost to the base . . . . *EVOLVULUS*

2' Corolla exceeding 1 cm in length. Style 1 or 2-cleft at the top.

3 Flowers pink-purple. Stigma 1 on a single style . . . . *IPOMOEA*

3' Flowers white to pale pink. Stigmas 2 on a cleft style . . . . *CONVOLVULUS*

CONVOLVULUS L.

1 Bracts beneath the calyx minute, usually less than 5 mm long. Corollas 1.5-2.5 cm long, pinkish. Plants with prostrate stems from deep, perennial roots. Leaves 2.5 cm long, their bases sagittate. A very persistent weed of fields and waste places which is difficult to eradicate. Common throughout the state, especially in the eastern part. July-Aug. Field Bindweed . . . . *Convolvulus arvensis* L.

1' Bracts beneath the calyx 1-2 cm long, almost hiding the calyx. Corollas exceeding 3 cm, usually 5-7 cm long, white. Leaves

long-petioled, the blades hastate to sagittate, about 7 cm
long. Stems twining on fences or other vegetation. Frequent
to common in waste places and low moist areas over the state.
June–Aug. Hedge Bindweed . . . . *Convolvulus sepium* L.

CUSCUTA L.

1 Bracts, usually 3–5, below each flower. Flowers sessile or
almost so in clusters, the twining stem with flower clusters
forming dense, ropey masses around its host. Flowers with
obtuse corolla lobes. Fruiting capsule globose with a very
short neck. Rare on shrubs of alluvial areas in Clay and
Union counties in the extreme southeast part. July–Sept . . .
. . . . *Cuscuta glomerata* Choisy

1' Bracts lacking below the flowers. Flowers in clusters but not
forming dense ropey masses. Sepals united at their bases.

   2 Corolla lobes 4, the flowers generally 4-parted and usually
   distinctly pediceled. Fruiting capsule globose-depressed.

      3 Lobes of the corolla spreading or recurved, obtuse.
      Fruiting capsule 2–2.5 mm across. Stems thread-like,
      the flowers in loose clusters. Flowers about 3 mm long,
      the calyx shorter than the corolla tube. Rare on herbs
      and shrubs in the eastern part. Aug–Sept . . . . . .
      . . . . *Cuscuta cephalantha* Engelm.

      3' Lobes of the corolla incurved and pointed. Fruiting
      capsule 3–4 mm across. Stems coarse, becoming yellow-
      orange with age, the flowers in dense to loose clusters.
      Flowers about 2.5 mm long, the calyx about one-half as
      long as the corolla. Rare on small trees and shrubs of
      alluvial places in the southern part. July–Sept . . . .
      . . . . *Cuscuta coryli* Engelm.

   2' Corolla lobes 5, the flowers generally 5-parted, sessile to
   short-pediceled. Fruiting capsules ovoid to globose.

      4 Lobes of the corolla obtuse, the tips spreading. Stems
      coarse, the flowers in sessile or subsessile clusters.
      Flowers 2.5–4 mm long, the calyx only one-half the
      length of the corolla tube. Fruiting capsule ovoid,
      3–5 mm across. Rare on coarse herbs or shrubs over
      the state. Aug–Sept . . . . . *Cuscuta gronovii* Willd.

      4' Lobes of the corolla acute, their tips incurved.

         5 Calyx lobes obtuse, almost as long as the corolla.
         Flowers usually less than 2 mm long. Corolla lobes
         spreading or reflexed, their surfaces not papillose.

The fruiting capsule depressed at its top. Frequent
on many herbaceous and woody hosts throughout the
state. {*C. arvensis* of auth. and *C. campestris*
Yuncker} Aug-Sept. Dodder . . . .
. . . . *Cuscuta pentagona* Engelm.

5' Calyx lobes acute, much shorter than the corolla.
Flowers 3-5 mm long, in irregular clusters. Corolla
papillose on its outer surfaces, the lobes erect.
Fruiting capsules not depressed at the tips. Frequent in low and moist alluvial or sandy places on
herbs and shrubs over the state. Aug-Sept . . . . .
. . . . *Cuscuta indecora* Choisy

EVOLVULUS L.

Evolvulus nuttallianus R. & S. Small perennial plants up to 2 dm
tall. Stems branching from the woody base. Leaves lanceolate,
sessile, up to 2 cm long, their surfaces silky-pubescent. Corolla rose-pink to purple, slightly less than 1 cm long, broadly rotund and funnelform. Fruiting capsule 2-4 valved with 1-several
seeds. Infrequent on dry plains of the southwestern part. {*E.
pilosus* Nutt.} June-Aug.

IPOMOEA L.

1 Plants with bushy stems, perennial from an exceptionally
large, fleshy taproot which may be 3-5 dm in diameter at the
top. Stems spreading, up to a meter tall, with entire, linear
leaves 5-8 cm long. Flowers pink to purple, the corollas 5-7
cm long. Frequent in dry, sandy prairie or on clay bluffs in
the southwest one third of the state. July-Aug. Bush Morning
Glory . . . . *Ipomoea leptophylla* Torr.

1' Plants with climbing stems up to 3 or more meters tall, annual.
Leaves broadly ovate in shape with cordate bases, mostly entire, 5-9 cm long. Inflorescence of 1-5 flowers on peduncles
arising from upper leaf axils. Corollas white to blue or purple, 5-7 cm long. Sepals hirsute towards their bases. Rare
in waste places in the eastern part. July-Aug. Common Morning Glory . . . . *Ipomoea purpurea* (L.) Roth.

Family Polemoniaceae

(Key to Genera)

1 Principal leaves alternate, simple or compound.

2 Plants annual.

   3 Leaves pinnatifid with ultimate segments with spinulose tips. Plants of small stature, usually less than 10 cm tall . . . . *NAVARRETIA*

   3' Leaves linear and simple. Plants often exceeding 10 cm . . . . *COLLOMIA*

2' Plants annual to perennial. Principal leaves pinnatifid with linear divisions. Stems usually taller than 10 cm.

   4 Flowers in clusters forming short, interrupted spikes . . . . *GILIA*

   4' Flowers in head-like clusters or open panicles . . . . *IPOMOPSIS*

1' Principal leaves opposite.

5 Plants perennial . . . . *PHLOX*

5' Plants annual.

   6 Leaves palmately dissected, the segments linear. Flowers long-pedicelled on capillary branches . . . . *LINANTHUS*

   6' Leaves simple, lanceolate. Flowers in axillary clusters at upper portions of stems . . . . *MICROSTERIS*

COLLOMIA Nutt.

Collomia linearis Nutt. Plants annual, 1-4 dm tall with mostly erect, unbranched stems. Leaves linear to lanceolate, sessile, 1-5 cm long, with entire margins. Flowers in terminal and upper axillary clusters, subtended by leaves that are wider than those below. Petals 8-12 mm long, white to bluish-pink. Fruiting capsules not rupturing the calyx tube at maturity, the calyx becoming chartaceous. Frequent in dry or sterile areas over the state. June-July.

GILIA R. & P.

Gilia spicata Nutt. Plants biennial or perennial from a taproot, the stems simple or occasionally branched, 1-4 dm tall. Principal leaves pinnatifid, 3 cm long, the lobes entire and linear. Upper leaves often simple and linear. Inflorescence of capitate clusters arranged in an irregular or interrupted spike. Corolla white, the petals 9-12 mm long. Fruiting capsule with many seeds. Rare on dry plains of the southwest part, including the Badlands. June-July.

IPOMOPSIS Michx.

1 Inflorescence paniculate and open, flat-topped. Corolla 3-5
cm long, their lobes up to 1 cm long, white. Plants annual
to weakly perennial, branched, from a taproot, the stems 2-4
dm tall. Leaves deeply pinnatifid, with narrow segments.
Calyx much shorter than the corolla tube. Rare on sandy
prairie in the sand hills area of Bennett and Todd counties
in the southern part. {*Gilia longiflora* (Torr.) G. Don}
July-Aug . . . . *Ipomopsis longiflora* (Torr.) G. Don

1' Inflorescence dense, the flowers in capitate head-like clusters. Corolla 6-8 mm long, the lobes about one-third the
length of the tube. Plants perennial, branched at the base
from a woody caudex, the stems 1-1.5 dm tall. Leaves variously pinnatifid with narrow segments, the ultimate divisions
spiny-tipped. Frequent on dry hillsides and prairies of the
western part. May-June. Ball-head . . . .
. . . . *Ipomopsis congesta* (Hook.) V. Grant

LINANTHUS Benth.

Linanthus septentrionalis Mason. Plants slender annuals with few
branches upward, the stems to 2 dm tall. Leaves sessile, palmately parted from near the base, the segments linear, to 1.5 cm
long. Flowers on long pedicels at the ends of branches, the corolla 2-5 mm long, white to pale lavender. Fruiting capsules
with 2-6 seeds in each locule. Rare on dry hillsides of the
Black Hills and Harding County. May-June.

MICROSTERIS Greene.

Microsteris gracilis (Hook.) Greene. Plants erect annuals with
few to several branches, the stems to 3 dm tall. Leaves lance-
shaped to narrowly elliptic, 2-5 cm long. Flowers in terminal
or upper axillary clusters, appearing sessile but often paired
on short pedicels. Corolla 5-12 mm long, white to pink-lavender.
Fruiting capsule with one seed per locule. Apparently rare in
meadows and on hillsides in the southwestern part. Ours var.
humilior (Hook.) Cronq. May-June.

NAVARRETIA R. & P.

Navarretia intertexta (Benth.) Hook. Plants erect annuals,
branched at the base or stems simple, 1-6 cm tall. Leaves pinnatifid, the segments filiform, pointed, up to 3 cm long. Flowers borne in dense, leafy bracted clusters at the ends of branches, the petals white, 5 mm long. Fruiting capsules with 1 or 2

seeds per locule. Rare at margins of temporary pools in sandy
soil of Harding County. Ours var. *propinqua* (Suskd.) Brand.
May-June.

PHLOX L.

1 Stems taller than 20 cm, not densely branched from a woody
   base.

   2 Leaves linear or linear-lanceolate, usually less than 1 cm
      wide. Stems erect, perennial, 2-5 dm tall, branched above.
      Inflorescence corymbose, the petals white to purple, 2-3 cm
      long, with broad lobes. Frequent in prairie remnants and
      along roadsides in the southeast part. Ours ssp. *fulgida*
      (Wherry) Wherry. June-July. Prairie Phlox . . . .
      . . . . *Phlox pilosa* L.

   2' Leaves elliptical, more than 1 cm wide and 2-5 cm long.
      Stems weak, not erect, many times decumbent. Flowers in
      loose, few-flowered corymbs, the petals lavender to bluish.
      Rare in alluvial and rich upland woods of Clay, Lincoln and
      Union counties in the southeast part. Ours ssp. *laphamii*
      (Wood) Wherry. May. Woods Phlox . . . *Phlox divaricata* L.

1' Stems less than 20 cm tall, tending to be branched from a
   woody, cespitose base.

   3 Leaves narrow-linear, pointed, not over 2 mm wide and
      rarely over 3 cm long.

      4 Corolla tube 12-15 mm long. Calyx 8 mm or more long.
         Longest leaves 1-3 cm long and 1-2 mm wide. Plants
         perennial, spreading in colonies, the stems 5-10 cm
         tall. Inflorescence 1-5 flowered, the flowers white.
         Corolla tube longer than the calyx. Frequent on dry
         prairies and plains in the western one-half. May-June.
         Moss Phlox . . . . *Phlox andicola* Nutt.

      4' Corolla tube 6-12 mm long. Calyx tube less than 8 mm
         long. Longest leaves less than 12 mm long and 0.5-1.5
         mm wide. Plants cespitose and tufted, the stems poorly
         developed, in small clustered mats. Inflorescence of
         solitary flowers at the ends of branches, the flowers
         white. Corolla tube about equaling the calyx. Infre-
         quent in dry soil of the western part. May-June . . . .
         . . . . *Phlox hoodii* Rich.

   3' Leaves 2-5 mm wide, subulate-pointed with whitened carti-
      laginous margins which are ciliate towards their bases.
      Plants perennial, the short, spreading branches with

fibrous, peeling bark. Flowers sessile, usually solitary,
the corolla white, with petals 1.5 cm long. Calyx 7-8 mm
long, shorter than the corolla tube. Frequent on slopes,
clay banks, and limestone ridges in the western part. May-
June . . . . *Phlox alyssifolia* Greene

Family Hydrophyllaceae

(Key to Genera)

1 Flowers solitary on axillary peduncles. Plants mostly less
than 4 dm tall . . . . *ELLISIA*

1' Flowers in definite terminal or axillary inflorescences.

   2 Petals rolled or twisted together while still in bud (convolute). Placentae enlarged, but not forming a partition-
   like structure. Plants perennial . . . . . . *HYDROPHYLLUM*

   2' Petals overlapped while still in bud (imbricate). Placentae not enlarged, but forming a partition-like structure. Plants annual or perennial . . . . . . *PHACELIA*

ELLISIA L.

Ellisia nyctelea L. Plants annual herbs with irregularly branched
stems up to 4 dm tall. Leaves pinnately separated, each with 3-5
pairs of leaflets oppositely arranged, the blades up to 5 cm long.
Flowers white to pale lavender, the petals 5-7 mm long. Fruiting
capsule of 2 carpels, the seeds mostly 4, finely reticulate. A
common weed in waste places and along roadsides and sandy railroad ballast over the state. June-July.

HYDROPHYLLUM L.

Hydrophyllum virginianum L. Plants perennial herbs 3-6 dm tall,
erect but branched in the upper part. Leaves pinnately divided
almost to the central vein, the segments coarse. The compound
blade 1-2 dm long and broadly triangular in outline. Inflorescence dense, almost capitate-like, on short pedicels. Petals
pale lavender, the stamens distinctly exsert. Ovary 2-valved,
with 4 seeds. Frequent in rich or alluvial woods in the eastern
part. May-June. Waterleaf.

PHACELIA Juss.

1 Plants annual, erect, up to 4 dm tall. Stamens slightly longer than the petals. Stems sparingly branched or simple.
Leaves linear, 2-6 cm long, the principal ones with two

lateral segments making the leaf appear trifoliolate. Flowers
white to blue-lavender, crowded in few-flowered axillary cymes.
Petals broad, about 1 cm long. Rare in western areas of the
Black Hills on sandy soils. More common westward in Wyoming
and Montana. Apr-June . . . . *Phacelia linearis* (Pursh) Holz.

1' Plants perennial, from a stout caudex and taproot, stems prostrate to sub-erect. Stamens distinctly longer than the petals.
   Stems covered with a dense grayish pubescence. Leaves simple
   or with a pair of lateral lobes near their bases, 5-12 cm long,
   oblanceolate. Inflorescence cymose or scorpioid, dense, the
   flowers dull white to lavender. Petals 4-7 mm long. Frequent
   to common in dry, open places in the western part. Ours var.
   leucophylla (Torr.) Cronq. June-July. Scorpion weed . . . .
   . . . . *Phacelia hastata* Dougl.

Family Boraginaceae

(Key to Genera)

1 Style terminal; ovary lacking a deep, four-lobed appearance.
   Corolla white, regular. Plants glabrous . . . . *HELIOTROPIUM*

1' Style arising from between the lobes of the ovary, not terminal.

   2 Corolla bilaterally symmetrical.

      3 Stamens unequal in length, extending beyond the corolla
         . . . . *ECHIUM*

      3' Stamens equal or nearly so, included within the corolla
         . . . . *LYCOPSIS*

   2' Corolla radially symmetrical or nearly so.

      4 Fruits with hooked barbs or bristles.

         5 Bristles uniformly covering the fruits, not confined
            to the margins. Fruits divergent . . . *CYNOGLOSSUM*

         5' Bristles or barbs at the margins of the fruits.
            Fruits erect.

            6 Plants annual. Fruiting pedicels erect or nearly
               so. Styles extending beyond the fruit . . . . . .
               . . . . *LAPPULA*

            6' Plants biennial or perennial. Fruiting pedicels
               curved. Styles shorter than the fruit . . . . . .
               . . . . *HACKELIA*

4' Fruits without hooked barbs or bristles.

  7 Flowers normally blue or purple.

    8 Plants glabrous . . . . *MERTENSIA*

    8' Plants pubescent or with coarse hairs and prickles.

      9 Stems trailing, armed with recurved prickles. Calyx in fruit enlarged and veiny, infolding the nutlets . . . . *ASPERUGO*

      9' Stems upright and branched. Calyx in fruit not enlarged or folded . . . . . . *MYOSOTIS*

  7' Flowers white, greenish-white to yellow or orange.

    10 Nutlets with a broad basal attachment.

      11 Corolla lobes rounded, spreading. Style not exsert from the corolla . . . . *LITHOSPERMUM*

      11' Corolla lobes pointed, erect. Style much longer than the corolla . . . . *ONOSMODIUM*

    10' Nutlets attached with a small non-basal area.

      12 Plants with sharp-pointed, rigid hairs. Fruits with a groove-scar running their length at the point of attachment . . . . *CRYPTANTHA*

      12' Plants with short, appressed hairs. Fruits with a keel along one-half the length at the point of attachment . . . . . *PLAGIOBOTHRYS*

ASPERUGO (Tourn.) L.

Asperugo procumbens L. Plants annual, the stems trailing and procumbent, with recurved, prickly hairs. Leaves oblong to spatulate, 3-5 cm long, opposite. Leaf blades hispid-hairy. Flowers solitary to few in axils of leaves, on recurved pedicels. Corolla 2-3 mm across, blue to purple, appearing radially symmetrical. Fruiting nutlets 4, attached laterally to an elongate receptacle. Calyx in fruit folded, 8-12 mm across, veiny and irregularly 5-lobed with smaller lobes in between. Rare in waste places in the western part. May-June. Catchweed.

CRYPTANTHA Lehm.

1 Plants annual. Flowers not very showy, white, rarely over 2.5 mm wide.

2 Nutlets alike, smooth and shiny, all four about the same
size, each oblong-ovate to lanceolate in shape, 1.5-2 mm
long. Plants with stems 1-5 dm tall, simple but with
branches above. Leaves narrow, oblanceolate, 2-4 cm
long. Flowers white, the corolla about 1 mm across. Inflorescence only slightly bracteate. Infrequent in dry
soil of the south central to the western part. June . . .
. . . . *Cryptantha fendleri* (Gray) Greene

2' Nutlets not alike, 3 of the 4 tuberculate, the fourth one
larger, finely granulate on the outer surface. Plants
small, 1-2 dm tall, the stems several and spreading.
Leaves oblanceolate, 1-3 cm long, reduced in shape towards
the upper part of the stem. Inflorescence bracteate with
reduced leaves. Petals 1-1.5 mm across. Nutlets to 1.5
mm long, the odd one 2.3 mm long. Rare in dry sandy soil
of the extreme southwest part. June-July . . . .
. . . . *Cryptantha minima* Rydb.

1' Plants perennial or at least biennial with coarse stems.
Flowers 5-12 mm across.

3 Nutlets smooth on their dorsal or outer surface. Stems to
3 dm tall, decumbent, branched from the base, with densely
set hairs on the upper part. Leaves linear to lanceolate,
3-8 cm long. Inflorescence of open cymes, the corolla tube
to 3 mm long, white. Nutlets 2-2.5 mm long, not all four
always maturing. Rare to infrequent on clay banks in the
Badlands and the extreme southwest part. June-Aug . . . .
. . . . *Cryptantha jamesii* (Torr.) Payson

3' Nutlets roughened or tuberculate on their dorsal or outer
surfaces. Inner surfaces smooth to rough.

4 Plants densely cespitose, the stems several from a much-
branched base, usually less than 1.5 dm tall. Leaves
narrowly oblanceolate, 2-6 cm long, with silky-strigose
hairs. Inflorescence narrow, without conspicuous bracts.
Petals white, the corolla tube 3-4 mm long. Fruits of 4
nutlets, usually only one maturing, this one up to 3 mm
long. Rare to infrequent on sandy soil in the southwest
part. June-Aug . . . . . *Cryptantha cana* (Nels.) Payson

4' Plants with stems simple or occasionally branched from
the base, 1.5-5 dm tall. Leaves not uniformly narrow
or silky-strigose.

5 Inflorescence narrow and tending to be glomerate.
Corolla limb 7-12 mm wide, the tube 3-7 mm long.
Plants with stems 1-several, 1-6 dm tall. Leaves

3-9 cm long, oblanceolate or spatulate. Nutlets lance-ovate, 3-5 mm long, 2 to 4 maturing, their outer surfaces rugose to tuberculate. A highly variable species common in the western one half of the state. {*C. bradburiana* Payson} June-July . . . . *Cryptantha celosioides* (Eastw.) Payson

5' Inflorescence broad and rounded in outline, especially towards maturity. Corolla limb 5-8 mm wide, the tube 3-4 mm long. Stems stout, 1-several from a biennial or perennial base, 2-4 dm tall. Leaves oblanceolate, 5-12 cm long. Nutlets ovate-lanceolate, 2-3 mm long, 2 to 4 maturing, their outer surfaces slightly rugose. A much rarer occurring species than the preceding with smaller flowers and flowering later. Rare in the extreme southwest part. July-Sept . . . . . *Cryptantha thrysiflora* (Greene) Payson

CYNOGLOSSUM L.

Cynoglossum officinale L. Plants herbaceous, biennial, with stems 4-8 dm tall, leafy. Basal leaves spatulate, petioled, the blades 1-3 dm long. Inflorescence of several many flowered racemes, the flowers red to purple, 7-10 mm long. Inflorescence branches elongating in fruit. Fruits separating into four divergent nutlets, each about 7 mm long, the outer surfaces scabrous-roughened. Infrequent in waste places in the western part of the state, less common eastward. June. Hounds-tongue.

ECHIUM L.

Echium vulgare L. Plants annual or weakly perennial, the stems hispid, leafy, up to 6 dm tall. Leaves 2-12 cm long, linear to oblong. Inflorescences in leafy bracted racemes in the upper one half of the stem. Flowers blue, irregular; the corolla 15-18 mm long. Fruits attached at the base, the 4 nutlets wrinkled or furrowed. Rare in waste places over the state. July-Aug. Blue Weed.

HACKELIA Opiz.

1 Leaves linear-oblong, gradually reduced upwards. Corolla exceeding 4 mm in width at the limb. The dorsal area of the nutlet 3-4 mm long along the longitudinal axis. Plants biennial or short lived perennials. Leaves 4-5 cm long, oblanceolate, petioled. The inflorescence lacking a pronounced spreading racemose appearance, mostly narrow. Petals bluish, the limb 4-7 mm wide. Infrequent in woods and thickets and

in brushy borders of the Black Hills. June-July . . . .
. . . . *Hackelia floribunda* (Lehm.) Johnst.

1' Leaves elliptic-oblong, with distinct petioles. Inflorescence
tending to spread from upper leaf axils. Corolla about 2 mm
wide. The dorsal area of the nutlet less than 3 mm long.

    2 Dorsal area of the nutlet with more than 5 erect bristles
as long, as the marginal ones. Plants up to 1 meter tall.
Leaves oblanceolate, the lower ones large, up to 15 cm
long, becoming smaller upward. Inflorescence spreading,
the flowers blue. Frequent in woods and thickets over the
state. {*Hackelia americana* (Gray) Fern.} June-July . . .
. . . . *Hackelia deflexa* (Wahl.) Opiz.

    2' Dorsal area of the nutlet smooth or pitted, lacking promi-
nent hooked bristles. Rarely with a few short bristles.
Stem up to 1 meter tall. Freely branched with a spreading
inflorescence upward. Leaves oblanceolate to oblong-
elliptic, petioled. Petals pale blue or white. Much like
the preceding. Frequent in rich upland woods over the
state. July-Aug . . . . . *Hackelia virginiana* (L.) Johnst.

HELIOTROPIUM L.

Heliotropium curassavicum L. Plants with stems diffuse and
branching, annual, 2-5 dm tall. Leaves spatulate, fleshy, mostly
less than 5 cm long. Flowers whitish, in one-sided scorpoid racemes. The corolla about 2 mm wide. Fruits ovoid and somewhat
depressed, about 2 mm long, soon separating into 4 nutlets. Infrequent in alkali soils over the state. {*H. spathulatum* Rydb.}
July-Aug. Heliotrope.

LAPPULA Gilib.

1 Pointed marginal prickles of the nutlets in 2 or more rows,
the prickles not swollen or fused to each other at their bases.

    2 Flowers about 3 mm across. Principal leaves linear to
spatulate, 2-6 cm long. Prickles shorter than the length
of the fruit. Plants up to 4 dm tall, much branched and
leafy. Flowers in several racemes, the corolla blue.
Fruiting nutlets 3-4 mm wide, granular-roughened on the
backs. A frequent weed of overgrazed or eroded soil over
the state. June-July . . . . . . *Lappula echinata* Gilib.

    2' Flowers less than 2 mm across. Principal leaves oblong to
ovate, 1-2 cm long. Prickles almost as long as the length
of the fruit. Plants 2-4 dm tall, much branched, the many
leaves strigose-hispid. Flowers in many racemes, the

corolla blue. Fruiting nutlets 4-5 mm long, papillose-
tubercled on the backs. Rare in dry soils in the western
part. Aug-Sept . . . . . . *Lappula cenchrusoides* A. Nels.

1' Pointed marginal prickles of the nutlets in one row, sometimes
swollen at the base and fused adjacently.

  3 Basal portions of the prickles inflated to form a cup
shaped nutlet. Stems branched, 1-3 dm tall. Leaves oblanceolate, 1 or more cm long, hirsute. Flowers in bracted
racemes, 2-3 mm long, blue. Fruits 3-4 mm long. Infrequent in dry soil of the western part. May-July . . . . .
. . . . *Lappula texana* (Scheele) Britt.

  3' Basal portions of the prickles not inflated; however, their
bases may be confluent or partially joined upward. Stems
branched above, 2-5 dm tall. Leaves with spreading hairs,
the lower ones spatulate, 2-4 cm long. Inflorescence of
several to many racemes, the flowers blue, 1.5-2 mm wide.
Fruit about 3 mm long. Common in disturbed soil over the
state. {*Lappula occidentalis* (S. Wats.) Greene} May-July
. . . . *Lappula redowskii* (Hornem.) Greene

LITHOSPERMUM L.

1 Plants annual. Flowers white or off-white. Fruits roughened
and gray to brown. Stems 1-6 dm tall, mostly simple, with
strigose hairs. Leaves linear to oblanceolate, the ones
lower on the stem larger, mostly 2-6 cm long. Flowers in the
axils of upper leaves, the corolla 5-8 mm long. A weedy species that occasionally is found in waste places in the western part. June-Aug . . . . *Lithospermum arvense* L.

1' Plants perennial. Flowers yellow or orange. Fruits smooth,
white or gray.

  2 Corolla tube less than 1.5 cm long, yellow-orange, the
petal lobes entire.

    3 Calyx lobes less than 7 mm long. Foliage leaves densely
canescent. Plants 1-3 dm tall, usually not branched.
Leaves lanceolate, 1-4 cm long. Flowers in leafy-
bracted cymes, the corolla orange. Nutlets smooth,
gray, about 3 mm long. Common in prairie over the
state. Apr-May. Hoary Puccoon . . . .
. . . . *Lithospermum canescens* (Michx.) Lehm.

    3' Calyx lobes 8-10 mm long. Foliage leaves rough-hispid.
Plants with erect stems 3-6 dm tall, simple to leafy.
Leaves linear to lanceolate, 3-6 cm long. Flowers

bright yellow to orange. Nutlets ivory-white, smooth,
about 4 mm long. Rare in sandy soil in the sandhills
area of the southern part. May-July . . . .
. . . . . . *Lithospermum carolinense* (Walt.) MacMill.

2' Corolla tube more than 1.5 cm long, yellow, the petal lobes
toothed or fringed. Plants from a woody taproot, the stems
up to 4 dm tall. Leaves linear, 1-5 cm long, usually not
more than 0.5 cm wide. Flowers in leafy racemes in upper
axils, bright yellow. Fruiting nutlets 3-4 mm long, shiny
white. Frequent in prairies over the state. May. Fringed
Puccoon . . . . *Lithospermum incisum* Lehm.

LYCOPSIS L.

Lycopsis arvensis L. Plants annual, the stems erect or ascending, 2-5 dm tall. Leaves narrowly oblong or wider, 3-7 cm long,
hairy. Flowers blue, slightly irregular, the corolla about 7 mm
long. The throat of the upper part of the corolla with 5 hairy
appendages. Nutlets ovoid. A rare weed in the central and eastern part of the state. June-Aug. Small Bugloss.

MERTENSIA Roth.

1 Plants more than 4 dm tall at maturity. Leaves with distinct
lateral veins, the blades of cauline leaves not petioled.
Basal leaves long-petioled, glabrous to strigose. Inflorescence branched and well-developed, the flowers blue. Corolla
10-15 mm long, the calyx 1-3 mm long, their lobes cleft nearly
to the base. Apparently rare in damp thickets in the western
part. June-July . . . . . *Mertensia ciliata* (James) D. Don

1' Plants less than 4 dm tall at maturity. Principal leaves
lacking prominent lateral veins.

2 Tube of the corolla longer than the limb. Leaves oblong to
lance-ovate, the blades glabrous to strigose-pustulate, 2-
12 cm long. Stems 1-4 dm tall, usually clustered from a
thick caudex. Flowers in small terminal cymes, the corolla
1-2 cm long. Infrequent in ravines and along streams in
the western part. May-June . . . .
. . . . *Mertensia oblongifolia* (Nutt.) G. Don

2' Tube of the corolla almost equal in length to the limb.
Principal leaves lanceolate, mostly glabrous. Stems sev-
eral from a deep-seated rootstock, 1-5 dm tall, glabrous.
Inflorescence cymose or paniculate, the flowers blue.
Petals about 13 mm long. Common in prairie and on open
hillsides over the state. May-June. Blue Bells . . . . .
. . . . *Mertensia lanceolata* (Pursh) A. DC.

MYOSOTIS L.

1   Plants annual or winter annuals, with fibrous roots, weedy.
    Flowers white. Stems 0.5-4 dm tall, branched, the vegetation
    hirsute. Leaves 1-5 cm long, oblanceolate to elliptic. In-
    florescence comprising the upper one third of the plant.
    Corolla inconspicuous, the limbs 1-2 mm wide. Rare in wet
    places in the sand hills area of the southern part. May-June
    · *Myosotis verna* Nutt.

1' Plants perennial, from a fibrous-stoloniferous base, not
    weedy. Flowers blue to purple-tinged.

    2   Calyx with spreading hairs at the base, the lobes much
        longer than the tube. Plants 0.5-4 dm tall, several stems
        from a short, fibrous-rooted base. Leaves oblanceolate to
        oblong, becoming smaller on the upper part of the stem.
        Inflorescence a scorpoid cyme, becoming lax and open to-
        wards maturity. Flowers blue, 4-8 mm wide. Rare to in-
        frequent in rich woods, on moist slopes, or along streams
        at higher altitudes in the Black Hills. June. Forget-me-
        not · · · · · · · · · · · · · · *Myosotis sylvatica* Hoffm.

    2' Calyx uniformly strigose, lacking spreading hairs, the
        lobes shorter than the tube. Plants 2-6 dm tall, the stems
        usually single or few from the base. Leaves oblanceolate
        on the lower part of the stem, becoming smaller upward.
        Inflorescence open, not leafy. Flowers blue, 5-10 mm wide.
        Apparently rare in moist soil in the northern Black Hills.
        June-July · · · · · · · · · · · · *Myosotis scorpoides* L.

ONOSMODIUM Michx.

Onosmodium molle Michx. Plants herbaceous perennials with several stems from woody rootstocks. Plants 3-8 dm tall. Leaves
lance-ovate, 2-7 cm long, with rough hairs, giving a hispid appearance. Inflorescence a leafy bracted cyme that is terminal,
the flowers greenish white, 11-15 mm long. Nutlets white-gray,
smooth and shiny, 3-5 mm long. Common in prairies and plains
over the state. {*Onosmodium occidentale* Mack.} Ours var.
hispidissimum (Mack.) Cronq. June-July. False Gromwell.

PLAGIOBOTHRYS F. & M.

Plagiobothrys scouleri (H. & A.) I. M. Johnst. Plants herbaceous
annuals with spreading, branched stems. Plants 5-15 cm tall,
with strigose hairs. Leaves linear, 1-5 cm long, roughly hairy.
Flowers in lax racemes that are leafy, white, the petals 1-2 mm
long. Fruits of 4 nutlets, 1-2 mm long. Infrequent in alkali

and saline soils in the west-central areas of the state. {*Allo-carya scopulorum* Greene} July-Aug. Popcorn Flower.

Family Verbenaceae

(Key to Genera)

1 Corolla with 4 lobes, the petals 2-lipped. Flowering spikes axillary . . . . *PHYLA*

1' Corolla with 5 lobes, the petals not obviously 2-lipped. Flowering spikes terminal . . . . *VERBENA*

PHYLA Lour.

1 Leaves lanceolate, 2-8 cm long, with 6-10 teeth on each blade margin. Leaves mostly petioled. Plants with weak, spreading stems, 1-4 dm long, rooting at the nodes. Flowers in dense, subglobose spikes up to 3 cm long, the petals white to pale purple, about 2.5 mm long. Frequent in low places of prairies and in swales or marshy areas in the eastern one half of the state. {*Lippia lanceolata* Michx.} June-Sept. Fog Fruit . . . . *Phyla lanceolata* (Michx.) Greene

1' Leaves cuneate to oblanceolate, 1-3 cm long, with 1-4 remote teeth on each blade margin. Leaves essentially sessile. Stems weak ascending to mostly prostrate, 2-9 dm long. Flowers in dense, fascicled axillary spikes, elongating to 2 cm in fruit. Petals white to purple, about 4 mm long. Infrequent in low places of prairies and in swales of prairie in the central and north central parts of the state. July-Sept . . . . *Phyla cuneifolia* (Torr.) Greene

VERBENA L.

1 Flowers with corollas exceeding 1 cm. The sterile style-lobe protruding beyond the stigmatic surface. Leaves bipinnatifid, 2-5 cm long. Plants prostrate, with stems weak ascending, 1-4 dm tall. Inflorescence densely bracted, the flowers rose-purple. Common in waste or sandy soil, especially in the western part. May-Aug . . . . . . *Verbena bipinnatifida* Nutt.

1' Flowers with corollas 4-9 mm long. The sterile style-lobe not protruding beyond the stigmatic surfaces.

2 Inflorescence of solitary spikes or in threes ending the branches or main stems.

3 Plants erect, 4-10 dm tall, perennial. Vegetation soft
pubescent. Leaves oval, serrated, 5-10 cm long. Stems
simple or branched above. Flowers vary from white to
deep purple, the petals with limbs 8-9 mm wide. Very
common in prairie, disturbed soil, and in waste places
throughout the state. June-Aug. Hoary Vervain . . . .
. . . . *Verbena stricta* Vent.

3' Plants mostly low or prostrate, branching from the base.
Leaf blades spatulate in outline, pinnately lobed, 1-4
cm long. Flowers in dense bracted spikes, pale blue to
purple, the bracts as long as or longer than the flow-
ers. Pubescence harsh, especially in the inflorescence.
Common in waste places and roadsides over the state.
{*V. bracteosa* Michx.} July-Sept. Prostrate Vervain . .
. . . . *Verbena bracteata* Lag. & Rodr.

2' Inflorescence of panicled spikes at the ends of branches
and stems.

4 Corolla white. Spikes very slender, interrupted. Plants
annual or perennial, up to 1.5 meters tall, single
stemmed but branched above. Leaves broadly lanceolate,
with coarse serrations, 8-18 cm long. Infrequent in alluvial woods and thickets, more common in the eastern
part. Rare western. July-Aug. White Vervain . . . . .
. . . . *Verbena urticifolia* L.

4' Corolla lavender. Spikes erect, crowded, densely panicled. Plants perennial, branched above, the stems up to
12 dm tall. Leaves broadly lanceolate, doubly serrate,
often hastate at the base, 4-18 cm long. Frequent in
moist thickets and waste places over the state. Aug-
Sept . . . . *Verbena hastata* L.

Family Lamiaceae

(Key to Genera)

1 Corolla radially symmetrical or nearly so, the petal lobes
almost equal.

2 Each flower with 2 stamens; plants lacking a minty odor
. . . . *LYCOPUS*

2' Each flower with 4 stamens. Plants with a more or less
minty odor.

3 Flowers 1-4 in each axil . . . . *ISANTHUS*

      3' Flowers numerous in each axil . . . . . . *MENTHA*

1' Corolla bilaterally symmetrical.

   4 The upper lip of the corolla small or lacking. The lower lip large . . . . *TEUCRIUM*

   4' The upper and lower lip of the corolla well-developed.

      5 Stamens 2.

         6 Flowers in compact axillary whorls or terminal clusters . . . . *MONARDA*

         6' Flowers in spikes or few in the axils of leaves.

            7 The upper lip of the corolla flat. Calyx not strongly two lipped . . . . *HEDEOMA*

            7' The upper lip of the corolla concave. Calyx strongly two lipped . . . . *SALVIA*

      5' Stamens 4.

         8 Calyx bearing a protuberance on the upper side. The bilabiate calyx smooth, without teeth . . . .
            . . . . *SCUTELLARIA*

         8' Calyx without a dorsal protuberance and with evident teeth.

            9 Inflorescence principally terminal in compact spikes or heads.

               10 Stamens extending out of the flower beyond the petals.

                  11 Leaves linear-lanceolate, entire, mostly sessile . . . . *PYCNANTHEMUM*

                  11' Leaves broader, toothed, with petioles . .
                    . . . . *AGASTACHE*

               10' Stamens within the curved portion of the upper petals.

                  12 Upper stamens longer than the lower.

                     13 Bracts beneath the flowers spinose-tipped. Calyx strongly irregular . .
                       . . . . *DRACOCEPHALUM*

                     13' Bracts beneath the flowers not spinose-tipped. Calyx only slightly irregular
                     . . . . *NEPETA*

12' Upper stamens shorter than the lower.

    14 Flowers borne individually in the axils of short bracts, the inflorescence a terminal raceme . . . . *PHYSOSTEGIA*

    14' Flowers clustered, verticillate in a dense terminal or interrupted inflorescence.

        15 Calyx strongly two-lipped . . . . *PRUNELLA*

        15' Calyx not strongly two-lipped, with 5 equal or sub-equal teeth . . . . *STACHYS*

9' Inflorescence principally axillary as dense verticillate arrangements of flowers or few in each upper axil.

    16 Teeth of the calyx lobes hooked or curved at the tips. Stamens mostly within the corolla . . . . *MARRUBIUM*

    16' Teeth of the calyx lobes not hooked or curved. Stamens projecting up and under the upper lip of the corolla.

        17 Upper stamens longer than the lower. Plants flattened, spreading, or growing close to soil . . . . *GLECOMA*

        17' Upper stamens shorter than the lower.

            18 Calyx teeth with rigid, spreading, awn-like points . . . . *LEONURUS*

            18' Calyx teeth not rigid.

                19 Leaves subtending the flower clusters sessile and clasping. Plants small, spreading . . *LAMIUM*

                19' Leaves subtending the flower clusters opposite, petioled. Plants upright or erect . . *GALEOPSIS*

AGASTACHE Clayt.

1 Petals blue to lavender.

  2 Undersurfaces of leaves pale green and vegetation mostly glabrous. Leaves ovate-lanceolate to ovate with coarse serrations. Plants perennial, up to 1 meter tall or more. Inflorescence of spike-like panicles, the flowers blue or blue-lavender. Petals 6-8 mm long. Rare to infrequent in upland woods in the eastern part. Aug-Sept . . . . *Agastache scrophulariaefolia* (Willd.) O. Ktze.

  2' Undersurfaces of leaves with a fine, white pubescence. Leaves ovate to deltoid-ovate, petioled, 5-8 cm long. Plants perennial, 5-10 dm tall, erect, with some branching above. Inflorescence of dense spikes up to 10 cm long, the flowers lavender. Calyx sparsely pubescent, bluish at the distal ends of the lobes. Frequent in woods and thickets over the state. July-Aug. Lavender Hyssop . . . . *Agastache foeniculum* (Pursh.) O. Ktze.

1' Petals yellow. Leaves thin, glabrous on the lower surfaces. Calyx glabrous. Plants perennial, a meter or more tall, with several ascending branches. Leaves petioled, narrowly ovate, with coarse serrations. Inflorescence of dense spikes 15 cm long and 2 cm thick, terminating erect branches. Rare in low alluvial woods in the southeastern part. Aug-Sept. Giant Hyssop . . . . *Agastache nepetoides* (L.) Ktze.

DRACOCEPHALUM L.

Dracocephalum parviflorum Nutt. Plants biennial from taproots, the stems principally simple and erect, 1.5-4 dm tall. Leaves lance-oblong, 2-6 cm long, the margins coarsely serrate with spine-tipped teeth. Flowers sessile or nearly so in the upper axils (and terminal) of bract-like leaves. Petals blue-purple, only slightly longer than the toothed calyx lobes. Frequent in valleys and in other mesic places in the Black Hills and in the northeast part. June-July. Dragonhead.

GALEOPSIS L.

Galeopsis tetrahit L. Plants annual, herbaceous, with opposite leaves, the stems 3-8 dm tall. Leaves petiolate, the blades ovate, coarsely serrated, the nodes of the stem swollen at the petiole base. Flowers in axillary verticils at the upper part, densely clustered. Petals 1.5-2.0 cm long, pink or lavender. Rarely escaped in waste places in the Black Hills. July-Aug. Hemp Nettle.

GLECOMA L.

Glecoma hederacea L. Plants perennial with fibrous roots, the
stems lax, stoloniferous, forming mats up to 1 meter in diameter.
Leaves petioled, opposite, the blades reniform, their distal margins toothed or crenate. Flowers few, pedicellate in the axils
of leaves, corolla narrow, 1-2.5 cm long, violet. Infrequent as
an escape in waste places and abandoned areas. Apr-May. Ground
Ivy.

HEDEOMA Pers.

1 Plants annual, the stems with long hairs. Leaves narrow.
   Stems much branched, up to 2.5 dm tall. Leaves on the main
   stems deciduous soon after their formation. Flowers borne
   along the entire length of the ascending branches, in axillary clusters forming a spikelike arrangement. Corolla blue,
   4-7 mm long. Common in sterile or sandy soil over the state.
   May-June. False Pennyroyal . . . . . . *Hedeoma hispida* Pursh

1' Plants perennial, the stems pubescent in the upper portions.
   The taproot slender, often branched with woody caudices.
   Plants 8-20 cm tall, diffusely branched. Leaves linear-
   oblong, 1-2 cm long. Flowers in small cymules in the axils
   of upper leaves, spike-like. Corolla rose-purple, 6.0-10 mm
   long. Frequent on dry exposed soil of the western part. {*H.
   camporum* Rydb.} June-July . . . . . *Hedeoma drummondii* Benth.

ISANTHUS Michx.

Isanthus brachiatus (L.) BSP. Plants annual, 2-4 dm tall, the
stems glandular-puberulent, with many branches. Leaves opposite,
elliptic to lanceolate, 1-4 cm long, mostly entire. Flowers in
axillary cymes, the entire inflorescence appearing like a large
panicle. Petals blue, 5-8 mm long, the corolla almost radially
symmetrical. Rare to infrequent in dry places in the eastern
part. Aug-Sept. False Pennyroyal.

LAMIUM L.

Lamium amplexicaule L. Plants annual from taproots, the stems
lax, spreading and ascending, 1-3 dm tall. Leaves rounded or
reniform, opposite, 1-3 cm across, the upper ones clasping the
stem, the lower ones petioled. Flowers axillary, sessile in
clusters of few to many. Corolla lavender, 10-15 mm long. Infrequent in waste or cultivated soils over the state. May-June.
Henbit.

LEONURUS L.

1 Calyx 5 angled and nerved. Upper corolla lobe long-hairy,
much exceeding the calyx. Plants perennial, from fibrous
roots, the stems sharply angled, 4.0-12 dm tall. Leaves opposite, petioled and lanceolate, with toothed lobes. Flowers
in dense, axillary clusters at the upper portion of the main
stem, the petals white or pale pink, 8-10 mm long. A common
weed thoroughly established in waste places in the eastern
and western parts of the state. June-July. Motherwort . . .
. . . . *Leonurus cardiaca* L.

1' Calyx not strongly angled, 10-nerved. Upper corolla lobe
short-pubescent, scarcely longer than the calyx. Plants biennial, the stems 6-10 dm tall, from fibrous roots. Leaves
ovate, petioled, the margins coarsely toothed. Flowers in
small axillary clusters, bracted, the petals white, naked on
the inner surface. Occasional in waste places in the eastern
part. July-Aug . . . . *Leonurus marrubiastrum* L.

LYCOPUS L.

1 Lobes of the calyx blunt, less than 1 mm long, not extending
beyond the nutlets. Plants perennial from a thickened tuberous rootstock, the stems 1-6 dm tall. Leaves lanceolate to
oblong, 2-5 cm long, with shallow teeth. Flowers crowded in
the axils of opposite leaves, the corolla 2-3 mm long, with
spreading lobes. Infrequent in moist places in the eastern
part. Aug-Sept . . . . *Lycopus uniflorus* Michx.

1' Lobes of the calyx acuminate-pointed, 2-3 mm long, longer
than the nutlets towards maturity.

  2 Leaves petioled and with pinnately divided incisions that
extend halfway or more to the main vein. Nutlets 1.0-1.5
mm long, the tops rounded. Plants perennial from elongate
rhizomes, the stems 2-6 dm tall, mostly simple, erect.
Leaves opposite, 3-6 cm long. Flowers verticillate in the
axils of upper leaves. Petals white, 2-3 mm long. Common
across the state in marshy areas or at aquatic margins.
June-Aug. Water Hoarhound . . . . *Lycopus americanus* Muhl.

  2' Leaves sessile, lanceolate, the margins coarsely serrate.
Nutlets 1.5-2.0 mm long, the tops flattened. Plants perennial from tuberous rhizomes, the stems 2-5 dm tall, sparingly branched if at all. Leaves 3-7 cm long. Flowers
axillary, the petals white, 3-5 mm long. Common in moist
places over the state. June-Aug . . . *Lycopus asper* Greene

MARRUBIUM L.

Marrubium vulgare L. Plants perennial from taproots, the stems
much-branched at the base, the branches prostrate or ascending,
3-6 dm long. Vegetative surfaces covered by soft, woolly hairs.
Leaves rounded, petioled, the blades 1-3 cm across. Flowers
densely crowded in the axils of upper leaves, the calyxes long-
hairy and with recurved, pointed lobes. Occasionally escaped
in waste places throughout the state. June-July. Horehound.

MENTHA L.

Mentha arvensis L. Plants perennial, herbaceous, 2-6 dm tall,
from a rhizome. Vegetation with a distinct minty odor. Stems
4-angled, simple, erect, pubescent on the angles. Leaves opposite, petioled, the blades elliptic or ovate, 1.5-6 cm long.
Flowers densely clustered in the axils of upper leaves. Corolla
whitish to bright pink, 4-6 mm long. Very common in moist or
marshy places over the state. July-Aug. Field Mint.

MONARDA L.

1 Plants perennial with flowers in dense, terminal clusters.
   Petals purple or rose-colored, 2.5-3.5 cm long, curved. Stems
   6-9 dm tall, pubescent. Leaves lance-ovate, pubescent, with
   sharply serrate margins. Vegetation with a strong minty odor.
   Very common in prairie and open woods over the state. Ours
   represented by the two varieties *fistulosa* and *menthaefolia*
   (Gray) Fern. July-Aug. Wild Bergamot . . . .
   . . . . *Monarda fistulosa* L.

1' Plants annual or weakly biennial or perennial; with flowers in
   the upper axils of opposite leaves. Petals white-yellow, 1-2
   cm long. Stems from a taproot, simple or branched from the
   base, 2-5 dm tall. Leaf blades lanceolate, petioled, 2-4 cm,
   the margins remotely dentate-crenate. Rare in sandy soil.
   Infrequent in the sandhills of Bennett and Todd counties in
   the south central part. June-July. Lemon Mint . . . .
   . . . . *Monarda pectinata* Nutt.

NEPETA L.

Nepeta cataria L. Plants perennial, herbaceous, the stem
branched upward, 3-8 dm tall. Stems and vegetation with a
dense, hairy pubescence. Leaves ovate to deltoid, petioled,
the blades coarsely crenate, 2-6 cm long. Inflorescence in
terminal interrupted clusters or verticils, appearing spike-
like. Petals bilabiate, whitish with purple dots, 10-12 mm
long. A common weed of disturbed and waste places over the
state. July-Aug. Catnip.

PHYSOSTEGIA Benth.

1 Corolla 1.5-3 cm long. Uppermost leaves below the inflorescence much reduced. Plants perennial, the stems mostly solitary, erect, 4-8 dm tall. Leaves opposite, lance-linear,
sharply serrate, 6-12 cm long. Inflorescence terminal, racemose, the flowers borne individually in the axils of small
bracts, lavender to rose-colored. Infrequent in thickets
and meadows in the eastern part. July-Aug. False Dragonhead
· · · · · · · · · · · · · · *Physostegia virginiana* (L.) Benth.

1' Corolla less than 1.5 cm long. Uppermost leaves not appreciably reduced. Plants perennial, the stems solitary, 4-10
dm tall. Leaves opposite, the principal upper ones lanceolate, broadest near their bases. Inflorescence terminal, up
to 10 cm long, the flowers in a dense raceme. Petals pink to
purple, about 1 cm long. Rare in moist thickets or alluvial
places in the eastern part. July-Aug · · · · · · · · · · · ·
· · · · · · · · · · · · · · · · *Physostegia parviflora* Nutt.

PRUNELLA L.

Prunella vulgaris L. Plants herbaceous, perennial, with spreading, ascending branches 1-5 dm tall. Leaves remote, petioled,
the blades broadly lanceolate with entire to irregularly toothed
margins 2-8 cm long. Inflorescence in axillary clusters and terminal, spike-like. Flowers 1-2 cm long, the petals pink to deep
lavender or blue. Infrequent in alluvial habitats in the eastern
part. June-July. Selfheal.

PYCNANTHEMUM Michx.

Pycnanthemum virginianum (L.) Durand and Jackson. Plants perennial, the stems much branched, ascending, 3-7 dm tall. Angles of
the stems short hairy. Leaves numerous, lance-linear, 2-5 cm
long, mostly glabrous above but sparsely puberulent on the lower
surfaces. Inflorescence cymose, the flowers clustered on terminal segments of branches, numerous. Corolla 6-7 mm long, white
or pink. Rare in moist prairie and thickets of the eastern part.
July-Aug. Mountain Mint.

SALVIA L.

1 Plants annual. Leaves lance-oblong, mostly entire, 2-6 cm
long. Upper calyx lip entire. Stems much branched, 1-4 dm
tall. Flowers in terminal, spike-like clusters, the petals
blue or purple, 6-8 mm long. Corolla deciduous shortly after
anthesis. A common weed in sandy or sterile soil over the

state. {*S. lanceolata* Willd.} June-Sept. Sage . . . . . .
. . . . *Salvia reflexa* Hornem.

1' Plants perennial. Leaves lanceolate-ovate, their margins remotely toothed, 5-8 cm long. Upper calyx lip toothed. Stems
   somewhat branched above, 3-6 dm tall. Flowers numerous in
   axile whorls at the upper part of the stem, appearing spike-
   like. Petals violet, 9-13 mm long. Infrequent in fields and
   waste places in the eastern part. June-July . . . .
   . . . . *Salvia nemorosa* L.

SCUTELLARIA L.

1 Flowers in terminal or axillary racemes 3-6 cm long. Stems
   solitary from rhizomes, branched above, 2-8 dm tall. Leaves
   ovate-lanceolate, 2-5 cm long, thin, with serrated margins.
   Corolla blue to white, about 6 mm long. Common in moist or
   marshy places over the state. July-Aug. Skullcap . . . . . .
   . . . . *Scutellaria lateriflora* L.

1' Flowers solitary or paired in the leaf axils.

   2 Petals 1.5-2.0 cm long. Leaves lance-ovate, 2-5 cm long.
      Plants perennial from rhizomes, the stems ascending and
      branched, 2-6 dm tall. Flowers solitary (appearing paired
      but in opposite axils) in the axils of upper leaves, the
      corolla blue with white. Infrequent in moist soil in the
      eastern and western part. July-Aug . . . .
      . . . . *Scutellaria galericulata* L.

   2' Petals 7-9 mm long, blue. Leaves ovate to rounded, 1.0-1.5
      cm long and almost as wide. Plants perennial from a thin
      rhizome system, the stems branched, up to 2 dm tall. Flow-
      ers mostly solitary in the axils of upper leaves, their
      calyxes glandular and hairy. Infrequent in dry prairie or
      sandy soil in the eastern part. Ours var. *leonardi* (Epl.)
      Fern. May-June . . . . *Scutellaria parvula* Michx.

STACHYS L.

Stachys palustris L. Plants perennial, herbaceous, the stems
simple, 3-8 dm tall. Leaves lanceolate-ovate, sessile or nearly
so, with coarsely toothed margins, 4-8 cm long. Vegetative surfaces densely pubescent. Inflorescence 4-6 flowered verticils,
bracted, appearing spike-like. Corolla rose-white, the petals
10-14 mm long, strongly bilabiate. A somewhat variable species
common throughout the state in mesic prairie and in marshes.
Ours var. *pilosa* (Nutt.) Fern. July-Aug. Hedgenettle.

TEUCRIUM L.

Teucrium canadense L. Plants perennial from rhizomes, the stems
solitary, simple, up to 8 dm tall. Leaves ovate to oblong, 6-10
cm long, the margins toothed. Flowers in compacted verticils,
on short pedicels, giving a terminal spiciform appearance.
Petals pink-purple, 10-18 mm long. Frequent to common in low
moist thickets and at the edges of alluvial woods over the state.
Ours var. *occidentale* (Gray) McCl. & Epl. July-Aug. Wood Sage.

Family Solanaceae

(Key to Genera)

1 Plants shrubby, the stems woody, often with short thorns or
   prickles. Corolla lavender. Fruit a berry . . *LYCIUM*

1' Plants herbaceous.

   2 Corolla exceeding 2 cm in length. Fruit a dehiscent cap-
      sule.

      3 Flowers sessile or nearly so. Corolla 2-6 cm long.
         Fruits not spiny . . . . *HYOSCYAMUS*

      3' Flowers pediceled. Corolla 6-10 cm long. Fruits with
         fleshy spines . . . . *DATURA*

   2' Corolla less than 2 cm long. Fruit a berry, often enclosed
      within the inflated or spiny calyx.

      4 Flowers with anthers longer than the filaments, the
         anthers opening by terminal pores or slits. Flowers
         usually with a rotate corolla . . . . . *SOLANUM*

      4' Flowers with anthers about equaling the filaments, open-
         ing longitudinally. Corollas campanulate or funnelform
         . . . . *PHYSALIS*

DATURA L.

Datura stramonium L. Plants coarse, herbaceous annuals with
stems branched, 1 meter or more tall. Leaves petioled, the
blades irregularly toothed and lobed, the principal leaves 10-15
cm long. Flowers terminal and usually solitary, but with additional growth of the stem, appearing axillary. Corolla white,
funnelform, 6-10 cm long. Fruits 2-carpelled, the capsule spiny,
3-5 cm long. Infrequent to rare in waste places in the central
to eastern parts. June-July. Jimson Weed.

HYOSCYAMUS L.

Hyoscyamus niger L.   A strong scented, coarse annual or biennial
with a branched stem to 1 meter tall.   Leaves alternate, sessile,
the blades pinnately lobed, 5-15 cm long.   Flowers borne in terminal, leafy bracted racemes.   Corolla funnelform, 5-lobed, 3-4
cm long, green-yellow with purplish lines.   Calyx persistent,
becoming urceolate in fruit, 2.5 cm long.   Infrequent on roadsides and in waste places over the state, a little more frequent
in the Black Hills.   June-July.   Henbane.

LYCIUM L.

Lycium halimifolium Mill.   Plants woody shrubs with arched
branches, the stems 1-3 meters tall, glabrous but occasionally
with thorns.   Leaves short-petioled, the blades elliptic-lanceo-
late, entire, 3-7 cm long.   Flowers axillary, 1-several in upper
locations, the corolla tubular to funnelform, 4-6 lobed, 9-13 mm
long, lavender.   Fruits red, the berries ovoid, 1-2 cm long, several to many seeded.   Widely planted and persisting over the
state in a variety of habitats.   July-Aug.   Matrimony Vine.

PHYSALIS L.

1 Flowers in groups of 2-4 at a node, their corollas white,
   broadly campanulate to rotate, 3-4 cm in diameter.   Plants
   annual, 3-7 dm tall, branched above.   Leaves ovate to lanceolate, their margins irregular.   Vegetation covered with a
   viscid pubescence.   Fruiting calyx ovoid, open at the distal
   end, to 1.5 cm long, almost completely filled with the berry.
   Rare in sandy soil of the Black Hills.   July-Aug . . . .
   . . . . *Physalis grandiflora* Hook.

1' Flowers solitary, the corollas yellow to green or brown, less
   than 3 cm across.   Plants perennial.

  2 Pubescence of the upper stems and leaves with long and
     short hairs.   Leaves rounded or rhombic in outline.

    3 Leaf blades 5 cm or more long, rhombic with irregular
       lobes, their bases cordate.   Pubescence of stem and
       leaves dense, with short hairs, and including soft,
       spreading hairs.   Plants perennial, the stems spread-
       ing, 3-8 dm tall.   Flowers yellow, the corolla 1.5-2.0
       cm long, with purple centers.   Staminal filaments great-
       ly expanded in width.   Fruiting berry yellow.   Frequent
       to common in dry or sandy soils over the state.   July-
       Sept.   Ground Cherry . . . . *Physalis heterophylla* Nees.

3' Leaf blades less than 5 cm long, rounded-ovate to rhombic in outline. Pubescence of stems and leaves mostly short, if long hairs present, these mostly on the calyx. Plants perennial, spreading to erect, the stems 2-5 dm tall. Flowers yellow, the corolla about 1.5 cm long, with brown centers. Fruiting calyx ovoid, 3-4 cm long, the berry yellow. Similar to the preceding but smaller. Rare in dry, sandy soils of the southwest part. Ours var. *comata* (Rydb.) Waterfall. July-Sept . . . . *Physalis hederaefolia* Gray

2' Pubescence of the upper stem and leaves of short stiff hairs or almost glabrous. Leaves ovate to lanceolate, thin, entire to coarsely toothed. Plants perennial, the stems mostly erect and branching above, 4-8 dm tall. Flowers yellowish, 1-2.5 cm across. Fruiting calyx glabrous to pubescent, the berry yellowish to red. Common in prairie throughout the state. A highly variable species. Represented in South Dakota by the following 4 varieties: *virginiana*, common; *hispida* Waterfall, rare; *subglabrata* (Mack. & Bush.) Waterfall, rare; and *sonorae* (Torr.) Waterfall, common. July-Sept . . . . *Physalis virginiana* Mill.

SOLANUM L.

1 Stems prickly or spiny. Leaves with a stellate pubescence.

2 Leaves pinnately lobed. Calyx spiny. Plants annual, branched, 3-6 dm tall. Leaf blades 3-8 cm long, their petioles also spiny. Flowers yellow, the corolla 2-3 cm across when fully expanded. Calyx expanded and armed with spines, enclosing a berry. Very common in waste soil and over grazed pastures over the state. July-Sept. Buffalo Bur . . . . *Solanum rostratum* Dunal.

2' Leaves irregularly ovate in outline with a few coarse lobes. Calyx occasionally spiny. Plants perennial, the stems erect, branched, 4-8 dm tall. Leaf blades 4-8 cm long. Inflorescence several flowered, appearing in the upper axils, becoming racemiform. Flowers white to violet, up to 2 cm across. Fruit a berry 1 cm or more in diameter, not enclosed by the calyx. Infrequent in waste areas and on roadsides in the eastern part. July-Aug. Horse Nettle . . . . *Solanum carolinense* L.

1' Stems not prickly or spiny. Leaves lacking a stellate pubescence.

3 Plants perennial with spreading or climbing stems up to 1 meter or more tall. Flowers blue or lavender. Leaves

petioled, the blades broadly ovate, often with a pair of
basal lobes, the entire leaf 4-8 cm long. Flowers in dichotomously branched inflorescences in upper axils or opposite the leaves, the corollas 1 cm across, pale blue or
violet. Fruits red, 8-10 mm long. A mildly poisonous
plant of thickets and low moist areas found infrequently
in the state. June-July. European Bittersweet . . . . .
. . . . *Solanum dulcamara* L.

3' Plants annual with branched or spreading stems. Flowers
  white.

  4 Leaves entire to sinuately dentate, the blades ovate-
    lanceolate in outline. Vegetation lacking a fetid odor.

    5 Stems and leaves glabrous to strigose. Fruiting
      berry black at maturity. Plants 1-7 dm tall, irregu-
      larly branched. Leaves petioled, bright green, ap-
      pearing glabrous. Flowers clustered on peduncles,
      umbel-like, from the upper nodes. Corolla 6-10 mm
      across, white to pale violet. Fruits 8 mm in diam-
      eter. Common weed of fields and disturbed areas over
      the state. Cases of livestock poisoning due to eat-
      ing the immature plant have been reported. {*S.
      nigrum*, in part} June-Sept. Black Nightshade . . .
      . . . . *Solanum americanum* Mill.

    5' Stems and leaves villous to viscid-hairy, gland-
      tipped. Fruiting berry yellowish at maturity and
      its pedicel expanding at maturity at the junction
      with the berry. Plants 1-7 dm tall, irregularly
      branched. Leaves bright to dark green, softly hairy.
      Corolla 6-10 mm across, white. Fruits 5-8 mm in di-
      ameter, yellow. Much like the preceding and consid-
      ered previously to be part of it. Frequent to common
      in waste places over the state. {*S. nigrum*, in part}
      June-Sept. Nightshade . . . . *Solanum villosum* Mill.

  4' Leaves pinnately lobed, on well-developed petioles.
    Vegetation with a fetid odor. Plants 1-4 dm tall, the
    stems branched from the base. Leaf blades 2-5 cm long,
    pale green. Flowers on stout peduncles, borne laterally
    at upper nodes. Corolla 5-8 mm across. Fruits green,
    8-12 mm in diameter, the pedicels deflexed in fruit.
    Frequent in dry areas of the western part, especially
    in disturbed soils. July-Aug . . . .
    . . . . *Solanum triflorum* Nutt.

Family Scrophulariaceae

(Key to Genera)

1 Fertile stamens 4 in each flower (an apparent 5th one is a sterile filament).

  2 Leaves alternate.

    3 Corolla spurred . . . . *LINARIA*

    3' Corolla not spurred.

      4 Leaves pinnatifid . . . . *PEDICULARIS*

      4' Leaves entire or pinnately trifoliolate.

        5 Corolla twice the length or more than the calyx . . . . *ORTHOCARPUS*

        5' Corolla not more than twice the length of the calyx . . . . *CASTILLEJA*

  2' Leaves opposite or whorled.

    6 Flowers greenish-brown, in a terminal panicle . . . . *SCROPHULARIA*

    6' Flowers other than greenish-brown.

      7 Corolla regular or almost so. Plants of aquatic or marshy places.

        8 Plants with floating or creeping stems. Flowers yellow, in the axils of obovate rounded leaves . . . . *BACOPA*

        8' Plants small, stemless. Flowers white or blue, inconspicuous, on peduncles from the basal rosette of leaves . . . . *LIMOSELLA*

      7' Corolla bi-labiate or otherwise irregular. Plants from a variety of habitats.

        9 Central lobe of the lower corolla lip folded, forming a pouch which encloses the stamens . . . . *COLLINSIA*

        9' Central lobe of the lower corolla lip arched or flattened but not forming a pouch.

          10 Flowers with a sterile filament in addition to the 4 normal stamens . . . . . . *PENSTEMON*

          10' Flowers without a sterile filament.

           11   Corolla almost closed due to the arching
                of the lower lip. Leaves wider than 2 mm
                . . . . *MIMULUS*

           11' Corolla with an open throat. Leaves lin-
                ear, 1-2 mm wide . . . . . *AGALINIS*

1' Fertile stamens 5 or 2.

   12 Stamens 5 . . . . *VERBASCUM*

   12' Stamens 2 (sterile filaments may be present in addition).

      13 Flowers in terminal spike-like racemes. Leaves
          chiefly basal or whorled.

         14 Corolla present, white. Leaves whorled . . . . .
             . . . . *VERONICASTRUM*

         14' Corolla lacking, flowers greenish-brown. Leaves
             chiefly basal . . . . *BESSEYA*

      13' Flowers mostly in the axils of upper leaves. Leaves
          alternate or opposite.

         15 Sepals 4 or 4-lobed . . . . *VERONICA*

         15' Sepals 5 or 5-lobed.

            16 Corolla white or yellowish. Sterile filaments
                lacking or vestigial . . . . . . *GRATIOLA*

            16' Corolla blue-violet. Sterile filaments pres-
                ent . . . . *LINDERNIA*

                       AGALINIS Raf.

1 Corolla 2-2.5 cm long, the flowers on stout, almost erect
   pedicels. Fruiting capsules oblong, 7-10 mm long. Plants
   of prairies, annual, the stems 1-5 dm tall, branched above.
   Leaves linear, scabrous, 1-3 cm long. Flowers solitary in
   upper leaf axils, the corolla rose-purple. Infrequent in
   prairies and on plains over the state. {*Gerardia aspera*
   Dougl.} Aug-Sept . . . . . *Agalinis aspera* (Benth.) Britt.

1' Corolla 1-1.5 cm long, the flowers on slender, diverging pedicels. Fruiting capsules globose, 3-7 mm long. Plants of
   moist or marshy places, the stems much branched, 1-6 dm tall.
   Leaves linear, 1-4 cm long. Flowers solitary or in axillary
   racemes with up to 15 flowers. Petals light to dark purple.
   Frequent in moist soil of pond margins or shores over the
   state. {*Gerardia tenuifolia* Vahl.} Aug-Sept . . . .
   . . . . *Agalinis tenuifolia* (Vahl.) Raf.

BACOPA Aubl.

Bacopa rotundifolia (Michx.) Wetts. Plants perennial herbs of marshy or aquatic areas, the stems to 5 dm long, mostly submersed with the tips floating or exposed. Leaves opposite, the blades thin, obovate rounded, 1-3 cm across, with slightly toothed margins. Flowers axillary, solitary on short peduncles, the petals yellow, 6-8 mm long. Fruiting capsules ovoid, many seeded. Frequent in springs and shallow water of ponds over the state. July-Aug. Water Hyssop.

BESSEYA Rydb.

Besseya wyomingensis (A. Nels.) Rydb. Plants perennial herbs from a short caudex, the stems 2-4 dm tall, white hairy. Leaves mostly basal, the blades ovate to oblong, with coarse teeth, 2-6 cm long. Stem leaves much reduced. Inflorescence a spike-like raceme, the flowers without petals, the 2 stamens exsert. Fruiting capsule compressed ovoid, 5-6 mm long. Infrequent on open slopes and hillsides in the Black Hills. {*B. cinerea* (Raf.) Pennell.} May-June. Kitten-tails.

CASTILLEJA L.

1 Principal leaves trifoliolately pinnatifid. Corolla 4-5 cm long. Plants perennial, the stems clustered from a stout vertical caudex. Stems 1-3 dm tall, densely villous, gray-green. Leaves 4-5 cm long, the divisions linear. Inflorescence a terminal leafy spike, the flowers greenish yellow, extending beyond their bracts. Frequent to common on dry hillsides of plains and prairies over the state. May-June. Indian Paintbrush . . . . . . *Castilleja sessiliflora* Pursh

1' Principal leaves entire, some dissected into linear lobes. Corolla 1.5-2.5 cm long. Stems 1-several, mostly erect, 3-5 dm tall, from a perennial woody base, not villous-gray. Leaves lanceolate, mostly glabrous, 4-5 cm long. Inflorescence compact, terminally spikate with leafy bracts. Flowers green, not conspicuously distinguished from the bracts. Frequent in meadows and open wooded hillsides at medium to upper elevations in the Black Hills. June-July . . . .
. . . . *Castilleja sulphurea* Rydb.

COLLINSIA Nutt.

Collinsia parviflora Lindl. Plants annual, herbaceous, with weak, narrow stems 1-3 dm tall, simple to diffusely branched. Leaves opposite, oblong to linear, entire, 1-5 cm long, the ones subtending the flowers often whorled. Flowers solitary or clustered in upper axils, the corolla blue, 4-6 mm long, 2-lipped. Infrequent but may be locally abundant under trees on slopes in the Black Hills. May-June. Blue-Lips.

GRATIOLA L.

Gratiola neglecta Torr. Plants herbaceous annuals from fibrous roots, the stems simple to diffusely branched, 1-3 dm tall. Leaves 1-4 cm long, linear to oblong, the margins entire to denticulate, mostly glabrous. Flowers solitary on axillary peduncles, the petals yellow or white, 8-10 mm long, bilabiate. Fruiting capsules ovoid to globose, 5-7 mm long. Frequent on muddy shores of ponds, lakes and streams over the state. June-Aug. Hedge Hyssop.

LIMOSELLA L.

Limosella aquatica L. Plants herbaceous, annual or weak perennials essentially stemless with basal leaves. Leaves in a rosette-like arrangement, 5-20 mm long, the blades spatulate on slender petioles. Flowers solitary on slender scapes 1-3 cm long. Petals 2-3 mm long, white or pink. Fruiting capsule 2-4 mm long with many seeds. Rare in moist soil at margins of lakes and ponds over the state. June-Aug. Mudwort.

LINARIA Mill.

1   Flowers blue or blue with white. Plants annual or weakly perennial, the stems essentially simple, erect, 1-4 dm tall. Many times a basal rosette of prostrate branches is formed. Leaves linear, 1-4 cm long, opposite or sub-opposite. Flowers in terminal racemes, the petals 8-12 mm long, including the spur. Fruiting capsules 3-4 mm long, almost globose. Infrequent on moist or mesic sandy prairie in the western part of the state. Ours var. *texana* (Sch.) Penn. May-June . . . . .
. . . . *Linaria canadensis* (L.) Dumont.

1' Flowers yellow or yellow with red-orange. Plants perennial.

2   Principal leaves ovate to broadly lanceolate, clasping the stem, 3-7 cm long. Stems 7-12 dm tall, robust, glabrous to glaucous. Inflorescence terminal or paniculate, the flowers yellow. Petals 2-4 cm long, not including the spur. Fruiting capsule with seeds that are not winged. Apparently escaped from cultivation and spreading in the western part. June-July. Toadflax . . . .
. . . . *Linaria dalmatica* (L.) Mill.

2' Principal leaves linear to narrowly lanceolate, 2-6 cm long. Stems 1-6 dm tall, glabrous, from a dense, perennial

rhizome system. Flowers in compact terminal racemes, the
corolla, including the spur, 2-3 cm long. The throat of
the corolla with a red-orange palate. Fruiting capsules
up to 1 cm long, with many flattened seeds. Commonly
naturalized in waste places and on roadsides by abandoned
dwellings over the state. June-Aug. Butter and Eggs . . .
. . . . *Linaria vulgaris* Hill

LINDERNIA All.

1 Lower stem leaves rounded at their bases. Flowering pedicels
1.5-2.5 cm long, slender, conspicuously surpassing the subtending leaves. Plants annual, the stems branched, up to 2
dm tall. Leaves entire to remotely dentate, the blades 5-15
mm long, glabrous. Flowers solitary in the axils of upper
leaves, the corolla blue or purple, 6-8 mm long. Fruiting
capsules ovoid, 3-7 mm long, with many seeds. Rare in muddy
areas along ponds and lakes in the eastern part. July-Aug.
False Pimpernel . . . . . *Lindernia anagallidea* (Michx.) Penn.

1' Lower stem leaves tapered to their bases. Flowering pedicels
1-1.5 cm long, shorter or only slightly longer than the subtending leaves. Plants annual, the stems branched, 1-2 dm
tall. Leaves oblong to ovate, 1-3 cm long, remotely dentate.
Flowers solitary in the axils of upper leaves, the corolla
blue, 9-10 mm long. Fruiting capsules ovoid, 3-7 mm long,
with wrinkled seeds. Rare in muddy or wet, sandy places in
the eastern part. July-Aug . . . . *Lindernia dubia* (L.) Penn.

MIMULUS L.

1' Flowers blue or rarely white. Plants perennial, from a creeping rootstock. Stems 3-8 dm tall, simple, erect, glabrous.
Leaves opposite, sessile, lanceolate to narrowly oblong, 3-8
cm long, the margins irregularly crenate to entire. Flowers
solitary on pedicels in the upper axils, the corolla bilabiate, 2-3 cm long. Frequent in moist alluvial woods and along
waterways and in swamps in the eastern part. July-Aug.
Monkey-flower . . . . *Mimulus ringens* L.

1' Flowers yellow.

2 Plants perennial. Corolla bilabiate, more than 1 cm long.

3 Stems weak and decumbent or creeping, 1-4 dm long, often
rooting at the nodes. Corolla 1-2 cm long, the throat
open. Plants perennial from a rhizome. Leaves with
short petioles or sessile, the blades round-ovate to
reniform with dentate margins. Flowers on pedicels

over 1 cm long, the calyx teeth very blunt. Infrequent
in spring creeks and ponds in the eastern and western
parts. Ours var. *fremontii* (Benth.) Grant. June-July.
. . . . *Mimulus glabratus* HBK.

3' Stems erect or occasionally creeping, 1-6 dm tall.
   Corolla 1-4 cm long, the throat nearly closed by the
   upraised palate. Plants perennial, by stolons from
   rhizomes, glabrous to pubescent. Leaves ovate to rotund, irregularly dentate, variable in size. Flowers
   in upper axils of leaves or in terminal racemes. Calyx
   lobes acute, the upper one the largest. Infrequent in
   moist places and shallow water over the state. July-
   Aug . . . . *Mimulus guttatus* DC.

2' Plants annual. Corolla regular or only slightly bilabiate,
   less than 1 cm long. Stems and leaves usually with a soft,
   viscid pubescence. Plants erect to diffusely branched, the
   stems 1-3 dm tall. Leaves opposite, ovate to broadly deltoid with coarse, triangular teeth. Blades 5-30 mm long.
   Flowers in the upper leaf axils, usually solitary, the
   corolla 7-10 mm long, rarely to 13 mm, yellow, streaked
   with red. Rare in marshy bottoms of rich wooded ravines
   in the Black Hills. June-July . . . .
   . . . . *Mimulus floribundus* Dougl.

ORTHOCARPUS Nutt.

Orthocarpus luteus Nutt. Plants herbaceous annuals with simple
or rarely branched stems 1 4 dm tall. Leaves usually of two
kinds. The lower linear, 2-4 cm long, the upper ones usually
three-parted. Vegetation with soft, spreading hairs. Inflorescence a terminal leafy spike, the flowers perfect, corolla two-
lipped, bright yellow. Petals 1-1.5 cm long. Common in upland
and sandy prairie over the state. June-July. Owl's Clover.

PEDICULARIS L.

1 Leaves pinnately divided almost to the midrib, the segments
   with sharp serrations. Flowers 2.5-3 cm long. Plants perennial, the stems 5-15 dm tall. Leaves 2-4 dm long, with large
   lanceolate divisions. Inflorescence a terminal leafy spike,
   the flowers conspicuously yellow with pink or red streaks.
   Rare on shaded hillsides in the Black Hills. July-Aug . . . .
   . . . . *Pedicularis grayii* A. Nels.

1' Leaves pinnately lobed, the lobes rounded. Flowers smaller,
   2-2.5 cm long.

2 Upper corolla lip with a lateral tooth on each side near
   the apex. The lower lip of the corolla much shorter than
   the upper. Stems usually clustered from the base, 1-4 dm
   tall. Leaves pinnately lobed, 4-12 cm long. Flowering
   spike compact, the petals bright yellow, 2-2.5 cm long.
   Frequent in mesic prairies in the eastern part. May-June.
   Lousewort . . . . *Pedicularis canadensis* L.

2' Upper corolla lip lacking a pair of lateral teeth. The
   lower lip of the corolla almost as long as the upper.
   Stems usually single, simple or rarely branched above,
   3-7 dm tall. Leaves usually opposite, 5-10 cm long, with
   crenately toothed lobes. Inflorescence a terminal spike,
   leafy-bracted, compact. Petals yellow, 2-2.5 cm long.
   Infrequent in swampy or marshy soil at the base of springy
   hills in the eastern part. Aug-Sept . . . .
   . . . . *Pedicularis lanceolata* Michx.

PENSTEMON Mitch.

1 Corolla 3 cm or more long. Stems stout and glabrous, usually
   exceeding 4 dm.

   2 Flowers with the sterile filament glabrous or sparsely
      puberulent. Corolla usually deep blue, about 3 cm long.
      Leaves oblong to lanceolate, 5-10 cm long. Plants 3-6 dm
      tall, perennial. Frequent in dry prairie over the state,
      more common in the western one half. June-July. Smooth
      Beardtongue . . . . *Penstemon glaber* Pursh

   2' Flowers with the sterile filament long bearded. Corolla
      pink to deep lavender, 3-5 cm long. Leaves ovate to sub-
      circular, 3-8 cm long, glaucous. Plants with stout stems
      up to 8 dm tall. Flowers in a loose, terminal raceme with
      leafy bracts. Very common in prairies over the state.
      May-June. Large Beardtongue . . . .
      . . . . *Penstemon grandiflorus* Nutt.

1' Corolla usually less than 3 cm long. Stems usually not exceeding 4 dm.

   3 Plants glabrous throughout. Flowers usually blue. Stems
      1-3 dm tall, clustered from a perennial base, glabrous.
      Leaves linear-lanceolate, waxy green, 5-8 cm long. Inflo-
      rescence dense, the compact raceme often interrupted.
      Flowers rarely white, the petals 1.5-2 cm long. Lobes of
      the upper lip smaller than the lower. Frequent on dry
      prairie and exposed clayey slopes of the western part.
      Represented in South Dakota by the two subspecies

angustifolius and *caudatus* (Heller) Keck. May-June. Narrow Beardtongue . . . . *Penstemon angustifolius* Pursh

3' Plants pubescent or pubescent-glandular, at least in the inflorescence. Flowers lavender-pink to white.

    4 Throat of the corolla crested with many coarse yellow hairs which project from the throat. Sterile filament also long bearded with yellow hairs. Plants, including the leaves, villous pubescent, the stems stout, perennial, 1-4 dm tall. Leaves oblong to linear-lanceolate, varying much in length. Inflorescence dense and leafy-bracted, the corollas lavender. Infrequent in dry prairie in the western two thirds. May-June. Crested Beardtongue . . . . *Penstemon eriantherus* Pursh

    4' Throat of the corolla cylindric, not crested or with yellow hairs projected. Plants not densely villous pubescent.

        5 Flowers essentially white with a few blue purple spots, the corolla lobes flaring. The sterile filament short bearded. Plants perennial, the stems 1-4 dm tall, puberulent-sticky. Leaves lanceolate, 3-8 cm long, puberulent. Inflorescence dense, the corollas about 2 cm long. Common on dry prairies over the state. May-June. White Beardtongue . . . . *Penstemon albidus* Nutt.

        5' Flowers lavender, the corolla lobes ascending. The sterile filament long-bearded on the upper part. Plants perennial, the stems slender, 2-6 dm tall, glabrous below. Leaves lanceolate, 5-8 cm long, with fine teeth. Inflorescence an open raceme, the flowers 1.5-2 cm long. Frequent in prairies and dry places over the state. June-July. Slender Beardtongue . . . . *Penstemon gracilis* Nutt.

SCROPHULARIA L.

1 Sterile filament greenish-yellow, as wide as or wider than long. Corolla 7-11 mm long. Plants perennial from a thickened root, the stems 1-several, 5-15 dm tall. Leaves coarsely toothed and broadly lanceolate, 5-15 cm long, becoming smaller towards the upper part of the stem. Inflorescence terminal, branched and open, the flowers reddish brown. Fruiting capsule dull brown, 6-9 mm long. Frequent in low, moist openings and thickets over the state. June-July. Figwort . . . . *Scrophularia lanceolata* Pursh

1' Sterile filament brown or purple, longer than wide. Corolla
   5-8 mm long, reddish-brown. Stems erect, up to 2 meters tall,
   perennial from a deep rootstock. Leaves ovate to lanceolate,
   the margins coarsely serrate, 5-20 cm long. Inflorescence
   loosely branched and open. Fruiting capsule 4-7 mm long,
   shiny brown. Rare in thickets in the southeast part. July.
   . . . . *Scrophularia marilandica* L.

VERBASCUM L.

Verbascum thapsus L. Plants stout, biennial herbs with simple
stems to 1.5 meters. Vegetation densely covered with gray, tomentose hairs. The first year the plant produces a basal rosette of
large, oblanceolate leaves up to 3 dm long. The second year the
principal leaves are smaller and decurrent on the flowering stem.
Inflorescence a spike-like raceme, the flowers numerous. Corolla
yellow, up to 2 cm across. Fruiting capsules ellipsoid with septicidal dehiscence. Common as a weed of overgrazed pastures and
on roadsides, especially in the eastern part. July-Sept. Mul-
lein.

VERONICA L.

1 Stems with inflorescences in the axils of upper leaves, the
   main stem not terminating in an inflorescence. Leaves opposite throughout. Plants perennial.

   2 Plants glabrous. Leaves with entire or almost entire mar-
      gins. Flowers light blue. Plants of aquatic or semi-
      aquatic habitats.

      3 Principal leaves short-petioled. Plants 1-5 dm tall,
         from rhizomes, the stems ascending to erect. Leaf
         blades lanceolate to elliptic, 1-6 cm long, the margins
         usually serrated. Flowers in axillary racemes, not
         crowded. Fruiting capsule compressed, apically notched,
         usually wider than long. Quite common in streams and
         shallow water of the northeast part and in the western
         one-half. May-July. Brooklime . . . .
         . . . . *Veronica americana* Schw.

      3' Principal leaves sessile.

         4 Leaves over 3 times as long as wide. Maturing pedi-
            cels spreading. Plants perennial, the stems sprawl-
            ing to weakly erect, 2-6 dm tall. Leaves ovate to
            lance-shaped, 2-12 cm long, entire to finely serrate,
            often clasping. Inflorescence a spreading raceme,
            flowers white to blue. Infrequent in moist places

and shallow water in the eastern and western parts.
June-July . . . . *Veronica catenata* Penn.

4' Leaves less than 3 times as long as wide. Maturing
pedicels ascending or curved upwards. Plants 2-8 dm
tall, from spreading rhizomes, the stems sprawling
to decumbent. Leaves oblanceolate to oblong, widest
above the middle, the margins entire to serrate. Inflorescence racemose, many-flowered. Fruiting capsules not compressed or apically notched. Infrequent
in aquatic habitats in the west and eastern parts.
June-July . . . . . . *Veronica anagallis-aquatica* L.

2' Plants pubescent. Leaves with finely toothed margins.
Stems rooted at the lower portions, creeping or decumbent,
5-25 cm long. Leaves elliptic to obovate, 2-5 cm long,
petioled. Flowers in axillary racemes, the petals blue,
3-6 mm across. Fruiting capsule 3-4 mm across, broadly
triangular. Rare in disturbed sites and on roadsides in
the eastern part. May-July . . . . *Veronica officinalis* L.

1' Stems terminating in inflorescences with additional inflorescences in axillary positions. Lower leaves opposite, upper
leaves alternate to sub-opposite. Plants annual.

5 Fruiting pedicels short, less than 2 mm long, even in
fruit.

6 Corolla white. Principal leaves linear-oblong to oblong, up to 2.5 cm long. Plants with stems 5-30 cm
long, simple or more often branched below. Leaves irregularly toothed or rarely entire, 1-3 cm long. Fruiting capsule broader than long, deeply notched, 4 mm long.
Infrequent on hillsides, in valleys and along streams
over the state. Ours var. *xalapensis* (H.B.K.) St. John
and Warren. May-Sept . . . . . . *Veronica peregrina* L.

6' Corolla blue. Principal leaves ovate to elliptic,
toothed, not over 1.5 cm long. Stems erect or ascending, 5-20 cm tall, simple to occasionally branched,
pubescent. Leaves 5-15 mm long, petioled, the upper
ones often sessile. Fruiting capsule obcordate, 2 mm
long. Rare in waste places of the eastern part. Apr-
May . . . . *Veronica arvensis* L.

5' Fruiting pedicels longer, 1-4 cm long. Plants lax, loosely
ascending, the stems finely to coarsely pubescent, 1-4 dm
long, often rooting at the lower nodes. Leaves opposite
to alternate, petioled, the blades ovate, 1-2 cm long.
Flowers blue, the petals up to 1 cm wide. Fruiting

capsules 6-8 mm wide, notched at the apex, not as long as wide. Rare in moist places in the Black Hills at middle altitudes. Apr-May . . . . *Veronica persica* Poir.

VERONICASTRUM Fabr.

Veronicastrum virginicum (L.) Farw. Plants perennial herbs with several stems, 5-15 dm tall. Principal leaves whorled, less commonly opposite, on short petioles, the blades lanceolate, 6-14 cm long, the margins finely serrate. Inflorescence a slender terminal spike 5-15 cm long, crowded with white to pink flowers. Petals 7-9 mm long, stamens projecting from the flowers. Rare to infrequent in low meadows and moist thickets in the eastern part. June-July. Culver's Root.

Family Orobanchaceae

OROBANCHE L.

1 Flowers solitary on pedicels 3-6 cm long, lacking bracts at the base of the calyx. Plants yellow to brown, lacking chlorophyll and parasitic on roots of various species but especially those of *Artemisia*. Stems up to 1 dm above soil level. Leaves bract-like and soon deciduous, non-functional. Flowers purplish or brown, the petals 1.5-2.5 cm long. Frequent in prairies of the central and western parts. June-July . . . . *Orobanche fasciculata* Nutt.

1' Flowers in a dense spicate inflorescence, sessile, with bracts at the base of each calyx. Plants 1-2.5 dm tall, parasitic on *Artemisia* and *Ambrosia* in our area. Flowers purple with yellow bearded areas in the throat of the corolla tube, the petals 1.5-2 cm long. Frequent in dry prairies over the state. July-Aug. Broom-rape . . . . . *Orobanche ludoviciana* Nutt.

Family Bignoniaceae

(Key to Genera)

1 Leaves pinnately compound. Woody vines . . . . *CAMPSIS*

1' Leaves simple, whorled. Trees . . . . *CATALPA*

CAMPSIS Lour.

Campsis radicans (L.) Seem. Plants perennial climbing vines, woody, the stems up to 8 meters. Leaves pinnately compound,

leaflets 7-11, each one 4-8 cm long, with coarsely toothed margins. Inflorescence clustered in panicles at the ends of
branches, the flowers red-orange, 6-8 cm long. Corolla funnelform, slightly irregular. Fruiting capsules flattened, oblong,
10-15 cm long. Occasionally persisting in fencerows and at
abandoned farmyards in the eastern part. July-Aug. Trumpet-
creeper.

CATALPA Scop.

Catalpa speciosa Warder. Trees to 25 meters, the trunk well
developed and branched above. Leaves whorled, usually 3 at a
node, petioled. Leaf blades broadly ovate to rounded with acute
tips 10-25 cm long, softly hairy. Inflorescence a large panicle
of white with purple spotted flowers, each 3-5 cm long. Fruits
elongate capsules 2-5 dm long with winged seeds. Infrequently
planted and persisting. June-July. Catalpa.

Family Lentibulariaceae

UTRICULARIA

1 Flowers with the corolla up to 6 mm across; the lower lip 4-8
   mm long, about twice as long as the upper one. The spur at
   the base of the lower part of the corolla a small protuberance. Leaves numerous, up to 1 cm long, their bases three-
   parted and their divisions becoming unequally branched. Each
   leaf with several bladders. Stems rooted in the mud in shallow water, the flowering peduncles emergent. Infrequent in
   shallow water of ponds in the northeastern part. July-Aug . .
   . . . . *Utricularia minor* L.

1' Flowers with the corolla up to 12 mm across; the lower lip
   10-20 mm long, not much longer than the upper one. The spur
   at the base of the lower part of the corolla conspicuously
   developed. Leaves 1-3 cm long or longer, 2 parted at their
   bases and their divisions becoming pinnately divided. Each
   leaf with many bladders. Flowers 1-several on peduncles that
   emerge from the water. Stems creeping on the bottom and ascending. Frequent to common in shallow water of ponds and
   lakes over the state. July-Aug. Bladderwort . . . .
   . . . . *Utricularia vulgaris* L.

Family Martyniaceae

PROBOSCIDEA Schmid.

Proboscidea louisianica (Mill.) Thell. Plants annual or weak
biennials with branched stems 2-7 dm tall. Leaves ovate to
rounded, the blades 5-25 cm across, petioled. Vegetation with
dense, coarse hairs. Flowers in irregular clusters at the upper
portion of the stem, the petals white, tinged with rose. Corolla
bilaterally symmetrical, 9-14 cm long. Fruiting pod tapered,
curved from a large base, 8-16 cm long, 2-valved. An occasional
or rare weed in waste places over the state. July-Aug. Unicorn
Plant.

Family Phrymaceae

PHRYMA L.

Phryma leptostachya L. Plants perennial herbs with leafy stems
2-8 dm tall. Leaves opposite, petioled, the blades ovate-
elliptic, 4-12 cm long, with serrated margins. Inflorescence
a spike or racemiform, terminal, the flowers up to 1 cm long.
Petals white, tinged rose or lavender, bilaterally symmetrical.
Fruiting achenes closely appressed downward on the spike, each
4-6 mm long. Frequent in alluvial and rich upland woods of the
eastern part and in the Black Hills. June-July. Lopseed.

Family Plantaginaceae

PLANTAGO L.

1   Leaves linear, less than 1 cm across at the widest part.

    2   Spikes 3 mm wide or narrower. Leaves seldom more than 1 mm
       wide. Plants annual, only occasionally more than 15 cm
       tall. Flowers with 2 stamens, dioecious or polygamous.
       Corolla lobes less than 1 mm long. Fruiting capsules 2-3
       mm long, each with 4-6 seeds. Rare to infrequent in wet
       meadows and prairie swales over the state. June . . . . .
       . . . . *Plantago elongata* Pursh

    2' Spikes 4 mm or more wide. Leaves more than 2 mm wide.
       Bracts subtending the flowers longer than the calyx, promi-
       nent. Plants annual to weakly perennial, from a short tap-
       root. Flowering scapes and leaves densely villous to white-
       pubescent. Flowers with 4 stamens, the sepals 2-3 mm long.
       Fruiting capsules oblong, with 2 seeds. Common in dry

soils over the state. Represented in South Dakota by the
two varieties *patagonica* and *spinulosa* (Dcne) Gray. {P.
purshii R. & S.} June–July. Buckhorn . . . .
. . . . *Plantago patagonica* Jacq.

1' Leaves wider than 1 cm at the widest part.

    3 Leaf blades lanceolate, gradually tapering into the petiole.

        4 Spikes slender, 7 cm or longer, sparsely flowered at the
lower part with the spike interrupted. Plants densely
brown-woolly at the base of the petioles, perennial, the
rootstocks long. Flowering scapes 10-40 cm tall. Flowers perfect, the corolla lobes 1-2 mm long. Fruiting
capsules 3-4 mm long, with 2-4 seeds. Infrequent in
prairie, swales and alkali marshes over the state.
June–July . . . . *Plantago eriopoda* Torr.

        4' Spikes less than 7 cm long, densely flowered.

           5 Plants commonly 4 dm or more tall, the flowering
scapes slender and flexuous. Spikes tapered, up to
1 cm in diameter at the base. Bracts and sepals glabrous to inconspicuously pubescent. Bracts longer
than the sepals. Leaves with prominent parallel
veins. Flowers with 4 stamens exsert and conspicuous. Fruiting capsules with 1-2 seeds, black and
shiny. Frequent as a weed in moist places, occurring sporadically over the state. May–July. English
Plantain . . . . *Plantago lanceolata* L.

           5' Plants 20 cm or less tall, the scapes erect. Bracts
and sepals hirsute. Petals erect and pointed, exceeding by twice the length of the sepals after the
capsule begins to mature. Bracts shorter than the
calyx. Plants annual or biennial, the scapes to 1.5
dm tall. A rare weed in dry and sandy soil in the
eastern part. May–June . . . . *Plantago virginica* L.

    3' Leaf blades broadly ovate, their bases abruptly tapering
to the petiole. Bracts and sepals elevated to form a
prominent keel.

        6 Fruiting capsules 3-6 mm long, slender, the circum-
scissile dehiscence line below the middle. Bracts
narrowly triangular. Plants perennial, the leaf blades
broadly elliptic, 5-15 cm long and over half as wide.
Spikes 5-20 cm long, loosely flowered. Capsules with
4-10 seeds, black. Infrequent in lawns, gardens and
disturbed areas. July–Aug. Plantain . . . .
. . . . *Plantago rugelii* Decne.

6' Fruiting capsules 2-5 mm long, stout, the circumscissile
dehiscence line at about the middle. Bracts broadly
ovate to obtuse. Plants perennial, the leaf blades
broadly elliptic with petioles hairy at the base.
Spikes about as long as the preceding, the flowers
mostly sessile. Fruits ellipsoid, with 6-25 seeds,
each about 1 mm long. A common weed naturalized from
Europe and now common in lawns, gardens and waste areas
over the state. June-July. Common Plantain . . . . . .
. . . . *Plantago major* L.

Family Rubiaceae

GALIUM L.

1 Ovary and fruit glabrous.

2 Flowers yellow, numerous in panicles. Leaves linear, usually 6 in a whorl, sharply pointed. Plants 3-10 dm tall,
erect. Stems puberulent, at least in the upper part. Occasionally escaped and persisting as a weed at abandoned
farmsteads and on roadsides. Rare in the eastern part.
June. Yellow Bedstraw . . . . *Galium verum* L.

2' Flowers white, solitary or in 2's or 3's. Leaves linear-
ovate, not sharply pointed.

3 Pedicels beneath the flowers scabrous and curved, from
5-10 mm long. Flowers solitary. Corolla 3-lobed.
Leaves 1-2 mm wide, usually in 6's at a whorl. Stems
diffuse, slender, 2-8 dm tall, prostrate to ascending
with recurved scabrous hooks on the angles. Rare in
moist soil in thickets and openings of low places in
Day and Roberts counties of the northeast and in the
Black Hills. July-Aug . . . . *Galium trifidum* L.

3' Pedicels beneath the flowers smooth, not curved, and
2-6 mm long. Flowers in 2's or 3's, the corolla 4-
lobed. Leaves 2-3 mm wide, the principal ones in
whorls of 4, 1-3 cm long. Stems prostrate-spreading
to erect, often diffusely branched, 2-7 dm tall. Fruits
smooth, 3 mm across. Infrequent in low meadows and
along streams and ponds over the state. May-June . . .
. . . . *Galium obtusum* Bigel.

1' Ovary and fruit pubescent, or at least with hairs.

4 Leaves 4 in each whorl, the tips not sharply pointed.
Plants perennial, the stems erect, to 5 dm tall, mostly

glabrous. Flowers many in dense panicles, the corolla
white. Petals 4-7 mm long. Fruits 2 mm wide, mostly
pubescent or with hairs. Frequent to common on rocky
ledges, along streams and in low meadows or thickets
over the state. May-June. Northern Bedstraw . . . . . .
. . . . *Galium boreale* L.

4' Leaves 6 or more in each whorl, the tips sharply pointed.
Stems usually weak and decumbent.

    5 Stems with stiff, downward pointed hairs. Leaves usu-
ally not much more than 5 mm at their widest part.
Plants annual, the stems weak and spreading. Flowers
white, 3-5 on axillary peduncles. Fruits 2-4 mm across.
Common in woods, thickets and alluvial areas over the
state. Apr-May. Cleavers . . . . . . *Galium aparine* L.

    5' Stems smooth, without pointed hairs. Leaves narrowly
oval, to 1 cm wide. Leaf blades 2-6 cm long. Plants
perennial, the stems spreading or scrambling, 3-6 dm
long. Inflorescence terminal or axillary, 3 or more
flowered, the petals greenish-white. Fruits 2-3 mm
wide. Frequent in woods and moist thickets over the
state, especially in the northeast and the Black Hills.
July-Aug. Sweet-scented Bedstraw . . . .
. . . . *Galium triflorum* Michx.

Family Caprifoliaceae

(Key to Genera)

1 Style short or lacking. Inflorescence densely branched, with
many flowers. Leaves simple or compound.

    2 Leaves pinnately compound. Fruits with 3-5 seeds . . . .
. . . . *SAMBUCUS*

    2' Leaves simple. Fruits with 1 seed . . . . . *VIBURNUM*

1' Style obvious, elongate. Inflorescence few flowered, variously shaped. Leaves simple.

    3 Plants trailing or stems creeping, with evergreen leaves.
Flowers in pairs, nodding . . . . *LINNAEA*

    3' Plants with erect or twining stems.

        4 Corolla less than 8 mm long. Flowers radially sym-
metrical . . . . *SYMPHORICARPOS*

4' Corolla more than 1 cm long. Flowers bilaterally symmetrical to nearly radially symmetrical . . *LONICERA*

LINNAEA L.

Linnaea borealis L. Perennial shrubby plants with twining or
trailing stems up to 5 dm long. Leaves evergreen, opposite, the
blades broadly ovate, 1-2 cm long. Flowers in pairs on axillary
peduncles 3-6 cm long. Corolla white to rose-pink, 1 cm long or
slightly longer, funnelform. Frequent in rich woods of ravines
and hillsides at medium altitudes in the Black Hills and Harding
County. {*L. americana* Forbes} July-Aug. Twinflower.

LONICERA L.

1 Flowers clustered at terminal portions of stems, each cluster
   subtended by a pair of basally fused leaves. Stems shrubby,
   trailing or twining, 2 or more meters long. Leaves broadly
   elliptic to rounded, 4-8 cm long, glaucous. Flowers yellow
   to red, the corolla to 3 cm long, funnelform, with a gibbous
   base. Infrequent on rocky or rich wooded slopes of the eastern part and in the Black Hills. {*Lonicera glaucescens* Rydb.}
   Ours var. *glaucescens* (Rydb.) Butters. May-June. Wild Honeysuckle . . . . *Lonicera dioica* L.

1' Flowers paired at terminal and axillary locations, not subtended by a pair of fused leaves. Plants shrubby, branched,
   to 4 meters tall. Leaves ovate to oblong, 3-5 cm long, tending to have blades with subcordate bases. Flowers on axillary
   peduncles, the corolla white to pink or red, about 2 mm long,
   glabrous. Fruits red, the berries 5-7 mm across. Widely
   planted as hedges over the state and occasionally escaped.
   May-June. Tatarian Honeysuckle . . . . . *Lonicera tatarica* L.

SAMBUCUS L.

1 Inflorescence flat topped or nearly so with the central axis
   not extending up into the panicle, up to 8 cm wide. The pith
   of the older twigs whitish. Woody shrubs to 4 meters, forming
   thickets. Leaves with 5-11 leaflets with 7 the common number.
   Each leaflet lance-ovate, sharply serrate, 5-12 cm long.
   Flowers white, the corolla rotate. Fruits purple at maturity,
   4-5 mm in diameter, edible. Frequent to common in alluvial
   areas and along fencerows in the eastern one-half. May-June.
   Elderberry . . . . *Sambucus canadensis* L.

1' Inflorescence pyramidal in outline, the central axis extending
   up beyond the lower branches, less than 8 cm wide. The pith
   of older twigs brown. Woody shrubs 2-4 meters tall. Leaflets

5-7, lanceolate to ovate, narrowed and oblique at their bases.
Flowering cyme 5-6 cm long, flowers white, turning brown.
Fruits amber or reddish at maturity, 4-6 mm in diameter, not
edible. Frequent in valleys, ravines and along streams in
the Black Hills. {*S. pubens* Michx. and *S. melanocarpa* Gray}
Ours ssp. *pubens* (Michx.) House. May-June . . . .
. . . . *Sambucus racemosa* L.

SYMPHORICARPOS Duham.

1 Stamens exsert, definitely longer than the corolla. Petals
5-9 mm long, rose-pink. Style exceeding 4 mm, also exsert
from the corolla. Sparingly branched shrub usually over 5
dm tall. Leaves usually exceeding 3 cm, ovate, with entire
to undulate or lobed margins. Fruits white, 6-9 mm long.
Common throughout the state in a variety of habitats, especially overgrazed pastures and hillsides. June-July. Wolf-
berry . . . . *Symphoricarpos occidentalis* Hook.

1' Stamens included in the corolla, not projecting from the
throat. Petals 4-6 mm long, white or pink. Style 2-3 mm
long. Plants shrubby, branched, 2-6 dm tall or taller.
Leaves 1-3 cm long, ovate to oblong, pale green beneath with
a dense pubescence. Flowers in short spikes or in pairs,
pedicelled. Corolla 5-8 mm long, the lobes conspicuous.
Fruits white, 6-9 mm in diameter. Infrequent in ravines and
on shaded hillsides in the Black Hills. {*Sambucus racemosus*
var. *pauciflorus* Robbins} May-June. Snowberry . . . .
. . . . *Symphoricarpos albus* (L.) Blake

VIBURNUM L.

1 Leaves not lobed. Leaf margins finely to coarsely serrate.

  2 Margin of leaf with 1 serration or tooth per millimeter.
Leaf blades ovate to oblong, 5-8 cm long, tapering to an
abruptly acuminate apex. Shrubs or shrubby trees to 6
meters tall. Inflorescence a cyme 5-8 cm across, the
flowers white, 4-8 mm wide. Fruits blue to black, glaucous, 6-10 mm across, ellipsoid. Infrequent in rich wooded
hillsides and in ravines in the eastern part and in the
Black Hills. June-July. Nannyberry . . . .
. . . . *Viburnum lentago* L.

  2' Margin of leaf with 1-3 serrations or teeth per centimeter.
Leaf blades lance-ovate to rounded, 3-7 cm long, with 7-10
teeth on each side. Plants shrubby, to 1 meter tall. Inflorescence of short branched cymes, the flowers creamy
white. Fruits black, up to 1 cm long, ellipsoid. Infre-

quent to rare in rich, loamy woods of the southeast part.
Ours var. *affine* (Bush) House.   May–June . . . .
. . . . *Viburnum rafinesquianum* Schultes

1' Leaves shallowly to deeply lobed.   Leaf margins variable, not
regularly serrated.

    3 Petioles with conspicuous glands at the summit immediately
below the base of the blade.   Leaves lobed to a depth of 2
cm, the lobes entire or with remote serrations.   Plants
shrubby, to 2 meters or more tall.   Inflorescence cymose,
flat topped, the peripheral or marginal flowers often enlarged and neutral.   Frequent in rich wooded ravines in
the northern Black Hills.   {*Viburnum trilobum* Marsh}   Ours
var. *americanum* Ait.   June–July.   Highbush Cranberry . . .
. . . . *Viburnum opulus* L.

    3' Petioles lacking conspicuous glands at their summit.   Leaves
shallowly lobed, with coarse serrations.   Leaf blades 5–8
cm long and almost as wide, hirsute beneath.   Small shrubs
usually less than 1 meter tall.   Inflorescence a few flowered cyme with a pair of leaves subtending it.   Corolla
whitish, 4–6 mm long, the flowers all alike.   Fruits red
or orange, about 1 cm long.   Rare in rich wooded ravines
in the northern Black Hills.   {*Viburnum eradiatum* (Oakes)
House}   June–July.   Squashberry . . . .
. . . . *Viburnum edule* (Michx.) Raf.

Family Adoxaceae

ADOXA L.

Adoxa moschatellina L.   Plants small perennial herbs with tuberous rootstocks, the stems 5–12 cm tall.   Leaves ternately compound, mostly basal, the blades glabrous, 0.5–2.0 cm across, with
a musky odor.   Flowers greenish, 3–6 in small terminal clusters,
each 5–8 mm wide.   Petals 4–6 lobed.   Rare in moist wooded canyons with cold air drainage.   Rare to infrequent in canyons of
the Black Hills.   May–July.   Moschatel.

Family Valerianaceae

VALERIANA L.

1 Plants with a stout taproot, the stems 2–6 dm tall, glabrous.
Lower leaves numerous, oblanceolate or spatulate, mostly basal,
the blades gradually tapering to a petiole base.   Cauline

leaves smaller than the basal leaves, pinnatifid into 3-7
lateral divisions. Inflorescence at maturity 10-40 cm long,
panicled. Petals 1-3 mm long, off white to yellow. Achenes
usually short-hairy, 3-5 mm long. Frequent in damp meadows
and hillsides in the Black Hills. June-July . . . .
. . . . *Valeriana edulis* Nutt.

1' Plants with a short caudex and many fibrous roots. Lower
leaves differentiated into blades and petioles. Plants generally smaller in stature than the preceding, 1-4 dm tall.

   2 Corolla mostly 2-4 mm long. Inflorescence in fruit elon-
      gate and paniculiform. Cauline leaves sessile, pinnatifid,
      with 1-7 pairs of lateral lobes. Blades of basal leaves up
      to 7 cm long, mostly undivided. Corolla white, the stamens
      exsert. Achenes lanceolate, glabrous, 2-5 mm long. Infre-
      quent in moist woods and valleys in the Black Hills. {*V.*
      septentrionalis Rydb.} June-July . . . *Valeriana dioica* L.

   2' Corolla longer, 4-7 mm long. Inflorescence in flower and
      fruit compact, somewhat capitate. Cauline leaves subses-
      sile, pinnatifid, with a few reduced lateral lobes. Basal
      leaves well developed, the blade up to 8 cm long. Corolla
      white, the petals 4-7 mm long. Achenes short-hairy to gla-
      brous, 3-6 mm long, linear-oblong. Infrequent to rare in
      moist woods of valley floors in the Black Hills. June-
      July . . . . *Valeriana acutiloba* Rydb.

Family Dipsacaceae

(Key to Genera)

1 Stem prickly. Receptacular bracts stiff, awn-sharpened,
persistent. Plants often exceeding 1 meter . . *DIPSACUS*

1' Stem not prickly. Receptable densely hairy, lacking awn-like
bracts. Plants less than 1 meter tall . . . . . *KNAUTIA*

DIPSACUS L.

Dipsacus sylvestris Huds. Plants stout biennials or weak perennials up to 2 meters tall, the stem with numerous prickles above.
Leaves large, lanceolate, with crenate margins. Flowers in terminal heads subtended by several upward arching bracts. Corolla
slender, with a white tube, almost hidden between the awn-like
bracts. Often confused as a member of the Asteraceae. Rare as
an escape in the eastern part. July-Sept. Teasel.

KNAUTIA L.

Knautia arvensis (L.) Duby. Plants annual herbs 4-7 dm tall, sparingly branched. Leaves pinnately divided, the principal ones on the lower one half of the stem, becoming reduced upwards. Flowers in dense terminal heads with small involucral bracts, the heads 2-3 cm across. Flowers pale lavender, the corolla 8-10 mm long. Receptacle hairy. Rare as an escape along fields and roadsides and in waste places. June-Sept. Blue Buttons.

Family Cucurbitaceae

(Key to Genera)

1 Corolla of male flowers with 6 lobes. Fruits inflated, up to 4 cm long; several seeded . . . . *ECHINOCYSTIS*

1' Corolla of male flowers with 5 lobes. Fruits not inflated, approximately 2 cm long; with one seed . . . . *SICYOS*

ECHINOCYSTIS T. & G.

Echinocystis lobata (Michx.) T. & G. Plants annual, climbing herbs, often to 6 meters tall. Leaves 5 lobed, angular, 7-10 cm across, with tendrils opposite the leaves. Plants monoecious, the male flowers in erect racemes, flowers up to 1 cm across. Female flowers sessile or short, peduncled in axils, the fruits becoming inflated to 4 cm long, prickly. Common in low or moist alluvial woods and thickets over the state. July-Aug. Wild Cucumber.

SICYOS L.

Sicyos angulatus L. Plants annual climbing herbs with branched tendrils, the stems 1-6 meters long. Stems often growing on other vegetation. Leaves broadly angular, shallowly 3-5 lobed, up to 10 cm wide. Plants monoecious, the male flowers on short peduncles. Female flowers clustered, the maturing fruits eventually on long peduncles. Fruits 1-3 cm long, prickly but not inflated. Infrequent in alluvial areas of the eastern part. July-Aug. Bur Cucumber.

Family Campanulaceae

(Key to Genera)

1 Plants perennial. Flowers in a terminal inflorescence or solitary, pedicelled . . . . *CAMPANULA*

1' Plants annual. Flowers axillary, sessile or almost so . . . . *TRIODANIS*

CAMPANULA L.

1 Flowers in terminal spikes or racemes. Upper leaves obovate to lanceolate.

  2 Corolla rotate. Style curved. Leaves thin, evenly serrate, the blades ovate, petioled, 7-15 cm long. Stems erect, usually solitary, many times more than 1 meter tall. Flowers in leafy bracted spikes at the upper part of the stem, the corolla spreading, about 2.5 cm across. Infrequent in alluvial woods in the southeast part. July-Aug. Tall Bellflower . . . . *Campanula americana* L.

  2' Corolla campanulate. Style erect. Leaves coarse, unevenly serrate, the blades ovate-cordate, long petioled, 6-13 cm long, becoming smaller upwards on the stem. Inflorescence spicate, the flowers on peduncles, nodding, subtended by bracts. Corolla blue-lavender, 2-3 cm long. Occasionally escaped from cultivation and persisting over the state. June July . . . . *Campanula rapunculoides* L.

1' Flowers solitary or in a loosely panicled cluster. Upper leaves linear or narrowly lanceolate.

  3 Stems erect, 1-4 dm tall. Leaves 2-6 cm long, the principal ones long-petioled, their blades ovate with coarse dentations. Flowers blue, the corolla exceeding 1 cm, drooping in a loosely panicled cluster. Frequent on sandy ridges, wooded hillsides and in meadows of the western one-half. {*C. petiolata* A. DC.} June-Aug. Blue Bell . . . . *Campanula rotundifolia* L.

  3' Stems weak, almost filiform, 0.5-3 dm tall. Leaves linear to narrowly lanceolate, 1-4 cm long. Flowers white to pale blue, solitary on the ends of slender pedicels. Corolla open, campanulate, 5-8 mm long. Rare in marshy or boggy places in the northeast and in the sandhills of the south and western part. July-Aug. Marsh Bellflower . . . . *Campanula aparinoides* L.

TRIODANIS Raf.

1 Leaves narrow, several times longer than wide. Capsules becoming prominent towards maturity, up to 2 cm long. Stems
   simple, erect, up to 3 dm tall. Leaves sessile, mostly lanceolate, 1-3 cm long. Inflorescence spicate, the flowers 8-10
   mm long, lavender or blue. Sepals subulate, 3-5, persistent.
   Frequent in dry soil of prairies and open places over the
   state. {*Specularia leptocarpa* (Nutt.) Gray} June-July . . .
   . . . . *Triodanis leptocarpa* (Nutt.) Nieuwl.

1' Leaves broad, rounded and sessile, appearing to clasp the
   stem and hiding the maturing capsules. Capsules not becoming
   elongate, usually less than 7 mm long. Stems simple, erect,
   2-5 dm tall, mostly hispid or scabrous. Leaves ovate to cordate, 1-3 cm across, as wide as or wider than long, toothed.
   Flowers lavender to deep purple, the corolla 8-12 mm long on
   open flowers. Fruiting capsule opening at about the middle.
   Frequent in waste or disturbed soil over the state. {*Specularia perfoliata* (L.) A. DC.} June-July . . . .
   . . . . *Triodanis perfoliata* (L.) Nieuwl.

Family Lobeliaceae

LOBELIA L.

1 Flowers 1.5-2 cm or longer, blue. Inflorescence terminal,
   racemose. Stems simple, erect, 2-8 dm tall. Leaves obovate
   to oblong, 8-12 cm long, with decurrent bases. Upper leaves
   becoming bract-like in the inflorescence. Flowers strongly
   bilaterally symmetrical. Fruits capsular, 7-10 mm long,
   partly inferior. Common in marshy areas, margins of ponds
   and in alluvial thickets over the state. Aug-Sept. Blue
   Cardinal Flower . . . . *Lobelia siphilitica* L.

1' Flowers less than 1 cm long, pale blue to white. Inflorescence a terminal spike or narrow raceme with narrow, leafy
   bracts. Leaves oblanceolate or lanceolate, the principal
   ones 4-8 cm long and up to 1 cm wide, the upper ones reduced.
   Flowers irregular, on short pedicels or almost sessile.
   Fruits short, 2-6 mm long, partly inferior. Frequent in
   moist prairies, prairie swales or meadows over the state.
   July-Aug. Spiked Lobelia . . . . *Lobelia spicata* L.

Family Asteraceae

(Key to Genera)

1 Flower heads composed of perfect ray flowers only; often times the stems and leaves with white milky juice (Liguliflorae).

  2 Pappus only of simple capillary bristles.

    3 Leaves all basal. Flower heads solitary on naked scapes from soil level.

      4 Margin of leaves entire, linear . . . . *AGOSERIS*

      4' Margin of leaves deeply lobed to lacinate . . . . *TARAXACUM*

    3' Leaves cauline as well as basal. Flowering heads 1-several on the main stem or on leafy branches.

      5 Achenes flattened. The leaf margins or midribs with sharp prickles.

        6 Heads few-flowered (less than 50), the achenes with a constricted beak to which the pappus is attached . . . . *LACTUCA*

        6' Heads many-flowered (more than 50), the achenes beakless . . . . *SONCHUS*

      5' Achenes not strongly flattened.

        7 Principal leaves linear, much reduced. Flowers rose-pink to purple . . . . *LYGODESMIA*

        7' Principal leaves not reduced or linear. Flowers variously colored.

          8 Heads nodding. Flowers cream-colored, greenish or pink . . . . *PRENANTHES*

          8' Heads erect. Flowers yellow or white.

            9 Pappus white. Achenes tapering upwards. Flowers yellow . . . . *CREPIS*

            9' Pappus sordid yellow to brown. Achenes not tapering gradually upwards. Flowers yellow but white in one species . . . *HIERACIUM*

  2' Pappus of simple bristles *and* scales, of plumose bristles, or of scales only.

    10 Flowers blue. Pappus of chaffy scales . . . . *CICHORIUM*

10' Flowers yellow.

    11 Pappus of bristles and narrow scales . . . . *MICROSERIS*

    11' Pappus plumose, the branches laced and webbed . . . . *TRAGOPOGON*

1' Flower heads *not* composed of ray flowers only (ray and disks present, or disk flowers only). (Tubuliflorae)

  12 Heads of ray and disk flowers.

    13 Ray flowers yellow or yellow-orange.

      14 Receptacle surface with chaffy scales at the base of the achenes . . . . *GROUP 1*   page 406

      14' Receptacle surface naked, lacking chaffy scales at the base of the achenes . . . *GROUP 2*   page 408

    13' Ray flowers white, pink, red, blue or purple colored, or some color intermediate between any of these . . . . *GROUP 3*   page 409

  12' Heads of disk flowers only (those with very small and inconspicuous ray flowers are included here as well as above).

    15 Receptacle surface with chaffy scales or bristles at the base of the achenes . . . . . . *GROUP 4*   page 410

    15' Receptacle surface naked; lacking scales or bristles at the base of the achenes . . . . *GROUP 5*   page 411

GROUP 1 (Heads with yellow to yellow-orange rays. Receptacle chaffy.)

1 Principal leaves alternate or basal.

  2 Receptacle chaffy only near the periphery . . *MADIA*

  2' Receptacle chaffy or bristly over its entire surface.

    3 Leaves all or some of them lobed or deeply divided.

      4 Receptacle merely bristly . . . . *GAILLARDIA*

      4' Receptacle definitely chaffy.

        5 Surface of receptacle conical or columnar.

          6 Achenes flattened . . . . *RATIBIDA*

          6' Achenes 4-sided . . . . *RUDBECKIA*

        5' Surface of receptacle flattened or convex.

       7 Ray flowers yellow, 3-5 cm long, in 2 or more series . . . . *SILPHIUM*

      7' Ray flowers golden yellow, 1 cm long, in 1 series . . . . *ENGELMANNIA*

3' Leaves not deeply divided.

    8 Plants annual, leaf-blades rhombic-ovate . . . . . .
. . . . *VERBESINA*

8' Plants perennial, leaves various.

    9 Principal leaves cordate or heart-shaped, petioled . . . . *BALSAMORHIZA*

9' Principal leaves lanceolate with tapered petioles.

    10 Receptacle merely bristly . . . *GAILLARDIA*

    10' Receptacle chaffy . . . . *RUDBECKIA*

1' Principal leaves opposite or whorled; some of the upper ones, however, may be alternate.

11 Pappus on the achenes of 2 or more barbed awns . . . . . .
. . . . *BIDENS*

11' Pappus on the achenes otherwise than with barbed awns.

    12 Principal leaves simple, not deeply lobed or dissected.

       13 Achenes flattened with thin edges.

         14 Leaves sessile. Involucral bracts broadly ovate to elliptic . . . . *SILPHIUM*

         14' Leaves petioled. Involucral bracts lanceolate with hirsute margins . . . *HELIANTHELLA*

       13' Achenes not flattened or with thin edges.

         15 Involucral bracts obtuse at their apex. Ray flowers fertile, pistillate, forming achenes
. . . . *HELIOPSIS*

         15' Involucral bracts acute to acuminate. Ray flowers sterile . . . . *HELIANTHUS*

    12' Principal leaves deeply lobed or dissected.

       16 Inner involucral bracts not joined on their margins . . . . *COREOPSIS*

       16' Inner involucral bracts joined on their margins one-half or more from the base . . *THELESPERMA*

GROUP 2 (Heads with yellow to yellow-orange rays.   Receptacle naked.)

1   Leaves deeply lobed or pinnately dissected.

    2   Achenes without pappus. Plants odoriferous . . . .
    . . . . *TANACETUM*

    2' Achenes with pappus; of bristles, awns or scales.

        3   Pappus of capillary bristles; outer scales may also be present.

            4   Involucral bracts in 1 series . . . . . *SENECIO*

            4' Involucral bracts in 2 or more series, imbricate . .
            . . . . *HAPLOPAPPUS*

        3' Pappus of several scales; these may be bristly at their tops.

            5   Involucral bracts with many resinous dots. Plants ill-smelling . . . . *DYSSODIA*

            5' Involucral bracts thin, lacking dots. Plants not odoriferous . . . . *PICRADENIOPSIS*

1' Leaves entire to toothed, not deeply lobed or pinnately dissected.

    6   Pappus principally or capillary bristles; some outer scales may also be present.

        7   Leaves opposite . . . . *ARNICA*

        7' Leaves alternate or basal.

            8   Involucral bracts in 1 series . . . . . *SENECIO*

            8' Involucral bracts in 2 or more series; imbricate.

                9   Plants shrubby with a stout taproot . . . .
                . . . . *HAPLOPAPPUS*

                9' Plants herbaceous.

                    10 Pappus simple in 1 series . . . *SOLIDAGO*

                    10' Pappus double, in 2 series . . *CHRYSOPSIS*

    6' Pappus of scales or awns.

        11 Leaves all basal . . . . *HYMENOXYS*

        11' Leaves cauline as well as basal.

            12 Involucral bracts gummy resinous and deflexed . . .
            . . . . *GRINDELIA*

12' Involucral bracts not gummy or resinous.

    13   Heads with disks less than 5 mm across . . . .
. . . . *GUTIERREZIA*

        14   Leaf bases decurrent on the stem as wings
. . . . *HELENIUM*

        14' Leaf bases not decurrent on the stem . . .
. . . . *GAILLARDIA*

GROUP 3 (Heads with ray flowers other than yellow or yellow-orange.)

1   Receptacle chaffy or bristly.

    2   Leaves opposite. Ray flowers white, small.

        3   Leaf blades lanceolate, sessile . . . . . *ECLIPTA*

        3' Leaf blades ovate, petioled . . . . *GALINSOGA*

    2' Leaves alternate.

        4   Involucral bracts spine-tipped or with pectinate margins
. . . . *CENTAUREA*

        4' Involucral bracts not pectinate or spine-tipped.

            5   Leaves dissected or toothed, receptacle not conical.

                6   Heads many, each one with 4-7 ray flowers . . . .
. . . . *ACHILLEA*

                6' Heads fewer, each one with more than 10 ray
flowers . . . . *ANTHEMIS*

            5' Leaves entire to slightly dentate, receptacle conical
. . . . *ECHINACEA*

1' Receptacle essentially naked.

    7   Pappus of capillary bristles.

        8   Rays short, hardly surpassing the disk flowers; needle-like, many . . . . *CONZYA*

        8' Rays normally developed, conspicuous.

            9   Involucral bracts well-developed with evident green
tips. Rays broad. Plants usually blooming in late
summer.

                10   Plants with a stout taproot or branched woody
caudex. The underground portions lacking many
fibrous roots . . . . *MACHAERANTHERA*

10' Plants fibrous rooted, lacking a taproot or
woody caudex . . . . *ASTER*

9' Involucral bracts narrow, lacking evident green tips.
Rays narrow. Plants usually blooming in early summer
. . . . *ERIGERON*

7' Pappus of scales, awns, or none.

11 Leaves conspicuously pinnatifid, pappus a short crown
or lacking . . . . *CHRYSANTHEMUM*

11' Leaves entire or toothed.

12 Plants low, less than 2 dm tall; leaves and flower
heads from essentially a stemless crown . . . . . .
. . . . *TOWNSENDIA*

12' Plants leafy stemmed, exceeding 2 dm in height . .
. . . . *BOLTONIA*

GROUP 4 (Heads of disk flowers only. Receptacle chaffy.)

1 Pappus of capillary bristles.

2 Leaves not prickly-spiny, narrow; less than 5 mm wide.
Plants less than 1 dm tall . . . . *EVAX*

2' Leaves prickly-spiny; plants much exceeding 1 dm in height.

3 Pappus barbellate, not plumose . . . . . . *CARDUUS*

3' Pappus plumose . . . . *CIRSIUM*

1' Pappus of scales, awns, or lacking.

4 Achenes with pappus lacking.

5 Plants with flowers perfect, of 1 kind; with leaves
entire, less than 5 mm wide . . . . *MADIA*

5' Plants with flowers of 2 kinds. These may be in different heads on the same plant or on different plants.
Leaves wider than 5 mm.

6 Staminate and pistillate flowers in the same heads;
involucre not spiny or tuberculate . . *IVA*

6' Staminate and pistillate flowers in separate heads;
the pistillate ones below; involucre of the pistillate heads bur-like with spines or tubercles.

7 Involucre bur-like, the spines with hooked
prickles . . . . *XANTHIUM*

7' Involucre with tubercles or straight spines
. . . . *AMBROSIA*

4' Achenes with scales or awns.

    8 Leaves opposite. Achenes variously awned.

        9 Involucral bracts separate . . . . . . *BIDENS*

        9' Involucral bracts joined their basal one-half or
          more . . . . *THELESPERMA*

    8' Leaves alternate. Involucral bracts forming soft,
      hooked spines . . . . *ARCTIUM*

GROUP 5 (Heads of disk flowers only. Receptacle naked.)

1 Plants woody shrubs . . . . *CHRYSOTHAMNUS*

1' Plants herbaceous.

    2 Achenes with a pappus of capillary bristles.

        3 Vegetation more or less white-woolly, at least on lower
        leaf surfaces.

            4 Plants scapose, the principal leaves basal and long
            petioled, with a large cordate or sagittate blade up
            to 30 cm long, developing after flowering . . . . .
            . . . . *PETASITES*

            4' Plants not scapose, the principal leaves not basal
            or long petioled.

                5 Flowering occurring in spring; basal leaves
                conspicuous . . . . *ANTENNARIA*

                5' Flowering occurring in middle or late summer.

                    6 Plants dioecious. Female plants may have a
                    few staminate flowers in the center of the
                    heads . . . . *ANAPHALIS*

                    6' Plants not dioecious. Flowers all fertile but
                    the outer ones may be pistillate . . . .
                    . . . . *GNAPHALIUM*

        3' Vegetation not white-woolly.

            7 Flowers at the outer edge of the heads pistillate.

                8 Involucral bracts essentially in one series . . .
                . . . . *ERECHTITES*

                8' Involucral bracts in several series.

                    9 Plants biennial or perennial (of the Black
                    Hills) . . . . *ERIGERON*

                  9' Plants annual.

          10 Involucral bracts green tipped, herbaceous
          . . . . *ASTER*

          10' Involucral bracts greenish but not herba-
          ceous or green-tipped . . . . *CONYZA*

   7' Flowers all perfect.

      11 Leaves all opposite or whorled . . *EUPATORIUM*

      11' Leaves alternate.

          12 Flowers pink to purple.

             13 Inflorescence in corymbiform cymes.
             Achenes with a pappus in 2 series . . .
             . . . . *VERNONIA*

             13' Inflorescence racemiform or spike-like.
             Achenes with a pappus in 1 series . . .
             . . . . *LIATRIS*

          12' Flowers cream or yellow, not purplish.

             14 Pappus bristles plumose. Flowers cream-
             colored . . . . *KUHNIA*

             14' Pappus bristles simple or merely barbel-
             lulate.

                15 Plants 6-10 dm tall, from a thick
                tuberous root. Stem angled and
                grooved. Leaves long-petioled with
                lance-ovate blades. Inflorescence
                an expanded, flat-topped corymb . .
                . . . . *CACALIA*

                15' Plants not as above.

                   16 Disk 1 cm broad or more; plant
                   perennial . . . . . *MACHAERANTHERA*

                   16' Disk less than 1 cm broad or, if
                   up to 1 cm broad, plant annual
                   . . . . *SENECIO*

 2' Achenes with pappus of scales or pappus lacking.

   17 Leaves opposite. Involucral bracts glandular . . . .
   . . . . *DYSSODIA*

   17' Leaves alternate.

      18 Flowers cream-yellow to white. Pappus of scales.

19 Involucral bracts petaloid. Leaves compound
    with linear segments . . . . . *HYMENOPAPPUS*

19' Involucral bracts not petaloid. Leaves pin-
    natifid . . . . *CHAENACTIS*

18' Flowers green or greenish. Pappus a short crown
    or lacking.

20 Heads few. Receptacles conical. Plant emit-
    ting a pineapple odor . . . . . . *MATRICARIA*

20' Heads numerous. Receptacles not conical.
    Plants bitter aromatic . . . . . *ARTEMISIA*

ACHILLEA L.

Achillea millefolium L. Plants perennial, the stems 3-5 dm tall,
several from a rhizome. Leaves several, pinnately dissected with
many divisions and appearing fern-like, the blades 4-8 cm long.
Vegetation aromatic. Inflorescence corymbiform with 10-40 heads.
The ray-flowers 4-10, white, fertile. Disk flowers pale white or
greenish. Very common over the state in prairies and disturbed
soils. {*A. lanulosa* Nutt.} Ours ssp. *lanulosa* (Nutt.) Piper.
June-July. Yarrow.

AGOSERIS Raf.

Agoseris glauca (Pursh) D. Dietr. Plants perennial from a taproot, lactiferous, the leaves glaucous, linear. Plants scapose,
the flowering scapes exceeding the leaves, 2-6 dm high. Flowers
all ligulate and perfect, yellow-orange. Flowering heads 2-4 cm
across, the involucre vase-like. Frequent in moist meadows and
near prairie swales of the northeast and in the meadows of the
western part. May-June. False Dandelion.

AMBROSIA L.

1 Pistillate involucres with 2-several rows of spines.

2 Plants annual. The spines of the fruit flat, not hooked.
   Stem erect, much branched, 2-5 dm high. Leaves pinnatifid,
   petioled. Fertile heads 1-flowered, the fruit 4-10 mm long,
   with 1 beak. Rare in swales of prairies of the central and
   western part. {*Franseria acanthicarpa* (Hook.) Cov.} Aug-
   Sept. Annual Bursage . . . . . *Ambrosia acanthicarpa* Hook.

2' Plants perennial from creeping rootstocks. The spines of
   the fruit rounded and hooked. Stems 1-4 dm tall. Leaves
   pinnatifid, tomentose below. Fertile heads 2-flowered,

the fruit 4-6 mm long, with 2 beaks. Rare in dry soils or
sandy disturbed areas of the southwestern part. {*Franseria
discolor* Nutt.} July-Aug. Perennial Bursage . . . .
. . . . *Ambrosia tomentosa* Nutt.

1' Pistillate involucres with 1 row of tuberculate spines.

    3 Principal leaves 3-5 lobed, some of the smaller upper
       leaves not lobed, opposite. Plants coarse, up to 2 meters
       tall. Flowers unisexual, the staminate borne in terminal
       spikes; the pistillate in axils of upper leaves. A very
       common weed in alluvial soil or waste places, its pollen
       helping to cause discomfort to hay fever sufferers. Aug-
       Oct. Giant Ragweed . . . . *Ambrosia trifida* L.

    3' Principal leaves pinnatifid.

        4 Plants annual, commonly branched above, 2-4 dm tall.
           Leaves once or twice pinnatifid, scabrous pubescent.
           Staminate flowers in terminal spikes, the individual
           flowers nodding, about 3 mm across. Pistillate flowers
           without a corolla. Fruits obovoid, 2-3 mm long, with
           5-7 spines. A common weed of waste places over the
           state. Aug-Sept. Small Ragweed . . . .
           . . . . *Ambrosia artemisiifolia* L.

        4' Plants perennial from a creeping rootstock. Stems
           sparingly branched, 2-3 dm tall. Leaves usually once-
           pinnatifid. The leaves thicker than in the preceding
           species but similar in other respects. Frequent in
           sandy soil and prairies throughout the state. Aug-Sept
           . . . . *Ambrosia psilostachya* DC.

ANAPHALIS DC.

Anaphalis margaritacea (L.) B. & H. Plants perennial from rhizomes, the stems 3-7 dm tall. Leaves linear-lanceolate, entire;
white-woolly pubescent below, green above. Inflorescence branched
above, many-headed, forming a rounded outline. Plants dioecious
or polygamodioecious. Heads 1 cm or less across, the bracts pearl
white. Infrequent along roadsides and in valleys of the Black
Hills. July-Aug. Pearly Everlasting.

ANTENNARIA Gaertn.

1 Leaves densely tomentose on upper and lower surfaces. Principal leaves usually less than 1 cm wide and less than 2.5 cm
   long.

    2 Pistillate heads 7-11 mm high. Flowering stems seldom over
       15 cm tall. Plants perennial, forming mats by spreading

stolons. Basal leaves spatulate, the upper ones linear and
bracteate. Inflorescence of 1-4 heads, sessile, clustered.
Outer involucral bracts of pistillate plants with obtuse
tips. Infrequent on dry plains over the state, more frequent westward. May-June . . . *Antennaria parviflora* Nutt.

2' Pistillate heads 4-7 mm high. Flowering stems commonly
over 15 cm tall, stoloniferous, forming mats. Involucral
bracts of pistillate plants white to bright pink. Basal
leaves narrowly spatulate, acutely tipped, 1-2 cm long.
Upper leaves narrow, erect. Heads several in a capitate,
dense cyme. Infrequent on dry or grassy hillsides over
the state. May-June . . . . *Antennaria rosea* Greene

1' Leaves green above, not densely white-hairy towards maturity.
Principal leaves more than 1 cm wide and 3 cm long.

3 Basal leaves 1-nerved, usually not more than 1.5 cm at the
widest part. Flowering stems varying considerably in height
from 0.5-4.0 dm tall, with long stolons and mat-forming.
Upper leaf surfaces thinly tomentose when young. Heads
several in a dense, cymose cluster. Staminate plants very
infrequent. Common in dry places and on prairie hillsides
over the state. Apr-June . . . *Antennaria neglecta* Greene

3' Basal leaves 3-nerved, usually exceeding 1.5 cm at the
widest part. Flowering stems tending to be taller than
the preceding, to 3 dm. Leaves becoming glabrate with age,
the basal ones 3-7 cm long, spatulate. Inflorescence of
several heads, on short peduncles. Frequent in moist
places in prairies and on the edges of woods over the state.
May-June . . . . . . *Antennaria plantaginifolia* (L.) Rich.

ANTHEMIS L.

Anthemis cotula L. Plants ill-smelling annuals with irregular
branches, 2-6 dm tall. Leaves twice or more pinnatifid, 2-6 cm
long, the ultimate segments narrow and pointed. Heads 7-12 mm
across, solitary at the ends of naked peduncles. Ray flowers
white, neutral, 8-10 mm long, becoming reflexed. Disk flowers
yellow. Achenes cylindrical, 10-ribbed, with glands. Sporadic
in waste places and on roadsides over the state. June-July.
Dogfennel.

ARCTIUM L.

1 Heads borne on long peduncles, forming a corymbose-like inflorescence. Principal lower leaves rounded at the distal ends.

2   Flowering heads 2-3 cm across, the involucral bracts often
    tomentose-hairy, usually not exceeding the corollas.
    Plants up to 1 meter tall, the inflorescence branched
    above. Leaves round-ovate, the petioles of the principal
    ones hollow. Heads pale to deep purple. Apparently rare
    in the western part. July-Sept . . . .
    . . . . *Arctium tomentosum* Mill.

2' Flowering heads 3-4 cm across, the involucral bracts gla-
    brous, exceeding the corollas. Plants robust, to 1.5
    meters or more, branched above, the inflorescence a broad
    corymb. Leaves round-ovate, the petioles of the principal
    ones solid. Heads pale purple, the involucral bracts dense
    and hook-like, forming the bur. Infrequent in moist waste
    places in the eastern and western parts. Aug-Sept. Great
    Burdock . . . . *Arctium lappa* L.

1' Heads borne in a racemose fashion at the upper part of the
    plant, with short peduncles or subsessile. Principal lower
    leaves tapered to the distal ends. Plants biennial, up to
    1.5 meter tall, simple stemmed or sparingly branched. Leaf
    blades narrow to broadly ovate, 4 dm across to 6 dm long.
    Inflorescence branched, racemiform. Heads pale purple.
    Locally abundant in thickets and alluvial woods of the east-
    ern part. Less common westward and in the Black Hills. Aug-
    Sept. Common Burdock . . . . *Arctium minus* Schkuhr.

ARNICA L.

1   Basal and lower cauline leaves strongly cordate, more than 10
    cm wide. The upper leaves smaller, ovate, tending to be ses-
    sile. Margins of leaves coarsely toothed. Plants perennial,
    2-5 dm tall, stamens solitary or few clustered together. Heads
    yellow, the raw flowers pistillate and fertile. Involucre 15-
    18 mm high. Involucral bracts with long, white hairs. Infre-
    quent in dry woods or ravines in Lawrence and Pennington coun-
    ties of the Black Hills. May-June . . *Arnica cordifolia* Hook.

1' Basal and lower cauline leaves not strongly cordate; lower
    leaves less than 10 cm wide.

2   Ray flowers 4-6, involucral bracts 9-11 mm long. Plants
    2-5 dm tall, from short rhizomes. Stems hairy as well as
    the leaves. Cauline leaves opposite, broadly lanceolate,
    3-8 cm long. Ray flowers yellow-orange. Involucre 10-12
    mm high. A subalpine species which is rare at higher alti-
    tudes of the Black Hills. June-July . . . .
    . . . . *Arnica rydbergii* Greene

2' Ray flowers 12-20, involucral bracts 10-15 mm long.

3   Stems and leaves densely pilose, the leaves not conspicuously toothed. Principal leaves basal, linear to
spatulate in shape. Plants perennial from scaly rhizomes, 2-6 dm tall. Basal leaves with brown tufts of
woolly hair. Principal leaves broadly oblanceolate.
Ray flowers 1.5 cm long, orange. Involucres 12-15 mm
high. Scattered in open woods of the Black Hills and
in Harding County. June-July . . . *Arnica fulgens* Pursh

3' Stems and leaves glabrous to moderately pubescent.
Basal leaves with toothed margins. Plants perennial
from dark, branching rhizomes. Stems mostly solitary,
3-5 dm tall. Principal leaves lance-ovate, the marginal
teeth 5-10 mm apart. Upper part of stem appearing almost naked, with 1-2 pairs of small opposite leaves
below the inflorescence of 3-9 heads. Ray flowers yellow orange. Involucres 9-10 mm high. Frequent in open
woods at higher altitudes of the Black Hills. {*A. arnoglossa* Greene} Ours ssp. *arnoglossa* (Greene) Maguire.
June-July . . . . *Arnica lonchophylla* Greene

ARTEMISIA L.

1 Plants with white hairs on the involucral bracts or between
the flowers of the head. Foliage usually with an obvious
whitish pubescence as well.

2   Ultimate divisions of the leaf less than 2 mm wide. Leaf
segments mostly filiform.

3   Principal leaves 2 to 3 times parted or dissected, the
ultimate segments usually less than 5 mm long. Plants
perennial, from 1-5 dm tall. Vegetation with a white-
silvery appearance, the plants forming mats with numerous upright stems. Common in prairies and on plains
over the state except for the extreme southeast part.
July-Aug. Pasture Sage-brush . . . .
. . . . *Artemisia frigida* Willd.

3' Principal leaves entire to two to three parted into
linear segments up to 2 cm long. Plants perennial, up
to 10 dm tall, their bases woody. Stems freely branched.
Leaves grayish-green. Heads in dense, narrow panicles
among the leaves on upright stems. Infrequent on dry
plains and hillsides of the western one-half. Aug.
Silvery Wormwood . . . . . . *Artemisia filifolia* Torr.

2' Ultimate divisions of the leaf exceeding 2 mm in width.*

    4 Shrubs or sub-shrubs with woody bases. Leaves entire or three-cleft at the apex. Flowers perfect and fertile.

       5 Principal leaves 3-5 cm long and mostly entire. The lower leaves sometimes with 2-3 teeth. Plants shrubby, branched, up to 1.5 meters tall, the twigs and leaves with a grayish-green canescence. Flowering heads in a large, leafy panicle. Flowers yellow, the heads 5-9 flowered. Frequent on dry prairies and plains over the state. Aug-Sept . . . . *Artemisia cana* Pursh

       5' Principal leaves less than 3 cm long with the apex divided into three distinct segments. Plants much branched, up to 3 meters tall, the foliage silvery-canescent. Lower part of plant woody, the bark shredding. Flowering heads yellow to brown, in large panicles, each head 5-8 flowered. Infrequent on dry hillsides and plains in the western one half of the state. July-Sept. Sagebrush . . . . *Artemisia tridentata* Nutt.

   4' Herbaceous or, if woody at the base, leaves pinnately dissected.

       6 Receptacle hairy. Leaves pinnately dissected, 3-5 cm wide, the ultimate segments 2-4 mm wide. Foliage fragrant. Plants 4-10 dm tall, shrubby. Heads many in a large panicle. Involucral heads 4-5 mm across. Frequent in waste places and in disturbed prairie over the state. Aug-Sept. Wormwood . . . . *Artemisia absinthium* L.

       6' Receptacle lacking hairs. Leaves entire or shallowly toothed and mostly simple with dense, white hairs.

          7 Involucre 4-5 mm high. Leaves narrow, entire. Plants with stems clustered from a caudex, 2-6 dm tall. Leaves tomentose below, tending to be involute with greenish to glabrate above. Flowering heads in a narrow inflorescence, the heads 5-6 mm across. Infrequent in dry, open prairie

*(The large, woody based *Artemisia vulgaris* L. may be found in the eastern part as a persisting escape from cultivation. It is aromatic with pinnately dissected leaves, the stems up to 2 meters tall.)

or eroded soil in the southern and western part.
Aug-Sept . . . . *Artemisia longifolia* Nutt.

   7' Involucre 3-4 mm high. Leaves entire to toothed,
simple with dense tomentum on the lower surface,
usually less than 1 cm wide. Plants 3-6 dm tall,
sometimes taller, from a cluster. Inflorescence
of numerous heads in panicles, the involucres 2-3
mm across. Very common in prairies, on roadsides,
and disturbed areas over the state. Ours var.
ludoviciana. July-Sept. White Sage . . . . . .
. . . . *Artemisia ludoviciana* Nutt.

1' Plants with involucral bracts lacking white woolly hairs or
dense pubescence. Foliage not at once obviously whitish-
canescent.

   8 Plants annual or biennial. The leaves pinnately dissected,
their ultimate divisions not linear or filiform. Disk
flowers fertile, the heads crowded into almost sessile
clusters at the upper part of the stem giving a spike-like
inflorescence. Plants 0.5-1.5 meters tall, with a stout
taproot. Infrequent in moist waste places and stream banks
throughout the state. Aug-Sept . . . .
. . . . *Artemisia biennis* Willd.

   8' Plants perennial. The leaves simple or pinnately divided,
their ultimate segments linear or filiform. The disk-
flowers sterile.

      9 Leaves simple, entire, linear, except some lower ones
with several linear segments. Foliage bright green,
mostly glabrous. Plants 30-70 cm tall, the upper part
of the stem often dull reddish-brown. Stems usually
not branched at the base but there may be several stems
from a root crown. Frequent in sandy, waste soil or in
moist alluvial areas over the state. Aug-Sept. Silky
Wormwood . . . . *Artemisia dracunculus* L.

      9' Leaves, especially the lower, pinnately compound or
dissected, distinctly petioled, usually crowded at the
base. The upper leaves pinnatifid and less divided,
the ultimate segments commonly linear. Foliage gray-
green to bright green. The stems several from a
branched caudex, reddish. Plants 3-10 dm tall, generally not as tall as the preceding species. Common in
dry places over the state. {*A. caudata* Michx. of earlier authors} Ours ssp. *caudata* Hall. & Clem. July-
Aug. Western Sagebrush . . . . *Artemisia campestris* L.

ASTER L.

1 Annuals with short rays or ray flowers completely lacking.
Plants 1-4 dm tall, branched. Leaves entire, linear, 3-10 cm
long. Flowering heads several to many at the end of spiciform
branches. Involucres 6-8 mm high, the bracts narrowly linear.
Pappus soft, copious, obvious at maturity. Frequent in moist
or saline soils over the state, especially in sandy alluvium
along the Missouri River. {*Brachyactis angusta* Britt.} Aug-
Sept. Rayless Aster . . . . *Aster brachyactis* Blake

1' Annuals, biennials or perennials with well-developed ray
flowers.

2 Principal basal or lower cauline leaves with cordate-shaped
blades or with an obvious petiole.

3 Inflorescence narrow, many-flowered, heads usually from
50 to 100 or more. Lower leaves thick, cordate to sagittate, their margins serrate. Branches of the inflorescence conspicuously bracted. Plants 4-10 dm tall,
the vegetation essentially glabrous. Flowers pale
lilac to lavender, the involucral bracts narrow with
green tips. Frequent in moist areas and swamps of the
northeast part. July-Aug . . *Aster sagittifolius* Willd.

3' Inflorescence with not more than 50 heads, usually
fewer flowered.

4 Principal leaves toothed, thin, shallowly cordate.
Upper part of the plant variously ciliate or pilose,
especially when young. Plants 3-10 dm tall, with a
branched inflorescence that is corymbose and only
rarely bracted. Flowers blue, rarely lavender, the
rays 8-15 mm long. Frequent in moist woods and ravines of the Black Hills and in the northeast part.
Aug-Sept . . . . *Aster ciliolatus* Lindl.

4' Principal leaves entire, thick, subcordate and with
an obvious petiole. Vegetation scabrous to glabrous.
Plants stiffly upright, 3-10 dm tall, with an open,
paniculiform inflorescence with many bracts. Flowers
blue or rarely pink, the rays 5-12 mm long. Rare in
dry, open woods in the eastern part. Aug-Sept . . .
. . . . *Aster azureus* Lindl.

2' Principal basal or lower cauline leaves not cordate-shaped
nor with an obvious petiole.

5 Bases of principal leaves strongly clasping the stem
(auriculate).

6 Involucral bracts glandular. Leaves and stems
   scabrous-pubescent or glandular or both.

   7 Plants less than 5 dm tall. Leaves less than 1 cm
      wide, lance-ovate in shape. Flowers light purple,
      the heads up to 2.5 cm wide but usually not exceed-
      ing 2 cm. Inflorescence branches with many small
      bracteal leaves. Common in dry prairies and on
      eroded bluffs over the state. Sept. Prairie
      Aster . . . . *Aster oblongifolius* Nutt.

   7' Plants commonly to 1 meter tall, the stems clus-
      tered, stout. Leaves 3-12 cm long and up to 2 cm
      wide, uniformly arranged on the stem. Upper sur-
      faces of leaves with stiff hairs, the lower sur-
      faces less so. Flowers showy, bright purple with
      yellow centers, clustered in a leafy inflorescence.
      Heads 2-3 cm across. Frequent in woods or alluvial
      areas, especially in the eastern part. Sept. New
      England Aster . . . . . . *Aster novae-angliae* L.

6' Involucral bracts glabrous. Leaves and stems glabrous or pubescent.

   8 Stems and leaves irregularly to uniformly hairy,
      especially in the upper parts. Vegetation not
      leathery. Plants of moist places.

      9 Stems slender, seldom over 2 mm in diameter,
         with decurrent lines of pubescence from upper
         leaves downward. Involucral bracts acute,
         strongly imbricate, often with purple tips or
         margins. Stems 2-7 dm tall, with linear leaves
         having slightly auriculate-clasping bases. In-
         florescence of few heads, broad, almost flat-
         topped, the rays white to lavender. Rare in
         boggy places in the Black Hills and in the
         northeast part. Aug-Sept . . . .
         . . . . *Aster junciformis* Rydb.

      9' Stems coarse, uniformly hairy, especially in
         the upper parts. Involucral bracts with long,
         acuminate tips, scarcely or not at all imbri-
         cate. Stems up to 1.5 meters tall, with leaves
         narrow to broadly lanceolate. Inflorescence
         corymbiform to paniculate, the heads few to
         many in a leafy cluster, blue. Apparently rare
         in moist places in the Black Hills. Aug-Sept.
         . . . . *Aster puniceus* L.

8' Stems and leaves glaucous, the vegetation leathery. Involucral bracts acute with chartaceous whitish margins and bases. Leaves lanceolate to narrowly ovate, highly variable in length, their bases not completely encircling the stem. Inflorescence ascending, stiffly racemiform to narrowly paniculate. Flowers rose to purple, the heads up to 1.5 cm across. Common in open woods of the Black Hills and in open, dry places elsewhere over the state. Aug-Sept. Blue Aster . . . . . . *Aster laevis* L.

5' Bases of principal leaves not auriculate-clasping. (Some leaf bases may be attached to as much as forty percent of the stem diameter but the lobed or auriculate condition should be lacking.)

10 Principal cauline leaves at flowering time (excepting basal rosette leaves) less than 1 cm wide (some rarely will exceed 1 cm).

11 Foliage with silky, glistening hairs covering both leaf surfaces. Plants with stems brittle and wiry, much branched, usually low, not more than 4 dm tall. Heads with pale to bright purple ray flowers and yellow disks. Inflorescence paniculiform and widely branched. Frequent in prairies over the eastern part of the state, less frequent westward. Aug-Sept. Silky Aster . . . . *Aster sericeus* Vent.

11' Foliage without silky hairs on both leaf surfaces.

12 Inflorescence with 20-50 or more heads. Lower leaves mostly deciduous at flowering time. Plants from short rootstocks, branched above with recurved and ascending branches. Heads white to lavender, densely clustered, often secund on the branches. Leaves at flowering time short, scabrous. Very common over the state in prairies, on roadsides, and other open places. Represented in South Dakota by the varieties *ericoides*, *falcatus* (Lindl.) and *pansus* (Blake). {*A. multiflorus* Ait.} Aug-Sept. White Aster . . . . *Aster ericoides* L.

12' Inflorescence few-headed, usually not more than 10.

ERRATA

1. In the *Keys to the Dicotyledoneae*, beginning page 144, under Section 9, *Herbaceous Dicotyledons with Petals United*, line 13' should read, "Ovary not 4-lobed. Fruit a capsule or berry."

2. Page 422, 3rd line, bottom. *Aster ericoides* L. Paragraph 12 describing *Aster ericoides* L. should be changed to include the following sections.

   12 Inflorescence with 20-50 or more heads. Lower leaves mostly deciduous at flowering time.

   12a Plants with well developed rootstocks or rhizomes.

   12b Flowering heads many (50 or more), secund on recurved branches. Involucres 5 mm or less high. Ray flowers usually less than 25. Plants from short rootstocks, the stems 3-10 dm tall, branched above. Leaves short, scabrous. Flowers white to pink. Very common over the state. Ours var. *ericoides*. [*A. multiflorus* Ait.] Aug-Sept. White Aster . . . . *Aster ericoides* L.

   12b' Flowering heads fewer (less than 50), solitary or few clustered at the ends of branches, the branches not noticeably recurved. Involucres 5-7 mm high. Ray flowers 25 or more. Plants from well developed, creeping rhizomes, the stems mostly 3-6 dm tall. Leaves much like the preceding, typically with appressed pubescence. Flowers mostly white, rarely pinkish. Frequent to common in the western part. Aug-Sept . . . . *Aster falcatus* Linl.

   12a' Plants fibrous-rooted, from a very short caudex. Stems clustered, up to 8 dm or more tall. Branches several, irregularly diverging from the main stem and tending to be recurved. Leaves many, linear, with rough, spreading hairs. Flowering heads many, secund on the recurving branches. Ray flowers 4-7 mm long, on disks 4-6 mm across. Rays white, 15-25 per head. Frequent on prairie over the state. Aug-Sept . . . . *Aster pansus* (Blake) Cronq.

3. Page 424, 7th line, bottom. *Aster simplex* Willd. Paragraph 18 describing *Aster simplex* Willd. should be changed to include the following sections.

counties in the Black Hills. July-Aug.
Rough Aster . . . . *Aster conspicuus* Lindl.

16' Leaves entire or nearly so. Ray flowers
white. The involucre sparsely puberulent
but not glandular. Plants of low, moist or
boggy areas. Heads many, forming a flat-
topped inflorescence. Stems stout, up to
1.5 meters. Rare to infrequent in marshy
places in the northeast part. Aug-Sept . .
. . . . *Aster pubentior* Cron.

14' Heads not compacted at the top of the plant or restricted to it. Inflorescence composed additionally
of branches from the sides of the main axes.

17 Flowering heads in narrow, 1-sided racemes.
Stems slender, branched, up to 1 meter tall,
glabrous to variously pubescent. Leaves lanceolate, thin and soft, the midrib beneath pubescent. Heads white to lavender, the involucres
4-7 mm high, glabrous. Apparently rare in open
woods and at edges of thickets in the eastern
part. Aug-Sept . . . .
. . . . *Aster lateriflorus* (L.) Britt.

17' Flowering heads not primarily in narrow, 1-sided
racemes.

18 Stem and branchlets with decurrent lines of
pubescence running down from the leaf bases;
otherwise, these glabrous. Lobes of the
disk corollas comprising less than one half
of the limb. Plants up to a meter tall, the
stems stout. Leaves lanceolate, the veins
forming a conspicuous reticulum. Flowers
white to rose or lavender. Involucres mostly 3-4 mm high. Common in low places over
the state. Ours represented by the following varieties: *simplex*, *hesperius* (Gray)
and *ramosissimus* (T. & G.) Cron. Aug-Oct.
. . . . *Aster simplex* Willd.

18' Stem and branchlets with uniform pubescence.
Leaves usually softly hairy on the lower
surfaces. Lobes of the disk corollas one-
half or more of the limb. Plants coarse
stemmed, much branched, up to a meter tall.
Leaves soft, green, veins not forming a

conspicuous reticulation. Inflorescence irregularly branched, usually elongate. Heads
usually white or pink, the involucres 3-5 mm
high. Infrequent in alluvial places in the
eastern part. Aug-Sept . . . .
. . . . *Aster ontarionis* Wieg.

BALSAMORHIZA Hook.

Balsamorhiza sagittata (Pursh) Nutt. Plants perennial from a
stout taproot, the stems almost scapose, 3-6 dm tall. Leaves
chiefly basal, cordate with petioles; large, 20-30 cm long and
5-10 cm wide. Vegetation with dense white hairs. Flowers yellow, one to a few heads on a short stem, usually with leaves
lacking or much reduced. Rare in woods of Lawrence and Pennington counties. Apr-May. Balsam Root.

BIDENS L.

1 Ray flowers yellow, conspicuous when heads are in bloom.

  2 Leaves simple, opposite, lanceolate to ovate in shape,
    their bases connate, sessile. Achenes 4-angled with 4
    awns retrorsely barbed. Plants branched, annual, less
    than 7 dm tall. Heads up to 2.5 cm across, hemispheric,
    nodding in fruit. Common in low wet places over the state.
    {*B. laevis* (L.) BSP} Aug-Sept . . . . . *Bidens cernua* L.

  2' Leaves pinnately dissected or pinnatifid, the ultimate seg-
    ments lance-linear, opposite, petioled. Achenes flattened,
    oblong, with two awns up to 2 mm long. Plants branched,
    annual, up to a meter tall. Heads with disks 8-15 mm wide,
    the rays 1-2 cm long. Infrequent in wet places along
    streams in the southern part. Aug-Sept . . . .
    . . . . *Bidens coronata* (L.) Britt.

1' Ray flowers inconspicuous or entirely lacking when heads are
in bloom.

  3 Leaves simple, opposite, lanceolate in shape, the margins
    toothed.

    4 Achenes flattened, 8 mm long. Corollas of disk flowers
      with 4 teeth. Heads with 6-8 leaf-like outer bracts
      that exceed the disk. Plants with simple stems or with
      strongly ascending branches, 3-6 dm tall. Leaves 3-20
      cm long and up to 5 cm wide. Heads 2-3 cm across.
      Achenes with 2-4 awns, often 3. Common in wet soil of
      ponds and streams over the state. Sept-Oct . . . . . .
      . . . . *Bidens comosa* (Gray) Wieg.

4' Achenes angled, 4-6 mm long. Corollas of disk flowers
with 5 teeth. Heads with 4 or 5 outer bracts, usually
not exceeding the disk. Plants 2-6 dm tall, with loose,
ascending branches. Leaves short-petioled, the principal ones with connate bases. Heads variable, rarely
with short ray flowers. Achenes with awns varying from
2 to 5. Infrequent in wet areas in the eastern part.
Sept-Oct . . . . *Bidens connata* Muhl.

3' Leaves pinnately compound or pinnately dissected.

5 Bracts beneath heads 10 or less. Achenes becoming blackened towards maturity, 2-awned. Annual plants up to a
meter tall, branched, with slender stems. Leaves pinnately compound, with 3-5 serrately margined leaflets.
Heads hemispheric, sometimes small and appearing depauperate, 1 cm or more wide. Common in moist places
over the state. July-Aug . . . . . . *Bidens frondosa* L.

5' Bracts beneath the heads more than 10. Achenes brownish
towards maturity, 2-awned. Plants annual, branched, the
stems more robust than the preceding. Heads with disks
1-3 cm across, yellowish in flower. Leaves pinnately
compound with 3-5 serrate leaflets. Entire plant leaf-
ier than the preceding. Common in moist open places and
thickets in the eastern part, less frequent western.
Aug-Sept . . . . *Bidens vulgata* Greene

BOLTONIA L'Her.

Boltonia asteroides (L.) L'Her. Plants short-lived perennials
3-8 dm tall, the stems branched. Leaves linear to lance-shaped,
2-5 cm long, up to 1 cm wide. Inflorescence corymbiform, the
heads many. Ray flowers white to lavender, with yellow disk
flowers. Involucres 2-3 mm high and 8-10 mm across. Frequent
in low moist places in the eastern part and in the Black Hills.
Represented in South Dakota by the two varieties *latisquama*
(Gray) Cronq. and *recognita* (Fern. & Grisc.) Cronq.

CACALIA L.

Cacalia tuberosa Nutt. Perennial herbaceous plants with thick,
tuberous roots, the stems 6-15 dm tall, angled and deeply grooved.
Leaves with petioles conspicuously elongate. Blades lance-ovate,
6-20 cm long, green on both surfaces, with prominent longitudinal
nerves. Inflorescence a flat-topped corymb, the flowers white or
flesh-colored. Heads 5-flowered, each head with 5 involucral
bracts. Apparently rare in prairie in the northeast part. July.
Indian Plantain.

CARDUUS L.

1 Heads 3-6 cm across, usually solitary and nodding at the ends
   of branches. Involucral bracts broad, sharply reflexed with
   a pointed tip. Plants 1-2 meters tall, biennial. Leaves
   lanceolate, deeply pinnatifid, 7-15 cm long. Flowers bright
   purple, the heads discoid. Becoming frequent in waste places
   in the southern part, especially in overgrazed pastures. June-
   July. Musk Thistle . . . . *Carduus nutans* L.

1' Heads 1-2.5 cm across, clustered at the ends of branches,
   erect. Involucral bracts narrow, rarely over 2 mm wide, not
   sharply reflexed. Plants 1-1.5 meters tall, biennial. Leaves
   1-2 dm long, with spinulose, pinnatifid divisions. Flowers
   bright rose-violet, the heads discoid. Infrequent in waste
   places in the southeast part. June-July . . . .
   . . . . *Carduus acanthoides* L.

CENTAUREA L.

1 Plants perennial, from deep seated creeping rootstocks. Outer
   involucral bracts entire. Mature pappuses over 6 mm long.
   Stems branched above, 5-8 dm tall. Leaves narrow, the upper
   ones entire, the lower ones toothed or shallowly lobed. Heads
   about 2 cm high, the flowers purple. Infrequent in eastern
   and western counties but locally abundant and considered noxious. July-Aug. Russian Knapweed . . . . *Centaurea repens* L.

1' Plants annual or biennial, more or less taprooted, without
   creeping roots. Outer involucral bracts with lacerate or
   pectinate margins. Mature pappuses 1-3 mm long.

   2 Leaves narrow, entire or remotely toothed, less than 1 cm
      wide. Plants annual, 3-6 dm tall, branched above. Heads
      2-3 cm across, the flowers blue, purple or white. Occa-
      sionally escaping from gardens and persisting as a weed
      over the state. June-Sept. Bachelor's Button; Cornflower.
      . . . . *Centaurea cyanus* L.

   2' Leaves pinnatifid with linear lobes. Plants biennial, tap-
      rooted, the stems 3-8 dm tall, branched. Heads 1.5-2 cm
      high, the outer involucral bracts with dark, pectinate
      tips. Flowers lavender to white. Infrequent as a weed
      in waste places in the eastern one half of the state, rare
      westward. July-Aug. Spotted Knapweed . . . .
      . . . . *Centaurea maculosa* Lam.

CHAENACTIS DC.

Chaenactis douglasii (Hook.) H. & A. Plants short-lived perennials 2-4 dm tall, sparingly branched from the base. Leaves

dissected, from 1-3 pinnatifid, tomentose. Flowers creamy white
to pale pink, the heads 2-3 cm across, corymbiform. Apparently
very rare in dry, open places in the western part. Our only report from the Slim Buttes area of Harding County in the northwest part. June-July.

CHRYSANTHEMUM L.

Chrysanthemum leucanthemum L. Plants perennial, the stems 2-6 dm
tall. Stems solitary or sparingly branched above. Leaves coarsely toothed to shallowly lobed or pinnatifid, oblanceolate in general outline. Heads solitary at the ends of stems or branches,
the involucres 2.5-3.0 cm across. Ray flowers white, the disk
flowers yellow. Frequent in disturbed prairie, along roadsides
and in waste places over the state. May-July. Ox-eye Daisy.

CHRYSOPSIS Nutt.

1 Stems hispid, stiff. Involucres and leaves granular-resinous
   but sparingly hispid. Plants perennial with an erect stem 2-4
   dm tall. Leaves narrow, linear to oblanceolate, petioled.
   Heads almost sessile, subtended by several narrow leaves. Involucres 8 mm tall and about 1 cm across. Rare in dry soil of
   the southeast part. {*Heterotheca stenophylla* (Gray) Shinners}
   Aug-Sept . . . . *Chrysopsis stenophylla* (Gray) Greene

1' Stems strigose with dense, spreading hairs. Involucres pubescent but not granular-resinous. Plants with several stems
   from a taproot, 2-5 dm tall. Leaves many, elliptic-oblong,
   2-5 cm long. Heads several at the top of each stem, peduncled.
   Involucres about 1 cm tall and up to 2 cm across. Common in
   prairies throughout the state. Ours represented by the following varieties: *angustifolia* (Rydb.) Cronq., *hispida* (Hook.)
   Gray and *villosa*. {*Heterotheca villosa* (Nutt.) Harms} July-
   Sept. Gold Aster . . . . . *Chrysopsis villosa* (Pursh) Nutt.

CHRYSOTHAMNUS Nutt.

1 Inflorescence racemose. Outer involucral bracts foliaceous
   and prolonged into acuminate tips. Heads 1-1.5 cm high, the
   maturing achenes pubescent. Plants 2-6 dm tall, shrubby, the
   branches white-tomentose. Leaves linear-filiform, 4-7 cm long.
   Heads in terminal racemes, with 5-20 flowers in a head. Corollas about 1 cm long. Rare in dry soils of the southwest part.
   Ours ssp. *howardi* (Parry) H. & C. July-Sept . . . .
   . . . . *Chrysothamnus parryi* (Gray) Greene

1' Inflorescence cymose, the heads at the ends of branches. Outer
   involucral bracts with acute tips, not elongate. Heads up to

1 cm high, the maturing achenes glabrous to pubescent. Plants
perennial, much-branched, up to 1.5 meters. Leaves linear, 2-
5 cm long, with a dense tomentum, especially when young. Heads
with 5 yellow flowers, the corollas 6-12 mm long. Frequent in
dry prairies and on hillsides in the western one-half. Ours
represented by the two subspecies *nauseousus* and *graveolens*
(Nutt.) Piper. Aug-Sept. Rabbit Brush . . . .
. . . . *Chrysothamnus nauseosus* (Pall.) Britt.

CICHORIUM L.

Cichorium intybus L. Plants perennial from a long taproot. Stems
irregularly branched, 2-4 dm tall. Lower leaves toothed to pinnatifid, upper leaves reduced, simple. Flowers blue, occasionally white, all ligulate, 1.5-2.5 cm across. Heads borne in
groups of 1-3 on upright branches, subsessile. Infrequent in
waste places and on roadsides in the southern one-half. July-
Sept. Chicory.

CIRSIUM Mill.

1 Heads less than 2 cm in diameter, unisexual. Stems from deep
   seated roots, perennial, branched above, 3-8 dm tall. Leaves
   sessile, green on both surfaces, glabrous or slightly pubescent, pinnatifid. Common in waste places over the state, becoming locally abundant as a noxious weed in fields in the
   eastern part. June-Aug. Canada Thistle . . . .
   . . . . *Cirsium arvense* (L.) Scop.

1' Heads exceeding 2 cm in diameter. Flowers mostly perfect.

   2 Involucral bracts conspicuously spine-tipped. Stems winged
      from the decurrent leaf bases.

      3 Heads less than 4 cm high. Stems and leaves green,
         lacking a dense white tomentum. Plants biennial,
         branched above, 4-12 dm tall. Leaves pinnatifid,
         bristly on the upper surface, gray tomentose below.
         Flowers purple to rose-colored, fading with age. In-
         frequent in waste places, pastures and farm lots over
         the state. July-Sept. Bull Thistle . . . .
         . . . . *Cirsium vulgare* (Savi) Airy-Shaw

      3' Heads 4-7 cm high. Both leaf surfaces with white, tomen-
         tose hairs, somewhat more dense on the lower surface.

         4 Flowering heads white to yellow or cream-colored, 4-5
            cm high. Plants biennial, 3-8 dm tall. Leaves
            sinuate-dentate with yellowish spines up to 1 cm long.
            Inner involucral bracts with twisted, eroded tips, the

outer ones with spreading, yellow spines. Infrequent
in sandy prairie of the south and southwest parts.
{*C. plattense* (Rydb.) Cockerell} June-July . . . . .
. . . . *Cirsium canescens* Nutt.

4' Flowering heads purple or rose-colored, rarely white,
4-6 cm high. Plants biennial, 3-9 dm tall. Leaves
pinnatifid, the segments with yellow spinose tips up
to 12 mm. Upper leaves with bases conspicuously decurrent on the stem. Inner involucral bracts elongate, yellowish, with ridges on their dorsal sides.
Rare to infrequent in prairies of the southwest part.
July-Aug . . . . *Cirsium ochrocentrum* Gray

2' Involucral bracts without conspicuous spine tips. Stems
not winged from the decurrent leaf bases.

5 Stems and lower leaf surfaces with a persistent, white
tomentum. Upper leaf surfaces variously pubescent or
whitened.

6 Main stems slender, mostly less than 5 mm in diameter.
Plants spreading by creeping roots. Early rosette
leaves entire or remotely toothed. Mature and late
leaves tending to be deeply pinnatifid. Heads rose-
purple, the achenes yellow-brown, about 4 mm long,
with an apical yellow band. Common in swales of
prairies over the state. {*Cirsium dakoticus* (A.
Nels.) Over; nomen nudum} July-Aug . . . .
. . . . *Cirsium flodmani* (Rydb.) Arthur

6' Main stems stout, 5 mm or more in diameter. Plants
with a stout taproot, usually not spreading by creeping roots. Early rosette leaves lobed, the lobes
overlapping. Later leaves coarsely toothed with
ovate to triangular lobes, wavy. Heads rose-purple,
the achenes brown, about 6 mm long. Frequent in
prairies and on hillsides over the state, occurring
less than the preceding species and flowering somewhat earlier. July-Aug. Wavy-leaved Thistle . . . .
. . . . *Cirsium undulatum* (Nutt.) Spreng.

5' Stems and upper leaf surfaces variously pubescent but
not persistently white tomentose.

7 Plants seldom more than 5 dm tall. Stem 1 cm or more
in diameter, appearing coarsely succulent, taprooted.
Leaves light green, with numerous pinnatifid divisions. Heads rose to lavender or white, solitary or
clustered at the stem terminus, often overtopped by

the leaves. Involucre 2.5-4 cm high, the outer
bracts bristle-tipped. Frequent in meadows and open
valleys of the Black Hills. July-Aug . . . .
. . . . *Cirsium drummondii* T. & G.

7' Plants commonly 1 meter or more tall. Main stems
fibrously robust, usually less than 1 cm in diameter.
Leaves dark green on the upper surface.

8 Upper leaves mostly entire or remotely lobed.
Lower leaves may be pinnatifid. Under surfaces
of leaves densely white. Involucral bracts with
serrulate tips. Vegetation not quite as spiny as
in the following species. Plants 1-2 meters tall,
branched above, with many heads. Flowers rose-
purple. Frequent in alluvial woods and low, waste
places in the eastern part. {*C. iowense* (Pammel)
Fern.} Aug-Sept . . . .
. . . . *Cirsium altissimum* (L.) Spreng.

8' Upper leaves deeply pinnatifid, the lobes ending
in spiny projections. Under surfaces of leaves
white tomentose. Inner involucral bracts with
entire tips, not spiny or serrulate. Plants 1-2
meters tall, branched, with many rose-purple heads.
Involucre 2-3 cm high. Infrequent in alluvial
areas in the eastern part. Aug-Sept . . . .
. . . . *Cirsium discolor* (Muhl.) Spreng.

CONYZA L.

1 Plants with a principal main stem and few or no side branches
except near the inflorescence; 1-13 dm tall, annual. Leaves
tending to be linear, the margins toothed, the lower leaves
deciduous before the heads mature. Flowers inconspicuous,
without showy rays. A coarse weed of waste places over the
state, abundant in many localities. {*Erigeron canadensis* L.}
Aug-Oct. Horseweed . . . . . . *Conyza canadensis* (L.) Cronq.

1' Plants without a main stem, much branched from the base,
rarely over 2 dm tall, annual. Leaves narrowly linear, pubescent, usually less than 3 cm long. Heads many, inconspicuous.
A rare weed of moist and sandy waste places in the eastern
part, often overlooked. {*Erigeron divaricatum* (Michx.) Raf.}
Aug-Oct . . . . *Conyza ramosissima* Cronq.

COREOPSIS L.

1 Principal leaves pinnately divided, the segments linear.
Plants annual, 4-10 dm tall, freely branched. Heads 1-2 cm

across; ray flowers yellowish with brown bases, the disk flowers reddish-purple. Frequent to common on moist banks, roadsides, and waste places over the state. July-Aug . . . . *Coreopsis tinctoria* Nutt.

1' Principal leaves palmately divided into three segments, sessile. Plants perennial, 3-7 dm tall, seldom branched. Heads 2-5 cm across, solitary or few. Ray flowers yellow, the disk flowers red or purple. Rare in native mesic prairies of the extreme eastern part. June-July . . . *Coreopsis palmata* Nutt.

CREPIS L.

1 Plants mostly 3-5 dm tall. Principal leaves basal, obovate-linear; stem leaves reduced or lacking. Margins of the basal leaves entire to toothed or shallowly pinnatifid. Stems and leaf surfaces glabrous to slightly hirsute. Heads 3-7 in irregular cymes, the bracts glandular-pubescent. Frequent in prairie swales and open meadow areas in the eastern part and in the Black Hills. June-July. Hawk's Beard . . . . *Crepis runcinata* T. & G.

1' Plants 1-3 dm tall. Principal leaves basal and cauline, deeply pinnatifid, covered with a dense tomentum. Heads cylindric, 4-10 in corymbose cymes. Rare in moist areas of the plains in the western part from Shannon County in the southwest to Harding County in the northwest. June-July . . . . *Crepis occidentalis* Nutt.

DYSSODIA Cav.

Dyssodia papposa (Vent.) Hitchc. Plants annual with many branches, the stems up to 4 dm tall. Vegetation unpleasantly odoriferous, the glands obvious on the involucral bracts. Leaves opposite, simple to pinnatifid, the ultimate segments linear. Heads numerous, yellowish brown, at the ends of paniculate branches. Rays few, inconspicuous. Frequent across the state in waste areas and along roadsides. Aug-Sept. Dogweed, Fetid Marigold.

ECHINACEA Moench.

Echinacea angustifolia DC. Plants perennial with a heavy, vertical taproot. Stems 4-8 dm tall, with coarse, hispid hairs. Principal leaves basal, becoming smaller upwards on the stem, the blades linear-lanceolate. Heads solitary, on long peduncles. The rays rose to purple, drooping. Chaffy bracts of the disk flowers spinescent, persisting. Common on open hillsides and prairie over the state. Ours var. *angustifolia*. May-July. Purple Cone-flower.

ECLIPTA L.

Eclipta alba (L.) Hassk. Plants annual herbs with spreading stems 2-8 dm long, often rooting at the nodes. Leaves lance-linear, entire to toothed, 2-8 cm long, sessile. Inflorescence of 1-3 heads clustered at terminal locations or in the axils of upper leaves. Involucre 4-6 mm wide. Ray flowers white, only 1-2 mm long. Disk flowers perfect, fertile. Pappus lacking or of several teeth. Rare as a weed in waste places in the eastern part. July-Aug.

ENGELMANNIA T. & G.

Engelmannia pinnatifida T. & G. Plants perennial with leafy stems, up to 6 dm tall, branched above. Leaves alternate, pin-natifid, 5-12 cm long, hirsute-hispid. Heads with golden-yellow rays, disk flowers forming hemispheres, the receptacle flattened. Rare in plains and on hillsides. Our only specimen from Badlands National Monument headquarters, Jackson County. More common south and west of the state. July. Engelmann's Daisy.

ERECHTITES Raf.

Erechtites hieracifolia (L.) Raf. Plants herbaceous annuals up to 1.5 meters tall, stems erect, glabrous to slightly hairy. Leaves alternate, with sharp serrations, lanceolate, up to 15 cm long. Heads clustered irregularly at top of stem, the flowers all discoid, white or creamy with exsert pappus. Frequent in marshy areas in the southeastern part. Becoming common and moving westward. Aug-Sept. Fireweed.

ERIGERON L.

1 Involucral bracts in 3 or 4 series, the outer shorter and thickened on the back. Plants perennial with a cespitose caudex. Stems 1-2 dm high, densely canescent. Old leaf bases persistent and becoming fibrous. Leaves narrow to linear, also canescent. Heads solitary; ray flowers rose-purple to white. Frequent on dry plains and prairies of the western part. June-July . . . . *Erigeron canus* A. Gray

1' Involucral bracts in 1 or 2 series of almost equal length and not thickened on the back.

    2 Leaves dissected into 2 or 3 linear or oblong divisions. Plants perennial with a caudex, cespitose. Leaves mostly basal, dense. Stems scapose-like, 3-10 cm high, the heads solitary. Flowers white to purple, or sometimes the ray flowers lacking. Rare in rocky crevices in higher altitudes

in the Black Hills and Harding County. July-Aug . . . . .
. . . . *Erigeron compositus* Pursh.

2' Leaves entire, toothed or somewhat lobed.

   3 Ligulate flowers of heads in full bloom 4 mm or less
      long, very numerous and filiform; inconspicuous or
      lacking.

      4 Rayless pistillate flowers present between the ray
         and disk flowers. Inflorescence corymbiform or heads
         solitary. Plants biennial or perennial, stems 5-60
         cm tall, from a simple or branched caudex. Basal
         leaves oblanceolate, the cauline leaves lanceolate
         to linear oblong, hirsute. Rare in rocky places of
         higher altitudes in the Black Hills. Ours var. *aster-
         oides* (Anderz.) DC. July-Aug . . . *Erigeron acris* L.

      4' Rayless pistillate flowers absent between the ray and
         disk flowers. Inflorescence racemiform or heads soli-
         tary. Plants mostly biennial, weak-rooted, 5-40 cm
         tall. Basal leaves oblanceolate, tapered, up to 12
         cm long. The cauline leaves more linear with spread-
         ing hairs. Ray flowers mostly white. Infrequent in
         meadows and other moist soil in the western part, in-
         cluding the Black Hills. July-Aug . . . .
         . . . . *Erigeron lonchophyllus* Hook.

   3' Ligulate flowers of heads in full bloom exceeding 4 mm,
      the rays well developed and spreading.

      5 Pappus bristles of ray flowers lacking, those of the
         disk flowers present. Plants mostly annual, tending
         to be weedy.

         6 Foliage of stems well-developed. Pubescence of
            the stem long-spreading. Plants 6-12 dm tall.
            Basal leaves elliptic, coarsely toothed. Cauline
            leaves numerous, broadly lanceolate, toothed to
            entire. Heads several to numerous, whitish-pink
            to blue. Infrequent in moist areas and disturbed
            prairie over the state. May-Aug . . . .
            . . . . *Erigeron annuus* (L.) Pers.

         6' Foliage of stems reduced, the leaves linear, re-
            duced. Pubescence of the stem more appressed than
            in the preceding. Plants 3-7 dm tall. Basal
            leaves oblanceolate, toothed, soon deciduous.
            Heads numerous, mostly whitish but sometimes blue.
            Common in prairies and dry areas over the state.

{*E. ramosus* (Walt.) BSP} June–Aug. Daisy Flea-
bane . . . . *Erigeron strigosus* Muhl.

5' Pappus bristles of ray and disk flowers present.
Annuals, biennials or perennials.

  7 Cauline leaves linear to lanceolate. Plants usually not over 1–5 dm tall, not *aster*-like.

    8 Plants biennials or true perennials, with woody
caudexes.

      9 Stems several, cespitose, from a thick tap-
root, less than 3 dm tall. Ligulate flowers
white. Basal leaves linear-oblanceolate, 2–
8 cm long. Heads several, corymbose, the
involucral disks 12–17 mm across. Ray flowers 50 or more. Frequent to common on dry
prairies in the western one-half. May–July
. . . . *Erigeron pumilus* Nutt.

      9' Stems simple or when occasionally more than
one, these decumbent-ascending, up to 5 dm
tall. Ligulate flowers light pink to purple.
Lower leaves oblanceolate, up to 15 cm long,
the upper ones reduced, linear. Heads solitary or several, the disks 1–2 cm across.
Ray flowers numerous, up to 175. Frequent
in meadows in the north and western part.
May–June . . . . . *Erigeron glabellus* Nutt.

    8' Plants without woody caudexes, annuals or bien-
nials.

      10 Hairs on the stem ascending or appressed,
sometimes sparse. Stems branched at the
base, the branches becoming stolon-like and
rooting at the ends. Basal leaves spatulate, stem leaves linear. Peduncles beneath the head naked. Flowers pink to blue-
lavender. Infrequent on open slopes and in
meadows of the Black Hills. June–July . .
. . . . *Erigeron flagellaris* Gray

      10' Hairs on the stem short-spreading, dense.
Stems not trailing or becoming stolon-like.

        11 Pappus single, of capillary bristles.
Plants mostly annual, simple to branched
above, 1–5 dm tall. Leaves many, linear
to oblanceolate, 2–5 cm long. Several

to many heads, the involucres about 8
mm across. Rays 30-35, each about 5 mm
long, white to pink. Infrequent on dry,
sandy prairie in the south and western
parts. July-Aug . . . .
. . . . . *Erigeron bellidiastrum* Nutt.

11' Pappus double, the inner row of capillary bristles, the outer of setae.
Plants biennial, branched from the
base with ascending stems 2-4 dm tall.
Leaves spatulate to oblanceolate, 2-6
cm long, densely hirsute with short
hairs. Several to many heads, the involucres 8-10 mm across. Rays many, 5
mm long, lavender. Infrequent in dry
sandy prairie of the southwest and in
open woods of the Black Hills. Ours
var. *cinerus* Gray. May-June . . . . .
. . . . . *Erigeron divergens* T. & G.

7' Cauline leaves lanceolate to wider. Plants usually over 4 or 5 dm tall, *aster*-like.

12 Rayflowers numerous, their corollas much less
than 1 mm wide, pink to pale lavender. The
stem leaves oblanceolate, toothed, the bases
clasping the stem. Stems up to 7 dm tall,
softly-hairy. Frequent in moist disturbed
soil over the state. June-Aug . . . .
. . . . *Erigeron philadelphicus* L.

12' Rayflowers fewer (less than 100), their corollas up to 1 mm wide, pink to purple.

13 Vegetation glabrous, but leaf blades may
have ciliate margins. Involucral bracts
glandular as well as hairy. Plants 3-5 dm
tall, the lower principal leaves oblanceolate. Inflorescence corymbose, the heads
1-2 cm across. Ray flowers violet to blue,
exceeding 12 mm in length. Frequent to
common in woods throughout the Black Hills.
Ours var. *macranthus* (Nutt.) Cronq. July-
Aug . . . *Erigeron speciosus* (Lindl.) DC.

13' Vegetation hairy. Involucral bracts not
glandular, but with hairs. Plants 3-6 dm
tall, the principal leaves oblanceolate.

Inflorescence corymbiform, the heads 1-2 cm across. Ray flowers pink to blue, occasionally white, about 10 mm long. Frequent in woods throughout the Black Hills. July-Aug . . . *Erigeron subtrinervis* Rydb.

EUPATORIUM L.

1 Flowers purple or pink. Leaves whorled. Plants 6-15 dm tall, the stems speckled with purple. Leaves 4-5 in a whorl, lanceolate, the blades coarsely toothed. Undersides of leaves with a dense, short pubescence. Inflorescence corymbiform, tending to be flat-topped, the flowers with corollas 5-8 mm long. Frequent to common in moist, low soil that is open or in thickets over the state. Ours var. *bruneri* (Gray) Breitung. July-Sept. Joe-pye Weed . . . . *Eupatorium maculatum* L.

1' Flowers white. Leaves opposite.

2 Leaves sessile, their bases joined around the stem, lanceolate with coarse crenate-serrate margins, with long spreading hairs beneath. Stems 6-12 dm high; perennial, with stout rhizomes. Inflorescence flat-topped to rounded, heads with 9-20 flowers, the corollas white. Frequent in moist alluvium or swampy places in the extreme eastern part. July-Sept. Boneset . . . . . . *Eupatorium perfoliatum* L.

2' Leaves petioled, the blades broadly ovate to cordate, with sharp, coarse serrations. Stems 3-12 dm tall, erect to weakly ascending. Perennial with fibrous roots. Inflorescence flat-topped. Heads with 12-20 flowers, white. Plants poisonous to animals, the toxic principle transmissible to humans through the milk from cows who are poisoned. Frequent in low or upland woods of the eastern part. Aug-Sept. White Snakeroot . . . . *Eupatorium rugosum* Houtt.

EVAX Gaertn.

Evax prolifera Nutt. Plants annual, stems 5-12 cm tall, the vegetation woolly. Leaves spatulate, 1 cm long, alternate, covered with a dense tomentum. Flowering heads in terminal glomerules, not showy. Involucre cylindrical, the bracts few, scarious. Receptacle chaffy. Achenes lacking a pappus. Infrequent in dry prairies and on plains of the southwestern one-fourth. {*Filago prolifera* (Nutt.) Britt.} May-June.

GAILLARDIA Foug.

1 Plants annual. Ray flowers purple-red with yellow tips, the corollas 1-1.5 cm long. Stems sparingly branched, 2-4 dm high.

Leaves in upper part of stem spatulate or oblanceolate, those towards the base sinuately pinnatifid. Rare in dry plains of the western part. May-June . . . . *Gaillardia pulchella* Foug.

1' Plants perennial. Ray flowers yellow with reddish bases, the corollas 1.5-3 cm long. Disk flowers purple. Stem 2-6 dm tall, usually not branched. Leaves oblanceolate in outline, more or less hirsute. Showy plant of the prairie. Frequent to common over the state. June-July. Blanket Flower . . . . *Gaillardia aristata* Pursh

GALINSOGA R. & P.

Galinsoga ciliata (Raf.) Blake. Plants annual with weakly ascending stems, 1-4 dm tall, irregularly branched. Vegetation pubescent with spreading hairs. Leaves ovate, petioled, with irregular dentate margins. Heads 3-7 mm wide, the rays white, 1-1.5 mm long. An occasional garden and greenhouse weed that persists in protected areas. July-Sept.

GNAPHALIUM L.

1 Inflorescence paniculiform, branched at the top of the stems. Plants 2-6 dm tall, biennial. Leaves lance-linear, densely tomentose below but becoming green-glabrate above. Leaf bases decurrent 1-2 cm on the stem. Involucres glabrous, up to 5 mm high, whitish. Infrequent in open places at higher altitudes of Lawrence and Pennington counties. {*G. macounii* Greene} Aug-Sept . . . . *Gnaphalium viscosum* HBK

1' Inflorescence of several to many small axillary and terminal clusters. Leaves whitish on both surfaces, not decurrent. Plants annual.

  2 Principal leaves oblong spatulate, 3-8 mm wide, the hairs loosely floccose on both surfaces. Plants 5-20 cm tall, with diffusely spreading branches. Rare in moist soil, ours from sandy soil along the Missouri River in the southern part and in the Black Hills. June-July . . . . *Gnaphalium palustre* Nutt.

  2' Principal leaves narrowly oblanceolate, not much over 3 mm wide, the tomentum appressed on both surfaces. Plants 8-25 cm tall, stems simple to branched. Involucre 2-3 mm high, woolly at the base. Rare in moist places along creeks in Custer, Lawrence and Pennington counties in the Black Hills. Aug-Sept . . . . *Gnaphalium uliginosum* L.

GRINDELIA Willd.

Grindelia squarrosa (Pursh) Dunal. Plants biennial or short-
lived perennials, branched above, the stems 2-7 dm tall. Leaves
ovate-oblong, glabrous with resinous dots, their margins serrate.
Heads with bright, yellow ray flowers. Involucral bracts highly
resinous, the outer ones strongly deflexed. Common in prairies
and plains over the state. Represented in South Dakota by the
two varieties *squarrosa* and *quasiperennis* Lunnell. Aug-Sept.
Gumweed.

GUTIERREZIA Lag.

Gutierrezia sarothrae (Pursh) Britt. & Rusby. Plants perennial
subshrubs or becoming shrubby. Stems 2-6 dm tall, much branched.
Leaves linear, 2-4 cm long and 1-2 mm wide. Heads numerous,
cylindric, with short yellow ray flowers. Frequent on dry plains
and hillsides of the western two-thirds. July-Aug. Matchbrush.

HAPLOPAPPUS Cass.

1 Leaves pinnatifid or bi-pinnatifid, 1.0-5.0 cm long, with narrow lobes. Plants perennial, 1-5 dm tall, with several stems
   from a branched, woody caudex. Flowering heads terminating
   the branches at the upper part of the stem. Involucres 8-12
   mm across, the bracts bristly pointed. Ray flowers up to 10
   mm long. Common in prairies over the state, especially westward. Aug-Sept. Iron Plant . . . .
   . . . . *Haplopappus spinulosus* (Pursh) DC.

1' Leaves entire, oblanceolate to spatulate, 2-8 cm long and 3-5
   mm wide, mostly erect.

   2 Flowering heads 10-11 mm high and broad. Involucral bracts
      oval with obtuse tips, in 3-4 series. Ray flowers 10-12 mm
      long. Plants perennial from stout taproots, the stems sev-
      eral, 5-20 cm tall. Leaves 3-nerved. Heads solitary at
      the ends of branches. Ray flowers yellow, showy. Achenes
      densely silky-villous, with many soft, white pappus bris-
      tles. Rare in dry clay soils of eroded banks in the north-
      west part. May-June . . . .
      . . . . *Haplopappus armerioides* (Nutt.) Gray

   2' Flowering heads 7-10 mm high and broad. Involucral bracts
      acuminate-cuspidate pointed, imbricate. Ray flowers 5-8 mm
      long, yellow, showy. Plants dwarf, tufted from a woody
      caudex, the stems 2-8 cm tall. Leaves narrowly linear, 1-
      nerved. Heads solitary at the ends of branches. Achenes
      strigose with longitudinal striations. Pappus few rigid

bristles that are straw or tan colored. Rare in rocky or
clay soil of the extreme southwest part. June . . . . . .
. . . . *Haplopappus multicaulis* (Nutt.) Gray

HELENIUM L.

Helenium autumnale L. Plants perennial with fibrous roots, the
stems 4-10 dm high. Leaves lanceolate with irregularly undulate
margins, the bases decurrent on the stem. Heads yellow, 2-4 cm
across, the rays usually 3-lobed and drooping. Infrequent in alluvial woods and other moist areas of the eastern part. Aug-Sept.
Sneezeweed.

HELIANTHELLA T. & G.

Helianthella quinquenervis (Hook.) Gray. Plants perennial from
branching rootstocks. Stems usually several, 4-12 dm high.
Leaves simple, mostly entire, broadly lanceolate, subopposite.
Heads usually solitary, nodding, the disks 2-4 cm across, the
rays pale yellow, 2-3 cm long. Infrequent in meadows and moist
slopes at higher altitudes of Custer, Lawrence and Pennington
counties in the Black Hills. July-Aug.

HELIANTHUS L.

1 Plants annual; leaves alternate on the stems. Disk flowers
   brownish.

   2 Center of disk with chaff of whitish hairs, the disk usu-
      ally not over 1.5 cm across. Plants not over 1 meter tall.
      Leaves lanceolate, their margins almost entire. The bracts
      of the involucre not ciliate. Frequent to common in sandy
      soil or in non-fertile waste areas over the state. Aug-
      Sept . . . . *Helianthus petiolaris* Nutt.

   2' Center of disk lacking chaff of whitish hairs, the disk
      commonly over 2.5 cm across. Plants 1-3 meters tall.
      Leaves broad, lance-ovate, their margins dentate. The
      bracts of the involucre ciliate to hispid. Very common
      in disturbed soil, fields and roadsides over the state.
      Aug-Sept. Common Sunflower . . . . . *Helianthus annuus* L.

1' Plants perennial; leaves opposite or alternate. Disk flowers
   various.

   3 Involucral bracts broadly ovate with rounded tips, decided-
      ly imbricate and appearing to be in several series. Disk
      flowers brownish purple. Heads usually solitary at the
      ends of irregular branched stems lacking leaves or leaves
      much reduced. Leaves lanceolate, thick, scabrous, opposite.

Common in prairies or sandy soil over the state. Represented in South Dakota by the two subspecies *rigidus* and *subrhomboideus* (Rydb.) Heiser. Aug-Sept . . . .
. . . . *Helianthus rigidus* (Cass.) Desf.

3' Involucral bracts lance-shaped with attenuate tips; not evidently imbricate or in several series.

  4 Stems glabrous or nearly so at the middle.

    5 Leaves coarsely serrate, blades lance-shaped, and commonly over 4 cm wide. Upper leaves tending to be alternate. Involucral bracts much longer than the height of the disk. Stems 1-3 meters tall, several from the rootstock. Heads several, the disk 1.5-2.5 cm across. Infrequent in low moist or alluvial soil and along roadsides in the eastern and western parts of the state. {Hybrids resulting from the crossing of this species and *H. maximiliana* Schrad. are called *H.* x *intermedius* Long. These occur infrequently over the state.} Aug-Sept . . . .
. . . . *Helianthus grosseserratus* Martens

    5' Leaves entire to shallowly serrate, the blades lanceolate-linear, rarely over 4 cm wide. Involucral bracts usually not longer than the height of the disk. Stems up to 1.5 meters tall. Involucral bracts with ciliate margins. Infrequent in bottom lands and along streams over the state. Represented in South Dakota by the two subspecies *nuttallii* and *rydbergii* (Britt.) Long. Aug-Sept . . . .
. . . . *Helianthus nuttallii* T. & G.

  4' Stems scabrous to hirsute.

    6 Leaves lance-ovate, commonly over 4 cm wide, distinctly petiolate. Stems stout, up to 3 meters tall, the roots often bearing tuberous rhizomes which are edible. Heads several, the disk 1.5-2.5 cm wide. Frequent in low moist places and in open waste land over the state. Aug-Sept. Jerusalem Artichoke . . .
. . . . *Helianthus tuberosus* L.

    6' Leaves linear-elliptic, the blades folded conduplicately and recurved, with short petioles or sessile. Stems slender to stout, up to 3 meters tall, with dense pubescence in the upper part. Heads several, occurring on side branches at the top of the plant. Common in disturbed prairie, along roadsides, and in dry places over the state. Aug-Sept. Maximilian's Sunflower . . . *Helianthus maximiliana* Schrad.

HELIOPSIS Pers.

Heliopsis helianthoides (L.) Sweet. Plants perennial, the stems up to 1 meter or more tall. Leaves ovate, with serrations, distinctly petioled, usually opposite. Heads with yellow ray flowers, the disk up to 2.5 cm wide. Involucral bracts in 1-3 series, the outer ones leaf-like, obtuse. Common in prairie remnants, undisturbed roadsides and in otherwise dry soil in the eastern three fourths of the state. Ours var. *scabra* (Dun.) Fern. July-Aug. Ox-eye.

HIERACIUM L.

1 Basal leaves the principal ones and persisting; those upwards on the stem much reduced. Flowers white, terminal on paniculiform branches. Plants up to 8 dm tall. Leaves long-hairy. Apparently very rare in the Black Hills in meadows and open places. This species is common in the Rocky Mountains and west. July-Aug . . . . *Hieracium albiflorum* Hook.

1' Basal leaves small and deciduous early. The principal leaves upwards on the stem; the upper ones reduced.

   2 Leaf margins lacking subconic hairs. The stem at the lower portion with long spreading hairs. Plants 5-10 dm tall, perennial. Leaves 2-5 times as long as wide, with rounded tips and somewhat clasping at the base. The margins with sharp, irregularly spaced teeth. Heads in a loose, terminal inflorescence. Rare in moist prairie in the eastern part and in the Black Hills. Aug-Sept . . . . *Hieracium canadense* Michx.

   2' Leaf margins with subconic hairs. The stem lacking long spreading hairs. Plants 4-8 dm tall, perennial. Leaves 4-12 times as long as wide, with a few coarse, irregular teeth. Heads in a loose terminal inflorescence, not umbellate. Frequent on open wooded slopes of Custer, Lawrence and Pennington counties at middle altitudes in the Black Hills and in the northeast part. Aug-Sept . . . . *Hieracium umbellatum* L.

HYMENOPAPPUS L'Her.

1 Plants biennial, without a conspicuous caudex, the stems 3-6 dm tall. Stems leafy, not limited to clusters from the base. Leaves pinnately compound, the ultimate segments linear. The involucral bracts yellow-colored at the apex. Flowers cream-

yellow. Pappus over 1 mm long, not hidden by the hairs of the
achenes. Plains and prairies over the state except in the extreme eastern part. June . . . *Hymenopappus tenuifolius* Pursh

1' Plants perennial, from a branching caudex, the stems 2-4 dm
tall. Stems sparingly leafy, in clusters from the base.
Leaves much like the preceding, the ultimate segments filiform. Involucral bracts pale yellow, the flowers deeper yellow than the preceding. Pappus less than 1 mm long, covered
by the silky hairs of the achenes. Infrequent in dry plains
and on plateaus of the western two thirds of the state. Represented in South Dakota by the variety *polycephalus* (Osterh.)
Turn. June . . . . *Hymenopappus filifolius* Hook.

HYMENOXYS Cass.

Hymenoxys acaulis (Pursh) Parker. A stemless perennial from a
stout taproot which is sometimes branched. Leaves linear, basal,
up to 6 cm long and 6 mm wide. Flowering scapes unbranched, 8-20
cm tall. Heads with deep yellow ray flowers, the disk 1-1.5 cm
wide. Ligules 7 mm long, their distal ends 3-4 toothed. Although tetraploid, ours not as variable as farther west. Frequent on dry hills and prairies from central South Dakota westward. May-June.

IVA L.

1 Plants perennial from a creeping rootstock. Flowering heads
in the axils of upper leaves. Stems branched, up to 5 dm
tall. Leaves sessile, entire and oblong, 2-3 cm long. Heads
hemispheric, 4-5 mm across, becoming recurved towards maturity.
Corollas greenish-white, tubular. Frequent in alkali flats or
in saline soil in the western one-half. June-July . . . . . .
. . . . *Iva axillaris* Pursh

1' Plants annual. Inflorescence terminal, racemiform or paniculate.

2 Inflorescence a terminal spike, the heads in the axils of
bract-like leaves. Plants 5-15 dm tall, mostly simple-
stemmed. Leaves lance-ovate, with tapering petioles,
about 10-15 cm long. Flowering heads drooping or recurved,
the pistillate flowers with a persistent corolla. Rare in
waste soil in the east-central part. Common south of the
state. {*I. ciliata* Willd.} Aug-Sept . . . . *Iva annua* L.

2' Inflorescence paniculate, the heads not subtended by bract-
like leaves. Plants up to 1 meter tall, mostly unbranched.
Leaves opposite, petioled, the blades broadly ovate, 5-20
cm long. Flowering heads 4-5 mm across, with 5 obovate

bracts. The corollas of pistillate flowers rudimentary or lacking. Frequent to common in lowland waste or disturbed soils over the state. July-Aug. Marshelder . . . . *Iva xanthifolia* Nutt.

KUHNIA L.

Kuhnia eupatorioides L. Plants perennial, the stems branched, up to 7 dm tall. Leaves lanceolate, up to 3 cm wide, irregularly toothed, their surfaces finely pubescent. Flowering heads clustered at the ends of branches, flowers all discoid, creamy white to dull yellow. Achenes with up to 20 pappus bristles. Common on dry hillsides and prairies over the state. Ours var. *corymbulosa* T. & G. Aug-Sept. False Boneset.

LACTUCA L.

1 Flowers yellow or occasionally cream-yellow and becoming bluish with age. Achenes flattened with a slender beak, almost as long as or longer than the body.

  2 Heads with 10 flowers or less. When mature the achenes with several to 7 ribs on each face. Plants annual, up to 8 dm tall. Leaves with spinulose-toothed margins and midribs also somewhat spined. Leaves usually pinnatifid but sometimes almost entire. Naturalized from Europe. Very common in waste places and on roadsides over the state. {*L. scariola* L.} July-Sept. Prickly Lettuce . . . . *Lactuca serriola* L.

  2' Heads with 10 to many flowers. Achenes with 1 to 3 prominent veins or ribs on each face, becoming dark brown towards maturity.

    3 Leaves soft, pinnatifid to irregularly lobed with oblong segments. Heads usually with 20 flowers or less, the involucres in fruit 10-15 mm long. Plants biennial, with unbranched stems up to 2 meters. Infrequent on the edges of woods and thickets in the eastern and western parts. June-Aug. Wild Lettuce . . . . *Lactuca canadensis* L.

    3' Leaves with spinulose margins and midribs, with pinnatifid or sinuate margins. Heads 20-many flowers, the fruiting involucres 15-20 mm long. Plants up to a meter tall, usually biennial, the stems stout. Frequent in prairies and prairie remnants over the state. July-Aug. Prairie Lettuce . . . *Lactuca ludoviciana* (Nutt.) Riddell

1' Flowers lavender or bluish, rarely yellow. Achenes with short beaks or beakless.

4 Plants perennial from a deep spreading root system. Flower heads showy, lavender, up to 3 cm across. Leaves glabrous, pinnately lobed to almost entire. Plants up to a meter tall. Achenes with several nerves on each face, the tapered short beak often white. Common in prairies, meadows, and on roadsides over the state. {*L. pulchella* (Pursh) DC.} July-Aug. Blue Lettuce . . . . *Lactuca oblongifolia* Nutt.

4' Plants annual or biennial. Flower heads not large and showy.

5 Leaves pinnatifid with the terminal lobe triangular in shape. Pappus remaining white at maturity of the achenes. Plants up to a meter or more tall, glabrous. Heads in a terminal panicle, usually bluish, up to 15 flowers. Rare in alluvial woods along streams in the extreme eastern part. July-Aug . . . . *Lactuca floridana* (L.) Gaertn.

5' Leaves pinnatifid to sinuate in shape. Pappus becoming brownish at maturity of the achenes. Plants to 1.5 meters tall, glabrous. Leaves sometimes hairy on the midrib underneath and sagittate at their bases. Many heads in a narrow panicle at the top of the plant. Flowers blue, sometimes white or yellowish, usually over 15 per head. Infrequent in low moist places in the extreme northeast part and in the Black Hills. July-Aug . . . . *Lactuca biennis* (Moench.) Fern.

LIATRIS Schreb.

1 Principal leaves more than 5 mm wide, the basal leaves widest. The pappus bristles barbellate, the lateral hairs very short and narrow. Under low magnification the bristles appearing only scabrous.

2 Heads broadly hemispheric at maturity, each usually with more than 18 flowers. Involucral bracts purplish, with scarious borders.

3 Flowering heads uniform in size, each head with 18-35 flowers. As many as 30 heads per plant. Stems 3-10 dm tall, from a deep-seated corm. Leaves toward the base lance-shaped, petioled. Frequent in dry to moist prairies in the eastern part of the state. {*L. scariosa* of reports} Aug-Sept . . . . *Liatris aspera* Michx.

3' Flowering heads uniform except that the terminal one is usually larger than the lower. Each head with more than 30 flowers. Fewer heads, commonly not more than 10, as

compared with the preceding species. Stems 2-6 dm tall,
from a shallow corm. The lower leaves are broadly lanceolate, petioled. Frequent in moist meadows and open ravines in the Black Hills and in the eastern part. Aug-
Sept . . . . . *Liatris ligulistylis* (A. Nels.) K. Schum

2' Heads cylindrical, each with 4-18 flowers. Involucral
bracts greenish to green-purple, with acuminate to obtuse
tips.

4 Involucral bracts with acuminate tips and spreading outward, squarrose, greenish. Heads in dense spikes. Stems
commonly up to a meter tall, unbranched, erect. Leaves
many, lanceolate to linear, alternate on the stem, much
larger below and reduced upward. Flowers rose-purple,
4-8 in each head. Rare in moist meadows in the eastern
part and not very distinct from the following species.
July-Aug . . . . *Liatris pycnostachya* Michx.

4' Involucral bracts with obtuse tips, erect, the margins
becoming erose, purplish. Heads in dense spikes, 6-16
flowered, bright red-pink. Stems commonly over a meter
tall, stout and unbranched. Leaves tending to be slightly narrower and linear than the preceding species. Lower
leaves longer and larger than the ones upward, numerous
and separate on the stem. Frequent in mesic or low prairie in the southeast part. July-Aug . . . .
. . . . *Liatris lancifolia* (Greene) Kittell.

1' Principal leaves usually less than 5 mm wide, the lowermost
not much wider than the others. The pappus bristles plumose,
the lateral hairs evident under low magnification.

5 Heads with 8 or fewer flowers, cylindric-shaped, usually 10
or more per plant. The involucral bracts appressed with
acuminate tips. Stems several from a thickened caudex,
usually less than 6 dm tall, stiff. Leaves numerous and
punctate, the lowermost deciduous toward mid-summer. Upper
leaves becoming bract-like in the inflorescence. Common in
dry prairies over the state. July-Sept . . . .
. . . . *Liatris punctata* Hook.

5' Heads with 20-30 flowers, the heads almost as wide as high,
usually 1 to 5 per plant. The outer involucral bracts
sharply pointed, bent out abruptly. Stems solitary, 2-6
dm tall. Leaves narrow, linear, much fewer than in the
preceding species. Infrequent in dry sandy prairie of the
southern part. {*L. squarrosa* (L.) Michx.} July-Aug . . .
. . . . *Liatris glabrata* Rydb.

LYGODESMIA (Pursh.) D. Don.

1 Plants annual. Leaves narrowly linear, the principal ones
   much exceeding 5 cm. Usually more than 5 flowers per head.
   Stems with ascending branches, erect, to 7 dm tall. Flowers
   rose-pink, the heads many at the ends of narrow branches.
   Pappus a sordid white color when mature. Rare in dry sandy
   places, principally in the western part. Aug . . . .
   . . . . *Lygodesmia rostrata* A. Gray

1' Plants perennial from a deep-seated root system. Stems
   branched, slender and stiff, up to 4 dm tall. Many times
   the stems have swollen places the size of hazel nuts. These
   are galls caused by several species of insects. Leaves linear, the principal ones rarely 5 cm long. Upper leaves becoming scale-like. Usually only 5 flowers per head, these
   rose-pink to white. Pappus tinged with brown or at least
   stramineous. Common in dry to moist prairies over the state.
   July-Aug. Skeleton Plant . . *Lygodesmia juncea* (Pursh) D. Don

MACHAERANTHERA Nees.

1 Ray flowers rudimentary or lacking, heads essentially discoid.
   Plants perennial from a stout taproot, 1-3 dm tall. Leaves
   thick, oblanceolate with spinulose tipped serrations. Lower
   leaves soon deciduous, the upper ones persisting, 2-4 cm long.
   Heads about 1 cm across, yellow, about 6-8 mm high. Disk
   flowers fertile and bisexual. Infrequent on dry, clay banks
   of the western part. {*Haplopappus nuttallii* T. & G.} June-
   July. Golden Weed . . . .
   . . . . *Machaeranthera grindelioides* (Nutt.) Shinners

1' Ray flowers well developed, white to rose-purple. Plants annual, biennial or perennial.

   2 Plants perennial with a thick, woody caudex. Involucral
      bracts acuminate, tipped with spines, in 2 or 3 series.
      Leaves many, entire to spinulose-toothed, 1 nearly oblance-
      olate, 2-5 cm long. Heads 1-1.5 cm wide, solitary at the
      ends of stems. Ray flowers pistillate, off-white to rose,
      10-12 mm long. Rare in alkali flats in the northwest part.
      {*Aster xylorrhiza* T. & G.} May-June . . . .
      . . . . *Machaeranthera glabriuscula* (Nutt.) Cronq. & Keck.

   2' Plants annual or biennial, with taproots. Involucral
      bracts in several series, with spreading or reflexed tips.

      3 Leaves once or twice pinnatifid. Plants annual, much-
         branched, 1-4 dm tall. Foliage glandular-pubescent.
         Heads rose-purple, up to 2 cm across, 1-few on stem

branches. Involucral bracts with a whitish base, the green tips spreading, ending in a bristle. Rays 1 cm or longer. Infrequent in dry soils of Pennington and Fall River counties. {*Aster tanacetifolius* HBK.} July-Aug. Tansy Aster . . . . *Machaeranthera tanacetifolia* (H.B.K.) Nees.

3' Leaves entire to toothed, simple. Plants biennial or rarely short-lived perennials.

4 Principal cauline leaves less than 3 mm wide, densely canescent, inrolled and recurved. At flowering time the wider, dentate margined leaves usually deciduous. Plants 1-4 dm tall, simple stemmed to much branched. Flowers many on the ends of branches, the rays purple. Heads about 1 cm across, the involucral bracts green-tipped with prominent straw-colored bases. Frequent in dry soil in the western one-half. {*Aster canescens* Pursh.} July-Sept. Hoary Aster . . . . *Machaeranthera canescens* Pursh.

4' Principal cauline leaves up to 1 cm wide, glabrous with many spinulose-tipped dentations. The lower leaves not tending to fall before or at flowering time. Plants 3-6 dm tall, the stems simple to branched. Flowers purple, subsessile on short, terminal branches. Heads 1.5 cm across, the involucral bracts subulate-curled, not prominently exposing straw-colored bases. Rare in dry or sandy soil in the southwest part. {*Aster sessiliflora* Nutt.} July-Aug . . . *Machaeranthera linearis* Greene

MADIA Mol.

Madia glomerata Hook. Plants slender, annual, 3-6 dm tall, with simple stems or sparingly branched towards the upper part. Leaves narrow, lance-shaped, 2-4 cm long and 1-4 mm wide, hairy. Heads in glomerate clusters at the tips of branches, the involucres 7-9 mm high. Ray flowers, when present, yellow, not over 3 mm long, fertile. Infrequent in dry soils or sandy and open sterile areas in the western part. July-Aug. Tarweed.

MATRICARIA L.

Matricaria matricarioides (Less.) Porter. Plants annual, the stems 2-4 dm high, usually with several branches. Leaves compound with linear segments, 2 or more times pinnatifid. Heads with ray flowers lacking, the disks conic to hemispheric, less than 1 cm across. Involucral bracts whitish, glabrous, with scarious margins. An infrequent weed of waste places, roadsides and overgrazed pastures throughout the state. May-Aug. Pineapple weed.

MICROSERIS D. Don

Microseris cuspidata (Pursh) Schultz-Bip. Plants stemless perennials from narrow, deep taproots. Leaves linear-lanceolate, 3-7 cm long, the margins wavy and with a whitened villous row of hairs. Flowering scapes 5-25 cm long, the heads solitary. Flowers yellow, ligulate, the disk 2-3 cm across. Common on dry, sandy prairie and rocky soil over the state. May-July. False Dandelion.

PETASITES Mill.

Petasites sagittatus (Pursh) Gray. Plants perennial from thick, creeping rootstocks. Stems erect, 1-4 dm tall, with leaves reduced to narrow bracts. Principal leaves basal, on long petioles, the blades 1-3 dm long, cordate or sagittate with a crenate-dentate margin. Petioles and undersides of leaves white tomentose. Inflorescence corymbiform, the heads about 6 mm high, ray flowers present or lacking. Plants with fertile pistillate and staminate heads in the same inflorescence. Achenes ribbed with many capillary bristles. Infrequent to rare in moist places in the Black Hills. May-June. Sweet Coltsfoot.

PICRADENIOPSIS Rydb.

1 Achenes hairy. Pappus scales ending in excurrent awns. Plants perennial, woody at the base, the stems branched, up to 2 dm tall. Leaves opposite, 3-parted into linear segments. Vegetation with appressed hairs. Heads 5-7 mm high, the ray and disk flowers yellow. Rare in dry plains in the western part and very similar to the following species. June-July . . . . *Picradeniopsis woodhousei* (Gray) Rydb.

1' Achenes glandular but not hairy. Pappus scales obovate, not ending in awns. Plants perennial with woody bases. Stems much-branched, 6-20 cm tall. Leaves nearly all opposite, 3-5 parted into linear segments. Vegetation densely gray-canescent. Flowers yellow, with few rays, each ray 2-4 mm long. Frequent on dry plains in the western one-half. {*Bahia oppositifolia* (Nutt.) DC.} June-July . . . . *Picradeniopsis oppositifolia* (Nutt.) Rydb.

PRENANTHES L.

1 Involucral bracts glabrous. Stem and leaves glaucous. Leaves, especially the lower ones, long petioled, the blade incised to

form a sagittate to hastate shape. Flowers white to greenish, the heads nodding. Pappus becoming cinnamon-brown at maturity. Rare in rich woods of Grant, Marshall and Roberts counties in the northeast part. August. White Rattlesnake Root . . . . *Prenanthes alba* L.

1' Involucral bracts pubescent with coarse hairs. Leaves toothed to entire, not evidently lobed.

    2 Flowers cream-color. Leaves and stem variously pubescent; however, the upper leaf surfaces may be glabrous. Lower leaves obovate, tapering to a petiole. Plants 7-12 dm tall. Inflorescence narrow, elongate. Rare in moist prairies of the eastern part. Aug-Sept . . . . *Prenanthes aspera* Michx.

    2' Flowers purple. Leaves and stem glabrous. Lower leaves broadly oblanceolate, petioled. Plants 4-10 dm tall. Inflorescence elongate, heads crowded, mostly ascending but tending to nod towards maturity. Infrequent in moist soil of the eastern part and in the Black Hills. Ours subspecies *multiflora* Cronq. Aug-Sept . . . . *Prenanthes racemosa* Michx.

RATIBIDA Raf.

1 Ray flowers 3 cm or less long. Leaf segments linear.

    2 Disk in full flower about 1 cm long, oblong. Plants perennial from a taproot. Stems solitary to several, 2-3 dm tall. Leaves pinnately compound or divided, the 3-6 divisions with coarse hairs. Flowering heads with yellow rays, drooping, up to 1 cm long. Apparently rare in dry soil in the southwest part. Common in the southern plains. July-Aug . . . . *Ratibida tagetes* (James) Barnh.

    2' Disk in full flower 1.5-3 cm long, cylindrical, 1 cm in diameter. Plants perennial from a woody taproot, the stems solitary or branched, 2-8 dm tall. Leaves pinnately divided, with 6 or more linear segments. Ray flowers yellow to purple or purple-tinged, 1.5-2 cm long, drooping, oval in outline. Very common in dry prairies and on roadsides over the state. June-Aug. Coneflower . . . . *Ratibida columnifera* (Nutt.) Woot. & Standl.

1' Ray flowers 3-5 cm long. Leaf segments lanceolate. Plants perennial from a woody caudex, the stems branched, up to 1 meter or more tall. Leaves pinnately compound, mostly with 5-7 lanceolate segments, their surfaces harsh and scurfy. Disk in full flower hemispheric to ellipsoid, about 2 cm

high and 1.5 cm across. Rays pale yellow, drooping, spatulate
in shape. Infrequent in prairie remnants and on roadsides in
the southeast part. July–Aug . . . .
. . . . *Ratibida pinnata* (Vent.) Barnh.

RUDBECKIA L.

1   Leaves oblanceolate to lance-shaped, their surfaces harsh due
    to rough hairs. Plants short-lived perennials, 3–7 dm tall,
    the stems solitary or branched. Heads on long peduncles; the
    rays orange-yellow, becoming pale. The disks brown to purple,
    1–2 cm across. Very common on dry prairies and on roadsides
    over the state. July–Sept. Blackeyed Susan . . . .
    . . . . *Rudbeckia hirta* L.

1' Leaves pinnately divided to lacinate, their surfaces mostly
    glabrous. Plants perennial from a woody base, stems up to 2
    meters tall, several. Leaves large, coarsely toothed and
    lacinate. Heads with yellow disks 1–2 cm across, becoming
    gray towards maturity. Rays 3–5 cm long, yellow, drooping.
    Frequent in low, moist places in the east and the Black Hills.
    July–Sept. Goldenglow . . . . *Rudbeckia laciniata* L.

SENECIO L.

1   Ray flowers lacking when plants are in bloom.

    2   Plants annual. Leaves all pinnatifid or sinuately incised,
    2–6 cm long. Outer involucral bracts with blackened spine
    tips, the inner ones linear and scarious on their margins.
    Stems moderately branched from the base; plants up to 4 dm
    tall. An infrequent weed in moist disturbed soil in the
    eastern part. Probably not persisting. July–Sept. Ground-
    sel . . . . *Senecio vulgaris* L.

    2' Plants biennial or short-lived perennials with well devel-
    oped rootstocks. Leaves broadly oblanceolate with denticu-
    late margins, the lower ones with long petioles. Upper
    leaves sessile, clasping the stem. Stems mostly solitary,
    only sparingly branched above, the inflorescence panicled
    or corymbiform. Plants 4–10 dm tall. Frequent in moist
    soils at the bases of talus slopes and on hillsides in
    Custer, Lawrence and Pennington counties in the Black
    Hills. July–Aug . . . . *Senecio rapifolius* Nutt.

1' Ray flowers present and obvious when plants are in bloom.

    3   Middle and upper stem leaves on maturing plants not appre-
    ciably fewer or smaller than the lower leaves.

4 Principal leaves or leaf segments linear, less than 5
  mm wide.

  5 Leaves simple, the blades sometimes with small short
    lobes toward the base. Stems glabrous, often several
    in a cluster, from a perennial rootstock; 2-5 dm tall.
    Flowering heads in a corymbose inflorescence, the
    heads usually less than 8 mm wide. The ray flowers
    yellow, few. Rare in valleys in the Black Hills.
    July-Aug . . . . *Senecio spartioides* T. & G.

  5' Leaves pinnately dissected. Stems tomentose when
    young, becoming glabrate with age, several from a
    suffruticose base. Plants 4-8 dm tall. Inflores-
    cence corymbose, the heads usually over 8 mm wide.
    Ray flowers 10-15 mm long. Frequent in valleys, on
    flood plains and in the foothills of the southwestern
    part. Ours principally from the Black Hills. Aug-
    Sept . . . . *Senecio riddellii* T. & G.

4' Principal leaves not linearly dissected, the blades exceeding 1 cm in width.

  6 Margins of leaves toothed or denticulate, the blades
    5 to 40 mm wide and up to 20 cm long, mostly sessile
    along the stem. Plants 2-6 dm tall, the stems dense-
    ly hairy. Heads several in a congested inflorescence.
    The rays pale yellow, 4-7 mm long. Infrequent in
    swampy or marshy places in the eastern counties and
    in the extreme northeast. {*S. palustris* (L.) Hook.}
    July-Aug . . . . . . *Senecio congestus* (R. Br.) DC.

  6' Margins of leaves distinctly pinnatifid, the toothed
    segments giving the leaf a lacinate appearance.
    Leaves 1-2 dm long. Plants 3-10 dm tall, the stems
    mostly glabrous. Inflorescence of several to numer-
    ous heads corymbosely arranged. Rare in wet soil
    along roadsides and in ravines in the Black Hills.
    Aug-Sept . . . . *Senecio eremophilus* Rich.

3' Middle and upper stem leaves on maturing plants smaller and
   fewer than the lower or basal ones.

  7 Plants more or less tomentose or densely pubescent when
    mature.

    8 Stems with tufts of floccose hairs at the nodes.
      Basal leaves ovate, petioled. The cauline leaves
      usually pinnatifid. Stems 2-6 dm tall, several from
      a branched caudex. Heads several to numerous in a

corymbose cyme, the involucres 6-7 mm high. Ray
flowers bright yellow or orange, about 8 mm long.
Common in moist to dry prairie over the state.
May-June. Prairie Ragwort . . . .
. . . . *Senecio plattensis* Nutt.

8' Stems and leaves densely canescent. Basal leaves
linear to lanceolate. Cauline leaves much reduced,
only dentate or rarely pinnatifid. Stems 1-4 dm
tall, several from the branched caudex. Heads 10-
12 mm high, with narrow, linear bracts. Ray flowers
bright yellow, 8-10 mm long. Frequent on dry prairie
in the western part. {*S. purshianus* Nutt.} May-June
. . . . *Senecio canus* Hook.

7' Plants glabrous or glabrate when mature. If pubescent,
the hairs are long-jointed and irregular.

9 Lower leaves with lanceolate to cordate bases not
gradually tapering to the petioles. Upper leaves
toothed or pinnatifid. Plants with well developed
horizontal or ascending rootstocks.

10 Principal basal leaves with petioles 6-9 cm long,
the blades with cordate bases and coarse teeth.
Cauline leaves toothed or pinnatifid. Plants 3-6
dm tall, the stems mostly solitary. Heads several to numerous at the ends of slender peduncles,
the disk 5-10 mm across. Ray flowers about 7 mm
long. Infrequent in moist meadows and thickets
in the eastern part and in the Black Hills. Ours
var. *semi-cordatus* (Mack. & Bush) T. M. Barkl.
July-Aug . . . . . . *Senecio pseudoaureus* Rydb.

10' Principal basal leaves with petioles 2-5 cm long,
the blades crenate to serrate, abruptly tapering
to the petiole. Upper leaves sessile, more or
less pinnatifid. Plants 1-5 dm tall, the vegetation floccose when young. Stems simple, erect.
Heads few, the disks 5-10 mm across. Ray flowers
about 5 mm long. Infrequent in meadows and other
moist open places in the Black Hills and in the
eastern part. June-July . . . .
. . . . *Senecio pauperculus* Michx.

9' Lower leaves lanceolate, gradually tapering to petioles. Upper leaves only shallowly toothed if at all.
Plants with short-lived erect rootstocks, becoming
fibrous-rooted.

11 Stems of maturing plants over 7 dm tall. Plants
of boggy places. Perennial from vertical rootstocks, the stems hollow and succulent. Heads
rather numerous and crowded, the involucres 5-9
mm high. Rays few. Rare in swampy and boggy
places in higher altitudes of the Black Hills.
July-Sept . . . . *Senecio hydrophilus* Nutt.

11' Stems of maturing plants less than 7 dm tall.
Plants not restricted to boggy places.

12 Leaves soft, fleshy; mostly entire or remotely sinuate. Stems hollow and juicy, becoming
flattened when dry. Heads several flowered,
usually more than 5 per stem. Involucral
bracts acuminate, many times blackened, but
without bearded tips. Frequent to common in
moist open places or in thickets over the
state. More common in the northeast. June-
July . . . . *Senecio integerrimus* Nutt.

12' Leaves firm, glabrous; mostly entire to finely serrate. Stems solid. Heads few flowered,
less than 5 per stem. Involucral bracts
black-tipped with tufts of hairs at their
apex. Rare to infrequent in meadows of high
altitudes in the Black Hills. July-Aug . . .
. . . . *Senecio crassulus* A. Gray

SILPHIUM L.

1 Principal leaves pinnately dissected and alternate. Each leaf
turned so that the blade is vertical, many times in a north-
south direction. Cauline leaves alternate, much reduced.
Plants 1-2 meters tall, perennial. Heads mostly sessile in a
racemiform arrangement at the upper part of the stem. Rays
pale yellow. Infrequent in moist prairies in the eastern
part. July-Aug. Compass-plant . . . . *Silphium laciniatum* L.

1' Principal leaves opposite, entire or merely toothed.

2 Leaves large, over 4 inches long, their bases perfoliate,
clasping around the stem, forming a large cup. Stems
angled, somewhat branched above, perennial. Plants over
a meter tall. Heads several in an open inflorescence.
Rays yellow, the disks 1-2 cm across. Rare in alluvial
places in the eastern part of the state. Aug-Sept. Cup-
plant . . . . *Silphium perfoliatum* L.

2' Leaves smaller than 4 inches long, their bases opposite but not perfoliate; blades ovate to elliptic in outline. Stems rounded in cross section, usually only sparingly branched above if at all. Heads few in a rather close terminal inflorescence, the rays pale yellow. Apparently very rare in the eastern part. Our only specimen from Union County along the Big Sioux River. July-Aug . . . . *Silphium integrifolium* Michx.

SOLIDAGO L.

1 Inflorescence compactly corymbiform to flat-topped, terminal.

    2 Leaves linear, less than 1 cm wide; up to 6 cm long, numerous. Inflorescence broad and flat-topped, the individual heads usually less than 1 cm across. Plants 4-8 dm tall, sparingly branched above; perennial from a rhizome. Leaf surfaces glandular-punctate. Frequent in moist soil of open places in the eastern part and in the Black Hills. Represented in South Dakota by the following 4 varieties: *graminifolia*, *gymnospermoides* (Greene) Croat, *major* (Michx.) Fern., and *media* (Greene) Harris. Aug-Sept . . . . *Solidago graminifolia* (L.) Salisb.

    2' Leaves lanceolate to ovate, commonly up to 4 cm wide and 3-6 cm long. Inflorescence densely corymbiform, the individual heads 1 cm or more across. Plants up to a meter tall, unbranched, the vegetation with a dense, short pubescence. Of dry soils. Common in dry prairies and plains over the state. Represented in South Dakota by the two varieties *rigida* and *humilis* Porter. Aug-Sept. Rigid Goldenrod . . . . *Solidago rigida* L.

1' Inflorescence variously paniculiform or pedicellate, or appearing spike-like; not corymbiform.

    3 Flowering heads in small axillary clusters or in short racemes, the branches erect or at least not obviously nodding or recurved.

        4 Stems below the inflorescence densely pubescent with short hairs. Plants with stems 1-4 dm tall, usually several from the creeping rhizomes. Leaves at the base soon deciduous, the others crowded, reduced upwards, their blades 3-nerved, elliptic or obovate. The distal ends of the blades remotely toothed, with a rounded or acute apex. Leaves 3-6 cm long. Inflorescence terminal, dense, an occasional lower branch extending outwards. Common on dry prairie over the state. {*S. incana* T. & G.} Aug-Sept . . . . . . *Solidago mollis* Bartl.

4' Stems below the inflorescence glabrous to sparsely puberulent or scabrous.

    5 Heads primarily in axillary clusters towards the top of the stem with an occasional short raceme appearing terminal. Leaves ovate to elliptic, sharply toothed with acuminate tips; the petioles winged. Stems 3-10 dm tall, perennial from rhizomes. Rare in rich woods of the extreme northeast part and also from the southeast part. {*S. latifolia* L.} Aug-Sept . . . . *Solidago flexicaulis* L.

    5' Heads primarily in crowded short racemes that are dense and with stiffly ascending branches. Plants 4-12 dm tall, from a woody caudex, stems 1-several from the base. Leaves lanceolate, their margins entire. Principal leaves 4-8 cm long and 1-3 cm wide, gradually becoming reduced in the upper part of the stem. Frequent in open places in the extreme western part and rare in the eastern part. Ours var. *rigiduscula* T. & G. Aug-Sept . . . . *Solidago speciosa* Nutt.

3' Flowering heads in terminal panicles or racemes, their branches radiating, many times recurved-nodding.

  6 Stems glabrous below the inflorescence.

    7 Lower or basal leaves largest, usually persisting, those upwards on the stem gradually becoming smaller. Plants 2-8 dm tall, perennial from a creeping rhizome. Principal leaves oblanceolate, linear, tapering to a petiole. Margins slightly toothed to entire, the blades firm. Inflorescence paniculiform, with irregular recurved branches. Common in prairies and open places throughout the state. {*S. glaberrima* Martens} Aug-Sept. Prairie Goldenrod . . . . *Solidago missouriensis* Nutt.

    7' Lower or basal leaves smaller than those at the middle of the stem and usually soon deciduous. Plants 6-20 dm tall, perennial. Principal leaves lanceolate, tapered at both ends, 3-8 cm long, distinctly serrate. Inflorescence commonly a dense pyramidal panicle, the branches up to 8 or more cm long. Frequent to common in meadows, marshes and other low places over the state. Ours represented by the two varieties *gigantea* and *serotina* (O. Ktze.) Cronq. Aug-Oct. Late Goldenrod . . . . *Solidago gigantea* Ait.

6' Stems puberulent to densely canescent below the inflorescence.

   8 Plants 3-5 dm tall, the stems curved and commonly growing in clusters from a fibrous-rooted perennial caudex. Principal leaves oblanceolate, gradually tapering to a petiole; mostly less than 1 cm wide at their widest point. Upper leaves smaller and linear.

      9 Involucral bracts broadly linear with blunt tips. Pubescence of leaves with hairs pointing in all directions. Rootstocks without rhizomes. Plants perennial, 2-6 dm tall, often in clumps. Cauline leaves oblanceolate, long-petioled, subtending smaller ones in axillary tufts. Inflorescence with a 1-sided panicle, many times conspicuously wand-like. Common in dry prairies and plains over the state. July-Sept . . . . *Solidago nemoralis* Ait.

      9' Involucral bracts lanceolate with acute tips. Pubescence of leaves with hairs pointing towards the apex. Plants perennial, 3-6 dm tall, the stems decumbent on the lower part. Cauline leaves oblanceolate, 5-8 cm long, tending to be remotely arranged on the stem. Rootstocks with creeping rhizomes. Inflorescence obscurely to distinctly panicled. Rare in dry soils. A western plant reaching South Dakota in the southwestern part. Aug-Sept . . . . . . *Solidago sparciflora* A. Gray

   8' Plants 5-15 dm tall, the stems erect, with 1-several in a cluster. Principal leaves lanceolate, tapered at both ends, usually exceeding 1 cm at the widest point. Leaves mostly glabrous or scabrous above with pubescence variable below. Upper leaves smaller but not essentially different in shape. Inflorescence compact to large and pyramidal, the branches diverging or recurved. A variable species widely distributed along roadsides, at the edges of thickets, and in open prairies over the state. {*S. altissima* L.} Represented in South Dakota by the following 5 varieties: *canadensis* (rare), *gilvocanescens* Rydb. (common), *hargeri* Fern. (rare), *salebrosa* (Piper) M. E. Jones (rare), and *scabra* (Muhl.) T. & G. (common). Aug-Sept. Canada Goldenrod . . . . *Solidago canadensis* L.

SONCHUS L.

1 Plants perennial with deep roots as well as horizontal spreading roots. Flowering heads 3-5 cm in diameter; bright yellow. Involucral bracts up to 1.5 cm high.

    2 Involucral bracts glandular or pubescent or both. Stems 6-15 dm tall, the upper part glaucous. Leaves prickly-margined, pinnatifid, their bases auriculate. Heads several in a corymbiform inflorescence, the heads relatively large. Achenes slightly compressed, with 5 or more prominent ribs, 2.5-3.5 mm long. Frequent in waste areas and at the edge of fields in the northeast part and rare in the western part. July-Aug. Field Sow-thistle . . . . *Sonchus arvensis* L.

    2' Involucral bracts glabrous or with a sparse tomentum. Plants perennial with a creeping rootstock, the stems simple, 4-8 dm tall. Leaves prickly-margined, from shallowly lobed to merely toothed. Lower leaves runcinate, their bases clasping the stem. Inflorescence with heads not as large as in the preceding species. Achenes turgid, 2.5 mm long, with transverse lines. Similar to the preceding. Frequent in fields and waste places, especially in the eastern one-half. July-Aug . . . . *Sonchus uliginosus* Bieb.

1' Plants annual. Flowering heads less than 3 cm in diameter. Involucral bracts 1 cm long.

    3 Terminal leaf segment large and triangular. The bases of the leaves sagittate-shaped and clasping; the margins of the blade scarcely prickly. Achenes longitudinally striate with roughened papillae. Plants commonly 2-8 dm tall, mostly glabrous. Inflorescence of several corymbiformly arranged heads. Infrequent in waste places over the state but more common in the east. July-Sept. Common Sow-thistle . . . . *Sonchus oleraceus* L.

    3' Terminal leaf segment not conspicuously triangular. The leaf margins pinnatifid with harsh prickles. Bases of leaves auriculate, rounded. Achenes merely longitudinally ribbed. Plants 3-12 dm tall. Inflorescence few-flowered, in corymbose cymes. Infrequent in waste places and old fields over the state. July-Sept. Spiny Sow-thistle . . . . *Sonchus asper* (L.) Hill

TANACETUM L.

Tanacetum vulgare L. Plants perennial from a spreading rhizome system, the stems up to a meter tall. Leaves numerous, compound

and pinnatifid, on short petioles. Leaf surfaces punctate, aromatic. Heads numerous in a corymbose panicle, the disks and rays
yellow, from 20 to 100 on a single plant. Disks about 7 mm wide.
Ray flowers pistillate with an oblique 3-toothed limb. Frequently escaped from cultivation over the state, persisting on roadsides and at abandoned homesteads. July-Sept. Tansy.

TARAXACUM Zinn.

1 Achenes red-brown at maturity. Principal leaves deeply incised their entire length, many times lacking an enlarged
   terminal segment. The tips of the involucral bracts with
   small enlargements or appendages. Plants perennial with taproots. Frequent in fields, pastures and thickets, especially
   in the eastern part. This species tends to be more common in
   shaded places than the following. {*T. erythrospermum* Andrz.}
   Apr-July. Red-seeded Dandelion . . . .
   . . . . *Taraxacum laevigatum* (Willd.) DC

1' Achenes olive-gray at maturity. Principal leaves not deeply
   incised to the midrib, usually possessing an enlarged terminal
   lobe. Involucral bracts with tips simply pointed. Plants
   perennial with taproots. More robust and broader than the
   preceding species. It also tends to be more aggressive in
   lawns and open places. Very common in waste places and in
   lawns over the state. Apr-June. Gray-seeded Dandelion . . .
   . . . . *Taraxacum officinale* Weber

THELESPERMA Less.

1 Flowering heads with yellow, inconspicuous rays. Plants biennial to weakly perennial, branched above, 3-7 dm tall. Leaves
   once or twice pinnately divided into linear segments, the upper leaves reduced. Inflorescence of several to many heads at
   the ends of branches. Ray flowers up to 1 cm long. Rare to
   infrequent in sandy prairies of the southern part. {*T. intermedium* Rydb.} Ours var. *intermedium* (Rydb.) Shinners. July-
   Aug. Greenthread . . . . *Thelesperma filifolium* (Hook.) Gray

1' Flowering heads discoid, with rays lacking. Plants perennial,
   with a woody caudex. Stems branched and leafy, 3-8 dm tall.
   Leaves once or twice pinnatifid, the ultimate segments linear.
   Inflorescence branched, the heads 12-14 mm across. Rare in
   dry, sandy soil of the southwest part. {*T. gracile* (Torr.)
   Gray} June-Aug . . . .
   . . . . *Thelesperma megapotamicum* (Spreng.) O. Ktze.

TOWNSENDIA Hook.

1 Involucral bracts yellow, rigid, with acuminate points. Plants
   biennial with stems short, branched, and spreading, usually
   less than 5 cm high. Leaves up to 9 cm long, spatulate. Heads
   1.5-3.0 cm across, the involucral bracts in 4-7 series, ciliate with a broadly scarious margin. Ray corollas white with
   pink or purple stripes below, 1.0-2.5 cm long, the disks yellow. Rare in dry prairies and on hillsides in the southwest
   part. May-June . . . . *Townsendia grandiflora* Nutt.

1' Involucral bracts not rigid, their points acute to acuminate.

   2 Heads 1-2 cm across, the disk corollas about 5 mm long.
      Involucral bracts with tufts of tangled cilia at the
      acuminate apex. Leaves linear with acuminate tips, up
      to 5 cm long, usually not more than 3 mm wide. Plants
      perennial, the stems very short and branched, terminated
      by tufts of leaves. Ray corollas white with pink stripes.
      Rare in dry soil of the extreme southwest part. Apr-May
      . . . . *Townsendia hookeri* Beaman

   2' Heads 2-4 cm across, the disk corollas 7-10 mm long. In-
      volucral bracts without cilia, the apex acute. Leaves ob-
      lanceolate, up to 8 cm long and 6 mm wide. Plants peren-
      nial, taprooted, the stems short and branched below or at
      the soil level. Ray corollas white to pink, the disk co-
      rollas yellow with purple tips. Rare in dry soils of the
      south and southwest parts. {*T. sericea* Hook.} Apr-May
      . . . . *Townsendia exscapa* (Rich.) Porter

TRAGOPOGON L.

1 Flowers purple. Heads with approximately 8 involucral bracts,
   these longer than the flowers. Peduncles below the heads enlarged. Plants biennial, 4-6 dm tall, somewhat branched.
   Leaves elongate, tapering, sometimes crowded on the stem. An
   infrequent escape from cultivation along roadsides and in
   waste places over the state. June-July. Vegetable Oyster . .
   . . . . *Tragopogon porrifolius* L.

1' Flowers pale yellow. Heads usually with 10 or more involucral
   bracts, these longer than the flowers. Peduncles enlarged immediately below the heads. Plants 3-8 dm tall, biennial or
   sometimes annual, more or less branched. Very common in dry
   places and regularly occurring in prairies over the state.
   {*T. major* Jacq.} June-Aug. Yellow Goatsbeard . . . .
   . . . . *Tragopogon dubius* Scop.

VERBESINA

Verbesina encelioides (Cav.) Benth. & Hook. Plants annual, 2-5 dm tall, sparingly to much-branched. Vegetation and stems with dense appressed hairs. Leaves lanceolate, alternately arranged, their margins coarsely toothed. Heads long-peduncled in a loose inflorescence, the disks 1.5-2.0 cm across. Rays 10-15, yellow, lax and spreading, their distal ends appearing ragged and irregular. Infrequent as a weed in the eastern part but may be locally abundant when it occurs. Ours subspecies *exauriculata* (Robins. & Greenm.) Coleman. Aug-Sept.

VERNONIA Schred.

1 Leaves glabrous on their under surfaces. Small pits appear from obscure punctate dots when leaves are dried. Plants 5-10 dm tall, clustered from a perennial rootstock, the stems mostly glabrous. Leaves sharply serrate, their surfaces occasionally puberulent. Inflorescence flattened, dense, the flower heads purple. Infrequent in wet or low open places, especially in the eastern part. Represented in South Dakota by the two varieties *corymbosa* (Schwein.) Schub. (common) and *fasciculata* (rare). Aug-Sept. Ironweed . . . . *Vernonia fasciculata* Michx.

1' Leaves hairy or at least pubescent on their under surfaces. Stems 4-12 dm tall, with some pubescence. Leaves lance-ovate, with sharp serrations. Inflorescence of many purple heads, the inner involucral bracts acuminate with resinous surfaces. Rare to infrequent in dry areas of prairie and open thickets in the south and eastern part. Ours var. *interior* (Small.) Schub. Aug-Sept . . . . *Vernonia baldwini* Torr.

XANTHIUM L.

Xanthium strumarium L. Plants annual, 1-15 dm tall, generally much-branched. Leaves broadly ovate, irregularly lobed with sub-cordate bases, petioled. Surface of vegetation with glandular-puberulence. Flowers unisexual, the staminate ones small, many-flowered and located in the upper part of the plant. Pistillate flowers axillary, composed of an involucre which matures to form a conspicuous 2-chambered bur with hooked prickles. Very common weed of fields and alluvial areas. July-Sept. Cockelbur.

General References

Barker, W. T. 1969. Flora of the Kansas Flint Hills. Univ. Kans. Sci. Bull. 48:525-84.

Beal, E. O., and P. H. Monson. 1954. Marsh and aquatic angiosperms of Iowa. State Univ. Iowa Studies Nat. Hist. 19, No. 5.

Beck, R. 1963. Additions to the flora of south central Iowa. Proc. Iowa Acad. Sci. 70:51-52.

Booth, W. E., and E. Wright. 1966. Flora of Montana I, II. Montana State Univ., Bozeman.

Brooks, R. E. 1969. The ferns of the Black Hills of South Dakota. Trans. Kans. Acad. Sci. 72:109-36.

Carter, J. 1961. Preliminary report on the vascular flora of northwest Iowa. Proc. Iowa Acad. Sci. 68:146-52.

_____. 1962. Vascular flora of Cherokee County, Iowa. Proc. Iowa Acad. Sci. 69:60-70.

Cooperrider, T. S. 1958. The ferns and other pteridophytes of Iowa. State Univ. Iowa Studies Nat. Hist. 20:1-65.

Correll, D. S., and M. C. Johnston. 1970. Manual of the Vascular Plants of Texas. Texas Research Foundation, Renner.

Cronquist, A., A. H. Holmgren, N. H. Holmgren, and J. L. Reveal. 1972. Intermountain Flora: Vascular Plants of the Intermountain West, USA, vol. 1. Hafner Publishing Co., New York.

Davis, R. J. 1952. Flora of Idaho. Brigham Young Univ. Press.

Deam, C. C. 1940. Flora of Indiana. Indiana State Dept. Conserv., Div. Forestry.

Fassett, N. C. 1957. A Manual of Aquatic Plants, 2nd ed. Univ. of Wis. Press, Madison. A revision of the 1940 ed., McGraw-Hill Book Co., New York.

Fernald, M. L. 1950. Gray's Manual of Botany, 8th ed. American Book Co., New York.

Flint, R. F. 1955. Pleistocene geology of eastern South Dakota. U.S. Geol. Surv. Profess. Paper 262.

Gleason, H. A. 1952. The New Britton and Brown Illustrated Flora of the Northeastern United States and Adjacent Canada. 3 vols. New York Botanical Garden, New York.

Gleason, H. A., and A. Cronquist. 1963. Manual of Vascular Plants of Northeastern United States and Adjacent Canada. Van Nostrand Co., Princeton, N.J.

Graustein, J. E. 1967. Thomas Nuttall, Naturalist. Explorations in America 1808-1841. Harvard Univ. Press, Cambridge, Mass.

Gray, A. 1880. Botany of the Black Hills of Dakota. In Report of the Geology and the Resources of the Black Hills of South Dakota. U.S.G.P.O., Washington, D.C.

Harrington, H. K. 1954. Manual of the Plants of Colorado. Sage
    Books, Denver.
Hartley, T. G. 1966. Flora of the Driftless Area of Iowa and
    Wisconsin. Univ. Iowa Studies Nat. Hist. 21:1-174.
Hermann, F. J. 1970. Manual of the carices of the Rocky Moun-
    tains and Colorado basin. USDA Agr. Handbook 374:1-397.
Hitchcock, A. S., and A. Chase. 1950. Manual of the grasses of
    the United States. USDA Misc. Publ. 200.
Hitchcock, C. L., A. Cronquist, M. Ownbey, and J. W. Thompson.
    1955-1969. Vascular Plants of the Pacific Northwest.
    3 vols. Univ. Wash. Publ. Biol., vol. 17. Seattle.
Humphrey, H. B. 1961. Makers of North American Botany. Ronald
    Press Co., New York.
Isley, D. 1950. The Leguminosae of the North-Central United
    States. Iowa State Univ. Press, Ames.
Jones, G. N. 1950. A Flora of Illinois. American Midland
    Naturalist Monograph No. 5. Univ. of Notre Dame Press,
    South Bend, Indiana.
_____. 1963. Flora of Illinois, 3rd ed. Univ. of Notre Dame
    Press, South Bend, Indiana.
Jones, G. N., and G. D. Fuller. 1955. Vascular Plants of
    Illinois. Univ. of Ill. Press, State Museum, Springfield,
    Ill.
Kingsbury, J. M. 1964. Poisonous Plants of the United States
    and Canada. Prentice-Hall, Inc. Englewood Cliffs, N.J.
Little, E. L. 1953. Checklist of native and naturalized trees
    of the United States. USDA Handbook 41.
_____. 1971. Atlas of the United States Trees, vol. 1, Conifers
    and Important Hardwoods. USDA Forest Serv. Misc. Publ. 1146.
Mackenzie, K. K. 1940. North American Cariceae 2 vols. New
    York Botanical Garden, New York.
Over, W. H. 1923. Trees and shrubs of South Dakota. S. Dakota
    Geol. Nat. Hist. Surv. Circ. 11.
_____. 1932. The Flora of South Dakota. Univ. of S. Dakota,
    Vermillion.
Perisho, E., and S. S. Visher. 1912. The geography, geology
    and biology of southcentral South Dakota. S. Dakota State
    Geol. Biol. Surv. 5:84-108.
Petry, E. J. 1925. Additions to the South Dakota flora. Proc.
    S. Dakota Acad. Sci. 10:25-27.
Pohl, R. W. 1966. The Grasses of Iowa. Iowa State J. Sci.
    40:341-566.
Porter, C. L. 1962, and continuing. A Flora of Wyoming (in 8
    parts). Agr. Exp. Sta., Univ. of Wyoming, Laramie.
Rydberg, P. A. 1896. Flora of the Black Hills of South Dakota.
    Contrib. U.S. Nat. Herb. 3:463-523.

_____. 1932 (reprinted, 1965). Flora of the Prairies and Plains
    of Central North America. Hafner Publishing Co., New York.
Shantz, H. L. et al. 1944. A checklist of the native and natu-
    ralized trees of the United States, including Alaska. USDA
    Forest Serv. Publ. (Mimeo).
Stevens, O. S. 1950. Handbook of North Dakota Plants. State
    Agr. Coll., Fargo, N. Dakota.
Steyermark, J. A. 1963. Flora of Missouri. Iowa State Univ.
    Press, Ames.
Thilenius, J. F. 1971. Vascular plants of the Black Hills of
    South Dakota and adjacent Wyoming. USDA Forest Serv. Res.
    Paper RM-71.
Thornbury, W. D. 1965. Regional Geomorphology of the United
    States. John Wiley and Sons, Inc., New York.
Vanorny, P. M. 1970. A history of the South Dakota geological
    survey. S. Dakota State Geol. Surv. Bull., Educ. Ser. 4.
Visher, S. S. 1914. The biology of Harding County, South
    Dakota. S. Dakota State Geol. Surv. 6:32-68.
_____. 1914. The geography of South Dakota. S. Dakota State
    Geol. Nat. Hist. Surv. 8:13-14.
_____. 1954. Climatic Atlas of the United States. Harvard
    Univ. Press, Cambridge, Mass.
Voss, E. G. 1972. Michigan Flora. Part I, Gymnosperms and
    Monocots. Cranbrook Inst. Sci. and Univ. of Mich. Herb.
Weaver, J. E. 1965. Native Vegetation of Nebraska. Univ. of
    Nebr. Press, Lincoln.
Winter, J. M. 1936. An analysis of the flowering plants of
    Nebraska. Univ. of Nebr. Conserv. Surv. Div.
Winther, O. O. 1964. The Transportation Frontier: Trans-
    Mississippi West, 1865-1890. Holt, Rinehart, and Winston,
    Inc., New York.

Selected References of Taxa
Arranged alphabetically by family

Aceraceae

Fleak, S. 1967. Hybridization in *Acer saccharum* and *A. nigrum*.
    Trans. Mo. Acad. Sci. 1:12-16.

Alismataceae

Bogin, C. 1955. Revision of the genus *Sagittaria*. Mem. N.Y.
    Bot. Gard. 9:179-233.
Fassett, N. C. 1955. *Echinodorus* in the American tropics.
    Rhodora 57:133-133-56.
Rubtzoff, P. 1964. Notes on the genus *Alisma*. Leafl. West Bot.
    10:90-95.

Amaranthaceae

Sauer, J. 1955. Revision of the dioecious *Amaranthus*. Madrono 13:5-46.

Anacardiaceae

Gillis, W. T. 1971. The systematics and ecology of poison ivy and the poison oaks. Rhodora 73:72-540.

Apiaceae

Crawford, D. J. 1970. The Umbelliferae of Iowa. State Univ. Iowa Studies Nat. Hist. 21:1-37.

Mathias, M. E. 1930. A monograph of *Cymopterus* including a critical study of related genera. Ann. Mo. Bot. Gard. 17: 213-476.

Mathias, M. E., and L. Constance. 1941. The North American species of *Eryngium*. Am. Midl. Nat. 25:361-87.

_____. 1944-45. North American flora 28B:43-297 (Apiaceae).

Asclepiadaceae

Woodson, R. E. 1954. The North American species of *Asclepias*. Ann. Mo. Bot. Gard. 41:1-211.

Asteraceae

Barkley, T. M. 1960. A revision of *Senecio integerrimus* and allied species. Leafl. West. Bot. 19:97-113.

Barkley, T. M. 1962. A revision of *Senecio aureus* and allied species. Trans. Kans. Acad. Sci. 65:318-408.

Beaman, J. 1957. The systematics and evolution of *Townsendia*. Contrib. Gray Herb., Harvard Univ. 183:1-151.

Bierner, M. W. 1972. Taxonomy of *Helenium* sect. Tetrodus and a conspectus of North American *Helenium*. Brittonia 24:331-55.

Clewell, A. F., and J. W. Wooten. 1971. A revision of *Ageratina* from eastern North America. Brittonia 23:123-43.

Coleman, J. R. 1968. A cytotaxonomic study of *Verbesina*. Thodora 70:95-102.

Croat, T. B. 1970. Studies in *Solidago* I. The *S. graminifolia-S. gymnospermoides* complex. Ann. Mo. Bot. Gard. 57:250-51.

_____. 1972. *Solidago canadensis* complex of the Great Plains. Brittonia 24:317-26.

Cronquist, A. 1943. Revision of the western North American species of *Aster*. Am. Midl. Nat. 29:429-68.

_____. 1943. The separation of *Erigeron* from *Conyza*. Bull. Torrey Bot. Club 70:629-32.

_____. 1946-1950. Notes on the compositae of the northeast United States. Inuleae and Senecioneae. Rhodora 48:116-25. 1946; Heliantheae and Helenieae. Rhodora 47:396-403. 1945; Cichorieae, Eupatorieae, and Astereae. Rhodora 50:28-35. 1950.

_____. 1947. Revision of the North American species of *Erigeron* north of Mexico. Brittonia 6:121-302.

_____. 1955. Phylogeny and taxonomy of the compositae. Am. Midl. Nat. 53:478-511.

_____. 1968. A commentary on specific delimitation in *Antennaria*. Am. Midl. Nat. 79:513-14.

Cronquist, A., and D. D. Keck. 1957. A reconstitution of the genus *Machaeranthera*. Brittonia 9(1):231-39.

Cruise, J. E. 1964. Biosystematic studies of three species in the genus *Liatris*. Can. J. Bot. 42:1445-55.

Ediger, R. L. 1970. Revision of section Suffruticosa of the genus *Senecio*. Sida 3:504-24.

Ellison, W. L. 1964. A systematic study of the genus *Bahia*. Rhodora 66:67-86.

Fuller, M. J. 1969. The genus *Carduus* in Nebraska. Univ. Nebr. Studies II 39:1-57.

Gardner, R. C. 1974. Systematics of *Cirsium* in Wyoming. Madrono 22:239-65.

Hall, H. M. 1928. The genus *Haplopappus*. Carnegie Inst. Wash. Publ. 389.

Harms, V. L. 1968. Nomenclatural changes and taxonomic notes on *Heterotheca*, including *Chrysopsis*, in Texas and adjacent states. Wrightia 4:8-20.

Heiser, C. B., D. M. Smith, S. B. Clevenger, and W. C. Martin. 1969. The North American sunflowers (*Helianthus*). Mem. Torrey Bot. Club 22:1-218.

Hsieh, T., A. B. Schooler, A. Bell, and J. D. Jalewaja. 1972. Cytotaxonomy of three Sonchus species. Am. J. Bot. 59:789-96.

Jackson, R. C. 1960. A revision of the genus *Iva*. Univ. Kans. Sci. Bull. 41:793-876.

_____. 1967. Biosystematic studies of *Haplopappus*. Taxon 16:303-4.

Jones, S. B. 1972. A systematic study of the Fasciculatae group of *Vernonia*. Brittonia 24:28-45.

King, R. M., and H. Robertson. 1970. Studies in the Eupatorieae XIX. New combinations in *Ageratina*. Phytologia 19:208-29.

Love, D., and P. Dansereau. 1959. Biosystematics studies on *Xanthium*: Taxonomic appraisal and ecological status. Can. J. Bot. 37:173-208.

Maguire, B. 1943. A monograph of the genus *Arnica*. Brittonia 4:386-510.

McGregor, R. L. 1968. The taxonomy of the genus *Echinacea*.
    Univ. Kans. Bull. Sci. 48:113-42.
Moore, R. J. 1972. Distribution of native and introduced knap-
    weeds (*Centaurea*) in Canada and the United States. Rhodora
    74:331-46.
Moore, R. J., and C. Frankton. 1961. Cytotaxonomy, phylogeny,
    and Canadian distribution of *Cirsium undulatum* and *Cirsium
    flodmanii*. Can. J. Bot. 39:21-33.
_____. 1963. Cytotaxonomic notes on some *Cirsium* species of the
    western United States. Can. J. Bot. 41:1553-67.
_____. 1964. A clarification of *Cirsium foliosum* and *Cirsium
    drummondii*. Can. J. Bot. 42:451-61.
_____. 1966. An evaluation of the status of *Cirsium pumilum* and
    C. Hillii. Can. J. Bot. 44:581-95.
_____. 1967. Cytotaxonomy of the Foliose thistles of the west-
    ern North America. Can. J. Bot. 45:1733-49.
_____. 1969. Cytotaxonomy of some *Cirsium* species of the east-
    ern United States, with a key to eastern species. Can. J.
    Bot. 47:1257-75.
Morley, T. 1958. Note on the distinction between the broad- and
    narrow-leaved Antennarias of Minnesota. Rhodora 60(719):306.
Payne, W. W. 1964. A re-evaluation of the genus *Ambrosia*. J.
    Arn. Arb. 45:401-38.
Richards, E. L. 1968. A monograph of the genus *Ratibida*.
    Rhodora 70:348-93.
Sherff, E. E. 1936. Revision of the genus *Coreopsis*. Field Mus.
    Publ. Bot. 11:279-495.
_____. 1937. The genus *Bidens*. Field Mus. Nat. Hist. Bot. Ser.
    16:(I and II).
Shinners, L. H. 1964. *Evax* transferred to *Filago*. Sida 1:252-
    53.
Smith, E. B. 1965. Taxonomy of the *Haplopappus*, section Isopap-
    pus. Rhodora 67:217-38.
Turner, B. L., and D. Horne. 1964. Taxonomy of *Machaeranthera*.
    Brittonia 16:316-31.
Weedon, R. R. 1973. Taxonomy and distribution of the genus
    Bidens in the north-central plains states. Ph.D. diss.,
    Univ. of Kansas, Lawrence.

Berberidaceae

Ahrendt, L. W. 1961. *Berberis* and *Mahonia*: A taxonomic revi-
    sion. J. Linn. Soc. 57:1-140.

Betulaceae

Dugle, J. R. 1966. A taxonomic study of western Canadian spe-
    cies of *Betula*. Can. J. Bot. 44:929-1007.

Boraginaceae

Higgins, L. C. 1971. A revision of *Cryptantha*, subgenus Oreo-
    carya. Brigham Young Univ. Sci. Bull. 23:1-63.
Johnston, I. M. 1952. Studies in the Boraginaceae, 23. A
    survey of the genus *Lithospermum*. J. Arn. Arb. 33:299-366.

Brassicaceae

Detling, L. E. 1939. *Descurainia* in North America. Am. Midl.
    Nat. 22:481-520.
Dudley, T. R. 1964. Synopsis of the genus *Alyssum*. J. Arn.
    Arb. 45:358-73.
Hitchcock, C. L. 1936. The genus *Lepidium* in the United States.
    Madrono 3:265-320.
_____. 1941. A revision of the *Drabas* of western North America.
    Univ. Wash. Publ. Biol. 11:1-132.
Hopkins, M. 1937. *Arabis* in eastern and central North America.
    Rhodora 39:63-98, 106-48, 155-86.
Montgomery, F. H. 1955. Preliminary studies in the genus
    Dentaria in eastern North America. Rhodora 57:161-73.
Mulligan, G. A., and C. Frankton. 1962. Taxonomy of the genus
    Cardaria with particular reference to the species introduced
    into North America. Can. J. Bot. 40:1411-25.
Rollins, R. C. 1941. A monographic study of *Arabis* in western
    North America. Rhodora 43:289-325, 348-411, 425-81.
_____. 1973. The genus *Lesquerella* in North America. Harvard
    Univ. Press, Cambridge. 288 pp.
Rossbach, G. B. 1958. The genus *Erysimum* in North America.
    Madrono 14:261-67.
Stuckey, Ronald L. 1972. Taxonomy and distribution of the genus
    Rorippa in North America. Sida 4(4):279-430.

Callitrichaceae

Fassett, N. C. 1951. *Callitriche* in the New World. Rhodora
    53:137-55.

Campanulaceae

McVaugh, R. 1945. *Triodanis* and its relationships to *Specularia*
    and *Campanula*. Wrightia 1:13-52.
Shetler, S. G. 1963. A checklist and key to the species of
    Campanula native or commonly naturalized in North America.
    Rhodora 65:319-37.

Capparidaceae

Iltis, H. H. 1966. Studies in the Capparidaceae VIII. Rhodora
    68:41-47.

Caprifoliaceae

Jones, G. N.   1940.   A monograph of the genus *Symphoricarpos*.
J. Arn. Arb. 21:201-52.

Caryophyllaceae

Core, E. L.   1941.   The North American species of *Paronychia*.
Am. Midl. Nat. 26:369-97.

Hartman, R. L.   1974.   Rocky Mountain species of *Paronychia*, a
morphological and chemical study.   Brittonia 26:256-63.

Hitchcock, C. L., and B. Maguire.   A revision of the North
American species of *Silene*.   Univ. Wash. Publ. Biol. 13:
1-73.

Porsild, A. E.   1963.   *Stellaria longipes* and its allies in North
America.   Nat. Mus. Can. Bull. 186:1-35.

Celastraceae

Blakelock, R. A.   1952.   A synopsis of the genus *Euonymus*.   Kew
Bull. 1951:210-92.

Ceratophyllaceae

Fassett, N. C.   1953.   North American *Ceratophyllum*.   Comun.
Inst. Trop. Invest. Cient. (El Salvador) 2:25-46.

Chenopodiaceae

Aellen, P., and T. Just.   1943.   Key and synopsis of the American
species of the genus *Chenopodium*.   Am. Midl. Nat. 30:47-76.

Baranov, A. J.   1964.   On the perianth and seed characters of
Chenopodium hybridum and *C. gigantospermum*.   Rhodora 66:
168-71.

Beatley, J. C.   1973.   Russian-thistle species in western United
States.   Range Manage. 26:225-26.

Brown, G. D.   1956.   Taxonomy of American *Atriplex*.   Am. Midl.
Nat. 55:199-210.

Reveal, J. L., and N. H. Holmgren.   1972.   *Ceratoides*, an older
generic name for *Eurotia*.   Taxon 21:209.

Wahl, H. A.   1954.   A preliminary study of the genus *Chenopodium*
in North America.   Bartonia 27:1-46.

Cistaceae

Daoud, H. S., and R. S. Wilbur.   1965.   A revision of the North
American species of *Helianthemum*.   Thodora 67:63-82, 201-16,
255-312.

Commelinaceae

Brashier, C. K.   1966.   Revision of *Commelina* in the U.S.A.
     Bull. Torrey Bot. Club 93:1-19.

Convolvulaceae

Yuncker, T. G.   1932.   The genus *Cuscuta*.   Mem. Torrey Club 18:
     113-331.
_____.   1965.   *Cuscuta*.   North Am. Flora 2(4):1-51.

Cornaceae

Wilson, J. S.   1964.   Variation of three taxonomic complexes of
     the genus *Cornus* in eastern U.S.   Trans. Kans. Acad. Sci.
     67:747-817.

Cupressaceae

Van Haverbeke, D. F.   1968.   A population analysis of *Juniperus*
     in the Missouri River basin.   Univ. Nebr. Studies.   New
     Ser. 38.

Cyperaceae

Beetle, A. A.   1947.   *Scirpus*.   North Am. Flora 188:481-504.
Fernald, M. L.   1905.   The North American species of *Eriophorum*.
     Rhodora 7:81-92, 129-136.
Harms, L. J.   1968.   Cytotaxonomic studies in *Eleocharis* subser.
     Palustres: central U.S.   Am. J. Bot. 55:966-74.
Hermann, F. J.   1968.   Notes on Rocky Mountain carices.   Rhodora
     70:491-521.
Kolstad, O. A.   1966.   The genus *Carex* of the High Plains.   Ph.D.
     diss., Univ. of Kansas, Lawrence.
Swenson, H. K.   1957.   *Eleocharis*.   North Am. Flora 189:509-40.

Elatinaceae

Fassett, N. C.   1939.   *Elatine* and other aquatics.   Rhodora 41:
     367-77.

Ericaceae

Krisa, B.   1966.   Contributions to the taxonomy of the genus
     Pyrola in North America.   Bot. Jahrb. 85:612-37.
Thompson, H. J.   1953.   The biosystematics of *Dodecatheon*.
     Contrib. Dudley Herb. 4:73-154.

Equisetaceae

Hauke, R. L. 1960. The smooth scouring rush and its complexities. Am. Fern J. 50:185-93.
_____. 1963. A taxonomic monograph of the genus *Equisetum* subgenus Hippochaete. Beih. Nova Hedwigia 8:1-123.

Euphorbiaceae

Richardson, J. W. The genus *Euphorbia* of the high plains and prairies of Kansas, Nebraska, South Dakota and North Dakota. Univ. Kans. Sci. Bull. 48:45-112.
Wheeler, L. C. 1941. *Euphorbia* subgenus Chamaesyce in Canada and the United States. Rhodora 43:97-154, 168-205, 223-86.

Fabaceae

Barneby, R. C. 1952. A revision of the North American species of *Oxytropis*. Proc. Calif. Acad. Sci. 27(4):177-312.
_____. 1964. Atlas of North American *Astragalus*. Mem. N.Y. Bot. Garden 13:1-1188.
Clewell, A. F. 1966. Native North American species of *Lespedeza*. Rhodora 68:359-405.
Dunn, D. B. 1959. *Lupinus pusillus* and its relationship. Am. Midl. Nat. 62:500-510.
Hermann, F. J. 1960. Vetches in the United States: Native, naturalized, and cultivated (*Vicia*). USDA Handbook 168: 1-84.
Isley, D. 1951. The leguminosae of the north-central United States. I. Loteae and trifolieae. Iowa State J. Sci. 25:439-82.
_____. 1954. Keys to the sweet clovers (*Melilotus*). Proc. Iowa Acad. Sci. 61:119-31.
_____. 1955. Leguminosae of the north central United States. II. Hedysareae. Iowa State J. Sci. 30:33-118.
_____. 1958. Leguminosae of the north-central United States. III. Mimosoideae and Caesalpinioideae. Iowa State J. Sci. 32:355-93.
_____. 1962. Leguminosae of the north-central United States. IV. Iowa State J. Sci. 37:103-62.
_____. 1971. Legumes of the U.S. III. *Schrankia*. Sida 4: 232-45.
_____. 1971. Legumes of the U.S. IV. *Mimosa*. Am. Midl. Nat. 85:410-24.
Phillips, L. L. 1955. A revision of the perennial species of *Lupinus* of North America. Res. Studies State Coll. Wash. 23:161-201.

Rollins, R. C. 1940. Studies in the genus *Hedysarum* in North
    America. Rhodora 42:217-39.
Schubert, B. G. *Desmodium*: Preliminary studies. I. Contribb.
    Gray Herb. 129:3-31. 1940; II. Gray Herb. 135:78-115.
    1941; III. Rhodora 52:135-55. 1950; IV. J. Arnold Arb.
    44:284-97. 1963.
Shinners, L. H. 1956. Authorship and nomenclature of bur
    clovers, *Medicago*, found wild in the United States.
    Rhodora 58:1-13.
Turner, B. L., and O. S. Fearing. 1964. A taxonomic study of
    the genus *Amphicarpaea*. Southwest Nat. 9:207-18.
Welsh, S. L. 1960. Legumes of the north-central states:
    Galegeae. Iowa State J. Sci. 35:111-250.
Wemple, D. K. 1970. Revision of the genus *Petalostemon*. Iowa
    State J. Sci. 45(1):1-102.
Whiteman, W. C., and O. A. Stevens. 1952. Native legumes of
    North Dakota grassland. Proc. N. Dakota Acad. Sci. 6:73-78.
Wilbur, R. L. 1964. A revision of the dwarf species of *Amorpha*.
    J. Eliza Mitchell Soc. 80:51-65.
Windler, D. R. 1974. A systematic treatment of the native uni-
    foliate *Crotalaria* of North America. Rhodora 76:151-204.

Fagaceae

Maze, J. 1968. Past hybridization of *Quercus macrocarpa* and
    Q. gembelii. Brittonia 20:321-33.

Fumariaceae

Ownbey, G. B. 1947. Monograph of the North American species of
    Corydalis. Ann. Mo. Bot. Gard. 34:187-258.
Stern, K. R. 1961. Revision of *Dicentra*. Brittonia 13:1-57.

Gentianaceae

Iltis, H. H. 1965. The genus *Gentianopsis*. Transfers and
    phytogeographic comments. Sida 2:129-54.
Pringle, J. S. 1966. *Gentiana puberulenta*, a known but unnamed
    species of the North American prairies. Rhodora 68:209-14.
Shinners, L. H. 1957. Synopsis of the genus *Eustoma*. South-
    west. Nat. 2:38-43.
St. John, H. 1941. Revision of the genus *Swertia* of the Ameri-
    cas and the reduction of *Frasera*. Am. Midl. Nat. 26:1-29.

Geraniaceae

Fernald, M. L. 1935. *Geranium carolinianum* and allies of north-
    eastern North America. Rhodora 37:295-301.

Jones, G. N., and F. F. Jones. 1943. A revision of the perennial species of *Geranium* of the United States and Canada.
Rhodora 45:5-26, 32-53.

Hydrocharitaceae

St. John H. 1962. Monograph of the genus *Elodea* I. Res.
Studies State Coll. Wash. 302:19-44.
_____. 1965. II. Monograph of the genus *Elodea*, Summary.
Rhodora 67:155-80.
_____. 1965. Monograph of the genus *Elodea*. IV and Summary.
Rhodora 67:1-35.

Hydrophyllaceae

Constance, L. 1940. The genus *Ellisia*. Rhodora 42:33-39.

Juncaginaceae

Love, A., and D. Love. 1958. Biosystematics of *Triglochin
maritimum*. Ag. Nat. Can. 85:156-65.

Labiatae

Epling, C. 1942. The American species of *Scutellaria*. Univ.
Calif. Publ. Bot. 20:1-146.
Henderson, N. C. 1962. A taxonomic revision of the genus
Lycopus. Am. Midl. Nat. 68:95-138.
Lint, H., and C. Epling. 1945. A revision of the genus
Agastache. Am. Midl. Nat. 33:207-230.

Lemnaceae

den Hartog, C., and F. vander Plas. 1970. A synopsis of the
Lemnaceae. Blumea 18:355-68.

Lentibulariaceae

Rossbach, C. B. 1939. Aquatic Utricularias. Rhodora 41:113-28.

Liliaceae

Beetle, D. E. 1944. A monograph of the North American species
of *Fritillaria*. Madrono 7:133-59.
Bowin, B., and W. J. Cody. 1956. The variations of *Lilium cana-
dense*. Rhodora 58:14-20.
Coker, W. C. 1944. The woody *Smilaxes* of the United States.
J. Elisha Mitchell Sci. Soc. 60:27-69.
Jones, Q. 1951. A cytotaxonomic study of the genus *Disporum* in
North America. Contrib. Gray Herb. 173:1-39.

Ownbey, G. B., and H. Aase. 1955. Cytotaxonomic studies in
    Allium I. The *Allium canadense* alliance. Res. Studies
    State Coll. Wash. Monogr. Suppl. 1.
Ownbey, M. 1940. A monograph of the genus *Calochortus*. Ann.
    Mo. Bot. Gard. 27:371-560.
Webber, J. M. 1953. Yuccas of the southwest. USDA Monogr. 17.
Wilbur, R. L. 1963. A revision of the North American species
    of *Uvularia*. Rhodora 65:158-88.

Linaceae

Harris, B. D. 1968. Chromosome numbers and evolution in North
    American species of *Linum*. Am. J. Bot. 55:1197-1204.
Rogers, C. M. 1968. Yellow-flowered species of *Linum* in Central
    America and western North America. Brittonia 20:107-35.
_____. 1969. Relationships of the North American species of
    Linum. Bull. Torrey Bot. Club 96:176-90.

Lythraceae

Shinners, L. H. 1953. Synopsis of the United States species of
    Lythrum. Field Lab. 21.

Malvaceae

Dwayne, A. W., and M. Y. Menzell. 1971. Genetic affinities of
    the North American species of *Hibiscus*. Brittonia 23:425-37.
Kearney, T. H. 1951. The American genera of Malvaceae. Am.
    Midl. Nat. 46:93-131.
Malva. In Flora Europea, vol. 2, 1968, p. 249-251.

Moraceae

Schultes, R. E., W. Klein, T. Plownan, and T. W. Lockwood. 1974.
    Cannabis: An example of taxonomic neglect. Harvard Univ.
    Bot. Mus. Leafl. 23:337-67.
Stern, W. T. 1974. Typification of *Cannabis sativa* L. Harvard
    Univ. Bot. Mus. Leafl. 23:325-36.

Najadaceae

Clausen, R. T. 1936. Studies in the genus *Najas* in the northern
    United States. Rhodora 38:333-45.
Fernald, M. L. 1914. The genus *Ruppia* in eastern North America.
    Rhodora 16:119-27.
_____. 1932. The linear-leaved species of *Potamogeton*. Mem.
    Gray Herb. 3:1-183.
Ogden, E. C. 1943. The broad-leaved species of *Potamogeton*
    north of Mexico. Rhodora 45:57-105, 119-63, 171-214.

Nymphaeaceae

Beal, E. O. 1956. Taxonomic revision of the genus *Nuphar* of North America and Europe. J. Eliza Mitchell Sci. Soc. 72: 317-46.

Onagraceae

Lewis, H., and J. Szweykowski. 1941. The genus *Gayophytum*. Brittonia 16:343-91.

Munz, P. A. 1965. Onagraceae. North Am. Flora 2(5):1-278.

Raven, P. H. 1964. The generic subdivision of the Onagraceae, tribe Onagreae. Brittonia 16:276-288.

Raven, P. H., and D. M. Moore. 1965. A revision of *Boisduvalia* (Onagraceae). Brittonia 17:238-54.

Ophioglossaceae

Clausen, R. T. 1938. A monograph of the Ophioglossaceae. Mem. Torrey Bot. Club 192:1-177.

Orchidaceae

Correll, D. S. 1950. Native orchids of North America north of Mexico. Chronica Botanica Co., Waltham, Mass.

Luer, C. A. 1969. The genus *Cypripedium* in North America. Am. Orch. Soc. Bull. 38:903-8.

Magrath, L. K. 1973. The native orchids of the prairies and plains region of North America. Ph.D. diss., Univ. of Kansas, Lawrence.

Stoutamire, W. P. 1974. Relationships of the purple-fringed orchids *Plantanthera psycodes* and *P. grandiflora*. Brittonia 26:42-58.

Papaveraceae

Ownbey, G. B. 1958. Monograph of the genus *Argemone* for North America and the West Indies. Mem. Torrey Bot. Club 21:1-149.

Pinaceae

Thilenius, J. F. 1970. An isolated occurrence of limber pine (*Pinus flexilis*) in the Black Hills of South Dakota. Am. Midl. Nat. 84:411-17.

Plantaginaceae

Bassett, I. J. 1973. The plantains of Canada. Res. Branch, Can. Dept. Agr. Monogr. 7.

Poaceae

Anderson, D. E. 1961. Taxonomy and distribution of the genus
    Phalaris. Iowa State Coll. J. Sci. 36:1-96.
Bowden, W. M. 1960. The typification of *Elymus macounii* Vasey.
    X-Agrohordeum. Torrey Bot. Club 87:205-8.
_____. 1965. Cytotaxonomy of the species and interspecific
    hybrids of the genus *Agropyron* in Canada and heighboring
    areas. Can. J. Bot. 43:1421-48.
Boyle, W. S. 1945. A cytotaxonomic study of the North American
    species of *Melica*. Madrono 8:1-26.
Church, G. L. 1949. A cytotaxonomic study of *Glyceria* and *Puc-
    cinellia*. Am. J. Bot. 36:155-65.
DeLisle, D. G. 1963. Taxonomy and distribution of the genus
    Cenchrus. Iowa State Coll. J. Sci. 37:259-351.
Erdman, D. S. 1965. Taxonomy of the genus *Sphenopholis*. Iowa
    State J. Sci. 39:289-336.
Fassett, N. C. 1924. A study of the genus *Zizania*. Rhodora
    26:153-60.
Gould, F. W. 1967. The grass genus *Andropogon* in the United
    States. Brittonia 19:70-76.
_____. 1974. Nomenclatural changes in the Poaceae. Brittonia
    26:59-60.
Gould, F. W., M. A. Ali, and D. W. Fairbrothers. 1972. A revi-
    sion of *Echinochloa* in the United States. Am. Midl. Nat.
    87:36-59.
Gould, F. W., and Z. J. Kapodia. 1964. Biosystematics studies
    in the *Bouteloua curtipendula* complex. II. Taxonomy.
    Brittonia 16:182-207.
Harvey, L. H. 1954. New entities in North and Middle America:
    Eragrostis. Bull. Torrey Bot. Club 81:405-10.
Henrard, J. T. 1950. Monograph of the genus *Digitaria*. Leyden
    Univ. Press, Leyden.
Hitchcock, A. S. 1925. The North American species of *Stipa*.
    Contrib. U.S. Nat. Herb. 24:215-62.
Kalms, B. Specimens of North American grasses: Their evaluation
    for typification. Can. J. Bot. 45:1848-1947.
Kawano, S. 1963. Cytogeography and evolution of the *Deschampsia
    caespitosa* complex. Can. J. Bot. 41:719-42.
Komarov, V. L. 1963. Flora of the USSR, vol. 2, Poaceae.
    Israel Prog. Sci. Transl. Jerusalem.
Marsh, V. L. 1952. A taxonomic revision of the genus *Poa* of
    United States and southern Canada. Am. Midl. Nat. 47:202-50.
Mobberley, D. G. 1956. Taxonomy and distribution of the genus
    Spartina. Iowa State J. Sci. 30:471-574.
Monachino, J. 1959. The type of *Setaria faberii*. Rhodora 61:
    220-23.

Pohl, R. W. 1954. The grasses of Iowa. Iowa State J. Sci. 49(4):341-566.
_____. 1959. Morphology and cytology of some hybrids between *Elymus canadensis* and *E. virginicus*. Proc. Iowa Acad. Sci. 66:155-59.
_____. 1962. *Agropyron* hybrids and the status of *Agropyron pseudorepens*. Rhodora 64:143-47.
_____. 1962. Notes on *Setaria viridis* and *S. faberii*. Brittonia 14:210-13.
_____. 1966. X *Elyhordeum iowense*, a new intergeneric hybrid in the Triticeae. Brittonia 18:250-55.
_____. 1969. *Muhlenbergia* in North America. Am. Midl. Nat. 82:512-42.
Pyrah, G. L. 1969. Taxonomic and distributional studies in *Leersia*. Iowa State J. Sci. 44:215-70.
Rominger, J. M. 1962. Taxonomy of *Setaria* in North America. Ill. Biol. Monogr. 29:1-132.
Sarkar, P. 1956. The crested wheatgrass (*Agropyron*) complex. Can. J. Bot. 34:328-45.
Shinners, L. H. 1956. Illegitimacy of Persoon's species of *Koeleria*. Rhodora 58:93-96.
Terrell, E. E. 1967. Meadow fescue: *Festuca elatior* or *F. pratensis*? Brittonia 19:129-32.
_____. 1968. A taxonomic revision of the genus *Lolium*. USDA Bull. 1392:1-65.
Wilson, F. D. 1963. Revision of *Sitanion*. Brittonia 15:303-23.

Polemoniaceae

Crampton, B. 1954. Morphological and ecological considerations in the classification of *Navarretia*. Madrono 12:225-38.
Wherry, E. T. 1955. The genus *Phlox*. Morris Arb. Monogr. 3.
_____. 1964. New combinations in Texas Polemoniaceae. Sida 1:250.

Polygonaceae

Brooks, G. M., and T. R. Mertens. 1971. A biosystematic study of *Polygonum ramosissimum* and *Polygonum tenue*. Proc. Indiana Acad. Sci. 81:277-83.
Love, D., and F. J. Bernard. 1958. *Rumex stenophyllus* in North America. Rhodora 60:54-57.
Mertens, T. R., and P. H. Raven. 1965. Taxonomy of *Polygonum* (Sec. Aviculare) in North America. Madrono 18:85-92.
Reveal, J. L. 1967. Notes on *Eriogonum* III. on the status of *Eriogonum pauciflorum*. Great Basin Nat. 27:102-17.
_____. 1968. Some nomenclatural changes in *Eriogonum*. Taxon 17:531-33.

_____. 1971. Notes on *Eriogonum* VI, a revision of the *Eriogonum microthecum* complex. Brigham Young Univ. Sci. Bull. 13:1-44.
Stokes, S. G. 1936. The genus *Eriogonum*. J. H. Neblett Press Room, San Francisco. 128 pp.

Polypodiaceae

Blasdell, F. 1963. A monographic study of the fern genus *Cystopteris*. Mem. Torrey Bot. Club 21:1-102.
Brooks, R. 1968. *Polystichum munitum* in South Dakota. Am. Fern J. 58:92.
Brooks, R. 1975. Notes on the *Polypodium vulgare* complex in South Dakota. Am. Fern J. 65:1.
Brown, D. F. 1964. A monographic study of the fern genus *Woodsia*. Beih. Nova Hedwigia 16:1-154.
Lang, F. A. 1971. The *Polypodium vulgare* complex in the Pacific Northwest. Madrono 21:235-254.
Lloyd, R. M., and F. A. Lang. 1964. The *Polypodium vulgare* complex in North America. Brit. Fern Gaz. 9:168-77.
Tryon, A., and R. Tryon. 1973. *Thelypteris* in northeastern North America. Am. Fern J. 63:65-76.

Primulaceae

Beamish, K. L. 1955. Studies in the genus *Dodecatheon* of northwestern America. Bull. Torrey Bot. Club 82:357-66.
Ray, J. D. 1956. The genus *Lysimachia* in the New World. Ill. Biol. Monogr. 243(4):1-160.
Robbins, G. T. 1944. North American species of *Androsace*. Am. Midl. Nat. 32:137-63.

Pteridaceae

Tryon, A. F. 1957. A revision of the fern genus *Pellaea* section *Pellaea*. Ann. Mo. Bot. Gard. 44:125-93.
Tryon, R. M. 1941. A revision of the genus *Pteridium*. Rhodora 43:1-31, 37-67.

Ranunculaceae

Benson, L. A. 1948. A treatise on the North American *Ranunculi*. Am. Midl. Nat. 40:1-264.
_____. 1954. Supplement to a treatise on the North American *Ranunculi*. Am. Midl. Nat. 52:328-69.
Boivin, B. 1944. American *Thalictra* and their Old World allies. Rhodora 46:337-77, 391-445, 453-87.
_____. 1948. Key to Canadian species of Thalictra. Can. Field Nat. 62:169-70.

Boraich, G., and M. Heimburger. 1964. Cytotaxonomic studies on New World anemones with woody rootstocks. Can. J. Bot. 42: 891–922.

Campbell, G. R. 1952. The genus *Myosurus* in North America. Aliso 2:398–403.

Ewan, J. A. 1945. A synopsis of the North American species of *Delphinium*. Univ. Colo. Studies 2:55–244.

Pringle, J. S. 1971. Taxonomy and distribution of *Clematis*, sect. Atragene, in North America. Brittonia 23:361–93.

Rhamnaceae

Brizicky, G. K. 1964. Further note on *Ceanothus herbaceous* vs. *C. ovatus*. J. Arn. Arb. 45:471–73.

Rosaceae

Jones, G. N. 1946. American species of *Amelanchier*. Ill. Biol. Monogr. 20:1–126.

Kruschke, E. P. 1965. Contributions to the taxonomy of *Crataegus*. Milwaukee Public Mus. Publ. 3.

Lewis, W. H. 1958. A monograph of the genus *Rosa* in North America. II. *R. foliolosa*. Southwest. Nat. 3:145–53.

Martin, F. A. 1950. A revision of *Cercocarpus*. Brittonia 7: 91–111.

Nielson, E. L. 1939. A taxonomic study of the genus *Amelanchier* in Minnesota. Am. Midl. Nat. 22:160–203.

Salicaceae

Argus, G. W. The willows of Wyoming. Univ. Wyo. Publ. 21(1): 1–63. Illus. Maps. 1957.

Crawford, D. J. 1974. A morphological and chemical study of *Populus acuminata* Ryd. Brittonia 26:74–89.

Froiland, S. G. 1962. The genus *Salix* in the Black Hills of South Dakota. USDA Forest Serv. Tech. Bull. 1269.

Raup, H. M. 1959. The willows of boreal western America. Contrib. Gray Herb. 185:1–95.

Saxifragaceae

Rosendahl, C. O. 1936. A monograph of the genus *Heuchera*. Minn. Studies Plant Sci. 2:1–180.

Taylor, R. L. 1965. The genus *Lithophragma*. Univ. Calif. Publ. Bot. 37:1–89.

Santalaceae

Piehl, M. A. 1965. The natural history and taxonomy of *Comandra* (Santalaceae). Mem. Torrey Bot. Club 22:1–82.

Saxifragaceae

Rosendahl, C. O., F. K. Butters, and O. Lakela. 1936. A monograph on the genus *Heuchera*. Minn. Studies Plant Sci. 2:
1-180.

Scrophulariaceae

Crosswhite, F. S. 1965. Revision of *Penstemon*. Am. Midl. Nat.
74:429-42.
Grant, A. 1924. A monograph of the genus *Mimulus*. Ann. Mo. Bot.
Gard. 11:99-388.

Solanaceae

Waterfall, U. T. 1958. A taxonomic study of the genus *Physalis*
in North America north of Mexico. Rhodora 60:107-14, 128-
42, 152-73.

Selaginellaceae

Tryon, R. M. 1955. *Selaginella rupestris* and its allies. Ann.
Mo. Bot. Gard. 42:1-99.

Tiliaceae

Jones, G. N. 1968. Taxonomy of American species of Linden
(*Tilia*). Ill. Biol. Monogr. 39.

Typhaceae

Hotchkiss, N., and H. L. Dozier. 1949. Taxonomy and distribu-
tion of North American cat tails. Am. Midl. Nat. 41:237-54.
Smith, G. 1967. Experimental and natural hybrids of North
America *Typha*. Am. Midl. Nat. 78:257-87.

Urticaceae

Hermann, J. F. 1946. The perennial species of *Urtica* in the
United States east of the Rocky Mountains. Am. Midl. Nat.
35:773-78.

Valerianaceae

Meyer, F. G. 1951. *Valeriana* in North America and the West
Indies. Ann. Mo. Bot. Gard. 38:377-503.
Nielsen, S. D. 1949. Systematic studies in the Valerianaceae.
Am. Midl. Nat. 42:480-501.

Verbenaceae

Lewis, W. H., and R. Oliver. 1961. Cytogeography and phylogeny
of the North American species of *Verbena*. Am. J. Bot. 48:
638-43.
Perry, L. M. 1933. A revision of the North American species of
Verbena. Ann. Mo. Bot. Gard. 20:239-362.

Violaceae

Russell, N. H. 1965. Violets (*Viola*) of the central and eastern
U.S. An introductory survey. Sida 2:1-113.
_____. 1955. The taxonomy of the North American acaulescent
white violets. Am. Midl. Nat. 54(2):481-94.
_____. 1956. Regional variation patterns in the stemless white
violets. Am. Midl. Nat. 56(2):491-503.

Vitaceae

Smith, C. E. 1965. Species indistinctions in the genus *Vitis*.
Leafl. Wet. Bot. 10:143-50.

GLOSSARY

A.   Prefix meaning without.

ABAXIAL.   On the side away from the axis.

ABORTIVE.   Failure to develop or hardly developed.

ACAULESCENT.   Without a stem aboveground or apparently so.

ACHENE.   A one-seeded, indehiscent fruit with a relatively thin
   wall, dry, in which the seed-coat is not fused to the fruit
   wall.

ACICULAR.   Like a needle or needle-shaped.

ACTINOMORPHIC.   Radially symmetrical or regular, the arrangement
   of flower parts.

ACUMINATE.   Gradually tapering to a fine point.

ACUTE.   Forming an acute angle at the tip or base less than 90°.
   Sharp-pointed tip with relatively straight sides.

ADAXIAL.   On the side nearest the axis, or turned to the axis.

ADHERENT.   Staying or sticking together.

ADNATE.   Grown or attached together. Refers to the union of
   unlike parts as stamens adnate to the petals.

ADVENTITIOUS.   Not in the usual place. Development where it
   would not be expected.

ADVENTIVE.   Arriving lately. Not fully established or only
   locally established.

AESTIVAL.   Blooming or occurring in summer.

AGGREGATE.   Referring to fruit derived from two or more pistils
   in the same flower; e.g., a raspberry.

ALATE.   With thin margins, winged.

ALLOPLOID, ALLOPOLYPLOID.   A polyploid in which the chromosome
   pairs are derived from two or more distinct species.

ALTERNATE.   Located singly at each node, as in leaves. Also,
   situated in between, as stamens alternate with the petals.

ALVEOLATE.   Having a honeycombed surface with regular, angular
   cavities (alveoli).

AMENT.   A dense spike or raceme with many small, usually naked
   flowers. Also a catkin.

AMENTIFEROUS.   Bearing aments or catkins.

AMPHIBIOUS.   Living both on land and in water. Continuing growth
   on land after water has evaporated or receded.

ANASTOMOSING.   Branching and rebranching to form a network.

ANDRO-DIOECIOUS.   Having male flowers on one plant and perfect
   flowers on another.

ANDROECIUM.   All of the stamens of one flower considered together.

ANDROGYNOUS.   A spike with both staminate and pistillate flowers,
   the staminate *above* the pistillate ones (in *Carex*).

ANEMOPHILOUS.   Pollinated by the wind.

ANGIOSPERM. A member of the flowering plants characterized by having the ovules enclosed in the ovary.

ANNUAL. A plant that germinates from seed, flowers and matures its fruit in one growing season. A winter annual germinates in the fall and matures the following spring or summer; like winter wheat.

ANNULAR. In the form of a ring or marked with rings.

ANNULUS. The specialized group of cells that rings the sporangium of ferns.

ANTERIOR. Meaning in front of. In reference to a flower, the side adjacent to the bract and away from axis.

ANTHER. The pollen producing part of the stamen, usually consisting of one or two pollen sacs.

ANTHESIS. The period of time when the flower is open and functional.

ANTHOCYANIN. A water-soluble pigment ranging from red through blue to purple that usually colors the petals.

ANTRORSE. Directed upward or forward from the point of attachment.

APETALOUS. Without petals.

APHYLLOPODIC. Having the lower leaves lacking or rudimentary. The foliage leaves borne well above the base of the plant.

APHYLLOUS. Without leaves.

APICAL. At the top of, towards the apex.

APICULATE. Ending in an abrupt point that is very sharp.

APOCARPOUS. The carpels free from each other or with only one carpel.

APOGAMOUS, APOMICTIC. Setting seed without fertilization.

APPRESSED. Pressed flat or closely against another part.

AQUATIC. Living in or inhabiting the water.

ARACHNOID. Having long, soft hairs that form a cobwebby network.

ARBORESCENT. Tree-like or having the form of a tree.

AREOLE. A small, clearly bounded area or clear space.

ARGENTEOUS. Silvery white or gray in color.

ARIL. A fleshy outgrowth that is attached to a seed or a fleshy thickening of the seed coat.

ARISTATE. Tipped with an awn or sharp bristle.

ARTICULATE. Jointed with a definite separation place.

ASCENDING. Growing obliquely upward or curving upward.

ASEPALOUS. Without sepals.

ATTENUATE. Gradually tapering to a very elongate and pointed tip.

AURICLE. A small lobe or rounded projecting appendage, usually at the base of a leaf.

AUTOPLOID, AUTOPOLYPLOID. A polyploid with the chromosome pairs derived from the same species.

AUTOTROPHIC. Independent nutritionally, being able to obtain raw materials from the substrate and to synthesize them. To be able to make its own food, as in a green plant.

AWN.   A slender, almost hair-like bristle that is either terminal or arising from a dorsal position.

AXIL.   The area of the angle formed by the leaf or petiole and the stem from which it arises.

AXILE.   Located at or near the central axis of the ovary. Usually refers to placental location.

AXILLARY.   Arising from or located in the axil.

BACCATE.   Berry-like, a berry.

BANNER.   The upper enlarged petal of the papilionaceous flower (Fabaceae) also called the standard.

BARBED.   Having short sharp, reflexed points at the surface or margin.

BASAL.   At the base or near the base of the stem or another organ.

BASIFIXED.   Attached by the base. Usually refers to anthers or styles.

BEAK.   A prolonged or slender tip of a fruit or seed.

BEARDED.   Having a tuft or ring of long hairs.

BERRY.   A fleshy fruit, indehiscent, having several or more seeds, from a single pistil.

BIENNIAL.   A plant that completes its life cycle in two years. Usually the flowers and fruits are produced only in the second year.

BIFURCATE.   Having two forks or becoming two forked.

BILABIATE.   Two-lipped.

BIPINNATE.   Twice pinnate, the pinnae being again pinnate.

BISEXUAL.   Having both sexes in the same flower. Also, perfect.

BIVALVED or BIVALVATE.   Having two valves.

BLADE.   The expanded or terminal portion of a leaf.

BLOOM.   A white or waxy covering of a surface, such as a leaf or fruit.

BOREAL.   Northern, of the north.

BRACT.   A small or specialized leaf-like structure usually associated with or immediately below a flower or flower structure.

BRACTEOLE, BRACTLET.   A small bract or diminutive of bract.

BUD.   An undeveloped stem, leaf or flower. Stems and leaves are enclosed by bud scales. Flowers are enclosed by the sepals.

BULB.   A short, vertical underground stem which has thickened leaves or thickened leaf bases developed as food storage organs.

BULBIL, BULBLET.   A small bulb formed at the base of a bulb or in the axils of some leaves in place of flowers.

CADUCOUS.   Falling off very early.

CALCAREOUS.   Limy or rich in calcium carbonate.

CALLUS. The small, firm thickening at the base of the lemma in grasses.

CALYX. The sepals of a flower considered collectively.

CAMBIUM. The meristem (sheet) in the stem that produces the xylem and phloem resulting in the increase in girth and diameter.

CAMPANULATE. Shaped like a bell, usually descriptive of a corolla or calyx.

CANALICULATE. Having small grooves or parallel channels.

CANESCENT. Pale or gray colored because of a fine dense pubescence.

CAPILLARY. Hair-like, very fine and slender.

CAPITATE. In a dense head, head-like.

CAPSULE. A dry, dehiscent fruit with more than two seeds, formed from more than one carpel.

CARINATE. Having keels or ridges.

CARPEL. The fertile leaf (or sporophyll) of the angiosperm that bears the ovules. Two or more carpels join to form a compound pistil.

CARTILAGINOUS. Tough and firm, like cartilage.

CARUNCLE. The small appendage that attaches the seed to the placenta.

CARYOPSIS. The fruit of grasses. Different from an achene in that the seed coat is fused or adnate to the pericarp. Often called a grain.

CATKIN. A dense spike or raceme with many small, usually naked flowers. Also an ament.

CAUDATE. Tail-like. Having a tail-like appendage.

CAUDEX. A short, vertical persistent stem, usually woody, just beneath the surface of the soil from which new aerial stems grow each year.

CAULESCENT. Having an obvious leafy stem with nodes and internodes.

CAULINE. Pertaining to the stem.

CELL. The cavity or locule of an ovary as used in taxonomy.

CERNUOUS. Nodding or having a tendency to nod.

CESPITOSE. Growing in low, dense tufts or without runners.

CHAFF. Thin dry scales. Usually referring to those at the base of achenes in the heads of the Asteraceae.

CHARTACEOUS. Papery in texture.

CHLOROPHYLL. The essential green pigment of flowering plants.

CILIATE. Having a fringe of marginal hairs.

CINEREOUS. Ashy-gray in color due to a covering of short hairs.

CIRCINATE. Coiled in bud while in an immature condition as in young fern fronds.

CIRCUMBOREAL. Occurring some distance all around the north pole.

CIRCUMSCISSILE. Dehiscing or separating by an encircling line
  so that the top comes off as a lid or cap.
CLAVATE. Shaped like a baseball bat, thicker at the distal end.
CLAW. The narrow, lower part of some petals and sepals having a
  large, distal end.
CLEFT. Cut or incised about halfway to the base forming two por-
  tions or lobes.
CLEISTOGAMOUS. A flower that sets seed without opening; self-
  pollinating.
CLIMBING. Growing erect or nearly so without supporting its own
  weight, attached to some other structure.
CLONE. A plant or group of plants originating by vegetative
  multiplication from a single individual.
COALESCENT. Running together as a single unit.
COHERENT. Alike parts sticking together.
COLLATERAL. Side by side.
COLUMN. A group of united filaments or filaments and styles, as
  in the Malvaceae.
COMA. A tuft of soft, long hairs, usually on a seed.
COMMISSURE. The face by which two carpels fuse together, as in
  the Apiaceae.
COMOSE. Having a coma, or a tuft of long soft hairs.
COMPLETE. A flower with sepals, petals, stamens and pistil.
COMPOUND. More than one. Leaf--with two or more distinct leaf-
  lets. Ovary or pistil--composed of more than one carpel.
CONDUPLICATE. Folded together the long way with the two parts
  or folds about equal.
CONE. A cluster of sporophylls or scales on an axis as in the
  pines.
CONIC. Shaped like a cone.
CONNATE. Similar or alike parts grown together or attached.
CONNECTIVE. The tissue connecting two pollen sacs of the anther
  together.
CONNIVENT. Converging or coming together but not fused or grown
  together.
CONTIGUOUS. So close as to be adjoining or connected.
CONTORTED. Sharply twisted or bent back on itself.
CONVOLUTE, CONTORTED. Referring to petals. Arranged so that
  each petal has one edge exposed with the other edge covered.
CORDATE. Shaped like a Valentine heart with the notch at the
  base, refers to leaves.
CORIACEOUS. Leathery in texture.
CORM. A short, vertical underground stem thickened as a food-
  storing organ but lacking prominent leaves.
COROLLA. All of the petals considered collectively.
CORONA. A crown-like arrangement of appendages between the
  stamens and petals.

CORYMB. A simple, racemose inflorescence that is flat-topped. The outer pedicels are progressively longer than the inner.

CORYMBIFORM. Having the form, but not necessarily the structure of a corymb.

CORYMBOSE. In a corymb.

COTYLEDON. A leaf of the embryo of a seed.

COULEE. An elongate, V-shaped ravine in upland prairie, often with shrubby or wooded vegetation at the lower part.

CREEPING. Growing along the ground and having roots from the nodes.

CRENATE. Having rounded teeth.

CRESTED. With an elevated ridge or appendage on the top or back.

CROWN. The persisting base of an herbaceous perennial or the leafy top of a tree.

CRUCIFORM. Like a cross, cross-shaped or opposite.

CRUSTACEOUS. Brittle and hard.

CUCULLATE. Hooded or shaped like the prow of a boat.

CULM. The stem of a grass or sedge.

CULTIGEN. A plant that originated in cultivation.

CUNEATE. Wedge shaped with the narrow end at the point of attachment.

CUPULATE. Cup-shaped.

CUSP. An abrupt, sharp point.

CUTICLE. The waxy covering of the epidermis of stems, leaves or fruits.

CYATHIFORM. Cup-shaped, the form of a cyathium.

CYATHIUM. The false inflorescence-like cluster of flowers in *Euphorbia*, consisting of a central, naked pistillate flower surrounded by several staminate flowers each of a single stamen.

CYLINDRIC. Having the shape of a cylinder.

CYME. An inflorescence characterized by the terminal flower blooming before others on that branch; a determinate inflorescence.

CYMOSE. With flowers in cymes or groups of cymes.

DECIDUOUS. Falling or separating after completion of normal functioning, such as leaves or petals.

DECOMPOUND. Repeatedly compound and often irregularly so. Usually referring to compound leaves with many leaflets.

DECUMBENT. With the lower part prostrate or curved and the top erect or ascending, as in stems.

DECURRENT. With a ridge or wing extending down the stem or axis below the point of insertion.

DECUSSATE. Arranged strictly oppositely in pairs with each succeeding pair at right angles to the preceding one.

DEFLEXED.   Curved or bent downward.

DEHISCENT.   Opening at maturity along a definite suture.

DELTOID.   Broadly triangular.

DENTATE.   With spreading, pointed teeth, the points directed at right angles from the margin.

DENTICULATE.   With small teeth.

DEPAUPERATE.   Poorly developed because of an unfavorable environment.

DETERMINATE INFLORESCENCE.   A flowering cluster in which the terminal flower blooms first, essentially stopping the elongation of the axis because all of the meristematic tissue becomes differentiated.

DI.   Prefix meaning two.

DIADELPHOUS.   Fused into two groups.   One of the two groups may consist of a single member.

DICHOTOMOUS.   A forking arrangement in which each fork or branch is of equal size.

DIDYMOUS.   Developing in pairs, twinning.

DIDYNAMOUS.   With 4 stamens in 2 unequal pairs, as in the Lamiaceae.

DIFFUSE.   Much spreading or loosely branching.

DIGITATE.   Having parts diverging from a common base, like fingers on the hand.   Similar to palmate.

DILATED.   Enlarged as if inflated.

DIMORPHIC.   Having two forms.

DIOECIOUS.   Bearing staminate and pistillate flowers on different or separate plants.

DIPLOID.   Having two full chromosome complements per body cell.

DISCOID.   Resembling a disk; in the Asteraceae, with the flowers of a head all tubular and perfect.

DISK.   An outgrowth of the receptacle that surrounds the base of the ovary.   In the Asteraceae, the central part of the head composed of tubular flowers.

DISSECTED.   Deeply divided into numerous or slender parts.

DISTAL.   Toward or at the far end.

DIURNAL.   Pertaining to daytime, open during the day.

DIVARICATE.   Widely spreading from the axis or branch.

DORSAL.   On the back or upper surface.

DORSIVENTRAL.   Flattened with a back side and a belly side.

DRUPE.   A fleshy fruit, indehiscent, usually with a single seed as an olive or cherry.

ECHINATE.   Prickly.

ECOTYPE.   Those individuals of a species adapted to a particular environment or specific habitat.

EGG.   A nonmotile gamete which can fuse with a sperm to produce a zygote.

ELLIPSOID. Elliptic in long section and circular in cross-
   section (a three dimensional body).
ELLIPTIC. The shape of a geometrical ellipse (two dimensional).
EMARGINATE. With a small notch or indentation at the tip.
EMBRYO. A young plant in seed prior to germination.
ENDEMIC. Confined to a particular geographic area.
ENDOCARP. The inner part of the pericarp or fruit wall.
ENSIFORM. Sword-shaped.
ENTIRE. Usually referring to a leaf margin that is not toothed
   or otherwise broken.
ENTOMOPHILOUS. Pollinated by insects.
EPHEMERAL. Lasting for a short time.
EPIGYNOUS. With the ovary below the place where the sepals,
   petals, and stamens are attached; that is, the ovary is
   inferior.
EPIPETALOUS. Attached to the petals, usually referring to
   stamens.
EPIPHYTE. A plant growing attached to another plant but not
   deriving food or water from it.
EROSE. With an irregular margin appearing as if it was worn
   away.
ESCAPE. Escaping from cultivation and maintaining itself.
EVEN-PINNATE. Lacking a terminal leaflet so there is an even
   number of leaflets.
EXCURRENT. With a continuing central axis from which lateral
   branches arise.
EXFOLIATING. Peeling off in layers.
EXSERTED. Projecting beyond the opening, the opposite of being
   included.
EXSTIPULATE. Meaning without stipules.

FALCATE. Sickle-shaped or curved like a hawk's beak.
FARINOSE. Covered with a meal-like powder.
FASCICLE. A small bundle or cluster pertaining to a morpho-
   logical detail of arrangement.
FASCICULATE. In small clusters or bundles.
FERRUGINOUS. Rust-colored.
FERTILE. Capable of producing seeds or pollen.
FIBRILLOSE. Breaking down into fibers.
FILAMENT. The stalk of a stamen that supports the anther.
FILIFORM. Very slender and thread-like.
FIMBRIATE. Having a fringed margin.
FLACCID. Weak and lax, flabby, not supporting its own weight.
FLESHY. Thick and juicy.
FLOCCOSE. Covered with long, soft, fine hairs that are loosely
   spreading.

FLORET. A little flower, as in the grasses or on the heads of
  the Asteraceae.
FOLIACEOUS. Leafy in texture.
-FOLIATE. Suffix indicating the number of leaflets, e.g., tri-
  foliolate, with 3 leaflets.
FOLLICLE. A dry fruit composed of single carpel that dehisces
  along the ventral, seed-bearing suture, as in milkweeds.
FREE. Not united to other organs or units.
FREE-CENTRAL. Referring to placentation in an ovary consisting
  of a free-standing column arising from the base.
FROND. The leaf of a fern, or any large leaf.
FRUIT. A ripened ovary, along with any other structures ripening
  and forming a unit with it.
FRUTICOSE. Shrubby or having woody parts.
FUNNELFORM. Shaped like a funnel.
FUSIFORM. Spindle-shaped. Thickest near the middle and tapering
  to each end.

GALEA. The upper, strongly concave lip of corollas that arches
  over the rest of the flower, as in *Castilleja* of the Scrophu-
  lariaceae.
GAMO. Prefix meaning connate or fused together.
GENICULATE. Abruptly bent or twisted.
GIBBOUS. Swollen on one side, usually near the base.
GLABRATE. Nearly glabrous or becoming so with age.
GLABROUS. Smooth, without hairs or glands.
GLAND. A swelling, protuberance, or depression that produces
  any type of viscous fluid.
GLAUCESCENT. Becoming glaucous, usually with age.
GLAUCOUS. Covered with a fine, waxy material giving a white or
  bluish cast to the surface.
GLOBOSE. Nearly spherical in shape.
GLOMERATE. Densely compacted in clusters or heads along the axis.
GLOMERULE. A small or compact cyme or cluster of flowers or
  fruits.
GLUME. A type of bract in the Poaceae found at the base of the
  grass spikelet.
GLUTINOUS. With a firm, sticky, usually dark-colored substance
  covering the surface.
GYMNOSPERM. A member of the group of plants characterized by
  having ovules not enclosed in an ovary.
GYNAECANDROUS. The type of spike in *Carex* with both pistillate
  and staminate flowers, the staminate ones *below* the pistillate.
GYNANDROUS. The stamens fused or adnate to the pistil, as in
  the orchids.
GYNOBASIC STYLE. The style attached directly to the enlarged
  receptable base as well as to the individual carpels.

GYNOECIUM. The carpels of a flower considered collectively.
GYNOPHORE. A stalk, usually elongate, that bears the gynoecium
  in some flowers.

HABIT. The general appearance or manner of growth of a plant.
HABITAT. A particular place or kind of environment where a
  plant grows.
HALOPHYTE. A plant adapted to grow in saline soils.
HAPLOID. Having only one set of chromosomes or one half of the
  normal diploid set.
HASTATE. Shaped like an arrowhead, with the basal lobes di-
  verging.
HEAD. An inflorescence of sessile flowers crowded closely to-
  gether at the tip or a peduncle.
HEMI. Prefix meaning half.
HERB. A plant with the stems dying back to the soil surface
  each year at the end of the growing season. It may be annual,
  biennial, or perennial.
HERBACEOUS. Not woody.
HERMAPHRODITE. Having both sexes in the same individual or
  flower.
HETERO. Prefix meaning unlike or differing.
HETEROSTYLIC. With styles of different lengths in different
  individual flowers.
HETEROTROPHIC. Obtaining nutrient material in ways other than
  producing it, either parasitic or saprobic.
HEXA. Prefix meaning six.
HIRSUTE. Pubescent with rather coarse or stiff hairs.
HISPID. Pubescent with coarse and firm, and often pungent hairs.
HOMO. Prefix meaning alike or of the same sort.
HOMOSPOROUS. Producing one kind of spore.
HYALINE. Thin and transparent or almost so.
HYDROPHYTE. A plant adapted to life in water.
HYGROSCOPIC. Taking up water, water-loving.
HYPANTHIUM. A ring or cup around the ovary, formed by the
  fusion of petals, stamens and sepals.
HYPO. Prefix meaning beneath.
HYPOGYNOUS. A flower with the sepals, petals, and stems attached
  to the receptacle, thus, below the ovary, meaning superior
  ovary.

IMBRICATE. Overlapping, as shingles on a roof.
IMMERSED. Growing under water.
IMPERFECT. Referring to flowers. One that has stamens but not
  pistils, or pistils but not stamens, regardless of what other
  flower parts may be present.
INCISED. Deeply and sharply cut.

INCLUDED. Contained within the structure or envelope. Not projecting outward.

INDEHISCENT. Staying closed at maturity, not opening.

INDETERMINATE INFLORESCENCE. A cluster of flowers that begins blooming from below, or those that are first formed. The younger, unopened flowers are above or towards the inside.

INDURATE. Hardened, not easily indented.

INDUSIUM. An outgrowth of cells that covers the sori (mass of sporangia) in ferns.

INFERIOR OVARY. An ovary having the sepals, petals, and stamens attached above or at its summit. Same as epigynous.

INFLORESCENCE. The flower cluster of a plant or the arrangement of flowers on the stem axis.

INFRA. Meaning below or less than.

INNOVATION. A growth or offset at the base of a stem or at the crown near soil level.

INTER. A prefix meaning among or between.

INTERNODE. The section of stem between adjacent nodes.

INTRA. A prefix meaning within or inside.

INTRORSE. Turned toward the inside or turned inward.

INVOLUCEL. Meaning small involucre, of the second order.

INVOLUCRE. A whorl or cycle of bracts or modified leaves immediately below or at the base of a flower cluster.

INVOLUTE. Rolled inward so that the upper side is on the inside of the roll, opposite of revolute.

IRREGULAR FLOWER. Any flower that has parts oriented so that the flower lacks radial symmetry.

KEEL. A sharp ridge along the longitudinal axis of an organ or structure resulting in a boat shape.

LABELLUM. The lip or large petal of the orchid flower.

LABIATE. Meaning lipped. Often two or more petals may form an upper lip while other petals form a lower lip resulting in a bilaterally symmetrical flower.

LACERATE. Irregularly torn or jagged.

LACINIATE. Deeply cut into narrow segments.

LACTIFEROUS. Producing a milky latex.

LAMINAR. Thin and flat, as the expanded portion of a leaf.

LANATE. Woolly.

LANCEOLATE. Shaped like a lance. Elongate structure tapering to both ends.

LANUGINOUS. Meaning woolly or with thick, dense hairs.

LATERAL. To the side or on the side away from the central axis.

LATICIFER. A specialized vessel in leaves and stems producing latex.

LEAFLET.   One of the blade segments of a compound leaf, or the ultimate or smallest unit.

LEGUME.   The fruit of a member of the Fabaceae, composed of a single carpel but having two sutures and dehiscing at maturity along them.

LEMMA.   One of the leaf-like bracts that immediately surrounds the individual florets of a grass flower.   Usually the lemma is borne so that its keel or ridge is oriented laterally or to the outside.   The other bract, the palea, is immediately above the lemma and opposite to it.

LENTICULAR.   Shaped like a lens having convex surfaces on both sides.

LIANA.   A climbing, woody vine.

LIGULATE.   Possessing a ligule.   In the Asteraceae, referring to flowering heads solely composed of ligulate flowers.

LIGULE.   Meaning little tongue.   The flat, strap-shaped flowers on the margin (ray flowers) of the disk in the Asteraceae. Also, the small growth along the culm of grasses where the sheath joins the leaf blade.

LIMB.   The enlarged or expanded part of the corolla beyond the throat or tubular part.

LINEAR.   Long and narrow with the sides essentially parallel.

LIP.   The parts of a lobed corolla.   The extension or growth of an organ or plant part.

LOBE.   The projecting part of an organ with divisions less than one-half the distance to the base.

LOCULE.   The cell, opening, or seed cavity in a fruit or ovary.

LOCULICIDAL.   Separating along the midrib or middle of the locule, usually referring to opening of the fruit wall.

LODICULE.   A small scale or vestigial perianth part found at the base of the floret in grasses.

LOMENT.   A fruit of the Fabaceae that has one-seeded transverse sections instead of longitudinal sutures as in legumes.

LUNATE.   Crescent or moon-shaped.

LYRATE.   Pinnatifid, usually leaf, with the terminal section enlarged and rounded.

MACRO.   Meaning large.

MALPIGHIAN HAIRS.   Hairs with two ends like a pickax with the attachment at the middle.

MARITIME.   Growing under the influence of salt water or saline conditions.

MEGA.   Meaning large.

MERICARP.   One side or the individual carpel of the fruit in the Apiaceae.

-MEROUS.   Referring to the number of parts; e.g., trimerous (3), pentamerous (5).

MESIC. Of average moisture, neither dry nor wet.

MESOPHYTE. A plant adapted to average moisture conditions.

MICRO. Prefix meaning small.

MICROSPOROPHYLL. A sporophyll or modified leaf bearing the
  small or male sporangia.

MONADELPHOUS. Having the stamens grouped together or connate,
  usually forming a tube.

MONILIFORM. Like a necklace, constricted at intervals.

MONO. Prefix meaning one.

MONOECIOUS. With unisexual flowers. Plants with male and
  female flowers at different locations on the same plant.

MUCRONATE. Tipped with a sharp, slender point.

MULTIPLE FRUIT. A fruit produced from several or many flowers,
  such as a mulberry.

MURICATE. With many, sharp projections making the structure
  very roughened.

NAKED. Not covered, lacking various normal enveloping structures.

NATURALIZED. Originally coming from another region but now
  thoroughly established.

NECTARY. A flower part producing odoriferous material.

NERVE. A vein or visible strand through the tissue in question.

NET-VEINED. Referring to leaves having the vascular traces
  forming a network.

NEUTRAL FLOWER. A flower with perianth parts only, and lacking
  functional stamens and pistils.

NOCTURNAL. Active at night. In plants, opening at night, such
  as *Mentzelia* in the Loasaceae.

NODE. That portion of the stem where a leaf or bud is located or
  has been attached.

NODOSE. Having enlarged portions along the axis due to growths.

NOMENCLATURAL. Referring to the application of a name based on
  a type specimen.

NUT. A hard, indehiscent, dry fruit with a thick wall, usually
  one-seeded.

NUTLET. A hard walled, small, usually one-seeded fruit that is
  indehiscent.

OB. A prefix meaning opposite or in the other direction.

OBCORDATE. Opposite of cordate. Heart-shaped but having the
  notch at the tip instead of at the base.

OBLANCEOLATE. Opposite of lanceolate. Having the widest part
  above the middle and tapering to the narrower base.

OBLONG. Shaped like a track in running events (two-dimensional).

OBOVATE. Opposite of ovate with the larger part towards the
  distal or far end.

OBSOLETE. Either lacking or much reduced as to be hardly notice-
able.
OBTUSE. Very blunt, the sides coming to the point or apex at an
angle of more than 90°.
OCREA, OCHREA. A stipule at the base of leaves that completely
sheaths the stem, as in members of the Polygonaceae.
OCHROLEUCOUS. Having the color of yellow-off white.
ODD-PINNATE. Referring to leaves with pinnate leaflets and a
single terminal leaflet making an odd number of leaflets.
OFFSET. A short, perennating structure growing from the base of
the shoot near soil level.
OPERCULATE. Having an opening, usually rounded.
OPPOSITE. Located directly across, usually referring to leaf
arrangement of stems.
ORBICULAR. Nearly circular in outline.
ORIFICE. An opening that is mouth-like.
OVAL. Broadly elliptical.
OVARY. The part of the pistil, usually enlarged, that contains
the ovules which mature into seeds.
OVATE. Shaped in outline like a hen's egg with the larger part
towards the base (two-dimensional).
OVOID. Shaped like a hen's egg (three-dimensional).
OVULE. A not-yet-matured seed located in the ovary.

PALATE. A raised portion of the throat of a corolla near the
juncture of the tube and the lip.
PALEA. The inner or upper one of the pair of bracts enclosing
the grass floret. The palea has its ridge or keel(s) toward
the rachilla.
PALMATE. With several lobes or divisions coming from a common
base.
PANICLE. A branched inflorescence that is indeterminate, usually compound. Often used to describe any many-branched flower
cluster.
PAPILIONACEOUS FLOWER. A flower type in the Fabaceae having a
banner petal, two wing petals, and two partly fused keel petals.
PAPILLOSE. Covered with many small bumps or projections.
PAPPUS. A group of hairs, bristles or scales that crown the summit of the achene in the Asteraceae.
PARALLEL-VEINED. Having many veins or nerves running parallel
along the longitudinal axis of most Monocotyledons' leaves.
PARASITE. Obtaining nutrients and water from a host plant to
which it usually is attached.
PARENCHYMA. A tissue of thin-walled, relatively undifferentiated
cells.
PARIETAL. Along the inside of the distal walls, opposite the

axis or axile location. Usually referring to placenta location in ovaries and fruits.

PECTINATE. Like the teeth of a comb. Having many units of
similar size adjacently located and originating from a base
which is 90º to the teeth.

PEDICEL. The stalk or stem of a single flower in an inflorescence.

PEDUNCLE. The stalk or stem of a group of flowers or of a solitary flower if only one is present.

PELLUCID. Appearing transparent or nearly so.

PELTATE. Circular or shield-shaped and attached at the lower
surface.

PENDULOUS. Drooping down or hanging.

PENTA. Prefix meaning 5.

PENTAMEROUS. Having 5 parts.

PEPO. A fruit that is relatively hard, indehiscent and with a
fleshy interior with many seeds, as cucumbers and watermelons.

PERENNIAL. Any plant that lives more than two years.

PERFECT. A flower with both stamens and pistils, regardless of
whether there is corolla and calyx or not.

PERFOLIATE. A leaf that has its basal portions clasping the stem.

PERIANTH. The calyx and corolla considered collectively.

PERICARP. The fruit wall.

PERIGYNIUM. The bract that encloses the achene (fruit) of members of *Carex* in the Cyperaceae.

PERIGYNOUS. The cup that surrounds the ovary due to the basal
fusion of the calyx, corolla and stamens.

PETAL. A modified floral leaf, usually colored other than green,
located between the calyx and the stamens.

PETIOLATE. With a petiole.

PETIOLE. A stalk or stem of a leaf.

PHREATOPHYTE. A plant growing in mud that is more or less con-
tinuously saturated with water.

PHYLLARY. A bract of the involucre at the outside of the head
of flowers in the Asteraceae.

PHYLLOPODIC. The lower leaves of the stem well developed.

PILOSE. Having long, soft spreading hairs.

PINNA. One of the primary lateral divisions of a pinnately compound leaf.

PINNATE. Having two rows of lateral divisions along a main axis,
like barbs on a feather.

PINNATIFID. Deeply cut in a pinnate fashion but not entirely to
the main axis.

PINNULE. The ultimate leaflet of a compound leaf which is two or
more times compound.

PISTIL. The female organ of a flower, consisting of the ovary,
style and stigma.

PLACENTA.   The area of tissue within an ovary to which the ovules
   are attached.
PISTILLATE FLOWER.   A flower possessing one or more pistils but
   lacking stamens.
PLUMOSE.   Feathery, with a long pubescence or with pinnately
   arranged bristles.
POD.   A general term applied to any type of fruit that dehisces.
POLLEN.   The contents of an anther on their release.
POLLINATION.   The transfer of male gametophytes (pollen) to the
   stigmatic surface resulting in the growth of a pollen tube.
POLLINIUM.   A cluster of pollen grains that remains together at
   transfer, as in the Asclepiadaceae and the Orchidaceae.
POLY.   Prefix meaning many.
POLYCARPOUS.   Having several to many separate carpels.
POLYGAMO-DIOECIOUS.   A plant that is nearly dioecious but having
   some scattered perfect flowers.
POLYGAMOUS.   A plant having perfect flowers and unisexual flowers.
POLYMORPHIC.   Having several different forms.
POLYPETALOUS.   A flower with several to many separate petals.
POLYPLOID.   A cell having more than the normal two complements of
   chromosomes.
POLYSEPALOUS.   A flower having several to many separate sepals.
POME.   A fruit derived from an inferior ovary with a fleshy fruit
   wall and a bony or papery endocarp containing the seeds.
POPULATION.   A grouping of plants in a particular locale.
PORICIDAL.   A type of dehiscence in which a fruit disseminates
   the seeds through pores.   Also referring to the way in which
   some anthers release pollen grains.
POSTERIOR.   On the back side.   The posterior side of a flower is
   the side nearest the axis.
PRICKLE.   A sharp outgrowth of epidermal tissue and lacking a
   developed vessel system connected to the stem or branch where
   it is borne.
PROCUMBENT.   A trailing stem that roots at the nodes.
PROPHYLL.   One of the pair of small bracts (bracteoles) found at
   the immediate base of the flower in the species of *Juncus*.
PROSTRATE.   Lying flat on the ground or nearly so.
PROXIMAL.   Near the base or the end of attachment.
PSEUDO.   Prefix meaning false.
PUBERULENT.   Softly and sparsely hairy.   Usually a lesser degree
   of pubescence.
PUBESCENT.   Any surface bearing hairs.
PULVINATE.   Cushion-like.
PUNCTATE.   Having dots, usually with small pits that are glandu-
   lar.
PUNGENT.   With a sharp odor that penetrates, or a rough surface
   that is uncomfortable to the touch.

PUSTULAR, PUSTULATE. Having small blisters or eruptions.
PYRIFORM. Pear-shaped.
PYXIS. A type of capsular fruit that opens along the midline or
   nearly so.

RACEME. A flower cluster that is elongate with short lateral
   branches (pedicels) on which the individual flowers are borne.
   The older flowers are below, the younger ones above.
RACEMIFORM. Having the appearance of a raceme but not necessarily the specific structure.
RACEMOSE. An inflorescence that appears like a raceme and devel-
   ops in an indeterminate way, the older flowers below, the
   younger flowers above.
RACHILLA. The ultimate axis of the inflorescence of grasses and
   sedges on which the individual florets or spikelets are borne.
RACHIS. The main axis on which leaves or flowering structures
   are borne.
RADIATE. A term used to describe the manner in which flowers in
   the Asteraceae are arranged. The marginal flowers ligulate,
   the central or disk flowers tubular.
RAPHE. Ridge-like area permanently attached to the ovule or
   seed that connects the seed to the placenta.
RAY. The ligule or ligulate flower in the Asteraceae; also
   refers to the primary or secondary branches of a compound
   inflorescence.
RECEPTACLE. The upper end of the pedicel (stem) of a plant to
   which the flowering parts are attached.
RECURVED. Curved downward or backward.
REFLEXED. Bent or curved backward.
REGULAR FLOWER. A flower that has radial symmetry, with parts
   arranged in a circle.
RENIFORM. Kidney-shaped.
RESUPINATE. Upside down or twisted to be inverted.
RETICULATE. Forming a network.
RETRORSE. Directed backward or downward.
REVOLUTE. Rolled outward or downward. The opposite of involute.
RHIZOID. A simple structure functioning like a root but lacking
   the anatomical complexity.
RHIZOME. A creeping, underground stem.
RHOMBIC. Obliquely angled in outline.
RIB. One of the main longitudinal veins on a leaf or other
   structure.
RIPARIAN. Referring to or growing along stream banks.
ROOTSTOCK. Similar to rhizome but more like a storage organ
   than a stem.
ROSETTE. A cluster of leaves arranged in concentric circles
   usually growing from a crown at soil level.

ROSTRATE. Beaked with a short, stout appendage.
ROTATE. Flat and circular in shape or outline.
ROTUND. Rounded or nearly so.
RUDERAL. Meaning weedy, growing as a volunteer.
RUDIMENTARY. Not perfectly or completely developed.
RUGOSE. Wrinkled or the surface irregular.
RUGULOSE. Having small irregularities on the surface.
RUNCINATE. Having sharply pinnatifid sections that point down-
    ward or backward.
RUNNER. A stolon or long slender stem on the surface of the soil.

SACCATE. Having a small pouch or sac.
SAGITTATE. Arrowhead shaped with basal lobes extending straight
    down or nearly so.
SALVERFORM. With a narrow tube and a broadly expanded upper
    portion or limb.
SAMARA. A dry, indehiscent fruit which is one-seeded and winged
    in some way.
SAPROBE. An organism that receives its nutrients from dead
    organic matter.
SAPROPHYTIC. A plant that functions as a saprobe.
SCABROUS. Rough to the touch due to the presence of imperfec-
    tions on the surface or of short hairs.
SCALE. A small, flat structure that may or may not resemble a
    leaf.
SCAPE. A leafless stem or peduncle that arises from soil level
    and is crowned with the flower or flower cluster.
SCAPOSE. Having the flowers on a scape.
SCARIOUS. Thin and white-transluscent, not green.
SCHIZOCARP. The fruit of the Apiaceae that splits into two
    halves at maturity.
SCLEROPHYLL. A leaf with much hard tissue that does not wilt
    upon drying.
SCORPIOID CYME. A flower cluster with a zig-zag appearance due
    to the successive lateral branches arising on opposite sides.
SCURFY. Having small scales on the surface.
SECUND. With flowers on one side of the axis.
SELENIFEROUS. Soils that contain selenium or plants that accumu-
    late it, such as certain species of *Astragalus, Atriplex* and
    Machaeranthera.
SEMI. Prefix meaning half.
SENESCENT. Becoming aged, losing function.
SEPAL. The usually greenish outermost structure making up the
    floral envelope, leaflike in appearance.
SEPALOID. Like a sepal in color and texture.
SEPTATE. Having partitions.

SEPTICIDAL. Dehiscing along the septum which separates the
  fruit cavities.
SERICEOUS. Having a silky appearance due to long, slender hairs
  parallel to the surface.
SERRATE. Toothed margin with the teeth arranged forward or up-
  ward away from the base.
SESSILE. Without a stalk or stem.
SETA, SETACEOUS. Bristle-like.
SETOSE. Having bristles.
SHEATH. A leaf-like structure that surrounds an organ, usually
  a stem.
SHRUBBY. Having woody or hardened stems, at least below, and
  usually with several stems from the base.
SIGMOID. S-shaped.
SILICEOUS. Containing sand or silica.
SILICLE. A wide, short fruit composed of two carpels in the
  Brassicaceae.
SILIQUE. A long, capsular fruit of two carpels in the Brassi-
  caceae.
SIMPLE LEAF. A leaf with a single blade although the blade may
  be lobed.
SIMPLE PISTIL. A pistil composed of a single carpel.
SINUATE. With a wavy margin.
SINUS. The indentation between two lobes or segments.
SORUS. A cluster of sporangia, as in the ferns.
SPADIX. A fleshy axis on which many flowers are crowded. A term
  used with plants in the Monocotyledons.
SPATHE. A large, leaf-like bract that subtends or encloses the
  spadix.
SPATULATE. Shaped like a spatula, being broader above or distal-
  ly than below or proximal.
SPICATE. Arranged as in a spike.
SPICIFORM. Having the appearance but not necessarily the struc-
  ture of a spike.
SPIKE. An elongate flower cluster in which the individual flow-
  ers are sessile on the main axis.
SPIKELET. The ultimate flower cluster in the grasses or sedges
  along with their subtending bracts.
SPINE. A sharp, pointed structure, usually a modified leaf.
SPINULE. Literally, a smaller spine.
SPORANGIOPHORE. A structure which bears sporangia, as in
  Equisetum.
SPORANGIUM. The vessel in which spores are produced.
SPORE. A one-celled reproductive structure characteristic of
  the primitive vascular plants.
SPOROPHYLL. A leaf or leaf-like structure that bears sporangia.

SPUR. A hollowed sac or elongate modification of a petal or
  sepal.
SQUARROSE. Spreading abruptly below the tip.
SSP. Abbreviation for subspecies, a taxonomic category that is
  infra-specific.
STAMEN. The male organ of the flower, consisting of an anther,
  or pollen sac, and its supporting filament.
STAMINATE FLOWER. A flower possessing one or more stamens but
  lacking pistils.
STAMINODE. A modified stamen that is sterile. Often it may
  appear intermediate between a petal and a stamen.
STANDARD. The uppermost petal of a papilionaceous flower in the
  Fabaceae. Also known as a banner.
STELLATE. Star-shaped; usually referring to hairs with many
  branches from the base.
STERILE. Not fertile; not capable of reproducing.
STIGMA. The upper part of the pistil that is capable of re-
  ceiving the pollen.
STIPE. A generalized term indicating a stalk or stem that sup-
  ports a structure or organ.
STIPITATE. Having a stipe-like support.
STIPULE. One of a pair of small, leaf-like appendages found at
  the base of many leaves.
STOLON. An elongate stem on the surface of the soil.
STRAMINEUS. Having a straw-like color.
STRIATE. With fine, parallel lines.
STRIGOSE. A surface with straight hairs oriented in much the
  same direction.
STROBILUS. A cluster of sporangia, or sporophylls with sporangia,
  on an axis.
STYLE. The usually elongate structure of the pistil connecting
  the ovary to the stigma.
STYLOPODIUM. An enlarged disc-like structure at the base of the
  style and crowning the ovary, characteristic of the pistil in
  the Apiaceae.
SUB. A prefix meaning below or under.
SUBMERSED. Below water level.
SUBTEND. To be located directly below or immediately at the base.
SUBULATE. Sharply pointed, awl-shaped.
SUCCULENT. Juicy and waterholding.
SUFFRUTESCENT. Having a woody or shrubby base.
SULCATE. Having longitudinal grooves on the surface.
SUPERIOR OVARY. An ovary in which other floral parts are at-
  tached below its insertion on the receptacle. Also hypogynous
  or perigynous.
SUTURE. A line or seam where two parts fuse or join, usually a
  place where a fruit dehisces.

SWALE.   A pothole-like depression in open prairie, occasionally with water.

SYMPETALOUS.   A term applied to petals being fused or grown together.

SYMPODIAL.   A branching of stems or inflorescence parts in which the main axis is repeatedly branched.

SYN, SYM.   A prefix meaning joined or united.

SYNGENESIOUS.   A term used to describe the fusion of stamens or anthers around the style, as in the tubular flowers of the Asteraceae.

TAPROOT.   A vertical, elongate root axis with branches originating from it.

TAXON.   A term used to describe a taxonomic entity of any rank.

TAXONOMIC SYNONYM.   A name applied to an organism that is based on a different type and not as acceptable as the current, correct name.

TENDRIL.   A modified leaf or stem that coils and grasps its supporting structure.

TEPAL.   An undifferentiated perianth part which is either a petal or a sepal.

TERETE.   A structure that is rounded in cross section.

TERNATE.   In three parts or in threes.

TETRA.   A prefix meaning four.

TETRADYNAMOUS.   With four long stamens and two short stamens, characteristic of members of the Brassicaceae.

TETRAMEROUS.   Having 4 parts.

TETRAPLOID.   A plant that has twice the number of chromosome complements, 4 instead of 2.

THALLUS.   A plant body not differentiated into roots, stems and leaves as in the Lemnaceae.

THORN.   A sharp pointed, branched or unbranched, modified stem structure.

THROAT.   The opening of a sympetalous corolla at the juncture of the tube and the limb.

THYRSE.   A flower cluster of racemosely arranged cymes organized into an elongate panicle.

TOMENTOSE.   A surface covered with matted and tangled hairs.

TOMENTUM.   A covering of matted or tangled hairs.

TORUS.   The receptacle of a flower.

TRAILING.   A stem that is prostrate but not rooting at the nodes.

TRI.   A prefix meaning three.

TRIFOLIOLATE.   With 3 parts or usually 3 leaflets.

TRIGONOUS.   Having three angles.

TRILOCULAR.   Having three locules or cavities.

TRIMEROUS.   With three parts or 3-parted.

TRUNCATE. With the apex or point abruptly straight as if cut off
transversely.

TUBER. An underground stem or rhizome enlarged with food storage
materials, often functioning in vegetative reproduction.

TUBERCLE. A small projection or tuberous outgrowth from the surface of an organ or structure.

TUBERCULATE. Having tubercles.

TURGID. Expanded or swollen; appearing as if there is internal
pressure.

TUSSOCKS. Densely clumped plants of the same clone, usually
grasses or sedges.

TWINING. Twisting or growing up and around in a spiral on some
other plant or structure.

UMBEL. An indeterminate inflorescence with branches arising from
a common point, often flat-topped or nearly so. In a compound
umbel the primary branches are again branched.

UNDULATE. Having a wavy margin.

UNI. A prefix meaning one.

URCEOLATE. Shaped like an urn; becoming urn-like.

UTRICLE. A small, one-seeded fruit with a very thin wall, dehiscing by the breakdown of the thin wall.

VALVATE. Arranged with the margins together, not overlapping.

VALVE. One of the portions of a compound ovary, usually referring to one unit of a capsule.

VAR. An abbreviation for the word variety, an infra-specific
taxonomic category.

VASCULAR. Referring to translocation or conduction in plants.
A vascular plant is one that normally possesses xylem and
phloem.

VEIN. A unit of conduction, a vascular bundle, in a leaf or
other organ.

VENTRAL. The lower or belly side. The adaxial side of a leaf
is the ventral side.

VERSATILE. Attached at the back and freely movable, as in the
attachment of the filament to the anther in the grasses.

VERTICIL. A single whorl or cycle of leaves or flowers.

VERTICILLATE. Arranged in verticils.

VESTIGIAL. Much reduced or not functioning.

VILLOUS. A hairy pubescence with long, soft hairs but not matted
or pressed down.

VISCID. Sticky or clammy.

WEED. A plant where it is not wanted, out of place in terms of
cultivation practices.

WHORL.   A ring or cycle of 3 or more similar structures originating at the same level or point.
WING.   A thin and flat extension on the side or end of a structure.

XEROPHYTE.   A plant adapted to life in dry places.

ZYGOMORPHIC.   Irregular in shape.   Having bilateral symmetry so
  that mirror images or equal halves can be obtained by bisection
  in only one plane.

INDEX